T0220787

Lecture Notes in Computer Science 11306

Commenced Publication in 1973
Founding and Former Series Editors:
Gerhard Goos, Juris Hartmanis, and Jan van Leeuwen

More information about this series at http://www.springer.com/series/7407

Long Cheng · Andrew Chi Sing Leung
Seiichi Ozawa (Eds.)

Neural
Information Processing

25th International Conference, ICONIP 2018
Siem Reap, Cambodia, December 13–16, 2018
Proceedings, Part VI

 Springer

Editors
Long Cheng (iD)
The Chinese Academy of Sciences
Beijing, China

Seiichi Ozawa
Kobe University
Kobe, Japan

Andrew Chi Sing Leung
City University of Hong Kong
Kowloon, Hong Kong SAR, China

ISSN 0302-9743   ISSN 1611-3349   (electronic)
Lecture Notes in Computer Science
ISBN 978-3-030-04223-3   ISBN 978-3-030-04224-0   (eBook)
https://doi.org/10.1007/978-3-030-04224-0

Library of Congress Control Number: 2018960916

LNCS Sublibrary: SL1 – Theoretical Computer Science and General Issues

This Springer imprint is published by the registered company Springer Nature Switzerland AG
The registered company address is: Gewerbestrasse 11, 6330 Cham, Switzerland

Preface

The 25th International Conference on Neural Information Processing (ICONIP 2018), the annual conference of the Asia Pacific Neural Network Society (APNNS), was held in Siem Reap, Cambodia, during December 13–16, 2018. The ICONIP conference series started in 1994 in Seoul, which has now become a well-established and high-quality conference on neural networks around the world. Siem Reap is a gateway to Angkor Wat, which is one of the most important archaeological sites in Southeast Asia, the largest religious monument in the world. All participants of ICONIP 2018 had a technically rewarding experience as well as a memorable stay in this great city.

In recent years, the neural network has been significantly advanced with the great developments in neuroscience, computer science, cognitive science, and engineering. Many novel neural information processing techniques have been proposed as the solutions to complex, networked, and information-rich intelligent systems. To disseminate new findings, ICONIP 2018 provided a high-level international forum for scientists, engineers, and educators to present the state of the art of research and applications in all fields regarding neural networks.

With the growing popularity of neural networks in recent years, we have witnessed an increase in the number of submissions and in the quality of submissions. ICONIP 2018 received 575 submissions from 51 countries and regions across six continents. Based on a rigorous peer-review process, where each submission was reviewed by at least three experts, a total of 401 high-quality papers were selected for publication in the prestigious Springer series of *Lecture Notes in Computer Science*. The selected papers cover a wide range of subjects that address the emerging topics of theoretical research, empirical studies, and applications of neural information processing techniques across different domains.

In addition to the contributed papers, the ICONIP 2018 technical program also featured three plenary talks and two invited talks delivered by world-renowned scholars: Prof. Masashi Sugiyama (University of Tokyo and RIKEN Center for Advanced Intelligence Project), Prof. Marios M. Polycarpou (University of Cyprus), Prof. Qing-Long Han (Swinburne University of Technology), Prof. Cesare Alippi (Polytechnic of Milan), and Nikola K. Kasabov (Auckland University of Technology).

We would like to extend our sincere gratitude to all members of the ICONIP 2018 Advisory Committee for their support, the APNNS Governing Board for their guidance, the International Neural Network Society and Japanese Neural Network Society for their technical co-sponsorship, and all members of the Organizing Committee for all their great effort and time in organizing such an event. We would also like to take this opportunity to thank all the Technical Program Committee members and reviewers for their professional reviews that guaranteed the high quality of the conference proceedings. Furthermore, we would like to thank the publisher, Springer, for their sponsorship and cooperation in publishing the conference proceedings in seven volumes of *Lecture Notes in Computer Science*. Finally, we would like to thank all the speakers, authors, reviewers, volunteers, and participants for their contribution and support in making ICONIP 2018 a successful event.

October 2018

Jun Wang
Long Cheng
Andrew Chi Sing Leung
Seiichi Ozawa

ICONIP 2018 Organization

General Chair

Jun Wang   City University of Hong Kong,
Hong Kong SAR, China

Advisory Chairs

Akira Hirose   University of Tokyo, Tokyo, Japan
Soo-Young Lee   Korea Advanced Institute of Science and Technology,
South Korea
Derong Liu   Institute of Automation, Chinese Academy of Sciences,
China
Nikhil R. Pal   Indian Statistics Institute, India

Program Chairs

Long Cheng   Institute of Automation, Chinese Academy of Sciences,
China
Andrew C. S. Leung   City University of Hong Kong, Hong Kong SAR,
China
Seiichi Ozawa   Kobe University, Japan

Special Sessions Chairs

Shukai Duan   Southwest University, China
Kazushi Ikeda   Nara Institute of Science and Technology, Japan
Qinglai Wei   Institute of Automation, Chinese Academy of Sciences,
China
Hiroshi Yamakawa   Dwango Co. Ltd., Japan
Zhihui Zhan   South China University of Technology, China

Tutorial Chairs

Hiroaki Gomi   NTT Communication Science Laboratories, Japan
Takashi Morie   Kyushu Institute of Technology, Japan
Kay Chen Tan   City University of Hong Kong, Hong Kong SAR,
China
Dongbin Zhao   Institute of Automation, Chinese Academy of Sciences,
China

Publicity Chairs

Zeng-Guang Hou	Institute of Automation, Chinese Academy of Sciences, China
Tingwen Huang	Texas A&M University at Qatar, Qatar
Chia-Feng Juang	National Chung-Hsing University, Taiwan
Tomohiro Shibata	Kyushu Institute of Technology, Japan

Publication Chairs

Xinyi Le	Shanghai Jiao Tong University, China
Sitian Qin	Harbin Institute of Technology Weihai, China
Zheng Yan	University Technology Sydney, Australia
Shaofu Yang	Southeast University, China

Registration Chairs

Shenshen Gu	Shanghai University, China
Qingshan Liu	Southeast University, China
Ka Chun Wong	City University of Hong Kong, Hong Kong SAR, China

Conference Secretariat

Ying Qu	Dalian University of Technology, China

Program Committee

Hussein Abbass	University of New South Wales at Canberra, Australia
Choon Ki Ahn	Korea University, South Korea
Igor Aizenberg	Texas A&M University at Texarkana, USA
Shotaro Akaho	National Institute of Advanced Industrial Science and Technology, Japan
Abdulrazak Alhababi	UNIMAS, Malaysia
Cecilio Angulo	Universitat Politècnica de Catalunya, Spain
Sabri Arik	Istanbul University, Turkey
Mubasher Baig	National University of Computer and Emerging Sciences Lahore, India
Sang-Woo Ban	Dongguk University, South Korea
Tao Ban	National Institute of Information and Communications Technology, Japan
Boris Bačić	Auckland University of Technology, New Zealand
Xu Bin	Northwestern Polytechnical University, China
David Bong	Universiti Malaysia Sarawak, Malaysia
Salim Bouzerdoum	University of Wollongong, Australia
Ivo Bukovsky	Czech Technical University, Czech Republic

Ke-Cai Cao	Nanjing University of Posts and Telecommunications, China
Elisa Capecci	Auckland University of Technology, New Zealand
Rapeeporn Chamchong	Mahasarakham University, Thailand
Jonathan Chan	King Mongkut's University of Technology Thonburi, Thailand
Rosa Chan	City University of Hong Kong, Hong Kong SAR, China
Guoqing Chao	East China Normal University, China
He Chen	Nankai University, China
Mou Chen	Nanjing University of Aeronautics and Astronautics, China
Qiong Chen	South China University of Technology, China
Wei-Neng Chen	Sun Yat-Sen University, China
Xiaofeng Chen	Chongqing Jiaotong University, China
Ziran Chen	Bohai University, China
Jian Cheng	Chinese Academy of Sciences, China
Long Cheng	Chinese Academy of Sciences, China
Wu Chengwei	Bohai University, China
Zheru Chi	The Hong Kong Polytechnic University, SAR China
Sung-Bae Cho	Yonsei University, South Korea
Heeyoul Choi	Handong Global University, South Korea
Hyunsoek Choi	Kyungpook National University, South Korea
Supannada Chotipant	King Mongkut's Institute of Technology Ladkrabang, Thailand
Fengyu Cong	Dalian University of Technology, China
Jose Alfredo Ferreira Costa	Federal University of Rio Grande do Norte, Brazil
Ruxandra Liana Costea	Polytechnic University of Bucharest, Romania
Jean-Francois Couchot	University of Franche-Comté, France
Raphaël Couturier	University of Bourgogne Franche-Comté, France
Jisheng Dai	Jiangsu University, China
Justin Dauwels	Massachusetts Institute of Technology, USA
Dehua Zhang	Chinese Academy of Sciences, China
Mingcong Deng	Tokyo University of Agriculture and Technology, Japan
Zhaohong Deng	Jiangnan University, China
Jing Dong	Chinese Academy of Sciences, China
Qiulei Dong	Chinese Academy of Sciences, China
Kenji Doya	Okinawa Institute of Science and Technology, Japan
El-Sayed El-Alfy	King Fahd University of Petroleum and Minerals, Saudi Arabia
Mark Elshaw	Nottingham Trent International College, UK
Peter Erdi	Kalamazoo College, USA
Josafath Israel Espinosa Ramos	Auckland University of Technology, New Zealand
Issam Falih	Paris 13 University, France

Bo Fan	Zhejiang University, China
Yunsheng Fan	Dalian Maritime University, China
Hao Fang	Beijing Institute of Technology, China
Jinchao Feng	Beijing University of Technology, China
Francesco Ferracuti	Università Politecnica delle Marche, Italy
Chun Che Fung	Murdoch University, Australia
Wai-Keung Fung	Robert Gordon University, UK
Tetsuo Furukawa	Kyushu Institute of Technology, Japan
Hao Gao	Nanjing University of Posts and Telecommunications, China
Yabin Gao	Harbin Institute of Technology, China
Yongsheng Gao	Griffith University, Australia
Tom Gedeon	Australian National University, Australia
Ong Sing Goh	Universiti Teknikal Malaysia Melaka, Malaysia
Iqbal Gondal	Federation University Australia, Australia
Yue-Jiao Gong	Sun Yat-sen University, China
Shenshen Gu	Shanghai University, China
Chengan Guo	Dalian University of Technology, China
Ping Guo	Beijing Normal University, China
Shanqing Guo	Shandong University, China
Xiang-Gui Guo	University of Science and Technology Beijing, China
Zhishan Guo	University of Central Florida, USA
Christophe Guyeux	University of Franche-Comte, France
Masafumi Hagiwara	Keio University, Japan
Saman Halgamuge	The University of Melbourne, Australia
Tomoki Hamagami	Yokohama National University, Japan
Cheol Han	Korea University at Sejong, South Korea
Min Han	Dalian University of Technology, China
Takako Hashimoto	Chiba University of Commerce, Japan
Toshiharu Hatanaka	Osaka University, Japan
Wei He	University of Science and Technology Beijing, China
Xing He	Southwest University, China
Xiuyu He	University of Science and Technology Beijing, China
Akira Hirose	The University of Tokyo, Japan
Daniel Ho	City University of Hong Kong, Hong Kong SAR, China
Katsuhiro Honda	Osaka Prefecture University, Japan
Hongyi Li	Bohai University, China
Kazuhiro Hotta	Meijo University, Japan
Jin Hu	Chongqing Jiaotong University, China
Jinglu Hu	Waseda University, Japan
Xiaofang Hu	Southwest University, China
Xiaolin Hu	Tsinghua University, China
He Huang	Soochow University, China
Kaizhu Huang	Xi'an Jiaotong-Liverpool University, China
Long-Ting Huang	Wuhan University of Technology, China

Panfeng Huang	Northwestern Polytechnical University, China
Tingwen Huang	Texas A&M University, USA
Hitoshi Iima	Kyoto Institute of Technology, Japan
Kazushi Ikeda	Nara Institute of Science and Technology, Japan
Hayashi Isao	Kansai University, Japan
Teijiro Isokawa	University of Hyogo, Japan
Piyasak Jeatrakul	Mae Fah Luang University, Thailand
Jin-Tsong Jeng	National Formosa University, Taiwan
Sungmoon Jeong	Kyungpook National University Hospital, South Korea
Danchi Jiang	University of Tasmania, Australia
Min Jiang	Xiamen University, China
Yizhang Jiang	Jiangnan University, China
Xuguo Jiao	Zhejiang University, China
Keisuke Kameyama	University of Tsukuba, Japan
Shunshoku Kanae	Junshin Gakuen University, Japan
Hamid Reza Karimi	Politecnico di Milano, Italy
Nikola Kasabov	Auckland University of Technology, New Zealand
Abbas Khosravi	Deakin University, Australia
Rhee Man Kil	Sungkyunkwan University, South Korea
Daeeun Kim	Yonsei University, South Korea
Sangwook Kim	Kobe University, Japan
Lai Kin	Tunku Abdul Rahman University, Malaysia
Irwin King	The Chinese University of Hong Kong, Hong Kong SAR, China
Yasuharu Koike	Tokyo Institute of Technology, Japan
Ven Jyn Kok	National University of Malaysia, Malaysia
Ghosh Kuntal	Indian Statistical Institute, India
Shuichi Kurogi	Kyushu Institute of Technology, Japan
Susumu Kuroyanagi	Nagoya Institute of Technology, Japan
James Kwok	The Hong Kong University of Science and Technology, SAR China
Edmund Lai	Auckland University of Technology, New Zealand
Kittichai Lavangnananda	King Mongkut's University of Technology Thonburi, Thailand
Xinyi Le	Shanghai Jiao Tong University, China
Minho Lee	Kyungpook National University, South Korea
Nung Kion Lee	University Malaysia Sarawak, Malaysia
Andrew C. S. Leung	City University of Hong Kong, Hong Kong SAR, China
Baoquan Li	Tianjin Polytechnic University, China
Chengdong Li	Shandong Jianzhu University, China
Chuandong Li	Southwest University, China
Dazi Li	Beijing University of Chemical Technology, China
Li Li	Tsinghua University, China
Shengquan Li	Yangzhou University, China

Ya Li	Institute of Automation, Chinese Academy of Sciences, China
Yanan Li	University of Sussex, UK
Yongming Li	Liaoning University of Technology, China
Yuankai Li	University of Science and Technology of China, China
Jie Lian	Dalian University of Technology, China
Hualou Liang	Drexel University, USA
Jinling Liang	Southeast University, China
Xiao Liang	Nankai University, China
Alan Wee-Chung Liew	Griffith University, Australia
Honghai Liu	University of Portsmouth, UK
Huaping Liu	Tsinghua University, China
Huawen Liu	University of Texas at San Antonio, USA
Jing Liu	Chinese Academy of Sciences, China
Ju Liu	Shandong University, China
Qingshan Liu	Huazhong University of Science and Technology, China
Weifeng Liu	China University of Petroleum, China
Weiqiang Liu	Nanjing University of Aeronautics and Astronautics, China
Dome Lohpetch	King Mongkut's University of Technology North Bangoko, Thailand
Hongtao Lu	Shanghai Jiao Tong University, China
Wenlian Lu	Fudan University, China
Yao Lu	Beijing Institute of Technology, China
Jinwen Ma	Peking University, China
Qianli Ma	South China University of Technology, China
Sanparith Marukatat	Thailand's National Electronics and Computer Technology Center, Thailand
Tomasz Maszczyk	Nanyang Technological University, Singapore
Basarab Matei	LIPN Paris Nord University, France
Takashi Matsubara	Kobe University, Japan
Nobuyuki Matsui	University of Hyogo, Japan
P. Meesad	King Mongkut's University of Technology North Bangkok, Thailand
Gaofeng Meng	Chinese Academy of Sciences, China
Daisuke Miyamoto	University of Tokyo, Japan
Kazuteru Miyazaki	National Institution for Academic Degrees and Quality Enhancement of Higher Education, Japan
Seiji Miyoshi	Kansai University, Japan
J. Manuel Moreno	Universitat Politècnica de Catalunya, Spain
Naoki Mori	Osaka Prefecture University, Japan
Yoshitaka Morimura	Kyoto University, Japan
Chaoxu Mu	Tianjin University, China
Kazuyuki Murase	University of Fukui, Japan
Jun Nishii	Yamaguchi University, Japan

Haruhiko Nishimura	University of Hyogo, Japan
Grozavu Nistor	Paris 13 University, France
Yamaguchi Nobuhiko	Saga University, Japan
Stavros Ntalampiras	University of Milan, Italy
Takashi Omori	Tamagawa University, Japan
Toshiaki Omori	Kobe University, Japan
Seiichi Ozawa	Kobe University, Japan
Yingnan Pan	Northeastern University, China
Yunpeng Pan	JD Research Labs, China
Lie Meng Pang	Universiti Malaysia Sarawak, Malaysia
Shaoning Pang	Unitec Institute of Technology, New Zealand
Hyeyoung Park	Kyungpook National University, South Korea
Hyung-Min Park	Sogang University, South Korea
Seong-Bae Park	Kyungpook National University, South Korea
Kitsuchart Pasupa	King Mongkut's Institute of Technology Ladkrabang, Thailand
Yong Peng	Hangzhou Dianzi University, China
Somnuk Phon-Amnuaisuk	Universiti Teknologi Brunei, Brunei
Lukas Pichl	International Christian University, Japan
Geong Sen Poh	National University of Singapore, Singapore
Mahardhika Pratama	Nanyang Technological University, Singapore
Emanuele Principi	Università Politecnica elle Marche, Italy
Dianwei Qian	North China Electric Power University, China
Jiahu Qin	University of Science and Technology of China, China
Sitian Qin	Harbin Institute of Technology at Weihai, China
Mallipeddi Rammohan	Nanyang Technological University, Singapore
Yazhou Ren	University of Science and Technology of China, China
Ko Sakai	University of Tsukuba, Japan
Shunji Satoh	The University of Electro-Communications, Japan
Gerald Schaefer	Loughborough University, UK
Sachin Sen	Unitec Institute of Technology, New Zealand
Hamid Sharifzadeh	Unitec Institute of Technology, New Zealand
Nabin Sharma	University of Technology Sydney, Australia
Yin Sheng	Huazhong University of Science and Technology, China
Jin Shi	Nanjing University, China
Yuhui Shi	Southern University of Science and Technology, China
Hayaru Shouno	The University of Electro-Communications, Japan
Ferdous Sohel	Murdoch University, Australia
Jungsuk Song	Korea Institute of Science and Technology Information, South Korea
Andreas Stafylopatis	National Technical University of Athens, Greece
Jérémie Sublime	ISEP, France
Ponnuthurai Suganthan	Nanyang Technological University, Singapore
Fuchun Sun	Tsinghua University, China
Ning Sun	Nankai University, China

Norikazu Takahashi	Okayama University, Japan
Ken Takiyama	Tokyo University of Agriculture and Technology, Japan
Tomoya Tamei	Kobe University, Japan
Hakaru Tamukoh	Kyushu Institute of Technology, Japan
Choo Jun Tan	Wawasan Open University, Malaysia
Shing Chiang Tan	Multimedia University, Malaysia
Ying Tan	Peking University, China
Gouhei Tanaka	The University of Tokyo, Japan
Ke Tang	Southern University of Science and Technology, China
Xiao-Yu Tang	Zhejiang University, China
Yang Tang	East China University of Science and Technology, China
Qing Tao	Chinese Academy of Sciences, China
Katsumi Tateno	Kyushu Institute of Technology, Japan
Keiji Tatsumi	Osaka University, Japan
Kai Meng Tay	Universiti Malaysia Sarawak, Malaysia
Chee Siong Teh	Universiti Malaysia Sarawak, Malaysia
Andrew Teoh	Yonsei University, South Korea
Arit Thammano	King Mongkut's Institute of Technology Ladkrabang, Thailand
Christos Tjortjis	International Hellenic University, Greece
Shibata Tomohiro	Kyushu Institute of Technology, Japan
Seiki Ubukata	Osaka Prefecture University, Japan
Eiji Uchino	Yamaguchi University, Japan
Wataru Uemura	Ryukoku University, Japan
Michel Verleysen	Universite catholique de Louvain, Belgium
Brijesh Verma	Central Queensland University, Australia
Hiroaki Wagatsuma	Kyushu Institute of Technology, Japan
Nobuhiko Wagatsuma	Tokyo Denki University, Japan
Feng Wan	University of Macau, SAR China
Bin Wang	University of Jinan, China
Dianhui Wang	La Trobe University, Australia
Jing Wang	Beijing University of Chemical Technology, China
Jun-Wei Wang	University of Science and Technology Beijing, China
Junmin Wang	Beijing Institute of Technology, China
Lei Wang	Beihang University, China
Lidan Wang	Southwest University, China
Lipo Wang	Nanyang Technological University, Singapore
Qiu-Feng Wang	Xi'an Jiaotong-Liverpool University, China
Sheng Wang	Henan University, China
Bunthit Watanapa	King Mongkut's University of Technology, Thailand
Saowaluk Watanapa	Thammasat University, Thailand
Qinglai Wei	Chinese Academy of Sciences, China
Wei Wei	Beijing Technology and Business University, China
Yantao Wei	Central China Normal University, China

Guanghui Wen	Southeast University, China
Zhengqi Wen	Chinese Academy of Sciences, China
Hau San Wong	City University of Hong Kong, Hong Kong SAR, China
Kevin Wong	Murdoch University, Australia
P. K. Wong	University of Macau, SAR China
Kuntpong Woraratpanya	King Mongkut's Institute of Technology Chaokuntaharn Ladkrabang, Thailand
Dongrui Wu	Huazhong University of Science and Technology, China
Si Wu	Beijing Normal University, China
Si Wu	South China University of Technology, China
Zhengguang Wu	Zhejiang University, China
Tao Xiang	Chongqing University, China
Chao Xu	Zhejiang University, China
Zenglin Xu	University of Science and Technology of China, China
Zhaowen Xu	Zhejiang University, China
Tetsuya Yagi	Osaka University, Japan
Toshiyuki Yamane	IBM, Japan
Koichiro Yamauchi	Chubu University, Japan
Xiaohui Yan	Nanjing University of Aeronautics and Astronautics, China
Zheng Yan	University of Technology Sydney, Australia
Jinfu Yang	Beijing University of Technology, China
Jun Yang	Southeast University, China
Minghao Yang	Chinese Academy of Sciences, China
Qinmin Yang	Zhejiang University, China
Shaofu Yang	Southeast University, China
Xiong Yang	Tianjin University, China
Yang Yang	Nanjing University of Posts and Telecommunications, China
Yin Yang	Hamad Bin Khalifa University, Qatar
Yiyu Yao	University of Regina, Canada
Jianqiang Yi	Chinese Academy of Sciences, China
Chengpu Yu	Beijing Institute of Technology, China
Wen Yu	CINVESTAV, Mexico
Wenwu Yu	Southeast University, China
Zhaoyuan Yu	Nanjing Normal University, China
Xiaodong Yue	Shanghai University, China
Dan Zhang	Zhejiang University, China
Jie Zhang	Newcastle University, UK
Liqing Zhang	Shanghai Jiao Tong University, China
Nian Zhang	University of the District of Columbia, USA
Tengfei Zhang	Nanjing University of Posts and Telecommunications, China
Tianzhu Zhang	Chinese Academy of Sciences, China

Ying Zhang	Shandong University, China
Zhao Zhang	Soochow University, China
Zhaoxiang Zhang	Chinese Academy of Sciences, China
Dongbin Zhao	Chinese Academy of Sciences, China
Qiangfu Zhao	University of Aizu, Japan
Zhijia Zhao	Guangzhou University, China
Jinghui Zhong	South China University of Technology, China
Qi Zhou	University of Portsmouth, UK
Xiaojun Zhou	Central South University, China
Yingjiang Zhou	Nanjing University of Posts and Telecommunications, China
Haijiang Zhu	Beijing University of Chemical Technology, China
Hu Zhu	Nanjing University of Posts and Telecommunications, China
Lei Zhu	Unitec Institute of Technology, New Zealand
Pengefei Zhu	Tianjin University, China
Yue Zhu	Nanjing University, China
Zongyu Zuo	Beihang University, China

Contents – Part VI

Image and Signal Processing

Time-Series Analysis

Handling Concept Drift in Time-Series Data: Meta-cognitive Recurrent Recursive-Kernel OS-ELM

Zongying Liu[1], Chu Kiong Loo[1], and Kitsuchart Pasupa[2(✉)]

[1] Faculty of Computer Science and Information Technology, University of Malaya,
50603 Kuala Lumpur, Malaysia
liuzongying@siswa.um.edu.my, ckloo.um@um.edu.my
[2] Faculty of Information Technology, King Mongkut's Institute of Technology
Ladkrabang, Bangkok 10520, Thailand
kitsuchart@it.kmitl.ac.th

Abstract. This paper proposes a meta-cognitive recurrent multi-step-prediction model called Meta-cognitive Recurrent Recursive Kernel Online Sequential Extreme Learning Machine with a new modified Drift Detector Mechanism (Meta-RRKOS-ELM-DDM). This model combines the strengths of Recurrent Kernel Online Sequential Extreme Learning Machine (RKOS-ELM) with the recursive kernel method and a new meta-cognitive learning strategy. We apply Drift Detector Mechanism to solve concept drift problem. Recursive kernel method successfully replaces the normal kernel method in RKOS-ELM and generates a fixed reservoir with optimised information. The new meta-cognitive learning strategy can reduce the computational complexity. The experimental results show that Meta-RRKOS-ELM-DDM has a superior prediction ability in different predicting horizons than the others.

Keywords: Time series · Recursive kernel · Recurrent
Kernel Adaptive Filter · Concept drift · Meta-cognitive learning

1 Introduction

Time series prediction impacts on humans' daily life, e.g. weather forecasting, wind speed forecasting for wind power systems [1], and financial trend forecasting [2]. In 2015, Scardapane et al. proposed an algorithm called Online Sequential Extreme Learning Machine (OS-ELM) with Kernel that enables implicit feature mappings by utilising kernel method [3]. The problem of OS-ELM–the unstable prediction results–was solved by kernel method. Its unstable feature mapping in OS-ELM was replaced by the kernel matrix. This can solve the problem of prediction deterministic in model as it is known that Extreme Learning Machine (ELM) randomly chooses its parameters, thus its generalisation ability cannot be guaranteed. However, the algorithm requires an extensive computational resource in the learning process, especially when the large-scale data sets are considered.

© Springer Nature Switzerland AG 2018
L. Cheng et al. (Eds.): ICONIP 2018, LNCS 11306, pp. 3–13, 2018.
https://doi.org/10.1007/978-3-030-04224-0_1

Thus, Scardapane et. al. further employed Kernel Adaptive Filters (KAF) to reduce the size of hidden neurons in the process of their algorithm–Approximate Linear Dependency (ALD) and Fixed-Budget (FB). It was found that ALD can reduce computational complexity while enhance the overall performance. Furthermore, the majority of data sets are non-stationary that means the characteristics of data streams may change–underlying distribution changes over time. This directly causes concept drift problem. Although there are many methods that deal with concept drift [4–6], however, these methods are only focus on classification problem. Hence, this paper pays more attention to the concept drift problem in time series prediction.

Besides, the hidden nodes of Kernel OS-ELM (KOS-ELM) include all information of training data, which keep the old information and cannot be overwritten with time progress. Reservoir Computing (RC) is a framework for computation–an extension of neural networks–that is employed in time series prediction. It is a fixed dynamical system that absorbs the information from their recent input history. As time passes, the old information in the reservoir is dissipated and overwritten. There are two major types of RC–Liquid-State Machine (LSM) and Echo State Network (ESN). Again, due to the random weight selection, this causes the unstable of forecasting performance. Therefore, this paper defines infinitely large reservoir called recursive kernels. These kernels can be readily analysed in terms of dynamical stability. To deal with an infinitely large hidden states, kernel trick is applied. Thus, we end up with a kernel function that is the inner product of the hidden states of infinite networks.

In order to solve the limitations of KOS-ELM and generate the fixed dynamical reservoir in its learning part, we apply recurrent multi-step algorithm with recursive kernel to release the restriction of prediction horizon. Furthermore, the KAF is applied to reduce the extensive computational resources. Moreover, Drift Detector Mechanism (DDM) is also considered in order to enable the model to have a good generalisation for concept drift in time series prediction and improve the performance of prediction. Finally, the modified meta-cognitive strategy–a new strategy for regression prediction–is applied to decide when the coming data in the learning part need to be updated, retrained or discarded.

2   Methodology

This section explains the theory of recursive kernel method and modifies DDM in the prediction model of KOS-ELM. At the end of the section, we present how a new meta-cognitive strategy is applied in our proposed time series prediction model.

2.1   Data Transformation

Time series data is a series of data points listed in time order. Assuming that the data $X = [X_1, \ldots, X_N]$ is given where N is the number of time series data. Then, the data is transformed into a matrix form in order to achieve the goal of multi-step prediction. This technique was previously applied in [7,8]. This gives the following training input data and target data,

$$S = \begin{bmatrix} X_{1,1} & \cdots & X_{1,D} \\ \vdots & \ddots & \vdots \\ X_{L,1} & \cdots & X_{L,D} \end{bmatrix} ; \ Y_p = \begin{bmatrix} X_{1,D+p} \\ \vdots \\ X_{L,D+p} \end{bmatrix}, \tag{1}$$

where D is the size of time window, L is the number of training data, p is the p-th step in the prediction $(p = 1, 2, \ldots, P)$.

2.2 Recurrent Multi-step Algorithm

Although KOS-ELM can perform well in time series prediction [3], its size of prediction horizon is restricted to utilise in the real-world application. The recurrent multi-step algorithm is a feedback network which backward from the output of current step to the input of the next step. In order to extend the prediction horizon, we employ recurrent multi-step algorithm in the prediction model, more details can be found at [8]. Based on the theory of recurrent multi-step algorithm, the training input data at the $(p + 1)$-th becomes as follow,

$$S_{p+1} = \begin{bmatrix} X_{1,1+p} & \cdots & X_{1,D} & y_1^{-1} & \cdots & y_1^{-p} \\ \vdots & \ddots & \vdots & \vdots & \ddots & \vdots \\ X_{L,1+p} & \cdots & X_{L,D} & y_L^{-1} & \cdots & y_L^{-p} \end{bmatrix}, \tag{2}$$

where y^{-p} represents the predicting values of training data in the p-th step and $(D + 1) \geq P$.

2.3 Combining Modified DDM with Recurrent KOS-ELM

In the dynamically changing or non-stationary environments, the data distribution can change over time that results in the phenomenon of concept drift. Hence, this phenomenon should be taken into consideration in the learning part of the model. According to the comparison of different concept drift detectors in [9], DDM is the best contender in classification tasks. In this paper, we modify DDM to suit with the process of the time series prediction and apply the modified DDM in Recurrent KOS-ELM (RKOS-ELM) in order to enable concept drift detector in the model.

For a sufficiently large number of samples, if the Binomial distribution is closely approximated by a Normal distribution with the same mean and variance, the process is wide-sense stationary. Therefore, the distribution is stationary [10]. Under this situation, no concept drift appears and the error rate decreases. Therefore, the probability distribution is a significant flag for detecting the change of context.

In the process of detecting concept drift of time series prediction by DDM, each of the l-th sample error-rate $(ER_{l,p})$ and the standard deviation $(SD_{l,p})$ in the p-th step can be calculated by (3) and (4), respectively.

$$ER_{l,p} = |Y_{l,p} - \hat{Y}_{l,p}|/Y_{l,p}, \tag{3}$$

$$SD_{l,p} = \sqrt{(ER_{l,p} \times (1 - ER_{l,p}))/l}, \tag{4}$$

For each sample in time series stream, we have to update two registers in order to keep track of error rate, including the minimum value of error rate ER_{min} and standard deviation SD_{min}. In the learning process of RKOS-ELM, we define the error rate in the $1 - \alpha/2$ confidence interval. It is approximately $ER_{l,p} \pm \alpha \times SD_{l,p}$ when there are a large number of examples ($L \geq 30$). The α is the confidence level. At the same time, the initial ER_{min} and SD_{min} are defined as $ER_{1,p}$ and $SD_{1,p}$, when the first input samples come in. The following new samples of the p-th step coming in the learning part of RKOS-ELM is processed and updated ER_{min} and SD_{min}. For instance, in the learning process of RKOS-ELM in the first step, there is the l-th sample $S_{l,.}$ with corresponding $ER_{l,1}$ and $SD_{l,1}$. The confidence level for concept drift is set to 99%, that is, the concept drift problem appears if $ER_i + SD_i \geq ER_{min} + 3 \times SD_{min}$. Otherwise, no concept drift problem happens if $ER_i + SD_i \leq ER_{min} + SD_{min}$. Besides, ER_{min} and SD_{min} will be updated.

2.4   Recursive Kernel

Assuming that there is a kernel function $k(x, x')$ with data points x and x'. The kernel function can be defined by $k(x, x') = \phi(x)\phi(x')$, where ϕ is a feature map. A norm can be defined using inner product of inputs and the kernel function has the following property:

$$\phi(x)\phi(x') = k(x, x'). \tag{5}$$

If k has ϕ associated with it for which (5) is valid, this means that this same feature map can be applied for recurrent kernels. The recurrent kernel with input data x and x' can be written as follow:

$$r_{t,t'}(x, x') = k(k_{t-1}(x, x') + x(t)x'(t'), k_{t-1}(x, x) + x(t)x(t), k_{t'-1}(x', x') + x'(t')x'(t')). \tag{6}$$

A recurrent formula that can be applied for any kernel function with the form specified in (5) [11]. This recurrent formula also requires us to compute the recurrent kernels $k_{t-1}(x, x)$ and $k_{t'-1}(x', x')$. Generally, a kernel function is only a function of the inner product of its arguments, and not their quadratic norms. In this case the corresponding recurrent kernel can be simplified to the following form:

$$r_{t,t'}(x, x') = k(r_{t-1,t'-1}(x, x') + x(t)x'(t')), \tag{7}$$

In RKOS-ELM with DDM, Radial Basis Function (RBF), which is a popular kernel function, is considered, $k(x, x') = exp(-\frac{||x-x'||^2}{2\sigma^2})$, where σ is a kernel parameter. Although ALD plays a significant role in filtering input data and reduce the number of hidden nodes, the computing process of kernel matrix only pays attention to the kernel matrix between the current input data and the current memory. In order to deal with the infinitely large hidden states of RKOS-ELM with DDM, we apply recursive kernel–which is the inner product of the hidden states of infinite networks–to replace RBF kernel method in the learning part. Thus, the old information of kernel matrix can be dissipated and overwritten by the recursive kernel over the time passes. Therefore, the reservoir of predicting model that is made by the recursive kernel has the more efficient, completed, and variety of information than by the common kernel method. In this paper, we utilise the recursive RBF kernel method to construct a kernel matrix. The algorithm is called "Recurrent Recursive RBF Kernel Online Sequential Extreme Learning Machine with DDM" (RRKOS-ELM-DDM).

According to (6), the recursive RBF can be represented as:

$$r_t(x, x') = exp(-\frac{||x(t) - x'(t)||^2}{2\sigma^2})exp(-\frac{r_{t-1}(x, x') - 1}{\sigma'^2}), \qquad (8)$$

where, σ' is the recursive kernel parameter and t represents the time sequence of time series data. As l denotes the number of training data ($l = (1, \ldots, L)$), the symbol of time sequence is replaced by l in RRKOS-ELM-DDM. Therefore, according to (8), the kernel matrix (G_R) in RRKOS-ELM-DDM can be represented as:

$$G_{R,l} = [r(S_l, S_1), \ldots, r(S_l, S_{l-1}))]^T, \qquad (9)$$

In RRKOS-ELM-DDM, denote the training input data $S = \{S_{(1,\cdot)}, S_{(2,\cdot)}, \ldots, S_{(L,\cdot)}\}$) and the corresponding target Y ($Y = \{Y_1, Y_2, \ldots, Y_L\}$). In the beginning, the kernel matrix for the first input sample is $G_{R,1} = exp(-\frac{||x(t+1) - x(t+1)'||^2}{2\sigma^2})$, $ER_{1,p} = 0$, $SD_{1,p} = \sqrt{ER_{1,p}(1 - ER_{1,p})}$, $ER_{min} = ER_{1,p}$, $SD_{min} = SD_{1,p}$. In the updating phase, $l \in \{2, \ldots, L\}$, the kernel matrix can be calculated by (8) and the output weight (β_l) with interval coefficient Q_l can be calculated by the following equations:

$$\beta_l = \begin{bmatrix} \beta_{l-1} - z_l r_l^{-1} e_l \\ r_l^{-1} e_l \end{bmatrix}, \qquad (10)$$

where $z_l = Q_{l-1} G_{R,l}$, $r_l = C^{-1} + 1 - z_l^T G_{R,l}$, $e_l = Y_l - G_{R,l}^T \beta_{l-1}$, and

$$Q_l = \begin{bmatrix} Q_{l-1} r_l + z_l z_l^T & -z_l \\ -z_l^T & 1 \end{bmatrix}. \qquad (11)$$

C is a regularisation parameter that is defined by 1. It is noted that the initial output weight $\beta_1 = Q_1 Y_1$ with $Q_1 = 1$.

In the training phase of prediction, the l-th prediction value $\hat{y}_l$ can be calculated by the kernel matrix $G_{R,l}$ and output weight β_l as follow:

$$\hat{y}_l = G_{R,l} \beta_l. \qquad (12)$$

2.5   New Meta-cognitive Learning Strategy

Seeking the optimal ALD threshold directly leads to the growth of learning time. In order to solve this problem, a new learning strategy, which is called "meta-cognitive learning strategy", is considered in the prediction model. Not only decide to add, retain or discard neuron when there is a new coming data into the learning part, but also automatically define ALD threshold. This strategy contains four parts, including under-sampling, neuron addition, retain sample, and discard samples. The diagram of meta-cognitive learning for time series prediction model is shown in Fig. 1.

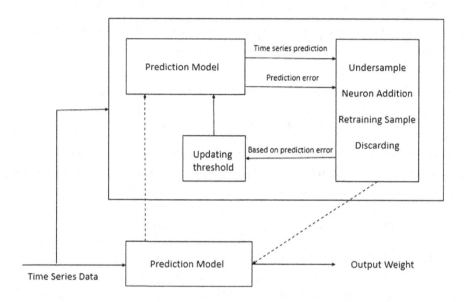

Fig. 1. Meta-cognitive learning for time series prediction model.

Under-sampling is the first phase–initialisation for the model–which requires the minimum number of hidden neurons. In the online learning models, the minimum number of hidden neurons is defined as one. At the same time, the initial threshold of ALD (φ) is equal to the current prediction error (e_1). The second phase is neuron addition, which contains φ and DDM criteria for new coming data. The hidden neuron will be increased by (10), and input data is added into the dictionary memory when the coming data fulfils the requirement of ALD and DDM. Then the current φ for the l-th input sample can be defined by the following equation:

$$\varphi_l = \lambda(\varphi_{l-1}) + (1 - \lambda)e_l, \tag{13}$$

where λ is the slope that controls the rate of self-adaptation and is set close to 1. If the coming sample data does not have concept drift problem or has the similar characteristics with dictionary memory by detecting from ALD filter, it will go to the retraining phase. The output weight will be retained by

$$\beta_{(l,p)} = \beta_{l-1,p} + Q_{l,p}Ge_{l,p}. \tag{14}$$

The last phase is discarding phase that defines the maximum number of hidden neurons to 1000 nodes. If the number of hidden neurons is more than 1000, the corresponding sample with the minimum error pattern that is determined by FB will be discarded from dictionary memory.

Finally, we apply meta-cognitive learning strategy in RKOS-ELM-DDM and RRKOS-ELM-DDM, which can be called as Meta-RKOS-ELM-DDM and Meta-RRKOS-ELM-DDM, respectively. The learning part of pseudo-code in RRKOS-ELM-DDM is shown in Algorithm 1.

3 Experimental Results and Analysis in Synthetic and Real-World Data Sets

We evaluate the performance of the proposed model–Meta-RKOS-ELM-DDM and Meta-RRKOS-ELM-DDM–with its conventional technique, RKOS-ELM, in synthetic and real-world data sets.

In the synthetic data sets, three time series data sets without concept drift problem–containing 20,035 values each set–are generated by autoregressive process as follows: (i) $X_t = 1.5X_{t-1} - 0.4X_{t-2} - 0.3X_{t-3} + 0.2X_{t-4} + w_t$ (TS$_1$), (ii) $X_t = -0.1X_{t-1} + 1.2X_{t-2} + 0.4X_{t-3} - 0.5X_{t-4} + w_t$ (TS$_2$), and (iii) $X_t = 0.9X_{t-1} + 0.8X_{t-2} - 0.6X_{t-3} + 0.2X_{t-4} - 0.5X_{t-5} - 0.2X_{t-6} + 0.4X_{t-7} + w_t$ (TS$_3$). Then these sets are combined to new time series with concept drift problem–the concept drift appears in the 10001-st value in the time series. TS$_4$ is a combination of TS$_1$ and TS$_2$ while TS$_5$ is a combination of TS$_3$ and TS$_1$. TS$_6$ is a combination of TS$_3$ and TS$_2$.

Furthermore, we employ two real-world data in the experiments, i.e., Shanghai Stock Exchange Composite Index (SSE) and ozone concentration of Toronto (Ozone). SSE data from 1 January 1991 to 6 January 2017 is retrieved from Yahoo Finance [12] and Ozone data is from 1 January 2003 to 31 December 2010 collected from the website of Ministry of the environment in Ontario [13]. We replace missing values by the mean value of their nearest values in all experiments. Here, we transform the real-world time series data sets into the matrices, including (6558 × 36) for SSE and (2887 × 36) for Ozone. In the experiments, Symmetric Mean Absolute Percentage Error (SMAPE) is employed to measure the performance of each model in the experiments, see [8] for more details on SMAPE. As it is required to tune a threshold φ of ALD in RKOS-ELM, we simply searched the optimal parameter uniformly in range of {0.00010, 0.00011, ..., 0.00100} based on SMAPE.

The performances of each contender in the four prediction horizons and the average values of four periods of prediction horizons are shown in Table 1. The table shows that the performances of Meta-RKOS-ELM-DMM and Meta-RRKOS-ELM-DMM are clearly better than RKOS-ELM, except in TS$_6$ data

---

Algorithm 1. Learning Phase of RRKOS-ELM-DDM

Require: Size of prediction horizon P; Time window size D; Number of training data
  L; Training data: S by (1) with its target; Kernel parameter $\sigma = 0.7$; Recursive
  kernel parameter σ'; Output weight of RKELM in the p-th step β_p; Prediction
  value in p-th step $\hat{y_p}$; Threshold of ALD (φ).
Ensure: Output weight β_p; Prediction value $\hat{y_p}$.
 1: **for** $p \in \{1, \dots, P\}$ **do**
 2:     Initialise:
 3:     $G_{R,1} = exp(-\frac{||x(t+1)-x(t+1)'||^2}{2\sigma^2})$;
 4:     $ER_{1,p} = 0$, $SD_{1,p} = \sqrt{ER_{1,p}(1 - ER_{1,p})}$;
 5:     $Q_{1,p} = 1$, $\beta_{1,p} = Q_1 Y_1$;
 6:     $mem_p = S_p(1)$;
 7:     $V_{1,p} = 1$;
 8:     $ER_{min} = ER_{1,p}$, $SD_{min} = SD_{1,p}$;
 9:     **for** $l \in \{2, \dots, L\}$ **do**
10:         $\Delta_{l,p} = k(S_{l+1,p}, S_{l+1,p}) - k(mem, S_{l+1,p})^T Q_{l,p} k(mem, S_{l+1,p})$;
11:         Compute $ER_{l,p}$, $SD_{l,p}$ by (3) and (4), respectively;
12:         Determine whether new coming data has concept drift problem or not;
13:         **if** The coming data has concept drift problem **then**
14:             $CD = 1$;
15:         **else**
16:             $CD = 0$;
17:         **end if**
18:         **if** $CD = 1$, and $\Delta_{l,p} \geq \varphi_l$ **then**
19:             Update $G_{R,l,p}$ by (9);
20:             Update $Q_{l,p}$ by (11);
21:             Update output weight ($\beta_{l,p}$) by (10);
22:             Add the current data into the memory dictionary (mem_p);
23:         **else**
24:             Update $V_{(l,p)} = V_{l-1,p} - G(z_{l-1,p}^T V_{l-1,p})$;
25:             Update output weight ($\beta_{p,out}$) by 14;
26:             $ER_{min} = ER_{l,p}$;
27:             $SD_{min} = SD_{l,p}$;
28:         **end if**
29:     **end for**
30:     Add prediction value into training data as new input data for next step by (2);
31: **end for**

---

set that Meta-RKOS-ELM-DMM is worse than RKOS-ELM. This prove that
meta-cognitive strategy and DDM can enhance the prediction performance and
solve the concept drift problem. Moreover, Meta-RRKOS-ELM-DMM is the best
contender.

We further compare the results between Meta-RKOS-ELM-DDM and Meta-
RRKOS-ELM-DDM in order to prove the ability of recursive kernel technique
in the learning process of Meta-RKOS-ELM-DDM. In recursive kernel method,
σ' has an impact on the prediction results. Therefore, σ' is tuned in the range
of 1.0 to 30.0 with 0.1 intervals. The optimal values are 14.0, 10.0, 17.0, 20.5,

Table 1. The performance comparison for multi-step-ahead prediction by RKOS-ELM, Meta-RKOS-ELM-DDM, and Meta-RRKOS-ELM-DDM.

Data sets	Algorithms	Predicting horizon (p)				Average predicting horizon			
		1	7	14	18	1–7	8–12	13–18	1–18
TS$_4$	RKOS-ELM	2.13	4.20	6.21	6.33	2.94	4.84	6.00	4.49
	Meta-RKOS-ELM-DDM	2.29	3.49	5.07	5.80	2.75	4.16	5.23	3.97
	Meta-RRKOS-ELM-DDM	**2.05**	**3.47**	**4.78**	**5.69**	**2.65**	**4.09**	**5.07**	**3.86**
TS$_5$	RKOS-ELM	1.96	6.66	8.59	8.77	4.15	7.26	8.03	6.31
	Meta-RKOS-ELM-DDM	**1.57**	5.02	**6.39**	7.53	3.45	5.63	6.99	5.24
	Meta-RRKOS-ELM-DDM	1.87	**3.27**	6.45	**7.21**	**2.80**	**5.12**	**6.81**	**4.78**
TS$_6$	RKOS-ELM	**1.27**	**2.80**	13.25	12.56	**2.74**	10.27	16.84	9.54
	Meta-RKOS-ELM-DDM	2.70	6.79	14.18	16.84	5.79	9.92	16.55	10.52
	Meta-RRKOS-ELM-DDM	3.52	6.38	**12.18**	**11.38**	5.59	**8.47**	**10.92**	**8.17**
SSE	RKOS-ELM	1.37	3.46	5.91	8.05	2.39	4.43	6.74	4.41
	Meta-RKOS-ELM-DDM	1.29	3.36	5.40	6.42	2.27	4.13	5.86	3.98
	Meta-RRKOS-ELM-DDM	**1.17**	**3.02**	**4.59**	**5.45**	**2.07**	**3.69**	**4.87**	**3.45**
Ozone	RKOS-ELM	4.26	5.65	6.93	7.07	5.21	**6.38**	6.98	6.12
	Meta-RKOS-ELM-DDM	**4.15**	5.74	6.61	6.92	5.10	**6.38**	6.80	6.02
	Meta-RRKOS-ELM-DDM	**4.15**	**5.18**	**6.02**	**6.30**	**4.87**	6.48	**6.18**	**5.48**

Note: The best performance is in boldface.

and 7.1 in TS$_4$, TS$_5$, TS$_6$, SSE, and Ozone, respectively. The results show that using recursive kernel technique can improve the overall performance in most of cases for all real-world data sets. Although, using recurrent kernel technique cannot improve SMAPE in the average values of 8 – 12 periods of prediction horizons of Ozone, but there is only a slight difference. This show that recursive kernel method is a good way to improve the performance in multi-step time series prediction.

4   Conclusions and Future Works

In this paper, we introduce an improvement of KOS-ELM named as Meta-RRKOS-ELM-DDM. According to the results of the experiments, Meta-RKOS-ELM-DDM can improve forecasting performance in different prediction horizons in both synthetic and real-world datasets. Because of meta-cognitive learning strategy, Meta-RRKOS-ELM-DDM can decide how to deal with the new coming data–when the model needs to add neuron, to retrain sample, or to discard sample. Moreover, the threshold of ALD can be automatically defined by meta-cognitive. This is not only save the learning time, but also solves the dependency of threshold of ALD. Furthermore, recursive RBF kernel successfully replaces the conventional RBF kernel in the model of Meta-RRKOS-ELM-DDM, which enhances the predicting performance of the different periods of prediction horizons. The results of Meta-RRKOS-ELM-DDM in are better than that of Meta-RKOS-ELM-DDM. The major benefits of Meta-RRKOS-ELM-DDM are as follows: the good generalisation model for solving concept drift problem can be achieved; the limitation of prediction horizon is released; and new modified DDM which is a method of deal with concept drift in time series prediction; improves the predicting performance in the multi-step prediction.

Meta-cognitive plays a significant role in the learning time reduction and dealing with parameter dependency while recursive kernel method generates a fixed reservoir with optimised information by dissipating and overwriting the information from the coming data in the learning part of the prediction model, which is helpful for improvement of forecasting performance. However, searching for the optimal recursive kernel parameter is computational extensive. In future work, we will pay more attention to find out the method to automatically define this parameter in order to reduce the computational complexity.

Acknowledgement. The authors would like to express special thanks of gratitude to UM Grand Challenge from the University of Malaya under Grant GC003A-14HTM, FRGS grant from MOHE FP069-2015A, and the Thailand Research Fund under grant agreement No. TRG5680090 which support our research.

References

1. Kaur, T., Kumar, S., Segal, R.: Application of artificial neural network for short term wind speed forecasting. In: Proceedings of 2016 Biennial International Conference on Power and Energy Systems: Towards Sustainable Energy, pp. 1–5 (2016)
2. Pradeepkumar, D., Ravi, V.: Forecasting financial time series volatility using particle swarm optimization trained quantile regression neural network. Appl. Soft Comput. **58**, 35–52 (2017)
3. Scardapane, S., Comminiello, D., Scarpiniti, M., Uncini, A.: Online sequential extreme learning machine with kernels. IEEE Trans. Neural Netw. Learn. Syst. **26**(9), 2214–2220 (2015)
4. Hammoodi, M., Stahl, F., Tennant, M.: Towards online concept drift detection with feature selection for data stream classification. In: Proceedings of the 22nd European Conference on Artificial Intelligence (2016)
5. Dehghan, M., Beigy, H., ZareMoodi, P.: A novel concept drift detection method in data streams using ensemble classifiers. Intell. Data Anal. **20**(6), 1329–1350 (2016)
6. Kulkarni, P., Ade, R.: Logistic regression learning model for handling concept drift with unbalanced data in credit card fraud detection system. In: Satapathy, S.C., Raju, K.S., Mandal, J.K., Bhateja, V. (eds.) Proceedings of the Second International Conference on Computer and Communication Technologies. AISC, vol. 380, pp. 681–689. Springer, New Delhi (2016). https://doi.org/10.1007/978-81-322-2523-2_66
7. Liu, Z., Loo, C.K., Masuyama, N., Pasupa, K.: Multiple steps time series prediction by a novel recurrent kernel extreme learning machine approach. In: Proceeding of the 9th International Conference on Information Technology and Electrical Engineering, p. SIG5.5 (2017)
8. Liu, Z., Loo, C.K., Masuyama, N., Pasupa, K.: Recurrent kernel extreme reservoir machine for time series prediction. IEEE Access **6**, 19583–19596 (2018)
9. Gonçalves, P.M., de Carvalho Santos, S.G., Barros, R.S., Vieira, D.C.: A comparative study on concept drift detectors. Expert Syst. Appl. **41**(18), 8144–8156 (2014)

10. Gama, J., Medas, P., Castillo, G., Rodrigues, P.: Learning with drift detection. In: Bazzan, A.L.C., Labidi, S. (eds.) SBIA 2004. LNCS (LNAI), vol. 3171, pp. 286–295. Springer, Heidelberg (2004). https://doi.org/10.1007/978-3-540-28645-5_29
11. Hermans, M.: Expanding the theoretical framework of reservoir computing. Ph.D. thesis, Ghent University (2012)
12. Yahoo: Shanghai stock exchange (2017). https://finance.yahoo.com/quote/000001.SS?p=000001.SS
13. Ministry of the Environment and Climate Change in Ontario: Air pollutant data (2000). http://www.airqualityontario.com/history

Analysis and Application of Step Size of RK4 for Performance Measure of Predictability Horizon of Chaotic Time Series

Shoya Matsuzaki, Kazuya Matsuo, and Shuichi Kurogi[✉]

Kyushu Institute of technology, Tobata, Kitakyushu, Fukuoka 804-8550, Japan
matsuzaki.shoya605@mail.kyutech.jp, {matsuo,kuro}@cntl.kyutech.ac.jp
http://kurolab.cntl.kyutech.ac.jp/

Abstract. So far, we have presented several methods for chaotic time series prediction, and shown performance improvement on predictability horizon. However, we could not have shown the comparison of the performance with other methods. In order to obtain general and absolute performance measure of predictability horizon, this paper analyzes to formulate the relationship between the mean predictability horizon and the step size of the fourth-order Runge-Kutta method, or RK4. By means of using the formula which we have obtained in this article, the step size of RK4 corresponding to the mean predictability horizon achieved by a learning machine can be obtained without executing RK4. We execute numerical experiment of the prediction by several learning machines, and compare the performance by means of the step size of RK4 corresponding to the mean horizon achieved by the learning machines, and we show the effectiveness of the present method.

Keywords: Chaotic time series prediction
Analysis and application of step size of RK4
Performance measure of predictability horizon

1 Introduction

So far, we have presented several prediction methods for chaotic time series and shown performance improvement on predictability horizon [1–4]. In our methods, we employ IOS (iterated one-step ahead) prediction, especially for Lorenz time series, and try to obtain longer predictability horizon. Here, the predictability horizon H [step] of a time series indicates the number of prediction steps after which the time series becomes unpredictable or the prediction error exceeds a certain threshold δ_y. Although we have shown our recent methods have obtained longer predictability horizon than our previous methods, there is no general performance measure of probability horizon to compare with the methods of other researches. Here, there are several problems. A problem is that no numerical

© Springer Nature Switzerland AG 2018
L. Cheng et al. (Eds.): ICONIP 2018, LNCS 11306, pp. 14–23, 2018.
https://doi.org/10.1007/978-3-030-04224-0_2

data have been provided for chaotic differential dynamics, such as Lorenz equations, mainly owing that numerical data can be obtained via various numerical methods. However, it is not so easy to obtain common data because time series is very sensitive to the initial state so that high precision data requiring large memory capacity will be necessary. Although several typical parameter values are provided, there are infinite number of parameter values theoretically. Furthermore, a number of sampling periods are possible. Another problem is that the predictability horizon H for different prediction start time changes largely. A solution may be the use of the mean of H. However, necessary and/or sufficient number of predictions and appropriate start points in time for obtaining the mean H are not so clear. Incidentally, the predictability horizon H [step] is almost the same as the critical time of decoupling T_c [LTU] (Lorenz time unit) [5–7], defined as the first point in time after which the state vector norm error exceeds a certain error tolerance δ. Here, T_c can be estimated for given small δ corresponding to Lipschitz constant, but we consider H for δ_y much larger than δ and the formula for estimating T_c shown in [5,6] could not be applicable to H by simply regarding T_c as HT_S with sampling period T_S [LTU/step].

This paper tries to obtain the formula of the mean H and presents a performance measure based on step size Δt of the fourth order Runge-Kutta method, or RK4. By means of numerical experiments on the predictability horizon of Lorenz time series, we approximate the formula of the function H of δ_y and Δt of RK4 in Sect. 2. By means of using the formula, the step size of RK4 corresponding to the mean predictability horizon achieved by a learning machine can be obtained without executing RK4. In Sect. 3, we show experimental result of the prediction by several learning machines, and compare the performance by means of the step size of RK4 corresponding to the mean horizon achieved by the learning machines, and we show the effectiveness of the present method.

2 Step Size of RK4 for Performance Measure of Predictability Horizon

2.1 Numerical Solution of Lorenz Equations and Predictability Horizon

We focus on Lorenz time series obtained from the differential dynamical system (or Lorenz equations) for state vector $\boldsymbol{x}_c = (x_c, y_c, z_c)^T$ and continuous time t_c given by

$$\frac{dx_c}{dt_c} = -\sigma x_c + \sigma y_c, \quad \frac{dy_c}{dt_c} = -x_c z_c + r x_c - y_c, \quad \frac{dz_c}{dt_c} = x_c y_c - b z_c, \quad (1)$$

where we employ $\sigma = 10$, $b = 8/3$, $r = 28$ and the initial state $\boldsymbol{x}_c(0) = (-8, 8, 27)^T$ (see [8] for properties of the dynamics). We have generated time series $y(t) = x_c(tT_S)$, where t [step] represents discrete time with the relationship $t_c = tT_S$ [LTU] for $t = 0, 1, 2, \cdots$ and sampling period $T_S = 0.025$ [LTU/step]. In the following, we use several time variables, such as predictability horizon

H, and discriminate the domain by denoting H [step] or H [LTU]. Since y_t is obtained not analytically but numerically, y_t involves an error owing to finite numerical precision, it is impossible to make a long-term prediction. As shown in [5–7], the critical time of decoupling T_c, defined as the first point in time after which the norm of state vector error exceeds a certain threshold, has mathematical relationship with the step-size Δt and the number of bits of precision, P. An important fact shown in [6] is that $T_c = 35$ LTU is achieved with double precision data ($P = 64$-bit), and $T_c = 10^4$ LTU is shown to be obtained by multiple precision method using high order Taylor expansion scheme. However, from [7] it is shown that 4000-digit multiple precision of initial condition is necessary to achieve $T_c = 1000$ LTU, which might have no physical meanings.

On the other hand, this paper focuses on the prediction of learning machines running on standard 64-bit Linux computer and evaluate the performance on predictability horizon. For simple expression in the following, let $y_{t_p:h_p} = y_{t_p} y_{t_p+1} \cdots y_{t_p+h_p-1}$ denote a time series of y_t with a start time $t = t_p$ and a horizon h_p. Furthermore, let us define predictability horizon $H = H\left(y_{t_p:h_p}; \delta_y\right)$ by

$$H\left(y_{t_p:h_p}; \delta_y\right) = \max\left\{h \mid \forall s < h \leq h_p; |y_{t_p+s} - y_{t_p+s}^{[\mathrm{gt}]}| \leq \delta_y\right\} \quad (2)$$

for a ground truth time series $y_{t_p:h_p}^{[\mathrm{gt}]}$ and an error threshold δ_y. Note that H [step] is almost the same as the critical time of decoupling $T_c \simeq HT_S$ [LTU], while the above error threshold δ_y is larger than the error tolerance $\delta = \|x_c - x_c^{[\mathrm{gt}]}\|$ satisfying Lipschitz condition for estimating T_c theoretically.

For evaluating H of the prediction of learning machines implemented on standard computer, we would like to generate a ground truth time series $y_{t_p:h_p}^{[\mathrm{gt}]}$ easily instead of using advanced methods as developed in [5–7]. So, we employ the classical fourth order Runge-Kutta method, or RK4, coded with GMP (GNU multi-precision library) implemented on standard 64-bit Linux computer. Furthermore, in the following, we regard a numerical solution obtained by RK4 with GMP as a prediction.

2.2 Predictability Horizon Obtained by RK4 and GMP

In order to see the performance of the numerical method using RK4 and GMP, we show numerical solutions obtained for several pairs of step size Δt and multiple precision P in Fig. 1. Here, we can see that larger critical time of decoupling T_c or larger predictability horizon H is achieved by smaller step size Δt from (a), and larger precision P from (b) for the prediction start time $t_p = 0$. From the result, we have decided to use the prediction with the smallest $\Delta t = 10^{-8}$ LTU and the largest $P = 256$ bit in the experiments as ground truth. From the predictability horizon for different prediction start time shown in Table 1, we can see that the above property holds for not each H but the mean H over different start time t_p. Namely, we can see that the mean H increases with the decrease of step size Δt for $P = 64$ bits and with the increase of P for $\Delta t = 10^{-8}$, although the variance of H seems large especially for small Δt. Incidentally, the result of the mean $H = 1363$ for $P = 64$ and $\Delta t = 10^{-8}$ indicates that $T_c \simeq 34.1$ LTU which seems consistent and compatible with $T_c = 35$ LTU achieved for double precision data ($P = 64$ bit) obtained with high order Taylor expansion scheme [7] as shown above.

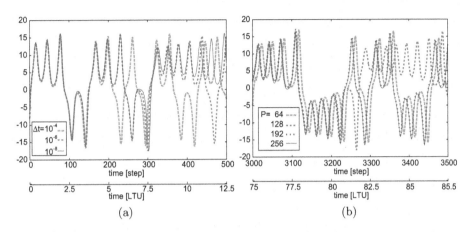

Fig. 1. (a) Prediction y_t ($t \in [0, 500]$) obtained for $\Delta t = 10^{-4}$, 10^{-6}, 10^{-8} LTU with $P = 256$ bit, and (b) prediction y_t ($t \in [3000, 3500]$) obtained for $P = 64$, 128, 192, 256 bit with $\Delta t = 10^{-8}$ LTU. The prediction start time is $t_P = 0$. Here, we cannot see the difference between the predictions for $P = 192$ and 256 until 5120 steps as shown in Table 1.

2.3   Formula of Mean Predictability Horizon

The mean predictability horizon H w.r.t. Δt and δ_y obtained for $P = 64$ bits is shown in Fig. 2. From the numerical result in (a), we have obtained a linear function H of $\ln(\Delta t)$ by means of applying the least square fitting method to the form given by

$$H = A \ln(\Delta t) + B \tag{3}$$

and we have obtained the values as

$$(A, B) = \begin{cases} (-40.0381, -149.529) & \text{for } \delta_y = 4, \\ (-39.6599, -135.641) & \text{for } \delta_y = 5, \\ (-39.3807, -118.594) & \text{for } \delta_y = 8, \\ (-40.0233, -114.167) & \text{for } \delta_y = 10, \\ (-41.9565, -115.719) & \text{for } \delta_y = 15. \end{cases} \tag{4}$$

We can see that a good approximation has been achieved from Fig. 2(b), although there are small approximation error. Here, we have excluded the data with

Table 1. Predictability horizon H for different prediction start time $t_p = 0, 1000, 2000, \cdots, 9000$ and the mean for several pairs of numerical precision P and step size Δt. The prediction $y_{0:10000}|_{P=256, \Delta t=10^{-8}}$ is assumed the ground truth prediction, and the error tolerance is $\delta_y = 15$.

P	64									128	192
Δt	$10^{-2}/2$	10^{-3}	$10^{-3}/2$	10^{-4}	10^{-5}	10^{-6}	10^{-7}	$10^{-7}/2$	10^{-8}	10^{-8}	
$t_p = 0$	71	161	192	225	377	460	640	611	1371	3252	5120
1000	118	154	156	273	338	458	551	583	1255	3170	5076
2000	74	135	138	254	348	500	478	481	1369	3218	5019
3000	70	102	105	247	343	459	498	510	1263	3227	5245
4000	49	165	198	261	352	444	564	596	1328	3366	5304
5000	119	197	197	327	444	507	626	656	1320	3283	5281
6000	74	164	168	230	353	490	480	483	1486	3430	5359
7000	403	368	427	460	584	645	710	713	1463	3446	5411
8000	63	128	156	244	339	401	467	553	1338	3414	5336
9000	60	151	152	214	279	418	572	599	1437	3430	5366
mean H [step]	110	173	189	274	376	478	559	579	1363	3320	5252
mean H [LTU]	2.8	4.3	4.7	6.8	9.4	12.0	14.0	14.5	34.1	83.0	131.3

$\Delta t = 10^{-8}$ for the fitting because they do not fit well to the above approximation lines (see below to examine the reason). Furthermore, we have tried to make a further approximation of A and B for all δ_y, and we finally have the form

$$H = (a_0 + a_1 \delta_y + a_2 \delta_y^2) \ln \Delta t + b_0 + b_1 \delta_y + \tau \ln \delta_y, \tag{5}$$

with $(a_0, a_1, a_2) = (-41.6501, 0.587778, -0.0406851)$, $(b_0, b_1) = (-7.4675, -239.551)$ and $\tau = 87.0274$ [step] $= 2.175685$ [LTU]. Here, note that τ represents time constant as shown below. This approximation seems good as shown in Fig. 2(c)–(e). Now, from (5), we have

$$\delta_y = \left[\Delta t^{-(a_0 + a_1 \delta_y + a_2 \delta_y^2)/\tau} \exp(-(b_0 - b_1 \delta_y)/\tau) \right] \exp(H/\tau). \tag{6}$$

Since this equation is recursive or the right hand side involves δ_y itself, it is not so easy to analyze. Here, note that δ_y basically represents the error threshold with the values 4, 5, 8, 10, 15, but δ_y can also be regarded as the error itself for given H and Δt. Following this context, let us derive the following equation:

$$\delta_y = \alpha \delta_{y0} \exp(H/\tau), \tag{7}$$

where

$$\delta_{y0} = \Delta t^{-(a_0 + a_1 + a_2)/\tau} \exp(-(b_0 + b_1)/\tau), \tag{8}$$

$$\alpha = \Delta t^{-(a_2(\delta_y - 1)^2 + (a_1 + 2a_2)(\delta_y - 1))/\tau} \exp\left(-b_1(\delta_y - 1)/\tau\right). \tag{9}$$

This equation shows that we have $\delta_y = \delta_{y0} \exp(H/\tau)$ for $\delta_y = 1$, which indicates that the error δ_y increases exponentially with time constant τ from $\delta_y = \delta_{y0}$ at the prediction start time $H = 0$.

Δt	$10^{-2}/2$	10^{-3}	$10^{-3}/2$	10^{-4}	10^{-5}	10^{-6}	10^{-7}	$10^{-7}/2$	10^{-8}
$\delta_y = 4$	69.0	135.3	157.7	193.0	316.0	400.8	498.0	528.3	1313.6
5	77.6	140.2	165.7	215.1	334.3	407.6	504.3	531.4	1321.7
8	93.0	159.6	169.3	241.2	342.9	422.2	515.3	544.7	1332.3
10	101.7	163.3	181.8	244.0	355.1	458.2	528.9	546.7	1335.2
15	110.1	172.5	188.9	273.5	375.7	478.2	558.6	578.5	1363.0

(a)

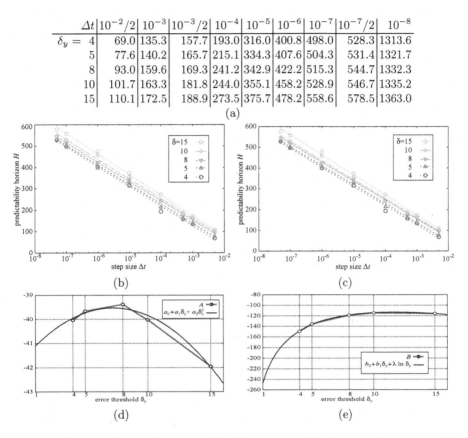

(b)   (c)

(d)   (e)

Fig. 2. Mean predictability horizon H w.r.t. step size Δt and error threshold δ_y. (a) Numerical result of mean H for Δt and δ_y, (b) linear approximation of H via (3), (c) further approximation of H via (5), (d) approximation of A by $(a_0 + a_1\delta_y + a_2\delta_y^2)\ln\Delta t$, and (e) approximation of B by $b_0 + b_1\delta_y + \tau\ln\delta_y$.

Next, let us examine the relationship between Eqs. (7)–(9) and the formula of the error tolerance, $\delta \approx \delta_0 \exp(\lambda T_c)$, shown in the theory of the critical time of decoupling [5,6]. Here, the initial error δ_0 consists of truncation error $\delta_{0t} \propto \Delta t^N$ and round-off error $\delta_{0r} \propto \Delta t^{-1/2}$, where N is the order of numerical scheme while we have employed RK4 with $N = 4$ in our experiment. Although $\delta_0 \propto \Delta t^2$ in [5] using $N = 2$, the above result $\Delta t^{-(a_0+a_1+a_2)/\tau}$ with $-(a_0 + a_1 + a_2)/\tau \approx 0.47$ seems very small with respect to $N = 4$ of RK4. The other parameter λ is estimated by the maximum Lyapunov exponent ($\lambda \approx 0.9$ for the Lorentz equations) in [5], while the corresponding $1/\tau \approx 0.46$ [LTU^{-1}] in the above result is smaller than $\lambda \approx 0.9$. Here, one of the differences between δ and δ_y is that T_c is estimated under a small δ satisfying Lipschitz condition but H has been obtained for δ_y larger than Lipschitz constant. Moreover, the mean $H = 34.1$[LTU] for

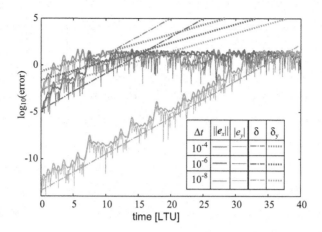

Fig. 3. Time evolution of state vector error $\|e_x\|$, prediction error $|e_y|$ and their approximation δ and δ_y for $\Delta t = 10^{-4}$, 10^{-6} and 10^{-8} for the prediction start time $t_p = 1$ step $= 0.025$ LTU.

$\Delta t = 10^{-8}$ compatible with $T_c = 35$ LTU as shown above seems to fit not the above equations but the estimation formula of T_c.

In order to examine these facts furthermore, we have obtained time evolution of the state norm error $\|e_x\| = \|x_c - x_c^{[gt]}\|$, the prediction error $|e_y| = |y_t - y_t^{[gt]}|$, the error $\delta = \delta_0 \exp(0.9t)$ for the approximation of $\|e_x\|$ and δ_y of Eqs. (7)–(9) for the approximation of the mean H as shown in Fig. 3. Here, we can see that $\|e_x\|$ changes as the upper envelope of $|e_y|$ as expected from their definitions. We can also see that δ approximates the ratio of the exponential increase of $\|e_x\|$ and $|e_y|$ well from the prediction start time. In contrast, δ_y seems to approximate the upper envelope of $|e_y|$ as well as $\|e_x\|$ well only for the points in time after $|e_y|$ increases larger than 1 (i.e. $\log_{10}|e_y| = 0$ in the figure) for $\Delta t = 10^{-4}$ and 10^{-6}. This is considered to be necessary and sufficient for approximating the mean H for the error threshold larger than 1. However, we can see that δ_y for approximating H will not work with $\Delta t = 10^{-8}$ which provided extremely small $\|e_x\|$ and $|e_y|$ compared with those for $\Delta t = 10^{-4}$ and 10^{-6}. We do not have clarified the reason so far, and we would like to examine it in our future research.

2.4 Step Size of RK4 for Performance Measure of Mean Predictability Horizon

From (5), we have

$$\Delta t = \exp\left(\frac{H - b_0 - b_1 \delta_y - \tau \ln \delta_y}{a_0 + a_1 \delta_y + a_2 \delta_y^2} \right). \tag{10}$$

With this equation, we can obtain the step size Δt of RK4 corresponding to the mean H for a threshold δ_y of the predictions achieved by a learning algorithm without executing RK4 for various Δt. Since RK4 is a standard and easily executable method with theoretically clear properties and the above smooth relationship among H, Δt and δ_y, we would like to use the step size of Δt for a performance measure of the mean H of learning machines.

3   Numerical Experiment on Predictability Horizon of Learning Machines

We have conducted numerical experiment of time series prediction with learning machines. We use a training time series $y_{0:20000}$ generated by RK4 with $P = 256$ bit and step size $\Delta t = 10^{-8}$, and sampling period $T_S = 0.025$ LTU. From time-embedding theory, we assume that y_t and $\boldsymbol{x}_t = (y_{t-1}, y_{t-2}, \cdots, y_{t-k})^T$ with embedding dimension k has a relationship given by

$$y_t = r(\boldsymbol{x}_t) + e(\boldsymbol{x}_t), \tag{11}$$

where $r(\boldsymbol{x}_t)$ is a nonlinear target function, and $e(\boldsymbol{x}_t)$ represents an error. The training set is constructed as $D^{[\text{train}]} = \{(\boldsymbol{x}_t, y_t) \mid t \in I^{[\text{train}]}\}$ for $I^{[\text{train}]} = \{k, k+1, \cdots, 1999\}$.

After the learning, the machine executes IOS (iterated one-step ahead) prediction by $\hat{y}_t = f(\hat{\boldsymbol{x}}_t)$ for $t = t_P, t_{P+1}, \cdots$, recursively. Here $f(\hat{\boldsymbol{x}}_t)$ denotes prediction function learned by the learning machine of $\hat{\boldsymbol{x}}_t = (\hat{x}_{t1}, \hat{x}_{t2}, \cdots, \hat{x}_{tk})$ whose elements are given by $\hat{x}_{tj} = y_{t-j}$ for $t - j < t_P$ and $\hat{x}_{tj} = \hat{y}_{t-j}$ for $t - j \geq t_P$. We have executed the prediction for $t_P = 2000, 2100, 2200, \cdots, 3000$.

As learning machines, we use bagging CAN2 (bagging competitive associative net 2), MLP (multi-layer perceptron), and LSTM (long-short term memory). For bagging CAN2, we use the same parameter values as shown in [3,4], i.e. the embedding dimension $k = 10$ and the number of units $N = 5i$ for $i = 1, 2, \cdots, 50$. For MLP implemented on the Chainer framework [9], we have optimized the parameters and obtained good results with two hidden layers involving 300 units with leaky ReLU, and dropout ratio being 0.001, which seems very small but provided best result in our experiments. We show the result for $k = 60$ which has achieved the best result among $k = 10, 20, 30, 40$ 50, 55, 60, 65, 70, 80 and 90. We have executed 10000 epochs with batchsize being 100 for learning. For LSTM also implemented on the Chainer framework, we have optimized the parameters and obtained good results with two hidden layers involving 200 units with ReLU, and dropout ratio with 0.3. We show the result for $k = 11$ which has achieved the best result among $k = 5, 7, 8, 9, 10, 11, 12, 13, 15, 20$ and 30. We have executed 10000 epochs for learning.

We show the achieved predictability horizon in Fig. 4. From (a), (b) and (c), the predictability horizon for different t_p changes largely. From (d), we can estimate the property on the performance of the predictability horizon by means of the step size Δt of RK corresponding to the mean H of each learning machine. Here, small Δt corresponds to large H as shown in Fig. 2(b) and (c). Although we can compare the performance of three learning machines by means of the mean H as a relative measure, i.e. CAN2 has the largest mean H, followed in

order by MLP and LSTM. On the other hand, the step size Δt of RK4 shown
in (d) indicates information much more as follows. Namely, we can see that
the performance of CAN2 for $\delta_y = 15$ is $\Delta t \approx 5.5 \times 10^{-4}$ which may be good
because we usually use RK4 with $\Delta t = 10^{-3} \sim 10^{-4}$ for a high numerical
precision. Furthermore, CAN2 has achieved the result with larger Δt for smaller
δ_y which indicates that CAN2 for smaller δ_y shows poor performance than RK4
with $\Delta t = 5.5 \times 10^{-4}$. This means that, of course, the mean H for smaller δ_y
of RK4 with the same Δt achieves smaller H as shown in Fig. 2(b) and (c), and
CAN2 has achieved much smaller mean H than RK4. This suggests that CAN2
may have a possibility to improve the performance for smaller δ_y somehow, e.g.
by means of improving learning algorithm, increasing the number of training and
test data, and so on, because RK4 with $\Delta t = 5.5 \times 10^{-4}$ has better performance
than CAN2. Ultimately and theoretically, it is expected that the performance of
a learning machine may be improved to the best performance of RK4 or other
numerical methods on a given computer with given precision bits.

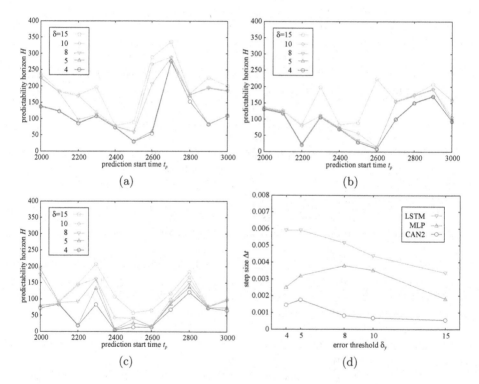

Fig. 4. Predictability horizon H achieved by (a) CAN2, (b) MLP and (c) LSTM, and
(d) the step size Δt of RK4 corresponding to the mean H of (a), (b) and (c). The mean
H for $\delta_y = 4, 5, 8, 10, 15$ is 112, 115, 163, 177, 199 for CAN2, 90, 92, 102, 111, 149 for
MLP, and 56, 67, 90, 103, 123 for LSTM, respectively. The mean for all δ_y is 153, 109
and 88 for CAN2, MLP and LSTM, respectively.

4   Conclusion

We have analyzed to formulate the relationship between the predictability horizon H, the step size Δt of RK4, and the prediction error threshold δ_y. By means of using the formula, Δt of RK4 corresponding to the mean H achieved by a learning machine can be obtained without executing RK4. With the mean H achieved by learning machines, CAN2, MLP and LSTM, the corresponding Δt of RK4 can be used as not relative but more absolute performance measure than the mean H of learning machines because RK4 is a standard and easily executable numerical method with clear properties. We have to clarify the obtained formula much more from the point of view of numerical and chaos theory. Especially, we could not have clarified the reason that the formula could not be applicable to H for $\Delta t = 10^{-8}$, which is for our future research.

References

1. Kurogi, S., Shigematsu, R., Ono, K.: Properties of direct multi-step ahead prediction of chaotic time series and out-of-bag estimate for model selection. In: Loo, C.K., Yap, K.S., Wong, K.W., Teoh, A., Huang, K. (eds.) ICONIP 2014, Part II. LNCS, vol. 8835, pp. 421–428. Springer, Cham (2014). https://doi.org/10.1007/978-3-319-12640-1_51
2. Kurogi, S., Toidani, M., Shigematsu, R., Matsuo, K.: Probabilistic prediction of chaotic time series using similarity of attractors and LOOCV predictable horizons for obtaining plausible predictions. In: Arik, S., Huang, T., Lai, W.K., Liu, Q. (eds.) ICONIP 2015. LNCS, vol. 9491, pp. 72–81. Springer, Cham (2015). https://doi.org/10.1007/978-3-319-26555-1_9
3. Kurogi, S., Toidani, M., Shigematsu, R., Matsuo, K.: Performance improvement via bagging in probabilistic prediction of chaotic time series using similarity of attractors and LOOCV predictable horizon. Neural Comput. Appl. J. (2017). https://doi.org/10.1007/s00521-017-3149-7
4. Kurogi, S., Shimoda, N., Matsuo, K.: Hierarchical clustering of ensemble prediction using LOOCV predictable horizon for chaotic time series. In: Proceedings of IEEE-SSCI 2017, pp. 1789–1795 (2017)
5. Teixeira, J., Reynolds, C.A., Judd, K.K.: Time step sensitivity of nonlinear atmospheric models: numerical convergence, truncation error growth, and ensemble design. J. Atmos. Sci. 64, 175–189 (2007)
6. Wang, P., Li, J., Li, Q.: Computational uncertainty and the application of a high-performance multiple precision scheme to obtaining the correct reference solution of Lorenz equations. Numer. Algorithms 59(1), 147–159 (2011)
7. Liao, S., Wang, P.: On the mathematically reliable long-term simulation of chaotic solutions of Lorenz equation in the interval [0, 1000]. Sci. China Phys. Mech. Astron. 57(2), 330–335 (2014)
8. Aihara, K.: Theories and Applications of Chaotic Time Series Analysis. Sangyo Tosho, Tokyo (2000)
9. Chainer Homepage. https://chainer.org/

Simultaneous Analysis of Subjective and Objective Data Using Coupled Tensor Self-organizing Maps: Wine Aroma Analysis with Sensory and Chemical Data

Keisuke Yoneda[1], Kimihiro Nakano[2], Keiichi Horio[1], and Tetsuo Furukawa[1(✉)]

[1] Kyushu Institute of Technology, Kitakyushu 808-0196, Japan
yoneda-keisuke@edu.brain.kyutech.ac.jp,
{horio,furukawa}@brain.kyutech.ac.jp
[2] Kuraray Co., Ltd., Tokyo 100-8115, Japan
Kirihiro.Nakano@kuraray.com

Abstract. In this paper, we propose a method for simultaneous analysis of subjective and objective data. The method, named coupled tensor self-organizing map (SOM), consists of two tensor SOMs, one of which learns the subjective data while the other learns the objective data. The coupled tensor SOM visualizes the dataset as three maps, namely, one target object map, and two survey item maps corresponding to the subjective and objective data. This method can be further extended to generate extra maps such as a map of attributes. In addition, the coupled tensor SOM also provides an interactive visualization of the relationship between the target objects and the survey items by coloring these three maps. We applied our proposed method to the wine aroma dataset. Our results indicate that this method facilitates an intuitive overview of the dataset.

Keywords: Subjective and objective data · Multi-view data
Coupled tensor decomposition · Multi-relational data · Tensor SOM
Self-organizing map

1 Introduction

Simultaneous analysis of subjective and objective data plays an important role in various fields pertaining to human senses, behavior, activity, and quality of life. Application examples include image/audio evaluation [11,19,22], engineering design [4], ergonomics [16], food testing [15], medical care and nursing [6], athlete coaching [18], and behavior analysis [23].

This work was supported by JSPS KAKENHI Grant Number 18K11472 and ZOZO Research.

L. Cheng et al. (Eds.): ICONIP 2018, LNCS 11306, pp. 24–35, 2018.
https://doi.org/10.1007/978-3-030-04224-0_3

In these fields, the aim of simultaneous surveys is not only analyzing the target objects, but also revealing the relationship between the subjective evaluations and the objective evidence. This typically requires examining combinations of subjective and objective survey items to determine their relationships. As the numbers of the survey items increases, the number of combinations grows extremely rapidly, hindering the feasibility of this analysis.

To solve this problem, we propose a method using a nonlinear dimensionality reduction approach. The proposed method not only visualizes the target objects by mapping them to a low-dimensional (usually two-dimensional) space, but it also maps the survey items to their own low-dimensional spaces. Thus, the method organizes three low-dimensional representations, corresponding to the target objects, the subjective survey items, and the objective survey items. As the result, each survey item is assigned to a low-dimensional coordinate, and the relationship between two survey items is translated to a correlation between two coordinates, which can be visualized by coloring the low-dimensional spaces.

For this purpose, we employed an extension of the self-organizing map (SOM) to tensorial data called the tensor SOM (TSOM) [12]. The proposed method consists of two TSOMs, one of which learns the subjective data, while the other learns the objective data. These two TSOMs are coupled so that they integrate the two datasets. The method, which we call coupled-TSOM (c-TSOM), organizes the following three maps: *the target object map*, *the subjective survey item map*, and *the objective survey item map*. In addition, it is further possible to organize an extra map for the attribute data by coupling a third TSOM, if necessary.

The structure of this paper is as follows. In Sect. 2, we formulate the problem. In Sect. 3, we overview related work. We introduce our c-TSOM algorithm in Sect. 4. Section 5 shows the results, and Sect. 6 concludes the paper.

2 Problem Formulation

We begin by defining two terms used in this paper, *view* and *mode*. The term *view* represents *an aspect of data observation*. A view corresponds to a data array constituting the entire dataset. Thus, when the target objects are measured over V views, then the dataset consists of V data arrays. The simultaneous survey dataset usually consists of two views: the *subjective view* and the *objective view*. A dataset may also have an additional view called the *attribute view*. In the case of the wine aroma dataset, the subjective and the objective views are the sensory and the chemical test data, respectively, and the attribute view corresponds to the grape varieties used in the wines. Note that the proposed method can be generalized to arbitrary V-view datasets.

The term *mode* represents *an aspect of data analysis*. A mode corresponds to a dimension of a data array. Simultaneous surveys typically consist of three modes: *the target object mode*, *the subjective survey item mode*, and *the objective survey item mode*. If the dataset has an attribute view, then we also have *the attribute item mode*. In the case of the wine datasets, there are four modes corresponding to the wines, the aroma types, the chemical substances, and the grape varieties. Note that the wine mode is shared by all of the views.

Suppose that we have N target objects to be analyzed, and they are surveyed by V-views, each of which consists of M_v survey items ($v = 1, \ldots, V$). Suppose we have M_1 subjective and M_2 objective survey items, respectively, and let $x_{nm_1}^{(1)}$ be the subjective measurement data from the n-th target object on the m_1-th survey item. Then, the subjective dataset becomes an array $\mathbf{X}_1 = \left(x_{nm_1}^{(1)}\right)$. Similarly, the objective data becomes an array $\mathbf{X}_2 = \left(x_{nm_2}^{(2)}\right)$, where $x_{nm_2}^{(2)}$ represents the measurement data from the n-th target objects on the m_2-th objective survey item. Finally, if necessary, the attribute data is given by $\mathbf{X}_3 = \left(x_{nm_3}^{(3)}\right)$, where $x_{nm_3}^{(3)}$ represents the degree to which the n-th object has the m_3-th attribute.

To visualize such a dataset, we assume the following data generation model. Let $z_n \in \mathcal{Z}$ be the low-dimensional (usually two-dimensional) latent variable of target object n, which represents intrinsic properties. Without loss of generality, we can assume that the latent space $\mathcal{Z}$ is a unit square with D_z dimension, (i.e., $\mathcal{Z} = [0,1]^{D_z}$), and the probability of z is uniform on $\mathcal{Z}$ (i.e., $p(z) = 1$). Similarly, $y_{m_v} \in \mathcal{Y}_v$ is the latent variable of the m_v-th survey item of the v-th view. Thus, $y_{m_1} \in \mathcal{Y}_1$ and $y_{m_2} \in \mathcal{Y}_2$ are the latent variables corresponding to the survey items of the subjective and the objective views, respectively. Then, the observed data are assumed to be generated by $x_{nm_1}^{(1)} = f^{(1)}\left(z_n, y_{m_1}\right) + \varepsilon_{nm_1}^{(1)}$ and $x_{nm_2}^{(2)} = f^{(2)}\left(z_n, y_{m_2}\right) + \varepsilon_{nm_2}^{(2)}$. Here, $f^{(v)} : \mathcal{Z} \times \mathcal{Y}_v \to \mathcal{X}$ is a smooth mapping from the product latent space to the observation space. The observation noise is represented by $\varepsilon_{nm_v}^{(v)} \sim \mathcal{N}\left(0, \beta_v(y_{m_v})\right)$, where $\beta_v(y_{m_v})$ indicates *the relevance* of survey item m_v. Similarly, the data of the attribute view is assumed to be represented by $x_{nm_3}^{(3)} = f^{(3)}\left(z_n, y_{m_3}\right) + \varepsilon_{nm_3}$, if necessary[1]. Under this assumption, the tasks consist of estimating the latent variables $\{z_n\}$, $\{y_{m_1}\}$, $\{y_{m_2}\}$, and the nonlinear mapping $f^{(1)}$, $f^{(2)}$, (and $\{y_{m_3}\}$, $f^{(3)}$, if necessary).

3   Related Work

In machine learning, the learning tasks presented above are referred to as *multi-view learning*. When the task involves estimating a low-dimensional subspace in an unsupervised manner, the task is referred to as *multi-view dimensionality reduction* (MVDR) [20,24]. The main aim of MVDR is to map the target objects to a low-dimensional space by integrating two (or more) views. The canonical correlation analysis (CCA) is the most basic method for MVDR [9,14]. While CCA is a linear method, some nonlinear methods using Gaussian processes have been also proposed [3,5,17]. However, these methods only map the target objects to a low-dimensional space, and they do not map the survey items. Therefore, CCA and its extensions cannot be applied to our case directly.

---

[1] To be precise, they must be treated as discrete variables, but in this paper we treat them as Gaussian random variables to simplify the explanation.

Aside from multi-view learning, similar studies have been developed in the field of *cross-domain data analysis*. In particular, cross-domain recommendation (e.g., recommending books from movie preference data, and vice versa) has become an important topic in cross-domain relational data analysis [10,25]. Though it appears different, the task of cross-domain recommendation is exactly the same as our task. In relation to the above example, users, books, movies, and user attributes are replaced by the wines, aroma types, chemical substances, and grape varieties, respectively. A representative linear method used for cross-domain recommendation is coupled tensor decomposition (CTD), which consists of multiple tensor decompositions coupled together [1].

Whereas most tensor decompositions are linear methods, the recently proposed TSOM [12] is a nonlinear tensor decomposition method. For an M-mode dataset, TSOM generates M maps corresponding to the visualization of the items of each mode. In addition, TSOM provides various interactive methods, enabling intuitive analysis based on the coloring of the maps. These properties are suitable for our task. Therefore, our aim is to extend CTD from linear to nonlinear modeling by replacing the linear tensor decomposition by the TSOM.

4 Theory and Algorithm

The objective function of SOM has been proposed in several past works [2,7, 8,21]. Thus it is given as the log-likelihood where the neighborhood function represents the posterior of the latent variables, as follows.

$$
\mathcal{L}_{\mathrm{SOM}} = \frac{1}{N} \sum_{n=1}^{N} \int h(\zeta, z_n) \left(-\frac{\beta}{2} \|\mathbf{x}_n - f(\zeta)\|^2 + \frac{D}{2} \ln \frac{\beta}{2\pi} \right) d\zeta, \tag{1}
$$

where $h(\zeta, z)$ is the neighborhood function. In the case of TSOM, the objective function (1) is extended to

$$
\mathcal{L}_{\mathrm{TSOM}} = \frac{1}{NM} \sum_{n=1}^{N} \sum_{m=1}^{M} \iint h_1(\zeta, z_n) h_2(\eta, y_m) \left(-\frac{\beta}{2} \|\mathbf{x}_{nm} - f(\zeta, \eta)\|^2 + \frac{D}{2} \ln \frac{\beta}{2\pi} \right) d\zeta \, d\eta, \tag{2}
$$

for two-mode case, where $h_1(\zeta, z)$ and $h_2(\eta, y)$ are the neighborhood functions, and z_n and y_m are the latent variables corresponding to mode 1 and 2 respectively [12].

By extending (2) for multi-view case, the objective function of c-TSOM is obtained as follows.

$$
\mathcal{L}_{\mathrm{c\text{-}TSOM}} = \sum_{v=1}^{V} \left\{ \frac{1}{NM_v} \sum_{n=1}^{N} \sum_{m_v=1}^{M_v} \right. \tag{3}
$$
$$
\left. \iint h_o(\zeta, z_n) h_v(\eta, y_{m_v}) \left(-\frac{\beta_v(\eta)}{2} \left(x_{nm_v}^{(v)} - f^{(v)}(\zeta, \eta) \right)^2 + \frac{1}{2} \ln \frac{\beta_v(\eta)}{2\pi} \right) d\zeta \, d\eta \right\},
$$

where $h_o(\zeta, z)$ and $h_v(\eta, y)$ are the neighborhood functions. Without loss of generality we assume that $\int h_o(\zeta, z) d\zeta = \int h_v(\eta, y) d\eta = 1$. Note that estimating

---

Algorithm 1 c-TSOM algorithm

Initialize $\{z_n\}$, $\{y_{m_v}\}$ randomly, and intialize $\beta_v(\eta) = 1$.
Calculate $\{f^{(v)}\}$ using Eq. (8), and $\{\varphi_n^{(v)}\}$ $\{\psi_{m_v}^{(v)}\}$ using Eqs. (5) (7).
repeat
    Determine $\{z_n\}$, $\{y_{m_v}\}$ by Eqs. (4) (6).
    Calculate $\{f^{(v)}\}$ using Eqs. (8), and $\{\varphi_n^{(v)}\}$ $\{\psi_{m_v}^{(v)}\}$ using Eqs. (5) (7).
    Calculate $\{\beta_v(\eta)\}$ using Eqs. (9) (10).
until the calculation converges.

Table 1. List of grape varieties.

1	Riesling	9	Syrah	17	Cabernet Sauvignon	25	Molinara	33	Zinfandel
2	Chardonnay	10	Sauvignon Blanc	18	Gamay	26	Nebbiolo	34	Petite Sirah
3	Pinot Noir	11	Terret Blanc	19	Arneis	27	Cabernet Franc	35	Semillon
4	Macabeo	12	Vermentino	20	Cortese	28	Sangiovese	36	Malvasia
5	Parellada	13	Merlot	21	Lagrein	29	Canaiolo	37	Mourvedre
6	Xarello	14	Montepulciano	22	Dolcetto	30	Malvasia Bianca	38	Koshu
7	Garnacha	15	Grenache Gris	23	Corvina	31	Sangiovese Grosso		
8	Cinsault	16	Carignan	24	Rondinella	32	Shiraz		

z_n without integrating the v-th view data can be accomplished by setting the relevance to $\beta_v(\eta) \equiv \beta_v$ and taking $\beta_v \to 0$.

The objective function $\mathcal{L}_{\text{c-TSOM}}$ is optimized with respect to $\{z_n\}$, $\{y_{m_v}\}$, $\{f^{(v)}\}$, and $\{\beta_v(\eta)\}$. They are updated iteratively until the estimated values converge. The latent variables $\{z_n\}$ are determined by

$$z_n \simeq \arg\min_{\zeta} \left\{ \sum_{v=1}^{V} \int \beta_v(\eta) \left(\varphi_n^{(v)}(\eta) - f^{(v)}(\zeta, \eta) \right)^2 p(\eta)\, d\eta \right\} \qquad (4)$$

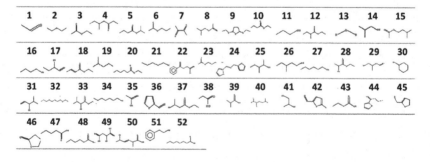

Fig. 1. List of chemical substances.

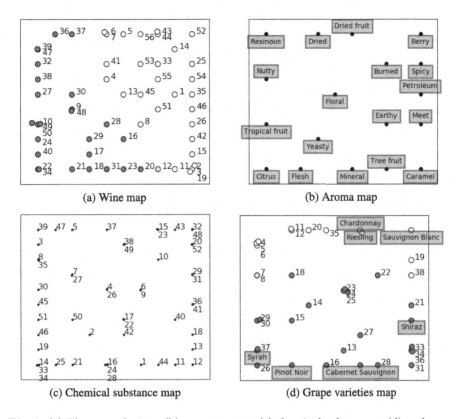

Fig. 2. (a) The map of wines, (b) aromas types, (c) chemical substances, (d) and grape varieties, generated by c-TSOM. (Color figure online)

where

$$\varphi_n^{(v)}(\eta) = \frac{1}{H_v(\eta)} \sum_{m_v=1}^{M_v} h_v\big(\eta, y_{m_v}\big) x_{nm_v}^{(v)}, \qquad H_v(\eta) = \sum_{m_v=1}^{M_v} h_v\big(\eta, y_{m_v}\big). \qquad (5)$$

To obtain (4), we introduce some approximations according to the original TSOM research [12]. Note that this approximation is generally used in the SOM family. In the actual program, the integral is evaluated numerically by discretizing the latent space to grid nodes. Similarly, the latent variables $\{y_{m_v}\}$ are determined by

$$y_{m_v} \simeq \arg\min_{\eta} \left\{ \beta_v(\eta) E_{m_v}^{(v)}(\eta) - \ln \beta_v(\eta) \right\} \qquad (6)$$

where

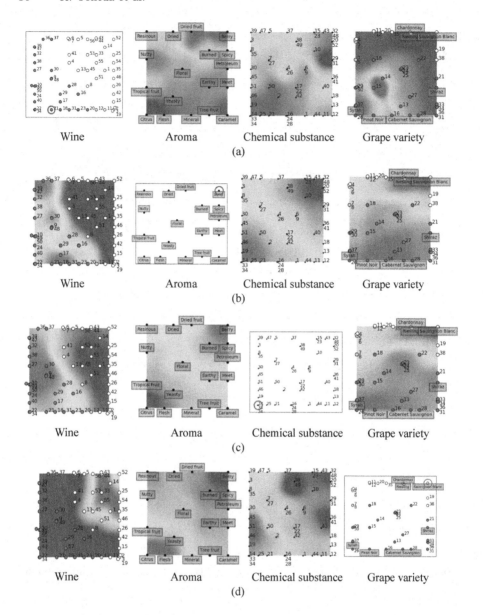

Fig. 3. Visualization using Conditional Component Plane (CCP). The red circles indicated in the white maps show the conditioned point. The red/blue regions denote the areas where the observed values are high/low with respect to the conditioned point. (Color figure online)

$$E_{m_v}^{(v)}(\eta) = \frac{1}{N} \sum_{n=1}^{N} \int h_o(\zeta, z_n) \left(x_{nm_v}^{(v)} - \psi_{m_v}^{(v)}(\zeta) \right)^2 d\zeta + \int \left(\psi_{m_v}^{(v)}(\zeta) - f^{(v)}(\zeta, \eta) \right)^2 p(\zeta) \, d\zeta$$

$$\psi_{m_v}^{(v)}(\zeta) = \frac{1}{H_o(\zeta)} \sum_{n=1}^{N} h_o(\zeta, z_n) x_{nm_v}^{(v)}, \qquad H_o(\zeta) = \sum_{n=1}^{N} h_o(\zeta, z_n). \tag{7}$$

After the latent variables are estimated, the mappings $\{f^{(v)}\}$ are estimated by

$$f^{(v)}(\zeta, \eta) = \frac{1}{H_o(\zeta) H_v(\eta)} \sum_{n=1}^{N} \sum_{m_v=1}^{M_v} h_o(\zeta, n) h_v(\eta, y_{m_v}) x_{nm_v}^{(v)}. \tag{8}$$

As for the ordinary SOM, $f^{(v)}$ is estimated for discretized grid nodes. Finally, the relevance values $\{\beta_v\}$ are updated by

$$\sigma_v^2(\eta) = \sum_{n=1}^{N} \sum_{m_v=1}^{M_v} \int h_o(\zeta, z_n) h_v(\eta, y_{m_v}) \left(x_{nm_v}^{(v)} - f^{(v)}(\zeta, \eta) \right)^2 d\zeta \, d\eta \tag{9}$$

$$\beta_{\text{new}}(\eta) = (1 - \varepsilon) \beta_{\text{old}}(\eta) + \varepsilon \sigma_v^{-2}(\eta), \tag{10}$$

where $0 < \varepsilon < 1$ is a small positive constant.

These calculations are iterated until they converge. During the iterations, the width of the neighborhood size is reduced, as in the ordinary SOM. Note that the objective function (3) is reduced in every step, and the convergence is assured (Table 1 and Fig. 1).

5   Application to Wine Aroma Analysis

We applied c-TSOM to visualize the wine aroma dataset [13]. The dataset consists of data obtained from 56 wines with 18 aromas as determined by sensory testing and 52 substances as determined by chemical measurement (1). In addition, the dataset also provides the grape varieties used in the wines (1).

Our results are shown in Fig. 2. To generate the wine map, the attribute view (i.e., grape varieties) was not used (i.e., $\beta_3 \rightarrow 0$). Thus, the wine map is determined only by the aromas and substances. In the wine map, the marker colors denote the wine types (i.e., white, red, and rosé). Although the wine types were not inputted to c-TSOM, they are separated correctly and clearly in the wine map. The same observation can also be seen in the grape variety map.

By using the Conditional Component Plane (CCP) method [12], it is further possible to analyze the relationship between different modes. Figure 3 shows examples of CCP. In Fig. 3(a), Wine-21[2] is the condition point in the wine map, and other three maps are colored indicating the observation values. The maps suggests that Wine-21 has "berry", "dried fruit", and some "spicy" aromas. It also estimates that Wine-21 made from Grenache Gris (Grape-15) or Montepulciano (Grape-14) (and some Carignan (Grape-16)).

In Fig. 3(b), the "berry" aroma is the condition point in the aroma map, and other three maps are colored indicating the observation values. Thus, the wines in the red region of the wine map have a berry-like aroma, and they contain the chemical substances in the red region of the substance map. It also shows that

---

[2] Wine-21 is a French red wine produced in Collioure, made from Grenache Gris and Carignan. This wine has aromas of "black cherry compote", "blackberry compote", "violet", and "spicy".

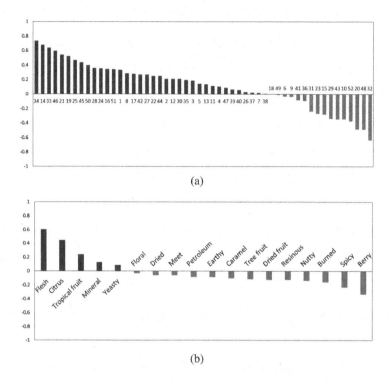

Fig. 4. (a) Correlation coefficients between aroma 'berry' and the chemical substances. (b) Correlation coefficients between aromas and grape variety 'Sauvignon Blanc'. (Color figure online)

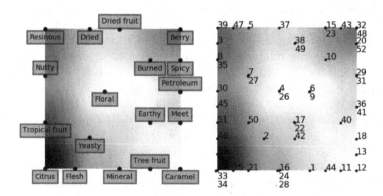

Fig. 5. Relevance β of aromas (left) and substances (right). Black/white indicates high/low relevance.

the aroma is related to the grape varieties in the red region of the grape variety map, for example, Pinot Noir and Cabernet Sauvignon. Similarly, Fig. 3(d) shows that wines made from Sauvignon Blanc have "citrus" and "flesh" aromas, and contain the substances in the red regions of the chemical substance map. These results are consistent to the correlation analysis (Fig. 4).

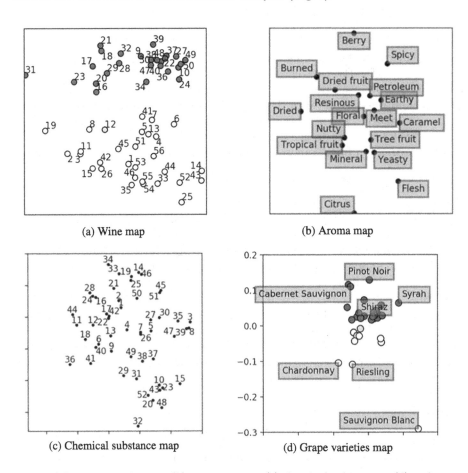

(a) Wine map   (b) Aroma map

(c) Chemical substance map   (d) Grape varieties map

Fig. 6. (a) The map of wines, (b) aromas types, (c) chemical substances, (d) and grape varieties, generated by the coupled tensor decomposition. (Color figure online)

Figure 5 shows the relevance of the survey items. The result shows that the aromas "citrus" and "flesh" in the bottom left corner, and "berry" in the top right corner are important for the consistent integration of the subjective and objective data. It also shows that the substances in the bottom left and the top right corners are important. Note that we can observe that the berry-like aroma is related the substances located at the bottom left corner in Fig. 3 (b) (c).

To examine the performance of c-TSOM, we compared c-TSOM and CTD, which would be the most popular method for the coupled tensor data analysis.

Table 2. Mutual information between views in latent variable estimation.

Methods	Mutual information
Coupled TSOM (Adaptive β_v)	**3.48 ± 0.02**
Coupled TSOM (Fixed β_v)	3.40 ± 0.02
Coupled tensor decomposition	0.70 ± 0.07

The mapping result of CTD is shown in Fig. 6. The obtained result was consistent with the result of c-TSOM, suggesting that they organized the maps appropriately. Furthermore, we assessed the view-integration performance by measuring the mutual information between the subjective and the objective views in latent variable estimation. The result is shown in Table. 2. The mutual information of the c-TSOM is much larger than CTD. In addition, it is also shown that relevance estimation further improves the mutual information.

6   Conclusion

In this paper, we have proposed a novel method called c-TSOM for simultaneous analysis of subjective and objective datasets. As shown in the experimental result on a wine aroma dataset, c-TSOM not only integrates multi-view datasets, but also allows for interactive viewing of data maps using the CCP method. This is useful for determining the relationship between the subjective and objective data items.

Because c-TSOM can be generalized to any V-view datasets, the application field of c-TSOM is not limited to subjective and objective data analysis. For example, cross-domain analysis of e-commerce data is another potential application field. It is likely possible to further generalize the method to any multi-view multi-mode dataset with complex data structures.

References

1. Acar, E., Nilsson, M., Saunders, M.: A flexible modeling framework for coupled matrix and tensor factorizations. In: European Signal Processing Conference, pp. 111–115, no. 1 (2014)
2. Cheng, Y.: Convergence and ordering of Kohonen's batch map. Neural Comput. **9**(8), 1667–1676 (1997)
3. Damianou, A.C., Ek, C.H., Titsias, M.K., Lawrence, N.D.: Manifold relevance determination. In: Proceedings of the 29th International Conference on Machine Learning, ICML 2012, vol. 1, pp. 145–152 (2012)
4. Dang, J., et al.: Optimal design of on-center steering force characteristic based on correlations between subjective and objective evaluations. SASAE Int. J. Passeng. Cars Mech. Syst. **7**(3) (2014)
5. Ek, C.H., Jaeckel, P., Campbell, N., Lawrence, N.D., Melhuish, C.: Shared Gaussian process latent variable models for handling ambiguous facial expressions. In: AIP Conference Proceedings, vol. 1107, pp. 147–153 (2009). https://doi.org/10.1063/1.3106464

6. Fitzpatrick, R., Davey, C., Buxton, M., Jones, D.: Evaluating patient-based outcome measures for use in clinical trials. Health Technol. Assess. **2**(14), i-74 (1998)
7. Graepel, T., Burger, M., Obermayer, K.: Self-organizing maps: generalizations and new optimization techniques. Neurocomputing **21**, 173–190 (1998)
8. Heskes, T., Spanjers, J.J., Wiegerinck, W.: EM algorithms for self-organizing maps. In: IJCNN, vol. 6, pp. 9–14 (2000)
9. Hotelling, H.: The relations of the newer multivariate statistical methods to factor analysis. Br. J. Stat. Psychol. **10**(2), 69–79 (1957). https://doi.org/10.1111/j.2044-8317.1957.tb00179.x
10. Hu, L., Cao, J., Xu, G., Wang, J., Gu, Z., Cao, L.: Cross-domain collaborative filtering via bilinear multilevel analysis. In: IJCAI International Joint Conference on Artificial Intelligence, pp. 2626–2632 (2013)
11. Hu, Y., Loizou, P.: Evaluation of objective quality measures for speech enhancement. IEEE Trans. Audio Speech Lang. Process. **16**(1), 229–238 (2008)
12. Iwasaki, T., Furukawa, T.: Tensor SOM and tensor GTM: nonlinear tensor analysis by topographic mappings. Neural Netw. **77**, 107–125 (2016). https://doi.org/10.1016/j.neunet.2016.01.013
13. Kirihiro, N.: Subjective analyses and their contrast to organoleptic and chemical detection of aroma components. Ph.D. thesis, Kyushu Institute of Technology (2016). (in Japanese)
14. Klami, A., Fi, A.K., Fi, S.J.V., Kaski, S., Fi, S.K.: Bayesian canonical correlation analysis. J. Mach. Learn. Res. **14**, 965–1003 (2013)
15. Kwan, W.O., Kowalski, B.: Correlation of objective chemical measurements and subjective sensory evaluations. Wines of vitis vinifera variety 'pinot noir' from France and the United States. Analytica Chimica Acta **122**(2), 215–222 (1980)
16. Oberg, T., Sandsjo, L., Kadefors, R.: Subjective and objective evaluation of shoulder muscle fatigue. Ergonomics **37**(8), 1323–1333 (1994)
17. Salzmann, M., Ek, C.H., Urtasun, R., Darrell, T.: Factorized orthogonal latent spaces. J. Mach. Learn. Res. Proc. Track **9**, 701–708 (2010)
18. Saw, A., Main, L., Gastin, P.: Monitoring the athlete training response: subjective self-reported measures trump commonly used objective measures: a systematic review. Br. J. Sport. Med. **50**(5), 281–291 (2016)
19. Seshadrinathan, K., Soundararajan, R., Bovik, A., Cormack, L.: Study of subjective and objective quality assessment of video. IEEE Trans. Image Process. **19**(6), 1427–1441 (2010)
20. Sun, S.: A survey of multi-view machine learning. Neural Comput. Appl. **23**(7), 2031–2038 (2013)
21. Verbeek, J., Vlassis, N., Krose, B.: Self-organizing mixture models. Neurocomputing **63**, 99–123 (2005)
22. Wang, Z., Bovik, A., Sheikh, H., Simoncelli, E.: Image quality assessment: from error visibility to structural similarity. IEEE Trans. Image Process. **13**(4), 600–612 (2004)
23. Weeden, J., Sabini, J.: Subjective and objective measures of attractiveness and their relation to sexual behavior and sexual attitudes in university students. Arch. Sex. Behav. **36**(1), 79–88 (2007)
24. Xu, C., Tao, D., Xu, C.: A survey on multi-view learning, pp. 1–59 (2013). ArXiv http://arxiv.org/abs/1304.5634
25. Xu, Z., Jiang, H., Kong, X., Kang, J., Wang, W., Xia, F.: Cross-domain item recommendation based on user similarity. Comput. Sci. Inf. Syst. **13**(2), 359–373 (2016). https://doi.org/10.2298/CSIS150730007Z

Marine Multiple Time Series Relevance Discovery Based on Complex Network

Lei Wang[1], Zongwen Huang[2(✉)], Suixiang Shi[1], Kuo Chen[3], Lingyu Xu[2],
and Gaowei Zhang[2]

[1] National Marine Data and Information Service, Tianjin, China
[2] Shanghai University, Shanghai, China
12716328@qq.com
[3] Tongji University, Shanghai, China

Abstract. Ocean measuring point is an important way to obtain many kinds of marine data. Reasonable layout of ocean measuring points can efficiently obtain marine data. At present, a marine measuring point can acquire multiple types of marine data, only by comprehensively using multiple types of ocean data we can more effectively discover the relationship between various ocean measuring points. This paper proposes a mapping method for fusion marine multiple time series into an image, and uses the similarity between different images to construct a complex network. Also, We build a complex network of marine multiple time series by selecting appropriate thresholds. Compared with the traditional method, the network constructed by our approach can find more accurate rules.

Keywords: Marine multivariate time series · Fusion image
Complex network · Relevance discovery

1 Introduction

Complex network theory has recently experienced a burst of activity, a large number of systems can be described by complex networks [14]. In a network, the vertices are elements in the system, and an edge connects two vertices. Many studies [3,9,13] have shown that complex networks are successfully used to describe various complex systems in reality, such as Internet [1], Collaboration networks [17], Transportation network [6], Biological network [16], Financial systems [5] and so on [7,12]. Due to the wide applications of network theory in many fields, ranging from financial times analysis [18], gene therapy [8] to data mining, the view of studying complex systems from the aspect of the evolution of network structure have been attracting more and more attention. Related statistical analyses have revealed plenty amazing structural features of real networks, such as high network transitivity, power-law degree distributions [2].

Moreover, one ocean measuring point can obtain many kinds of marine data, so many ocean measuring points can also establish a complex system, there may

© Springer Nature Switzerland AG 2018
L. Cheng et al. (Eds.): ICONIP 2018, LNCS 11306, pp. 36–45, 2018.
https://doi.org/10.1007/978-3-030-04224-0_4

be reciprocal ties in different Ocean measuring points of irregular number and weight, which create a highly connected structure with the features of a complex network. The interaction between marine data is also a complex system where the marine data in each Ocean measuring point contains a multivariate time series (MTS) [10]. We map the ocean measuring points as a network [15], of which the vertices are ocean measuring points and edges between vertices are relationships of shares. From a different angle of view, people put forward much network construction arithmetic, such as the minimum-cost spanning tree (MST) [11], the planar maximal filtering graph (PMFG) [19] and the correlation threshold method [4], and use traditional threshold method to construct a ocean measuring points network, the purpose is to study the network structure properties and topological stability.

There are many limitations to build marine multiple time series complex networks using traditional methods, this paper proposes a new method to build the network. Firstly, we convert the marine multiple time series of each ocean measuring point into a corresponding Gramian Angular Field (GAF) [20] grayscale image. Secondly, MGAF color images are combined by the multidimensional GAF gray images, and we use the image similarity to construct a multivariate ocean measuring point network. At last, using the method proposed our approach has higher connection efficiency than traditional method which is constructed by only one correlation threshold. We will introduce the method and experimental results in detail.

2   Constructing the Merged-Image Network

In this part, we will introduce the method of constructing ocean measuring points complex network in detail. The following is to show the process (Fig. 1):

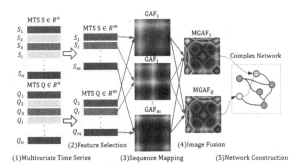

Fig. 1. The architecture proposed by our approach. The construction process of this method includes a total of four steps. (1) Select the corresponding subsequence from a marine multivariate time series; (2) Convert each subsequence into a corresponding image; (3) Fuse multiple marine source time series into one image; (4) Construct complex networks with similarity of fused images.

2.1 Feature Selection

One ocean measuring point is constituted by Multivariate Time Series (MTS). In order to reduce the computational complexity while keeping necessary properties, the method of pearson correlation coefficient is use to reduce dimension for time series with high similarity. The more similar the time series is, the higher the value of pearson correlation coefficient is. The formula for calculating the Pearson correlation coefficient is as Eq. 1.

$$\rho_{x_i,x_j} = \frac{cov\left(X_i, X_j\right)}{\sigma_{x_i}\sigma_{x_j}} \tag{1}$$

where X_i represents the time series of the number of i attribute of ocean measuring points X, σ_{x_i} represents the standard deviation of X_i, $cov\left(X_i, X_j\right)$ is the covariance of the X_i and X_j. The ρ_{x_i,x_j} are restricted to the interval [-1,1], where $\rho_{x_i,x_j} = 1$ defines perfect correlation, and $\rho_{x_i,x_j} = -1$ corresponding to perfect anti-correlation. $\rho_{x_i,x_j} = 0$ corresponds to uncorrelated pairs of ocean measuring points.

2.2 Image Fusion and Similarity Calculation

In this part we use the selected Features to fuse a comprehensive image, however,the Gramian Angular Field (GAF) converts time-series $X_i = \{x_{i1}, x_{i2}...x_{in}\}$ into grayscale images. Since the fused image is a time series, the time characteristic must be preserved when the image is processed, and the GAFS method can well preserve the time dependency.In addition,the $G_{(i,j||i-j|=k)}$ indicates the superposition of the time series in the uniform direction of the time interval k. However, one ocean measuring point has many multiple attributes, so we need fuse many multiple attributes into one image, in this paper, the multiple GAFs are used to be defined as Eq. 2.

$$MGAF_m\left[i,j\right] = [G_1\left[i,j\right], G_k\left[i,j\right], ..., G_m\left[i,j\right]], 0 \le i, j < n \tag{2}$$

where G_k is GAFs image converted from subsequences respectively and $[i,j]$ is one pixel of the $n \times n$ pixels image. At last m GAF images fuse into one image of $MGAF_m$.

The full pipeline for generating the MGAFs is illustrated in Fig. 2: illustration of the proposed encoding map of Multiple Gramian Angular Fields. From this picture we can see that, $s1$, $s2$, and $s3$ represent different time series respectively, then we respectively transform $s1$, $s2$, and $s3$ into a GAF image, at last, We merge the GAF images into a MGAF image. We can see that the MGAF image, which incorporates the many kinds of marine data, can well distinguish this situation. The advantages of multivariate time series conversion to fused images are well represented.

In data analysis and data mining and search engines, the similarity method can distinguish the correlation between different attributes. In order to better

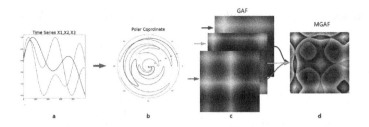

Fig. 2. The process of converting time series into images.

reflect the difference between images, we choose a more effective method structural similarity (SSIM [21]) to calculate the correlation between images, and this method is applied in the fields of super-resolution, image deblurring and the like. From the perspective of image fusion, the structural similarity index can reflect the attributes of the object structure in the scene, and has no relationship with the brightness and contrast of the image. The distortion is modeled as three combinations of brightness, contrast and structure. Average values are used as luminance estimates, standard deviations are used as contrast estimates, and we use the covariance to measure the structural similarity, Eq. 3 is used to assess the image similarity.

$$MSSIM\,[M_i, M_j] = \frac{1}{N} \sum_{n=1}^{N} SSIM\,[M_{in}, M_{jn}] \tag{3}$$

Where in the formula, we use the same way to divide M_i and M_j into N sub-images.

2.3 The Topological Features of Ocean Measuring Points Associated Networks

The ocean measuring point associated network is constracted by using complex network theory, the purpose is to better constructedfeatures of different ocean measurement points. Here we introduce the meaning of some complex network statistical attributes and the calculation method.

The clustering coefficient of network assumes that a node i has k_i edges that connect other nodes to it. Among the k_i nodes, there are at most $k_i\,(k_i - 1)\,/2$ edges, but the number of real edges between the k_i nodes is l_i. The ratio between l_i and $k_i\,(k_i - 1)\,/2$ is then defined as the clustering coefficient of node i, that is Eq. 4.

$$C_i = \frac{2l_i}{k_i\,(k_i - 1)} \tag{4}$$

The average clustering coefficient is the average clustering coefficient of all ocean measuring point, that is Eq. 5.

$$C = \frac{1}{n} \sum_{i-1}^{n} C_i \tag{5}$$

The efficiency of a pair of nodes is the multiplicative inverse of the shortest path distance between the nodes, the average efficiency is defined as Eq. 6.

$$E = \frac{2}{N(N-1)} \sum_{i \neq j} \frac{1}{d_{ij}} \tag{6}$$

Where N is the total number of ocean measuring point in a network and d_{ij} is the distance of ocean measuring point i and ocean measuring point j.

We use network density to represent the degree of similarity between different nodes, as shown in Eq. 7.

$$D = \frac{2m}{n(n-1)} \tag{7}$$

In formula, n means the number of nodes, m means the number of edges.

If there is a relationship between the two nodes v_i and v_j, it is represented as a connecting edge in the undirected graph G. A connected component in G is a maximally connected subgraph of an undirected graph G. The ratio between the maximum number of links between nodes and the number of ocean measuring points is called the coverage of the network, the formula is as Eq. 8.

$$F = \frac{S_{max}}{|V|}, S = \{|V_i|, V_i \subset V\} \tag{8}$$

where V is the collection of ocean measuring point in a network and V_i is the collection of the connected component of a ocean measuring point network.

The strength of linked nodes is represented by connection efficiency. The higher of the connectivity efficiency, the tighter the tightness between the two nodes. It is defined as Eq. 9.

$$CE = F - D \tag{9}$$

where D is the density of a network, F is the fraction of the coverage of the network.

3   Experiments and Discussion

3.1   Data

In this article, data is from fifteen ocean measuring points of East China Sea, they are belongs to four different provinces, dafeng, yangkou and lvsi belong to jiangsu province, sheshan, lvchao, tanxu and donghai belong to shanghai, daishan, zhenhai, wushashan and wenzhou belong to zhejiang province, pingtan, chongwu, jinjiang and longhai belong to fujian province. Every ocean measuring point contains five different factors, including air temperature, air pressure, humidit, wind speed and water temperature, the time is from April 1, 2016, to April 5, 2016, the frequency of obtaining data is 1 hours per time.

3.2   Attribute Selection

We use the method of Pearson correlation coefficient to indicate the properties' correlation between different ocean measuring points. In this paper, dimension reduction method is adopted when the value is greater than 0.7, through the Fig. 3, properties of an ocean measuring point are divided into five groups, then $m = 5$.

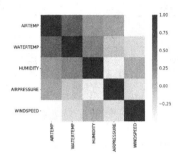

Fig. 3. The relationship of various attributes of a ocean measuring point.

The $MGAF_5$ formula is as Eq. 10.

$$MGAF_5\,[i,j] = [G_1\,[i,j]\,, G_2\,[i,j]\,, ...G_3\,[i,j]]\,, 0 \le i, j < n \qquad (10)$$

where $G_1, G_2, ...G_5$ is GAFs image converted from air temperature, air pressure, humidit, wind speed and water temperature respectively and $[i,j]$ is one pixel of the image of the $n \times n$ pixels. We can see that the fused images in Fig. 4 are calculated by corresponding five different ($m = 5$) time series in an ocean measuring point.

3.3   Results Under Different Thresholds

Different threshold networks are obtained under different thresholds. In order to obtain more accurate threshold networks, the selection of threshold is very important. Then empirical method is the most effective method to determine the threshold. In this paper we choose the networks from the threshold of 0.01 to 0.86, and the global properties of the system are counted by each restriction, which is showed in Fig. 5. From the characteristics of the network, each node plays a very different role in the network, and some key nodes play a leading role in the network. By comparing the central nodes under different thresholds, we select the more stable top ten central nodes.

From the Fig. 5, we find that with the increase of threshold, the obtained threshold network is more sparse, which means that the network density, average clustering coefficient and so on continue to decline. However, when the threshold is taken to a smaller value, many nodes in the network are almost connected. In order to remove the noise edge and keep the key nodes in the network as much

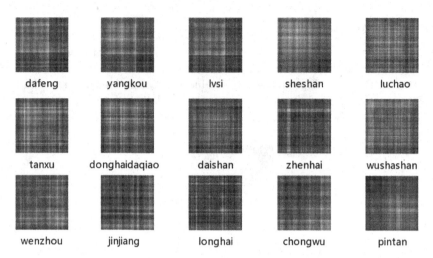

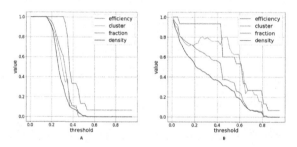

Fig. 4. The fused image.

Fig. 5. Fugure A is our method; Fugure B is results of the average of ocean measuring points correlation threshold method (traditional method).

as possible, we select the threshold when the link efficiency is the highest to obtain the network. From the Fig. 6, we can see that the connectivity efficiency has increased sharply in the threshold range, and has declined rapidly after rare extremes.

Then we calculate the image similarity $MSSIM$ and establish ocean measuring points associated networks models with different thresholds. The importance of nodes depends on node degree. From the Fig. 6, we can see that whether the network constructed by this method or the network constructed by the traditional method, the most suitable threshold network can be obtained when the threshold value is 0.4. Next, We choose the best threshold to get the network, in Fig. 7. The method proposed in this paper gets fewer edges, the relationship between the ocean measuring points in the same province is relatively close, and the relationship between the measuring points in different provinces is relatively loose, this phenomenon is in line with the actual situation. However, the network constructed by the traditional method gets many edges, although the

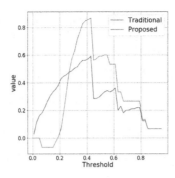

Fig. 6. The marine network connectivity efficiency.

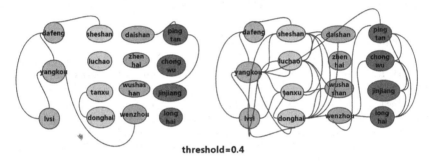

threshold=0.4

Fig. 7. The network.

ocean measuring points in the same province can get a strong relationship, the ocean measuring points in different provinces also have a strong relationship, this means the traditional method gets more meaningless edges. Concerning Fig. 6, we can see that the network constructed by our method can clearly and accurately express the relationship between different nodes, while the network constructed by the traditional method can not find the relationship between the important nodes.

4 Conclusion

In this paper, we presented a detailed study of the properties of the ocean measuring points correlation network. Finding components, cliques and independent sets in the network gives us a new tool for the analysis of the ocean measuring points structure by classifying the ocean measuring points into different groups. We selects the characteristics of marine multiple time teries, the time series of different attributes of each ocean measuring point are transformed into different images, and then the images of different attributes of each ocean measuring point are fused into an image by image fusion method. Then the similarity between the images is used to measure the relationship between the ocean measuring points, we use the edges to represent the relationship of different ocean measuring points, then the ocean measuring point network is constructed. At last, we choose different threshold networks by different thresholds, and choose the network under the optimal threshold by analysis, and through experiments prove that the network constructed by our method can clearly and accurately express the relationship between different nodes, while the network constructed by the traditional method can not find the relationship between the important nodes, so the network constructed by our approach is better than traditional method, it helps government department to layout the ocean measuring points more reasonable.

Acknowledgments. This thesis is supported by National Key R&D Program of China (2016YFC1403200)(2016YFC1401900), youth fund project of east china sea branch of state oceanic administration (201614).

References

1. Albert, R., Barabási, A.: Statistical mechanics of complex networks. Rev. Modern Phys. **74**(1), xii (2002)
2. Barabási, A.L., Albert, R.: Emergence of scaling in random networks. Science **286**(5439), 509–512 (1999)
3. Barthálemy, M., Barrat, A., Pastor-Satorras, R., Vespignani, A.: Velocity and hierarchical spread of epidemic outbreaks in scale-free networks. Phys. Rev. Lett. **92**(17), 178701 (2004)
4. Boginski, V., Butenko, S., Pardalos, P.M.: Mining market data: a network approach. Comput. Oper. Res. **33**(11), 3171–3184 (2006)
5. Caldarelli, G., Battiston, S., Garlaschelli, D., Catanzaro, M.: Emergence of complexity in financial networks. Lect. Notes Phys. **650**, 399–423 (2004)
6. Colizza, V., Barrat, A., Barthélemy, M., Vespignani, A.: The role of the airline transportation network in the prediction and predictability of global epidemics. Proc. Natl. Acad. Sci. U.S.A. **103**(7), 2015–2020 (2006)
7. Brockmann, D., Helbing, D.: The hidden geometry of complex, network-driven contagion phenomena. Science (New York, N.Y.) **342**(6164), 1337–1342 (2013)
8. Eliazar, I., Koren, T., Klafter, J.: Searching circular dna strands. J. Phys. Condens. Matter **19**(6), 160–164 (2007)
9. Gómezgardeñes, J., Latora, V., Moreno, Y., Profumo, E.: Spreading of sexually transmitted diseases in heterosexual populations. Proc. Natl. Acad. Sci. U.S.A. **105**(5), 1399–1404 (2008)
10. Karl, D.M., Michaels, A.F.: The Hawaiian ocean time-series (hot) and bermuda atlantic time-series study (bats). Deep. Sea Res. Part II Top. Stud. Ocean. **43**(2–3), 127–128 (1996)
11. Kobayashi, M., Okamoto, Y.: Submodularity of minimum-cost spanning tree games, pp. 231–238 (2014)
12. Maslov, S., Sneppen, K.: Specificity and stability in topology of protein networks. Science **296**(5569), 910–913 (2002)
13. Newman, M.E.: Spread of epidemic disease on networks. Phys. Rev. E Stat. Nonlinear Soft Matter Phys. **66**(1 Pt 2), 016128 (2002)
14. Newman, M.E.J.: The structure and function of complex networks. SIAM Rev. **45**(2), 167–256 (2003)

15. Nodoushan, E.J.: Monthly forecasting of water quality parameters within bayesian networks: A case study of honolulu, pacific ocean. Civil Eng. J. 4(1), 188 (2018)
16. Guimera, R., Amaral, L.A.N.: Functional cartography of complex metabolic networks. Nature 433(7028), 895 (2005)
17. Boccaletti, S., Latora, V., Moreno, Y., Chavezf, M., Hwang, D.-U.: Complex networks: structure and dynamics. Complex Syst. Complex. Sci. 424(4-5), 175–308 (2006)
18. Stanley, H.E., et al.: Self-organized complexity in economics and finance. Proc. Natl. Acad. Sci. U.S.A. 99(3), 2561–2565 (2002)
19. Tumminello, M., Matteo, T.D., Aste, T., Mantegna, R.N.: Correlation based networks of equity returns sampled at different time horizons. Eur. Phys. J. B 55(2), 209–217 (2007)
20. Wang, Z., Oates, T.: Imaging time-series to improve classification and imputation. arXiv preprint arXiv:1506.00327 (2015)
21. Wang, Z., Bovik, A.C., Sheikh, H.R., Simoncelli, E.P.: Image quality assessment: from error visibility to structural similarity, vol. 13, pp. 600–612. IEEE (2004)

An Effective Lazy Shapelet Discovery Algorithm for Time Series Classification

Wei Zhang, Zhihai Wang, Jidong Yuan$^{(\boxtimes)}$, and Shilei Hao

School of Computer and Information Technology,
Beijing Jiaotong University, Beijing, China
{zhangwei2102,yuanjd}@bjtu.edu.cn

Abstract. Shapelet is a primitive for time series classification. As a discriminative local characteristic, it has been studied widely. However, global shapelet-based models have some obvious drawbacks. First, the progress of shapelet extraction is time consuming. Second, the shapelets discovered are merely good on average for the training instances, while local features of each instance to be classified are neglected. For that, instance selection strategy is used to improve the efficiency of feature discovery, and a lazy model based on the local characteristics of each test instance is proposed. Different from the commonly used nearest neighbor models based on global similarity, our model alleviates the uncertainty of predicted class value using local similarity. Experimental results demonstrate that the proposed model is competitive to the benchmarks and can be effectively used to discover characteristics of each time series.

Keywords: Time series · Lazy learning · Local similarity · Shapelet
Instance selection

1 Introduction

In recent years, time series classification (TSC) problems have received great attention. Although a large number of TSC algorithms have been proposed, extensive experiments show that 1NN classifier combined with improved distance function is still a competitive model in many problem areas [1]. However, the interpretability of 1NN based on global similarity is insufficient. It cannot indicate the common characteristics of similar instances and the dissimilarity among different classes. In reality, except classification accuracy, the characteristics of distinct instances are our concern. These features can help us have a deeper understanding of data and improve the interpretability of the classification model [2,3].

Shapelet is a discriminative subsequence of time series [4]. Since the shapelet can be used to establish interpretable classification models, it has been widely concerned [5]. Shapelet-based classification models can be divided into two categories. One type method uses the top-k shapelets to create a transformed dataset, on which traditional classification algorithms can be applied [6,7]. The other

© Springer Nature Switzerland AG 2018
L. Cheng et al. (Eds.): ICONIP 2018, LNCS 11306, pp. 46–55, 2018.
https://doi.org/10.1007/978-3-030-04224-0_5

employs selected shapelets to build the classification model directly [4,8,9]. The shapelet extraction is always embedded in the model building process for the latter one, while for the former one, it is independent.

The high time complexity of extracting shapelets is the main defect of the shapelet-based algorithms. In order to reduce that, some methods have been presented to avoid retrieving the entire candidate space. These approaches fall into two main groups: subspace-based and approximation-based. The former one introduces heuristic or random search algorithms to find the most discriminative characteristics [9–11]. The latter one learns shapelet candidates through numerical optimization algorithms or some other approximation methods [8,12]. For example, Rakthanmanon et al. proposed a fast discovery algorithm based on symbolic aggregate approximation [8]. However, compared with Brute-Force search algorithm, it is inevitable that some meaningful characteristics may be neglected by these two categories of methods. In addition, these global models have the following disadvantages:

(1) The global model evaluates and selects shapelets on the whole training set, which ignores the information contained in the instance to be classified. Meanwhile, due to the influence of redundant instances and intra-class variability, the shapelets extracted are merely good on average for the training instances.

(2) The main focus of the presented studies is to search shapelets in a smaller candidate space or to find local characteristics by approximation algorithms. However, the candidate dataset used for shapelets evaluation has not been paid enough consideration.

In order to solve the limitations of global model, a data driven, shapelet-based model (Lazy Shapelet Classification Route, LSCR) is proposed, which combines lazy learning strategy with the feature extraction. Figure 1 shows the characteristics discovered by the classic Shapelets Decision Tree (SDT) proposed by

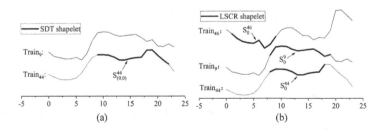

Fig. 1. (a) The shapelet found by SDT on ItalyPowerDemand dataset and its corresponding training instance. The black bold part indicates the time series shapelet discovered. (b) Three time series from ItalyPowerDemand dataset and their corresponding shapelets captured by LSCR. $Train_9$ and $Train_{44}$ (in blue and red separately) are instances with different classes, while $Train_9$ and $Train_{46}$ (in blue) are instances of the same type.

Ye et al. [4] and LSCR respectively. The symbol $Train_j$ denotes the jth training instance. $S^j_{(i,l)}$ represents the ith shapelet from the jth instance on the lth layer of the shapelets decision tree built by SDT. S^j_i is the ith shapelet of the jth instance discovered by LSCR.

From Fig. 1(b), we can find that not only the heterogeneous instances are different, there are also differences between the homogeneous instances. In addition, compared with SDT shapelet (as shown in Fig. 1(a)), our model may capture characteristics that SDT cannot discover. For example, the shapelet S^{46}_0 shown in Fig. 1(b) does not appear in the model built by SDT. Based on the shapelets discovered, our model can make correct predictions for the three instances. The main contributions of this paper are summarized as follows:

(1) Unlike the 1NN model based on global similarity, our model gives priority to local similarity.
(2) In order to reduce the huge number of redundant candidate shapelets generated by Brute-Force or approximate algorithm, a novel strategy that extracts candidate shapelets from the instance to be classified is proposed.
(3) Instances selection strategy is used in evaluating the discrimination of shapelets. The smaller training set can not only alleviate the interference of intra-class variation, but also reduce the computational complexity and improve the classification accuracy of our model.

This paper is organized as follows. Section 2 describes the model proposed and algorithm designed in detail. Section 3 presents the experimental analysis. Case study is presented in Sect. 4. Section 5 offers the conclusion of this paper.

2 Shapelet Extraction for Each Test Instance

This section outlines the algorithm proposed in this paper. In order to reduce the amount of distance calculation between candidate shapelets and time series, a smaller candidate shapelets collection is established for the instance to be classified at first. Then, a small part of the training instances is used to evaluate the candidate shapelets.

2.1 Candidate Shapelets Set for the Test Instance

In this paper, the symbol $\mathbf{D} = \{T_1, T_2, ..., T_n\}$ represents the dataset containing n time series, where T_i denotes a time series. The symbol $W_l(T_i)$ is used to indicate the set of subsequence of length l for the time series T_i. $\mathbf{D}_{node}$ is a dataset corresponding to an arbitrary tree node in the model SDT or LSCR, and $W_l(\mathbf{D}_{node})$ is the set of the candidate shapelets of length l. $W_l(\mathbf{D}_{node})$ in SDT could be represented:

$$W_l(\mathbf{D}_{node}) = \bigcup_{T_i \in \mathbf{D}_{node}} W_l(T_i) \tag{1}$$

However, the set of candidate shapelets of length l for each node in LSCR is:

$$W_l(\mathbf{D}_{node}) = W_l(T) \tag{2}$$

where T is the test instance.

Then, the whole candidate shapelets sets $W(\mathbf{D}_{node})$ for each node in SDT and LSCR can be get through the equation:

$$W(\mathbf{D}_{node}) = \bigcup_{l=min}^{max} W_l(\mathbf{D}_{node}) \tag{3}$$

where min and max are the minimum and maximum candidate length, respectively.

Hence, the whole candidate features set of each node in SDT contains $O(sm^2)$ elements, where s is the size of dataset D_{node} and m is the length of each time series. However, there are only $O(m^2)$ candidates need to be evaluated on each node in LSCR.

2.2   Building the Evaluation Dataset

The key idea of evaluation dataset is to sample the targeted training instances with different class values for shapelet evaluation. For the test instance T, we select k nearest neighbors with the same and different class value of its nearest neighbor respectively to make up two sub-datasets $\mathbf{D}_{same}$ and $\mathbf{D}_{diff}$. At last, the two subsets are combined to form the neighborhoods of the test instance.

$$\mathbf{D}_e = \mathbf{D}_{same}\left\{T_i | c_{T_i} = c_T\right\} \cup \mathbf{D}_{diff}\left\{T_j | c_{T_j} \neq c_T\right\} \tag{4}$$

where c_{T_i} is the class value of instance T_i, and c_T is the class value of the nearest neighbor instance of T.

In our model, same as the way proposed by Ye et al. [4], information gain is used to measure the discriminability of shapelets for the dataset $\mathbf{D}_e$. Meanwhile, a threshold δ of the shapelet would be determined.

2.3   Lazy Shapelet Classification Route

Instead of global similarity, local characteristics are used to reduce the uncertainty of prediction. When a new instance to be classified is coming, a customized classification route will be built by LSCR. The pseudo code of LSCR is shown in Algorithm 1.

The algorithm above shows the process of generating the classification route. At the beginning, the shapelet evaluation collection is established with Eq. 4 for the test instance (line 1). Then the terminating condition is judged (line 2). If the terminating condition is not satisfied, we extract the best shapelet based on the evaluation dataset $\mathbf{D}_e$ corresponding to the node (line 3). The candidate shapelets extracted through Eq. 3 are sorted according to the information gain. If the best shapelet is determined, the instance whose distance with the shapelet

---

Algorithm 1. LSCR(**D**, T, k)

Input: D, the training dataset; T, the test instance; k, the number of neighbor instances with the same and different class value
Output: the classification route of the test instance: $CRForT$
1: $\mathbf{D}_e \leftarrow$ DataSampling(**D**, T, k)
2: if no identification shapelet can be found in the dataset $\mathbf{D}_e$ corresponding to the node, and the majority class value is c, then return $CRForT$ and the predictive value is c; otherwise:
3: $best\_shapelet \leftarrow$ FindingBestShaplet(T, $\mathbf{D}_e$)
4: $\mathbf{D}'_e \leftarrow$ GeneratingNodeDataset($best\_shapelet$, $\mathbf{D}_e$, T)
5: Build the child node and repeat 2-5 on $\mathbf{D}'_e$ until the end

---

is larger than the split threshold in the dataset $\mathbf{D}_e$ would be excluded, and then the remaining neighbor instances constitute a new subset $\mathbf{D}'_e$ (line 4). After that we create a child node and repeat steps 2–5 until the terminating condition is met (line 5). At last, the shapelet-based classification route of the instance to be classified will be returned.

The predicted result obtained by model LSCR may be different from the class value of the nearest neighbor or the majority class in the initial dataset $\mathbf{D}_e$. This is the biggest difference between our model and the nearest neighbor models.

3   Experiment and Evaluation

Considering the time complexity of our algorithm, we present an experimental analysis on 44 datasets from the UEA & UCR time series repository [13].

3.1   Parameter Analysis and Selection

In order to study the effect of the neighborhood size on the discriminative evaluation of the features, we try to observe the accuracy trends of LSCR based DTW Distance (DTWLSCR) within the specified range over 16 datasets in Fig. 2.

From Fig. 2(a), we can see that most accuracy rates reach the maximum values when parameter $k = 5$, and then all the curves show a significant downward trend. Therefore, in the following experiments, the value k used in DTWLSCR on binary-class datasets is set to 5. From Fig. 2(b), we can find that, except Beef dataset, the accuracy rates on seven multi-class datasets have a growth trend as k rises. When k is greater than 6, the accuracy on each dataset tends to be stable. Hence, we set the parameter k of DTWLSCR to 6 on multi-class datasets.

3.2   Performance Analysis

In this section, DTWLSCR without instance selection is first compared with DTWLSCR ($k = 5/6$) on the 16 small datasets in Fig. 3. Setting parameter

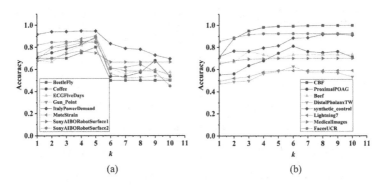

(a)   (b)

Fig. 2. The accuracy variation trends of DTWLSCR on different datasets when the value k increases. (a) Experiments on 8 binary-class datasets. (b) Experiments on 8 multi-class datasets.

k to 0 means instance selection is not carried out in the model LSCR. Considering the time complexity of SDT, we also compare LSCR with SDT on the 16 small datasets.

From Fig. 3(a), we can see that DTWLSCR($k = 5/6$) is significantly better than DTWLSCR ($k = 0$). Hence, in the following, we choose DTWLSCR combined with instance selection for further analysis. Meanwhile, in Fig. 3(b), we can see that the classification performance of DTWLSCR is better than SDT (10 of 16) over the 16 datasets. In order to compare the time complexity of DTWLSCR with SDT, we analysis the running time of DTWLSCR ($k = 5/6$) vs. SDT in Fig. 4.

In Fig. 4(a), it is clear that the running time of both models presents a linear increasing trend with respect to the number of instances, but SDT takes more time than DTWLSCR under the same condition. In Fig. 4(b), the runtime of SDT increases exponentially with respect to the length of instance. As the length

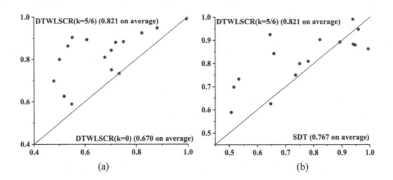

(a)   (b)

Fig. 3. Accuracy rates of DTWLSCR($k = 5/6$) compared with DTWLSCR without instances selection and SDT over 16 datasets. (a) DTWLSCR($k = 5/6$) vs. DTWLSCR ($k = 0$). (b) DTWLSCR vs. SDT

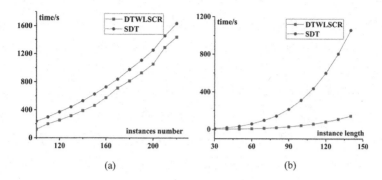

(a)   (b)

Fig. 4. The running time trends of DTWLSCR ($k = 5/6$) vs. SDT based on different variables. (a) The sizes of training and testing datasets increase by 10 at a time. The length of each instance remains 60. (b) The length of time series increases by 10 at a time. The sizes of training set and testing set both take 60.

of instance grows, SDT run significantly slower than DTWLSCR. However, the running time of DTWLSCR is not sensitive to the length of each time series. The above experiments indicate that, compared with SDT, DTWLSCR reduces the time complexity.

Furthermore, three benchmark classifiers based on global similarity are used to compare with DTWLSCR on 44 datasets, namely, 1NN model based on Euclidean distance (ED1NN), DTW-based 1NN (DTW1NN) and DTW-based kNN model (DTWkNN). From Fig. 5(a) and (b), we can see that DTWLSCR is better than the DTWkNN (30 of 44) and ED1NN (30 of 44). Especially, the average accuracy of DTWLSCR is nearly 10% higher than that of DTWkNN ($k = 10$ on binary-class datasets, $k = 12$ on multi-class datasets). DTWLSCR is slightly worse than DTW1NN (20 of 44) in Fig. 5(c).

Though DTWLSCR selects instances for test instance according to the class value of its nearest neighbor, the experimental results show that the prediction result is not necessarily the same as the class value of the nearest training

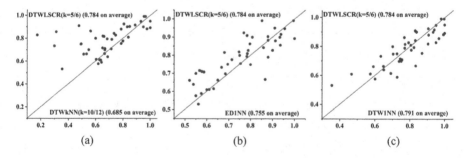

(a)   (b)   (c)

Fig. 5. Accuracy rates of DTWLSCR ($k = 5/6$) compared with three nearest neighbor models based on global similarity over 44 datasets. (a) DTWLSCR vs. DTWkNN ($k = 10/12$). (b) DTWLSCR vs. ED1NN. (c) DTWLSCR vs. DTW1NN

instance or the majority in the neighborhoods. In conclusion, the above accuracy comparison results suggest that narrowing the search space of class attributes continuously based on local similarity is an effective method.

4 Case Study

In this section, the interpretability of LSCR will be discussed on MoteStrain dataset. The classification performance of DTWLSCR on MoteStrain dataset is significantly better than the models used for comparison, it is also competitive to the best result provided by Bagnall et al. [13]. Figure 6 shows four instances and their shapelets extracted through DTWLSCR. Our model can correctly predict the class properties of these four instances, while SDT fails.

From Fig. 6, we can see that it is obvious that there are significant intra-class variations in similar cases. For example, the differences among the two instances with same class label $Test_{11}$ and $Test_{129}$ (shown in Fig. 6(a) and (b)) are noticeable, while there are no obvious common features. However, in the shapelet decision tree built by SDT, only a shapelet is found, which is not enough to distinguish these two classes. This is the reason why the shapelets decision tree is poorly performing on the MoteStrain dataset. Meanwhile, because the

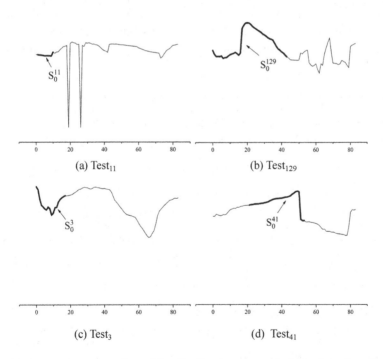

(a) $Test_{11}$

(b) $Test_{129}$

(c) $Test_3$

(d) $Test_{41}$

Fig. 6. Four test instances from MoteStrain dataset and their shapelets found by DTWLSCR. The top two instances (a–b) are from Class1 and the bottom two (c–d) from Class2.

characteristics of each test instance have been considered in our model, in face of this situation, we can achieve better classification results. In addition, based on the shapelet obtained by DTWLSCR, the prediction process of each instance can be explained. Since the shapelets extracted in LSCR come from the test instance, these shapelets can be a good explanation of what features determine its class value. For example, the reason why the 11th test instance belongs to Class1 is that the local feature S_0^{11} lies in its initial stage, while the local feature S_0^{41} of the 41th test instance in the middle part determines its predicted class label. Based on the characteristics captured, the other examples shown in Fig. 6 can also explain their prediction process in the same way.

5   Conclusion

In order to create a more pertinent and interpretable classification model, we propose to build a classification route for each time series using its own local characteristics directly. Meanwhile, the concept of shapelets evaluation dataset is put forward to reduce the consumption of distance calculation, and to weaken the impact of intra-class variation on the feature assessment. Our model can be used to answer the question: "What characteristics of the instance to be classified determine its class properties?". Extensive experiments and detailed case study demonstrate the rationality and feasibility of the proposed method. The main directions of our future work involve studying other generation and evaluation methods of local characteristics for the test instance, and making detailed statistical analysis of the acquired shapelets.

Acknowledgments. This work is supported by National Natural Science Foundation of China (No. 61672086, 61702030, 61771058), Beijing Natural Science Foundation (No. 4182052), China Postdoctoral Science Foundation (No. 2018M631328) and the Fundamental Research Funds for the Central Universities (No. 2017YJS036, 2018JBM014).

References

1. Lines, J., Bagnall, A.: Time series classification with ensembles of elastic distance measures. Data Min. Knowl. Discov. **29**(3), 565–592 (2015)
2. Nguyen, T.L., Gsponer, S., Ifrim, G.: Time series classification by sequence learning in all-subsequence space. In: 33th International Conference on Data Engineering, pp. 947–958. IEEE Press, San Diego (2017)
3. Senin, P., Malinchik, S.: SAX-VSM: interpretable time series classification using sax and vector space model. In: 13th International Conference on Data Mining, pp. 1175–1180. IEEE Press, Dallas (2013)
4. Ye, L., Keogh, E.: Time series shapelets: a new primitive for data mining. In: 15th ACM SIGKDD International Conference on Knowledge Discovery and Data Mining, pp. 947–956. ACM Press, Paris (2009)
5. Bagnall, A., Lines, J., Bostrom, A., Large, J., Keogh, E.: The great time series classification bake off: a review and experimental evaluation of recent algorithmic advances. Data Min. Knowl. Discov. **31**(3), 606–660 (2017)

6. Hills, J., Lines, J., Baranauskas, E., Mapp, J., Bagnall, A.: Classification of time series by shapelet trans-formation. Data Min. Knowl. Discov. **28**(4), 851–881 (2014)
7. Yuan, J.D., Wang, Z.H., Han, M.: A discriminative shapelets transformation for time series classification. Int. J. Pattern Recognit. Artif. Intell. **28**(6), 1–28 (2014)
8. Rakthanmanon, T., Keogh, E.: Fast shapelets: a scalable algorithm for discovering time series shapelets. In: 13th SIAM International Conference on Data Mining, pp. 668–676. SIAM Press, Austin (2013)
9. Karlsson, I., Papapetrou, P., Bostrom, H.: Generalized random shapelet forests. Data Min. Knowl. Discov. **30**(5), 1053–1085 (2016)
10. Gordon, D., Hendler, D., Rokach, L.: Fast and space-efficient shapelets-based time-series classification. Intell. Data Anal. **19**(5), 953–981 (2015)
11. Shi, M.H., Wang, Z.H., Yuan, J.D., Liu, H.Y.: Random pairwise shapelets forest. In: 22th Pacific-Asia Conference on Knowledge Discovery and Data Mining, pp. 68–80. Springer Press, Melbourne (2018)
12. Hou, L., Kwok, J.T., Zurada, J.M.: Efficient learning of timeseries shapelets. In: 30th AAAI Conference on Artificial Intelligence, pp. 1209–1215. AAAI Press, Phoenix (2016)
13. Bagnall, A., Lines, J., Vickers, W., Keogh, E.: The UEA & UCR Time Series Classification Repository. www.timeseriesclassification.com

Multivariate Chaotic Time Series Prediction: Broad Learning System Based on Sparse PCA

Weijie Li, Min Han$^{(\boxtimes)}$, and Shoubo Feng

Faculty of Electronic Information and Electrical Engineering,
Dalian University of Technology, Dalian 116024, Liaoning, China
minhan@dlut.edu.cn

Abstract. The sparse principle component analysis (SPCA) comprehensively considers the maximal variance of principal components and the sparseness of the load factor, thus making up for the defects of the traditional PCA. In this paper, we are committed to propose a novel approach based on broad learning system with sparse PCA named as SPCA-BLS for chaotic time series prediction. We also develop the incremental learning algorithms to rapidly rebuild the network without full retraining if the network is considered to be expanded. The core of the SPCA-BLS is that we achieved the dimensionality reduction and the features extraction of high-dimensional data and the dynamical reconstruction of the network without the entire retraining. The method has been simulated on an artificial and an actual data sets and the experimental results in regression accuracy confirm the characteristics and effectiveness of the SPCA-BLS.

Keywords: Chaotic time series prediction · Broad learning system
Sparse PCA

1 Introduction

The chaotic time series is a time sequence sampling of a dynamic system. In the real life, most meteorological, hydrological and financial systems are chaotic, and chaotic systems are difficult to model because of its chaotic nature and high dimensional and complex features [1]. Therefore, time series prediction has been a significant domains of research. So far, experts and scholars have developed many artificial neural networks to predict the time series more accurately [2–4].

Principal component analysis (PCA) is an important data processing method. In high-dimensional data analysis, it is an important data analysis and dimensionality reduction technology, and it plays a great role in the fields of biostatistics, social sciences, economics, and finance [5]. Despite its popularity, the main drawback of PCA is that the derived principal components (PCs) are a linear combination of all initial components [6]. Moreover, the coefficients of

L. Cheng et al. (Eds.): ICONIP 2018, LNCS 11306, pp. 56–66, 2018.
https://doi.org/10.1007/978-3-030-04224-0_6

all linear combinations are usually nonzero. Therefore, each obtained principal component lacks sparseness and may not have practical meaning [5,7].

Sparse representation is the normal requirement for data representation. In the statistical literature, the earliest attempts of sparsitying PCA includes simple axis rotations and component thresholding, whose basic goal is the subset selection based on the identification of principal variables. JoLiffe et al. proposed the first real computing technology called SCoLTASS [8] and provided an efficient optimization architecture using the lasso, however it was proved unpractical. Soon after, Zou et al. formulated sparse PCA (SPCA) [9] using the elastic network architecture for L1 penalized regression of conventional PCs, and the algorithm was effectively solved by using the minimum angle regression (LARS). The SPCA pursues an approximate sparse "eigenvectors" whose projections capture the maximal variance of the input.

Broad learning system (BLS) [10] is an efficient method in processing large-scale data. We proposed a novel prediction network named Broad Learning System based on Sparse PCA (SPCA-BLS) in this paper for feature extraction and prediction of large-scale data. We illustrate the excellent performances of the SPCA-BLS in two different data sets.

The paper is arranged as follows. We elaborate the architecture of the proposed SPCA-BLS method in Sect. 2. In Sect. 3, we introduce its incremental learning algorithm. In Sect. 4, we present the simulations and result analysis based on Lorenz time series and UCI time series about PM2.5. And Sect. 5 gives the conclusions.

2   Broad Learning System Based on Sparse PCA

The evolution of any time series is determined by other time series in the same chaotic system. Therefore, the evolution of the system in a certain moment contains all the information of the chaotic system. In actual prediction, we use the phase space reconstruction (PSR) [11] to represent the evolution characters of the chaotic time series in which the topological structure is equivalent to the original dynamic system.

2.1   Phase Space Reconstruction

A phase space can reflect a deterministic dynamical system which contains the all possible states of a system. By using the time delay method, a one-dimensional time series can be reconstructed into m-dimensional space where m is the embedding dimension [12]. Supposing that the chaotic network has N time series of the input, and assuming the embedding dimension as $\mathbf{m} = (m_1, m_2, ..., m_N)^T$, after PSR, we get

$$\mathbf{X} = \begin{pmatrix} x_{1,n} & x_{1,n-\tau_1} & \cdots & x_{1,n-(m_1-1)\tau_1} \\ x_{2,n} & x_{2,n-\tau_2} & & x_{2,n-(m_2-1)\tau_2} \\ \vdots & & \ddots & \\ x_{N,n} & x_{N,n-\tau_N} & \cdots & x_{N,n-(m_N-1)\tau_N} \end{pmatrix} \tag{1}$$

where $\tau_i (1 \leq i \leq N)$ is the time lag. We begin with an input matrix $\mathbf{X}_i \in \mathbf{R}^{N \times P}$, where P is the input dimension. And in this process, we set $\tau_1 = \tau_2 = \cdots = \tau_N = d$. Through PSR, we obtain the reconstructed input $\mathbf{X} \in \mathbf{R}^{N \times L}$, where $L = Pd$. The new input $\mathbf{X}$ will preserve all the intrinsic characteristics of the chaotic system after PSR.

2.2 Sparse PCA

The SPCA mainly concentrates on the maximal variance of the PCs or the minimum reconstruction error. We extract the features of high-dimensional data by SPCA [9] to obtain the PCs with sparse loadings based on L_1 penalties. For a single component, the SPCA solves

$$\hat{\beta} = \arg\min_{\theta, \mathbf{v}} \sum_{i=1}^{N} \|\mathbf{x}_i - \theta \mathbf{v}^T \mathbf{x}_i\|_2^2 + \lambda \|\mathbf{v}\|_2^2 + \alpha \|\mathbf{v}\|_1, \; subject \; to \; \|\theta\|_2 = 1 \quad (2)$$

where $\mathbf{x}_i$ is the ith component of $\mathbf{X}$, and $\|\mathbf{v}\|_1 = \sum_{i=1}^{N} |\mathbf{v}_i|$ is the 1-norm of $\mathbf{v}$. There are some remarks on this formulation: if $N > L$ and both λ and α_1 are zero, we get $\mathbf{v} = \theta$ and the direction of the largest principal component in this condition; the second penalty on $\mathbf{v}$ promotes the loadings sparseness.

For multiple components, the procedure of SPCA minimizes

$$\hat{\beta}_k = \arg\min_{\mathbf{v}_k} \sum_{i=1}^{N} \|\mathbf{x}_i - \Theta \mathbf{V}^T \mathbf{x}_i\|_2^2 + \lambda \sum_{k=1}^{K} \|\mathbf{v}_k\|_2^2 + \sum_{k=1}^{K} \alpha_k \|\mathbf{v}_k\|_1 \quad (3)$$

where $\Theta^T \Theta = \mathbf{I}_K$, $\Theta \in \mathbf{R}^{L \times K}$, and $\mathbf{v}_k$ is the columns of $\mathbf{V} \in \mathbf{R}^{L \times K}$. And apparently, K is the dimension after dimensionality reduction. $\mathbf{V}$ and Θ are not jointly convex in Eq. 3, but it is convex in each parameter with the other parameter fixed. Minimizing $\mathbf{V}$ with Θ fixed is a kind of K elastic net problem and can be solved easily. Similarly, minimizing Θ with $\mathbf{V}$ fixed is another version of the Procrustes problem which can be done by SVD. These steps are repeated until convergence. We define $\hat{\mathbf{V}}_k = \hat{\beta}_k / \|\hat{\beta}_k\|$ as the approximation of $\mathbf{V}_k$, and we get $\mathbf{X}\hat{\mathbf{V}}_k$, the kth approximated principal component. As we know, normalization can not affect the scaling factor of $\mathbf{V}_k$. Obviously, a large enough α_k generates a sparse $\hat{\beta}$ and a corresponding sparse $\mathbf{V}_k$. Given a fixed λ, Eq. 3 can be efficiently solved for all α_k by the LARS-EN algorithm [13]. Thus, we derive a sparse representation of the kth PC.

2.3 SPCA-BLS

After dimensionality reduction, we get $\mathbf{E} = \mathbf{XV}$, the sparse principal components as well as the extracted features. And then BLS maps the features to the m enhancement layers through a random non-linear mapping:

$$\mathbf{H}_j = \xi \left(\mathbf{EW}_{hj} + \beta_{hj} \right), j = 1, 2, \cdots, m \quad (4)$$

where the weights $\mathbf{W}_{hj}$ and bias β_{hj} are generated randomly. Then the output of enhancement layers is $\mathbf{H}^m = [\mathbf{H}_1, \mathbf{H}_2, \cdots, \mathbf{H}_m]$. Finally, the mapped features and the enhancement groups are jointly solved by the least square method on the target $\mathbf{Y}$. The output of SPCA-BLS is computed by:

$$\mathbf{Y} = [\mathbf{E}|\mathbf{H}^m]\,\mathbf{W} = \mathbf{A}_n^m \mathbf{W} \tag{5}$$

where $\mathbf{W}$ is the output weights of SPCA-BLS and $\mathbf{A}_n^m = [\mathbf{E}|\mathbf{H}^m]$. Then the output weight is calculated by ridge regression learning algorithm:

$$\mathbf{W} = (\delta \mathbf{I} + \mathbf{A}\mathbf{A}^T)^{-1}\mathbf{A}^T\mathbf{Y} \tag{6}$$

where δ is the regularization parameter denoting the further constraints and it always tends to be zero.

3   The Analysis and Incremental Learning of SPCA-BLS

SPCA achieves the feature extraction while reducing dimensions. Adding the L_1 penalty to the network can get a sparse representation of the results. However, this results still depend on the PCA. We get a self-contained method by adding the elastic net.

In addition, we give the incremental algorithm to update features dynamically, which avoids the entire retraining of the system. When the training accuracy cannot meet the requirement, we can implement dynamical expansion by adding new enhancement nodes. If we denote that there are p additional enhancement nodes, after adding the new representations, the output matrix is

$$\mathbf{A}_{n+1} = \left[\mathbf{A}_n^m | \xi(\mathbf{E}\mathbf{W}_{hm+1} + \beta_{hm+1})\right] \tag{7}$$

where $\mathbf{W}_{hm+1} \in \mathbf{R}^{K \times p}$ and $\beta_{hm+1} \in \mathbf{R}^p$ are regenerated randomly. We renew the joint features by the following operation:

$$\mathbf{A}_{n+1}^+ = \begin{bmatrix} \mathbf{A}_n^+ - \mathbf{D}\mathbf{B}^{\mathbf{T}} \\ \mathbf{B}^T \end{bmatrix} \tag{8}$$

where

$$\mathbf{D} = \mathbf{A}_n^+ \mathbf{A}_a$$
$$\mathbf{B} = \begin{cases} \mathbf{C}^+ & \mathbf{C} \neq 0 \\ (1 + \mathbf{D}^T\mathbf{D})^{-1}\mathbf{D}^T(\mathbf{A}_n^+) & \mathbf{C} = 0 \end{cases} \tag{9}$$
$$\mathbf{C} = \mathbf{A}_a - \mathbf{A}_n\mathbf{D}$$

and $\mathbf{A}_a = \xi(\mathbf{E}\mathbf{W}_{hm+1} + \beta_{hm+1})$. The new output weights is

$$\mathbf{W}_{n+1} = \begin{bmatrix} \mathbf{W} - \mathbf{D}\mathbf{B}^T\mathbf{Y} \\ \mathbf{B}^T\mathbf{Y} \end{bmatrix} \tag{10}$$

When m new samples arrive, we have

$$^x\mathbf{X} = \begin{bmatrix} \mathbf{X} \\ \mathbf{X}_a \end{bmatrix} \tag{11}$$

where $\mathbf{X}_a = [\mathbf{x}_{n+1}, \mathbf{x}_{n+2}, \cdots, \mathbf{x}_{n+m}]^T$, is the new data matrix added into the SPCA-BLS. And the other initial architecture conditions are the same as the above network. The corresponding feature of the new data is donated as ${}^x\mathbf{E} = \mathbf{X}_a\mathbf{V}$. After feature extraction and the mapping in enhancement layers, the joint matrix for the new data is formulated as follows:

$$\mathbf{A}_x = [\mathbf{E}|\xi(\mathbf{Z}_n^a\mathbf{W}_{h1} + \beta_{h1}, \cdots, \xi(\mathbf{Z}_n^a\mathbf{W}_{hm} + \beta_{hm})] \tag{12}$$

Therefore, the updating matrix for the new input can be donated as follows:

$$ {}^x\mathbf{A}_n^m = \begin{bmatrix} \mathbf{A}_n^m \\ \mathbf{A}_x^T \end{bmatrix} \tag{13}$$

Then, the updating algorithm calculating by pseudo inverse is:

$$({}^x\mathbf{A}_n^m)^+ = [(\mathbf{A}_n^m)^+ - \mathbf{B}\mathbf{D}^T|\mathbf{B}] \tag{14}$$

where

$$\mathbf{D^T} = \mathbf{A}_x^T\mathbf{A}_n^{m+}$$
$$\mathbf{B}^T = \begin{cases} \mathbf{C}^+ & \mathbf{C} \neq 0 \\ (1 + \mathbf{D}^T\mathbf{D})^{-1}\mathbf{A}_n^{m+}\mathbf{D} & \mathbf{C} = 0 \end{cases} \tag{15}$$
$$\mathbf{C} = \mathbf{A}_x^T - \mathbf{D}^T\mathbf{A}_n^m$$

Assuming that $\mathbf{Y}_a$ is the corresponding output of additional $\mathbf{X}_a$, so the new output weights are:

$${}^x\mathbf{W}_n^m = \mathbf{W}_n^m + \left(\mathbf{Y}_a{}^T - \mathbf{A}_x{}^T\mathbf{W}_n^m\right)\mathbf{B} \tag{16}$$

The above operations can handle multiple incremental learning cases without complete retraining of the model. It is undeniable that the algorithm accelerates the efficiency of the network.

4   Experiments and Analysis

We conduct the experiments and analyze the experimental performances in this section to verify the designed network. In order to validate the effectiveness of SPCA-BLS, we conduct simulations based on two large-scale time series.

4.1   Data Sets Description

The Lorenz [15] time series are generated by the following formulas:

$$\begin{cases} \dfrac{dx}{dt} = \alpha(y - x) \\ \dfrac{dy}{dt} = (\beta - z)x - y \\ \dfrac{dz}{dt} = xy - \gamma z \end{cases} \tag{17}$$

When $\alpha = 10$, $\gamma = 3/8$ and $\beta = 28$, the series present the chaotic characteristic. The Lorenz system is a nonlinear, aperiodic, three-dimensional and deterministic chaotic system. We generate Lorenz series using the fourth-order Runge-Kutta method. We set the initial state as $(x, y, z) = (1, 1, 1)$ and the time step as 0.01. In our simulation there are 50,001 sample groups, and each group has 3 variables. Lorenz is then reconstructed into a high-dimensional phase space with the embedding dimension setting as [20, 20, 20] and the time lag [1, 1, 1] for the x, y, z series respectively.

We add a real-life meteorological data in the experiments. The Beijing hourly air quality index(AQI) data set collected from Beijing Capital International Airport including the PM2.5 data is used in the simulation. It contains 43824 group samples and each group consists of eight variables including PM2.5 μg/μm. We choose the first five variables in the simulation. In the processing of reconstruction, we set the embedding dimensions as [40,40,40,40,40] and the delay times as [1,1,1,1,1] for five variables respectively. The two data sets are both normalized into the range $[-1, 1]$. We divide the two data sets into two parts separately: the former 90% for training and the rest for testing.

4.2 Experimental Results and Analysis

We carried out the simulations on MATLAB 2016a on a 64-bit Windows 7 system with an Intel-i3 CPU and 6 GB memory. In this section, we verify the designed method on two data sets. Simulations on the reconstructed typical chaotic time series-Lorenz time series and Beijing Air Quality Index data set are carried out to prove the effectiveness of SPCA-BLS. Some other methods are compared to the proposed method: extreme learning machine (ELM), echo state network (ESN), ESN with leaky integrator neuron (LIESN) [14] and BLS. The performances of the prediction are measured by the four following indexes such as Root Mean Square Error (RMSE), Normalized Root Mean Squared Error (NRMSE), Mean Absolute Error(MAE) and Symmetric Mean Absolute Percent Error (SMAPE).

Table 1. The $t + 1$ step prediction results on Lorenz time series

Index	RMSE ($\times10^{-4}$)			NRMSE ($\times10^{-5}$)			MAE ($\times10^{-4}$)			SMAPE($\times10^{-4}$)		
Methods	x	y	z	x	y	z	x	y	z	x	y	z
ELM(1000)	20.2	26.4	31.6	5.35	5.13	7.30	10.7	14.4	16.1	7.06	11.7	0.70
ELM(4000)	4.61	7.21	7.32	1.22	1.40	1.69	2.15	3.96	3.90	1.05	2.04	0.17
ESN	11.2	19.5	16.4	2.76	3.78	3.79	7.64	12.5	10.2	3.92	9.41	0.44
LIESN	12.9	32.5	26.5	3.43	6.30	6.13	8.54	21.9	18.2	4.85	10.0	0.84
BLS	4.91	11.3	10.2	1.30	2.18	2.35	3.08	6.27	5.40	1.25	2.91	0.22
SPCA − BLS	**8.22**	**4.65**	**3.93**	**2.20**	**0.90**	**0.90**	**5.12**	**3.04**	**2.42**	**2.28**	**1.38**	**0.10**

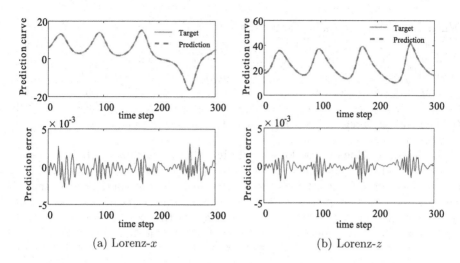

Fig. 1. $t + 1$ step prediction performances on Lorenz time series by SPCA-BLS.

The testing results of $t + 1$ step prediction comparisons of Lorenz are tabulated in Table 1. In Table 1, ELM (1000) and ELM (4000) represent that the hidden nodes of ELM is 1000 and 4000 respectively. It is clear that: for ELM when the hidden nodes is 1000, the RMSE of y sequence is 26.4×10^{-4} which is vary large actually; and when the hidden nodes is 4000, the RMSE of y sequence is just 7.21×10^{-4} which is still much larger than SPCA-BLS of it. It should be realized that with the increase of the hidden nodes of ELM, the operation of programs will become very slow, while the accuracy is not improved.

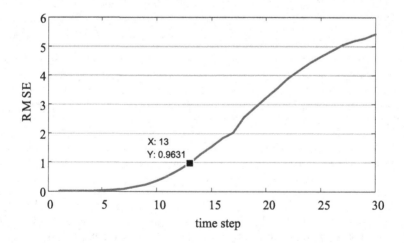

Fig. 2. RMSE of m-step-ahead prediction of Lorenz-z by SPCA-BLS.

Although the x sequence predicted by SPCA-BLS is not superior to all the methods, y and z columns predicted by it are much better than the results of other methods. And the $t+1$ step prediction results of Lorenz are presented in Fig. 1. Prediction on m-step-ahead ($1 \leq m \leq 30$) on Lorenz-z is conducted directly by SPCA-BLS, where the output is delayed m-step corresponding by the time window. The RMSE of m-step-ahead prediction performance conducted by SPCA-BLS is illustrated in Fig. 2. We can be see that in Fig. 2 the RMSE is less than 1 even if m has increased to 13 which thoroughly demonstrates the effectiveness of multi-step prediction of SPCA-BLS. As can be seen, the prediction of SPCA-BLS is accurate and effective, as all the RMSEs are less than 1. From fourteenth step, the RMSE increases fast, and the result becomes inaccurate.

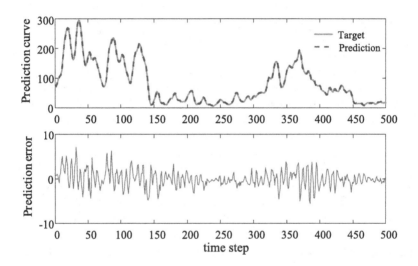

Fig. 3. $t+1$ step prediction performance on Beijing AQI by SPCA-BLS.

Table 2. The $t+1$ step prediction results on PM2.5 time series

Index	RMSE		NRMSE ($\times10^{-3}$)		MAE		SMAPE ($\times10^{-2}$)	
Methods	Train	Test	Train	Test	Train	Test	Train	Test
ELM	4.853	4.579	5.752	5.426	3.139	3.138	6.058	8.228
ESN	4.806	3.997	5.696	4.736	2.883	2.609	5.093	5.824
LIESN	4.808	3.994	5.698	4.734	2.878	2.607	5.071	5.762
BLS	6.052	4.815	7.172	5.707	3.822	3.314	7.268	8.404
SPCA-BLS	**3.284**	**2.670**	**3.892**	**3.165**	**2.177**	**1.863**	**3.959**	**4.644**

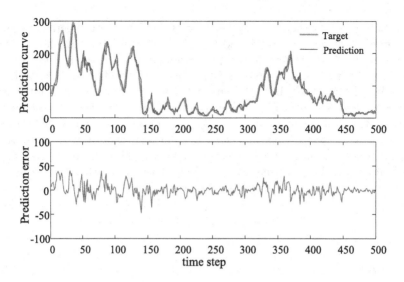

Fig. 4. $t + 3$ step prediction performance on Beijing AQI by SPCA-BLS.

Table 3. The running time comparison on 3-step-ahead predicting of Beijing Air Quality index

Time (/s)	Method						
	ELM (1000)	ELM (2000)	ELM (4000)	ESN	LIESN	BLS	SPCA-BLS
TraintimeofLorenz	6.3024	19.4533	77.4545	9.5785	9.3289	1.4329	**6.6924**
TesttimeofLorenz	0.2496	0.4992	1.014	0.8892	0.9048	0.0823	**0.0312**
TraintimeofPM2.5	5.6628	18.0649	65.7544	56.0044	55.3804	0.9400	**10.2493**
TesttimeofPM2.5	0.234	0.546	0.99841	5.7096	5.928	0.0936	**0.0624**

We also conducted a group of comparative simulations on a real world data. The predicting comparison results on Beijing AQI data set list in Table 2. In the table, all the indexes of SPCA-BLS have achieved the optimal value. The $t + 1$ step prediction results of PM2.5 are presented in Fig. 3. Besides, the $t + 3$ step prediction results of PM2.5 are presented in Fig. 4. Obviously, we can see that in the application of prediction of actual data, the performance of the SPCA-BLS is more prominent. To further illustrate the effectiveness of the proposed method, we compared the training time and test time of all methods. The comparison results are given in Table 3. In this table, ELM(1000) represents Extreme Learning Machine whose hidden nodes are 1000, and similarly ELM(2000) represents the ELM with 2000 hidden nodes, ELM(4000) represents the ELM with 4000 hidden nodes. The results show that the proposed method has a shorter training time and the shortest test time among all methods which strongly indicates that SPCA-BLS can balance the prediction accuracy and the running time.

Therefore, we can believe that SPCA-BLS can better extract the features and learn the evolutionary rules in actual complex meteorological data.

5   Conclusion

We propose an effective chaotic time series prediction method SPCA-BLS in this work, which achieves dynamical high-dimensional features extraction when new nodes are added. The SPCA-BLS has some advantages, such as efficient feature extraction and compression of input and dynamic expansion of the model. Its performances have been testified by two simulations. Compared with the mentioned comparison algorithms, the designed model achieves higher accuracy especially in the real-world data. This results ensure the effectiveness of SPCA-BLS for predicting multivariate real life time series.

Acknowledgment. This work is supported by the National Natural Science Foundation of China (Grant No. 61773087) and the Fundamental Research Funds for the Central Universities (DUT17ZD216).

References

1. Weigend, A.S.: Time series prediction: forecasting the future and understanding the past. Routledge (2018)
2. Han, M., Zhong, K., Qiu, T., Han, B.: Interval type-2 fuzzy neural networks for chaotic time series prediction: a concise overview. IEEE Trans. Cybern. (2018). https://doi.org/10.1109/TCYB.2018.2834356
3. Ak, R., Fink, O., Zio, E.: Two machine learning approaches for short-term wind speed time-series prediction. IEEE Trans. Neural Netw. Learn. Syst. **27**(8), 1734–1747 (2016)
4. Han, M., Ren, W., Xu, M., Qiu, T.: Nonuniform state space reconstruction for multivariate chaotic time series. IEEE Trans. Cybern. (2018). https://doi.org/10.1109/TCYB.2018.2816657
5. Jolliffe, I.T., Cadima, J.: Principal component analysis: a review and recent developments. Phil. Trans. R. Soc. A **374**(2065) (2016). 20150202
6. Hu, Z., Pan, G., Wang, Y., Wu, Z.: Sparse principal component analysis via rotation and truncation. IEEE Trans. Neural Netw. Learn. Syst. **27**(4), 875–890 (2016)
7. Zou, H., Xue, L.: A selective overview of sparse principal component analysis. Proc. IEEE **106**(8), 1311–1320 (2018)
8. Jolliffe, I.T., Trendafilov, N.T., Uddin, M.: A modified principal component technique based on the lasso. J. Comput. Graph. Stat. **12**(3), 531–547 (2003)
9. Zou, H., Hastie, T., Tibshirani, R.: Sparse principal component analysis. J. Comput. Graph. Stat. **15**(2), 265–286 (2006)
10. Chen, C.P., Liu, Z.: Broad learning system: an effective and efficient incremental learning system without the need for deep architecture. IEEE Trans. Neural Netw. Learn. Syst. **29**(1), 10–24 (2018)
11. Takens, F.: Detecting strange attractors in turbulence. In: Rand, D., Young, L.-S. (eds.) Dynamical Systems and Turbulence, Warwick 1980. LNM, vol. 898, pp. 366–381. Springer, Heidelberg (1981). https://doi.org/10.1007/BFb0091924

12. Huang, Y., Kou, G., Peng, Y.: Nonlinear manifold learning for early warnings in financial markets. Eur. J. Oper. Res. **258**(2), 692–702 (2017)
13. Zou, H., Hastie, T.: Regularization and variable selection via the elastic net. J. R. Stat. Soc.: Ser. B (Stat. Methodol.) **67**(2), 301–320 (2005)
14. Jaeger, H., Lukoševičius, M., Popovici, D., Siewert, U.: Optimization and applications of echo state networks with leaky-integrator neurons. Neural Netw. **20**(3), 335–352 (2007)
15. Lorenz, E.N.: Deterministic nonperiodic flow. J. Atmos. Sci. **20**(2), 130–141 (1963)

Dynamic Streaming Sensor Data Segmentation for Smart Environment Applications

Hela Sfar[(✉)] and Amel Bouzeghoub

CNRS Paris Saclay, Telecom SudParis, SAMOVAR, Évry, Essonne, France
{hela.sfar,amel.bouzeghoub}@telecom-sudparis.eu

Abstract. With the increasing availability of unobtrusive, and inexpensive sensors in smart environments, online sensor data segmentation becomes an important topic in reconstructing and understanding sensor data. Usually, in the literature, the segmentation is either performed by following a fixed or a dynamic time-window length. As stated in several works, static time-window length has several drawbacks while adjusting dynamically the window length is more appropriate. However, each of previous methods for dynamic data segmentation targets only a particular type of application. Hence, there is a need for a general method independent of applications providing high degree of usability. To achieve this aim, in this paper, we propose a novel method that dynamically adapts the time-window size. The proposal is designed to be applied in a wide range of applications (activity recognition, decision making, etc.) by combining statistical learning and semantic interpretation. This hybridization allows to analyze the incoming sensor data and choose the better time-window size. The presented approach has been implemented and evaluated in several experiments using the real dataset Aruba from the CASAS project.

Keywords: Clustering · Ontology · Segmentation
Smart environment

1 Introduction

Due to the rapid advances in sensing technology, we are witnessing a growing interest in smart environments, in which a variety of sensors are continuously sending data for processing and analysis in order to be used for different domain applications. To gain a meaningful data understanding from sensor data, one of the major tasks in this area is to divide the long sequence of sensing records into a set of individual segments. Each segment corresponds to a "specific" concept which can be interpreted differently according to the target application.

In the literature, the dynamic data segmentation has proven better results than the static one [4] since it identifies the time points in a more flexible way.

© Springer Nature Switzerland AG 2018
L. Cheng et al. (Eds.): ICONIP 2018, LNCS 11306, pp. 67–77, 2018.
https://doi.org/10.1007/978-3-030-04224-0_7

However, dynamic streaming sensor data segmentation is still a challenging problem. Generally, previous researches in this field [1–3], either assume that a pre-segmented dataset is available for learning the suitable time-window size or being conceived to be applied to a particular application. Accordingly, most related work suffer from generality issue. Hence, the segmentation should be done in a general manner. Afterwards, it can be adapted according to the application.

In this paper, in order to overcome the aforementioned issue, we propose a novel method which combines a clustering method, that is to say, a statistical learning method, and a logic-based method, that relies on high-level "symbolic" representation, in order to dynamically choose the best time-window size. Initially, to tackle the cold start problem that usually concerns statistical learning approaches, an ontology with a default classification is created. This task can be done by the designer after the setup of the environment. Afterwards, once a training dataset is acquired related to the target application[1], a clustering based classification of the dataset is done offline according to a defined features. The proposal dynamically updates offline the ontology whenever change occurs in the obtained clusters. During the online process, the method uses the ontology to decide which better size for the current time-window. The proposed method thanks to this hybridization neither requires a pre-segmented dataset nor being limited only for a particular type of application's use. In order to prove the accuracy of the proposal we apply it for an activity recognition application. In addition, we test it with Support Vector Machine (SVM) [12], a machine learning method, as an activity classifier module, with the real Aruba dataset from the CASAS project [11][2], and synthetic datasets. In the following, we summarize the main contributions of the paper:

– A new method for dynamic streaming sensor data segmentation. The proposal is flexible to be used in a wide range of applications.
– A novel marriage between knowledge oriented method and machine learning to provide a flexibility for the time-window size choice.
– The method is fully implemented and tested in the specific case of activity recognition application. SVM is used as an activity recognizer module with different datasets. The output is compared with SVM using fixed time-window size as well as state of the art work. The results are promising proving the high accuracy level.

2   Related Work

In this paper we classify previous works of dynamic streaming sensor data segmentation into three main classes: (1) metric-based methods, (2) learning-based methods, and (3) knowledge-based methods.

---

[1] Dataset of user activities for activity recognition application, for anomalies in case of anomaly detection application, etc.
[2] http://ailab.wsu.edu/casas/datasets/.

Metrics-Based methods: the general principle of these methods is that the window size is chosen according to the result of computed metrics such as mutual information [2,5]. For activity recognition process, the authors in [2] propose a segmentation method that consists of dividing the sensor events into chunks according to the incidence of activity. To achieve this aim, they use the Pearson Product Moment Correlation (PMC) metric to compute the correlation between any pair of sensor events in the window. The authors in [5] propose to compute the mutual information between each couple of sensor events within a window and calculate a feature's weight. Only highly related sensor events are supposed to be in the same activity. Another work [13] proposes a similarity measure to segment a motion stream. The proposed measure is highly related to the body motion attributes. For accelerometer streaming data segmentation, the authors in [6] propose to classify the data according to a set of features which are used as metrics such the variance, mean, etc. The method is specific for accelerometer data.

Knowledge-Based methods. This class applies mainly semantic representation (e.g ontology) and semantic reasoning (e.g logic inference) to derive dynamically the window size. For example, the work in [3] proposes an ontology representing the activities that can be realized by the residents, the types of installed sensors in the smart home, etc. The dynamic segmentation consists of either expanding or shrinking the time-window by queering the ontology at the same time as recognizing the activity. The work in [7] proposes a semantic based approach for segmenting sensor data series using ontologies to perform a terminology box (TBOX) and a assertion box (ABOX) reasoning, along with logical rules to infer whether the incoming sensor event is related to a given sequences of the activity. As in the previous work, the segmentation method is also integrated into the activity recognition method.

Learning-Based methods. Under this class, methods use machine learning in order to learn which suitable window size regarding the coming sensor data. Generally, a pre-segmented dataset is required for the training phase (i.e a dataset containing the ground truth of time window sizes for the segments). The authors in [1] propose to use a probabilistic data driven approach to identify, as a first step, the possible window size using a pre-segmented dataset. As a second step, they propose to learn the most likely window size for an activity based on the computed probabilities of the possible window sizes.

To the best of our knowledge, all the previous methods are designed to be used for a particular kind of application domain such as activity or gesture recognition. However, it is more useful to conceive a flexible method that can be applied in different applications.

3   Method

Through this section, we explain in details the whole process of the method for the dynamic streaming sensor data segmentation.

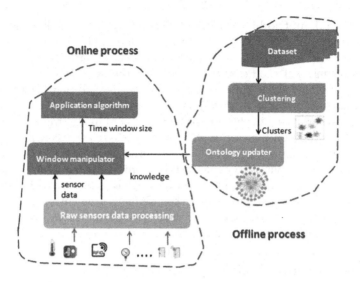

Fig. 1. Method architecture

As we can see in Fig. 1, the method holds mainly two processes: Offline and Online executions. The former starts by applying a `Clustering` algorithm in order to classify the given dataset into a set of classes based on given features. The dataset must concern one application and the features are defined accordingly. As mentioned before, the proposed method could be used with different application domains. The clustering is executed offline once a dataset is given or updated. Afterwards, the `Ontology updater` dynamically updates the ontology when the clusters are updated or new ones appear. The ontology is designed to contain the information that must be provided by the built clusters for each application. Next, when the dataset is ready and the set of clusters are obtained or changed, the ontology is dynamically updated. The online process dynamically segments the streaming sensor data into time-windows based on the knowledge provided by the ontology.

3.1 Clustering

The clustering aims, as a first step, to divide a given dataset into a set of clusters based on the given features. Each target application has its own features. For example, *activity durations* and *used sensors* are features corresponding to the activity recognition application while the *anomaly start time* can be considered as a feature for anomaly detection application.

3.2 Ontology Updater

A general ontology is designed to semantically represent the information provided by the clusters for each dataset. In order to allow a quick start of the

---

Algorithm 1. Ontology Updater

Input: Ontology O, Features set sf, Clusters C, Name of the dataset name, Application app
Output: Updated ontology O
1 **if** *(ExecuteRule(search, O, name)* $\neq \emptyset$*)* **then**
2 | ExecuteRule(insertInstance, Dataset, name);
3 **end**
4 appC ← GetClass(O, app);
5 **for** *(each cluster $c_i \in C$)* **do**
6 | **for** *(each element $el \in c_i$)* **do**
7 | | ExecuteRule(insertInstance, el, hasElement(O, appC));
8 | **end**
9 | **for** *(each feature $f_j \in sf$)* **do**
10 | | fC ← GetClass(O,f_j);
11 | | val ← GetValue(c_i, f_j);
12 | | ExecuteRule(insertInstance, val, fC);
13 | | setObj ← GetObjectProperties(hasElement(appC), fC);
14 | | **for** *(each objectProperty $obj \in setObj$)* **do**
15 | | | ExecuteRule(insertRelation, obj, c_i, val);
16 | | **end**
17 | **end**
18 **end**

---

system even if a dataset is not yet ready, the ontology initially contains similar information that should be provided by the clusters for each application. When the dataset is ready and the clusters are obtained, the ontology is dynamically updated with the content of the given clusters. The dynamic update is insured by using program and SPARQL queries. Algorithm 1 shows the operations performed by the ontology updater. Initially the ontology contains the three classes *Application*, *Dataset*, and *Element*. Moreover, it contains the applications that will be held by the method represented as subclasses of *Application* class. The class *Element* represents the main information provided by the application. For instance, for the activity recognition application the ontology must contain the class *Activity*, a subclass of *Element*, related to the class *activity recognition*, a subclass of *Application*, and related to the classes corresponding to the application's features (i.e Duration). After receiving the application name, the set of features and clusters, and the dataset name, Algorithm 1 starts by executing a rule named *search*, by calling the function *ExecuteRule*, which searches whether the given dataset is already in the ontology (Line 1 and 2). It is important to mention that the algorithm stores the corresponding classes in the ontology for each possible application and their corresponding features. Hence, the function *GetClass* (Line 4) returns the corresponding ontological class for the input application *app*. Afterwards, by applying a logic rule, named *insertInstance*, all the different elements belonging to the given clusters become instances of the corresponding subclass of *Element* related to the given application (Lines 5–8). For example, if the application is *Activity recognition* the ontology must hold the class *Activity* subclass of *Element*. In this case the activities in the clusters become instances of the *Activity* class. For a better insight see Fig. 2. Next, each feature value in the cluster becomes an instance of the corresponding feature's class in the ontology using a logic rule (Line 9–12). For example, 5 min is a value in the cluster of the feature Duration. Accordingly, 5 min becomes an instance

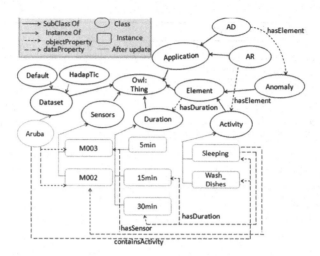

Fig. 2. An example of the ontology used in the experiments after being updated with Aruba dataset's clusters. Added elements are in red (Color figure online)

of the class *Duration* in the ontology (see Fig. 2). Finally, the algorithm extracts the set of objectProperties having the feature class as a range (Line 13). Then, it relates the instances corresponding to elements in the clusters with theirs features values using the adequate objectProperty. For example, in the ontology (Fig. 2), the feature's class *Duration* is the range of the objectProperty *hasDuration* where the Class *Activity* is its domain. Therefore, Algorithm 1 links the added activity's instances for each cluster with their corresponding durations.

3.3   Data Processing and Window Manipulator

Generally, a sensor sequence can be represented as $\{S1, S2...Sn\}$ where Si refers to the it^{th} sensor event, and each sensor event is encoded in the template of $\{date, time, sensorID, sensorValue\}$. Once the coming sensor data are represented in this template, the main aim of this step is to divide this streaming sensor data into a time windows with size "sw". Each chunk "sw" is chosen dynamically using the Window manipulator Algorithm 2.

Initially, the time window size is the minimum duration in the ontology regarding the given application and dataset. It is important to note here that Duration is defined as a features for all applications. For instance, if we target the activity recognition application and we set up the method for aruba dataset, then the algorithm sets the initial time window size as the minimum activity's durations belonging to Aruba. This duration should not be used before in the same session (Line 7). Afterwards, the algorithm keeps extracting the set of sensor data (*sensorsData*), that are in the streaming SD, occurred during the defined time window size (*actualSize*). This is achieved using the function *ReadOnline()* (Line 8). The algorithm stops the loop when *sensorsData* are included, in at least one of the feature vectors extracted from the ontology (Lines 12,18)

Algorithm 2. Window Manipulator

Input: ontology O, streaming sensor data SD, application app, application's features feats,
    Dataset's name dName
Output: set of possible results possRes, time window size sw

1 usedsizes← ∅;
2 sensorsData ← ∅;
3 keep ← true;
4 possRes ← Null ;
5 previousSize ← 0 ;
6 **while** *(keep)* **do**
7   actualSize ← getMinimumDuration(O, usedSizes, previousSize, app, dName);
8   sensorsData ← ReadOnline(SD, actualSize-previousSize, sensorsData);
9   EleFeatVects ← getElementsVectors(O, app, feat);
10   **for** *each appEl ∈ EleFeatVects* **do**
11   **if** *(testAppart(sensorData, appEl)* **then**
12   add(possRes, appEl);
13   **end**
14   **end**
15   previousSize ← actualSize;
16   Add(usedSizes, actualSize);
17   **if** *(possRes!= Null)* **then**
18   keep=false;
19   sw← actualSize;
20   **end**
21 **end**

---

using the *testAppart* function. In other words, based on the given application, the algorithm extracts the features vector for each given application's element. For example, in case of anomaly detection, it extracts for each anomaly its features vector holding the features' values. Then, it checks whether the data in the streaming are belonging to this vector. In the positive case, this element is added to the set of possible results *possRes* (Line 13). Finally, if the algorithm detects at least one possible element then it stops reading (Lines 19, 20). Otherwise, it extends the actual time-window size (*actualSize*) with the difference between it and the next minimum element's duration extracted from the ontology (Line (7)), the actual size becomes the previous size, and this process is repeated until possible results are detected (Line 17, 18).

4 Experiments and Discussion

To evaluate the usefulness of our proposal for dynamic streaming sensor data segmentation, we have applied it in case of activity recognition application. We tested it with 6 weeks over Aruba dataset and a synthetic dataset. On the one hand, Aruba was collected by the Center for Advanced Studies in Adaptive Systems (CASAS) [11]. The Aruba dataset contains ground-truthed activities of a home-bound person in a small apartment for 16 weeks. On the other hand, we have generated automatically two datasets that can be obtained from the GIS MADONAH[3]. For the activity recognizer method we applied the Support Vector Machine (SVM) algorithm. More details about the dataset and the experiments are given in the next subsections.

---

[3] http://www.bourges.univ-orleans.fr/madonah/.

4.1 Datasets

Data collected from Aruba dataset was obtained using 31 motion sensors, three door sensors, five temperature sensors, and three light sensors. 11 activities were performed for 220 days (7 months). The dataset is imbalanced, as some of the activities occur more frequently than others. Table 1 presents the statistics of the sensor events and activities performed in the 6 weeks over Aruba dataset. "Other activity" class contains events with missing labels. It covers 54% of the entire sensors events sequence.

The synthetic dataset contains sensor data values similar to that provided in GIS MADONAH. It is generated separately for half day (90 events) and one day (185 events) of an elderly like routine. It contains a set of presence, motions sensors and light detectors. The dataset holds principally four activities: Sleeping, Watching TV, Discomfort, and Eating.

These datasets were used in our method, on the one hand, as training dataset for SVM and, on the other hand, for the clustering.

Table 1. Statistics of six weeks over Aruba dataset

Id	Activity	Number of events
1	Bed_to_Toilet	266
2	Eating	3207
3	Enter_Home	404
4	Housekeeping	2117
5	Leave_Home	384
6	Meal_Preparation	57029
7	Relax	70917
8	Resperate	108
9	Sleeping	6536
10	Wash_Dishes	2092
11	Work	3264
12	Other activity	174 264

4.2 Evaluation Result

F-Score measures was used to evaluate our proposal with 6 weeks over Aruba dataset. Moreover, in order to discuss its performance regarding static time-window size, we firstly tested SVM with fixed time-window size (i.e. 5 min).[4] Afterwards we used SVM as an activity recognizer in the method. Furthermore, the results are compared with the dynamic data segmentation process, SWMIex [5]. SWMIex is a metrics-based method which uses the Mutual Information measure to compute sensor correlation. SWMIex, SVM with static time window, and our proposal are tested in same conditions (i.e dataset). For the process, we used three quarters of the dataset for the offline training and one quarter for the online testing. Figure 3 shows the different F-score values of the three methods for each activity belonging to Aruba dataset.

---

[4] After doing a set of experiments with different time window sizes, five minutes has shown best results for SVM using 6 weeks over Aruba dataset.

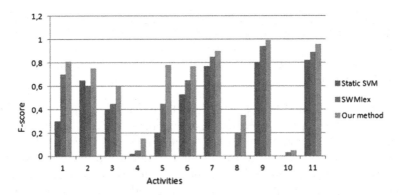

Fig. 3. F-score value for the different activities in Aruba dataset using only SVM with static time-window length (5 min), our and SWMIex methods

Table 2. Average F-score for all activities

Dataset /method	Static SVM	Our method
Half day synthetic dataset	0.39	0.42
One day synthetic dataset	0.42	0.53
Six weeks over Aruba	0.40	0.64

As it is shown in Fig. 3, our proposal outperforms static SVM and SWMIex in terms of F-score. From Fig. 3, we conclude that the three methods have unbalanced performance. In other words, they have higher efficiency to recognize some activities (i.e. 9. Sleeping, 11. Work) than to detect some others (4. HouseKeeping, 10. Wash_Dishes, 8. Resperate). This result, has two possible explanations. On the one hand, since also the dataset is unbalanced then infrequent activities have low chance to be recognized such as the *Resperate* activity. On the other hand, the *used sensor* is a feature for the three methods and some activities in the dataset can have different used sensors for each occurrence. The methods are then sometimes confused (i.e *HouseKeeping*).

The SVM with 5 min as time-window length has the lowest performance (F-score $\in [0..0.9]$) since it has a static segmentation. In fact, the duration of each activity varies according to the resident routine. For some activities five minutes is sufficient for the recognition and for others this length is either too long or too short. The proposal has the best performance (F-score $\in [0.38..0.99]$) thanks to its dynamic segmentation.

Table 2 shows the average F-score values of our method compared to Static SVM using the synthetic datasets and Aruba. Obviously, as we can see in the table, our method is more performing using Aruba than the two synthetic datasets and outperforms the Static SVM using the three datasets. The better F-score value obtained by our method using Aruba can be explained by the high sensors number used for Aruba. Indeed, the *used sensor* is considered as a feature in our method for classifying activities. Therefore, when the sensor number increased, our method differentiates better the activities and then provides better results. For One day synthetic dataset, our proposal has better F-score value than for that of half day because a richer training model is built from the former since it holds double data.

5   Conclusion

In this paper we proposed an hybrid method for dynamic streaming sensor data segmentation. The proposal combines clustering and ontology techniques to provide high flexibility in order to be used in different types of applications. As a matter of fact, the main advantage of our method regarding previous works are its ability to be adapted with different applications. Moreover, it is able to dynamically update the ontology whenever any update in the clusters occurs. In order to prove the efficiency of the proposal, we tested it in case of activity recognition application using a six weeks over Aruba dataset and two synthetic datasets and SVM as an activity recognizer. The evaluation result proves the proposal high efficiency level which is better than static method and a previous metric-based method.

References

1. Narayanan, C.K., Diane, J.C.: Activity recognition on streaming sensor data. Pervasive Mob. Comput. **10** (2014)
2. Jie, W., Michael, J.O., Gregory, M.P.O.: Dynamic sensor event segmentation for real-time activity recognition in a smart home context. Pers. Ubiquitous Comput. **19**, 287–301 (2015)
3. Georgo, O., Liming, C., Hui, W.: Dynamic sensor data segmentation for real-time knowledge-driven activity recognition. Pervasive Mob. Comput. **10**, 155–172 (2014)
4. Tak-chung, F.: A review on time series data mining. Eng. Appl. Artif. Intell. **24**, 164–181 (2011)
5. Nawala, Y., Belkacem, F., Anthony, F.: Towards improving feature extraction and classification for activity recognition on streaming data. J. Ambient. Intell. Hum. Comput. **8**, 177–189 (2016)
6. Tian, G., Zhixian, Y.I., Karl, A.: An adaptive approach for online segmentation. In: International Workshop on Data Engineering for Wireless and Mobile Access (2012)
7. Darpan, T., Liming, C., ChenZumin, W.: Semantic segmentation of real-time sensor data stream for complex activity recognition. Pers. Ubiquitous Comput. **21**, 411–425 (2017)
8. Warren, T.L.: Clustering of time series data-a survey. Pattern Recogn. **38**, 1857–1874 (2005)
9. Wilpon, J.G., Rabiner, L.R.: Modified k-means clustering algorithm for use in isolated word recognition. IEEE Trans. Acoust. Speech Signal Process. **33** (1985)
10. Tran, D., Wagner, M.: Fuzzy C-means clustering-based speaker verification. In: Pal, N.R., Sugeno, M. (eds.) AFSS 2002. LNCS (LNAI), vol. 2275, pp. 318–324. Springer, Heidelberg (2002). https://doi.org/10.1007/3-540-45631-7_42
11. Diane, J. C.: Learning setting-generalized activity models for smart spaces. IEEE Intell. Syst. **99** (2010)

12. Hearst, M.A., Dumais, S.T., Osuna, E., Platt, J., Scholkopf, B.: Support vector machines. IEEE Intell. Syst. Their Appl. **13** (1998)
13. Chuanjun, L., Prabhakaran, B.: A similarity measure for motion stream segmentation and recognition. In: International Workshop on Multimedia Data Mining: Mining Integrated Media and Complex Data (2005)

Gaussian Process Kernels for Noisy Time Series: Application to Housing Price Prediction

Juntao Wang[1(✉)], Wun Kwan Yam[1], Kin Long Fong[1], Siew Ann Cheong[2], and K. Y. Michael Wong[1(✉)]

[1] The Hong Kong University of Science and Technology, Hong Kong, China
{jwangcc,wkyam,klfong}@connect.ust.hk, phkywong@ust.hk
[2] Nanyang Technological University, 21 Nanyang Link, Singapore 637371, Singapore
cheongsa@ntu.edu.sg

Abstract. We study the prediction of time series using Gaussian processes as applied to realistic time series such as housing prices in Hong Kong. Since the performance of Gaussian processes prediction is strongly dependent on the functional form of the adopted kernel, we propose to determine the kernel based on the useful information extracted from the training data. However, the essential features of the time series are concealed by the presence of noises in the training data. By applying rolling linear regression, smooth and denoised time series of change rates of the data are obtained. Surprisingly, a periodic pattern emerges, enabling us to formulate an empirical kernel. We show that the empirical kernel performs better in predictions on quasi-periodic time series, compared to popular kernels such as spectral mixture, squared exponential, and rational quadratic kernels. We further justify the potential of the empirical kernel by applying it to predicting the yearly mean total sunspot number.

Keywords: Gaussian processes · Time series prediction
Housing price prediction · Empirical kernels

1 Introduction

In recent years, Gaussian processes (GPs) have become a useful technique for machine learning [1]. They are based on a principled approach to statistical inference as long as correlations among the data can be estimated *a priori*. Hence, it provides a Bayesian nonparametric approach to classification, regression and prediction, and excellent results were produced [2].

The prior of the inference embodies our natural expectation about the behaviors of the data, and many forms of kernels have been proposed [1,2]. For example, for quantities that have smooth variations in space or time, the Squared Exponential (SE) kernel is a suitable choice. For data with stronger correlation range, one may choose the Rational Quadratic (RQ) kernel. To capture the characteristics of different kernels, mixtures of kernels such as Spectral Mixture (SM) kernel [2] have also been used.

The importance of the appropriate choice of kernels was illustrated in GP time series prediction [2,3]. For example, while the SE kernel is suitable to describe smoothly varying data, it performs poorly when processing time series with periodic variations. For time series data with both long and short term trends, kernels such as RQ can capture the long term trend but not short one, whereas kernels such as SE performs oppositely. To improve the performance, kernels composed of a spectral mixture [2] was proposed, and was able to overcome the above drawbacks.

In real applications of time series prediction in science and economics, the data are often very noisy. The quasi-periodic components of the data may be masked by the noise. The prediction of housing prices considered in this work is a typical example. And for the quasi-periodic time series, traditional kernels such as SE and RQ may not be able to make meaningful predictions and sophisticated kernels such as SM can suffer from over-parameterization and overfitting.

Financial time series have been investigated using different methods, such as fuzzy time series [4], Support Vector Machine [5,6], and Artificial Neural Network (ANN) [7], in recent decades for the interest of maintaining stable and profitable market environment. However, most of the methods fail to be applicable to real time series [4] and they are too complex and lack the power of interpretation to capture the essential properties of the real time series.

Among various markets, the housing market might be the most appealing one, because it is closely related to the livelihood of individuals. For example, the Japanese housing bubble burst in 1991 exerted serious impact not only on the housing market but also the overall Japanese economy [8], leading to the sudden decline in both investment and consumption. However, because of the stochastic nature of financial markets [9], it is extremely difficult for scientists to accurately predict the subsequent observations based on the current information. On the other hand, many researchers are seeking indicators which could reveal the hidden states of the financial markets. For example, to prevent great depression caused by housing bubble bursts, there were many efforts to identify the explosive behavior, which is regarded as bubble inflation, in housing price time series. Several statistical techniques were used to confirm the existence of explosive behavior, such as root test, augmented Dickey Fuller test, and Cointegration test [10]. One popular method was introduced by [11], called PSY method, and gained its success in identifying multiple bubbles in historical housing data [12]. However, it lacks evidence that it is practical in predicting future trends in housing markets.

In this paper, we investigate the useful statistical information extracted from a real time series, namely, the Hong Kong housing price, and propose a comprehensive kernel using the prior knowledge obtained from the time series. By using this kernel, we can improve the predictive power of the GP models. In Sect. 2 we first review the theoretical framework of GPs. In Sect. 3, we describe how the Fourier spectrum of the auto-covariance of the time series can be used to construct the High Order Periodic (HOPE) kernel for Gaussian processes. In Sect. 4, we describe the construction of the kernel from the Hong Kong housing price data. In Sects. 5 and 6, we will present the predictions on housing price data and yearly mean total sunspot number given by the GP models using the HOPE kernel and some other alternative popular kernels. The paper is concluded in Sect. 7.

2   Gaussian Processes

Given a set of points $X = \{x_1, x_2, \ldots, x_t\}$, a Gaussian process [1] assumes that $f(x_i)$, the true value generated at point x_i, satisfies the joint Gaussian distribution

$$\boldsymbol{f} = [f(x_1) \ldots f(x_t)]^T \sim \mathcal{N}(\boldsymbol{\mu}, \boldsymbol{K}(X, X)), \tag{1}$$

with mean vector $\boldsymbol{\mu} = [\mu(x_1), \mu(x_2), \ldots, \mu(x_t)]^T$ and covariance matrix $\boldsymbol{K}$, where $K_{ij} = k(\boldsymbol{\theta}, x_i, x_j)$ is the kernel parameterized by hyperparameter $\boldsymbol{\theta}$. And $\mathcal{N}$ is the Gaussian distribution. Generally, the true value $f(x_i)$ cannot be known and only their noisy versions $y_i = f(x_i) + \epsilon$ can be observed, here ϵ is the independent identically distributed Gaussian noise with variance σ_n^2. Based on the given observations and prior distribution, the expected value and variance of the true value generated at a set of points X' can be inferred. Specifically, to estimate the true values $\boldsymbol{f'}$ at X' based on the known observations and prior, we have the joint distribution

$$\begin{bmatrix} \boldsymbol{y} \\ \boldsymbol{f'} \end{bmatrix} \sim \mathcal{N}\left(\begin{bmatrix} \boldsymbol{\mu} \\ \boldsymbol{\mu'} \end{bmatrix}, \begin{bmatrix} \boldsymbol{K}(X, X) + \sigma_n^2 I & \boldsymbol{K}(X', X) \\ \boldsymbol{K}(X, X') & \boldsymbol{K}(X', X') \end{bmatrix}\right), \tag{2}$$

where I is the identity matrix, and the posterior predictive distribution

$$p(\boldsymbol{f'}|\boldsymbol{y}, X, X') = \mathcal{N}(\boldsymbol{m}, \boldsymbol{C}), \tag{3}$$

where $\boldsymbol{m}$ is the estimated mean and $\boldsymbol{C}$ is the estimated variance at points X', which can be analytically derived as:

$$\boldsymbol{m} = \boldsymbol{\mu'} + \boldsymbol{K}(X, X')(\boldsymbol{K}(X, X) + \sigma_n^2 I)^{-1}(\boldsymbol{y} - \boldsymbol{\mu}), \tag{4}$$

$$\boldsymbol{C} = \boldsymbol{K}(X', X') - \boldsymbol{K}(X, X')(\boldsymbol{K}(X, X) + \sigma_n^2 I)^{-1}\boldsymbol{K}(X', X). \tag{5}$$

Therefore, specifying the kernel k, the variance σ_n^2, and the mean function μ determines the GP, and hence, the estimated mean value can be calculated by the model. To tune the hyperparameters of GP, we need to optimize the marginal log-likelihood $\mathcal{L}$ with respect to the unknown hyperparameters in the covariance matrix and the mean function:

$$\mathcal{L} = \log p(\boldsymbol{y}|\boldsymbol{\theta}, X) = -\frac{1}{2}(\boldsymbol{y} - \boldsymbol{\mu})^T (\boldsymbol{K} + \sigma_n^2 I)^{-1} (\boldsymbol{y} - \boldsymbol{\mu})$$

$$-\frac{1}{2}\log\det(\boldsymbol{K} + \sigma_n^2 I) - \frac{|X|}{2}\log 2\pi, \tag{6}$$

where X is the set of observation points, $\boldsymbol{y}$ is the corresponding vector of observed values at X, and $|X|$ is the number of observations. Thus, given the estimated hyperparameters, variance σ_n^2, and mean function that maximize the log-likelihood $\mathcal{L}$, we are able to predict the expected value and standard deviation of the unknown true value generated at points X' by the predictive distribution.

3   Construction of Kernels

In this section, we construct a new kernel for noisy realistic time series. Although empirical correlation can be questionable in reflecting the true theoretical underlying processes [2], we may show that it is still possible to improve predictions on noisy quasi-periodic time series by using the information extracted from statistical analysis, such as auto-covariance analysis and Fourier Transform.

The cross-covariance $\gamma_{FG}(l)$ for two finite discrete real time series $F = \{F_t : t \in T\}$ and $G = \{G_t : t \in T\}$ is defined as

$$\gamma_{FG}(l) = \sum_t F_{t+l}G_t, \tag{7}$$

where l is the time that G lags F. For the case that $F = G$, $\gamma_{FG} = \gamma_{FF}$ is the auto-covariance of the time series F. For the housing price data we adopt the definition that is without the subtraction of expected value of the time series.

Given a time series F with N observations, the auto-covariance $\gamma_{FF}(l)$ can characterize the similarity in the data up to lag $N - 1$. To construct the kernel function describing the similarities between arbitrary data, we transform the auto-covariance $\gamma_{FF}(l)$ into its frequency-domain:

$$X(f) = \sum_{l=0}^{N-1} \gamma_{FF}(l)e^{-i2\pi fl/N}, \tag{8}$$

and obtain the auto-covariance spectrum $\frac{|X(f)|}{N}$.

Therefore, we select the significant periodic components having large enough amplitudes to construct the kernel and ignore the other weaker components, since the components with small amplitudes may be just noises or redundancies and have little contribution to the true connections between the data. Thus we propose the one-dimensional High Order Periodic (HOPE) kernel with order n:

$$k_n(x_i, x_j) = \alpha e^{-(x_i-x_j)^2/(2\ell^2)} \sum_{m=1}^{n} A_m \cos\left(2\pi f_m \left|x_i - x_j\right|\right), \tag{9}$$

where $n \in \mathbb{Z}^+$, α, ℓ, and $f_m, m = 1,\ldots,n$, are hyperparameters, and the weights $A_m, m = 1,\ldots,n$ is the m-th largest amplitudes (peaks) of the components in the auto-covariance spectrum. The introduction of the exponential factor $e^{-(x_i-x_j)^2/(2\ell^2)}$ is to ensure that the kernel vanishes when the space (time) difference between two observations points approaches infinity. Intuitively, two data points well separated in space (time) should have weak correlations. The HOPE kernel can be considered as an extension of the Spectral Mixture (SM) kernel [2] to deal with noisy quasi-periodic data, whereby we determine a number of hyperparameters directly from the statistical features of the data to alleviate the problem of over-parameterization and overfitting, which will be shown in Sects. 5 and 6. Then by optimizing the log-likelihood (Eq. 6) with respect to these hyperparameters and the mean function μ, we obtain a trained GP regression model.

4   Hong Kong Housing Price

The data of the Hong Kong property transaction prices were obtained through negotiations from EPRC Limited, a wholly-owned subsidiary of the Hong Kong Economic Times that specializes in providing property information to market-related industries. The dataset contains records of transacted properties for the period 1992 to 2010 and amounts to 2,492,842 transaction records. Figure 1(a) illustrates the monthly average of real price index (MARPI) (in HKD per square foot normalized by the consumer price index) of Hong Kong housing transactions and the Prime Lending Rates (PLR) in Hong Kong from Jan 1992 to Dec 2010 (totally 228 data points). Since the data of MARPI are very noisy, to pre-process and denoise the data, we investigate the time series of MARPI rate of change over 12 months (a year), defined as the slope of the linear regression line for the past 12 months MARPI data. Similarly, the time series of PLR rate of change over 12 months can also be calculated. As a comparison, these two time series are illustrated in Fig. 1(b).

As shown in Fig. 1(b), quasi-periodic features exist in both the MARPI rate of change and the PLR rate of change. The quasi-periodicity is exhibited even more clearly in the auto-covariance (Eq. 7) of the rate of change as shown in Fig. 2(a).

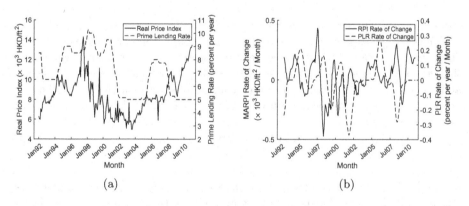

(a)   (b)

Fig. 1. (a) The monthly average real price index (MARPI) per square foot of Hong Kong housing and the prime lending rate (PLR). (b) The MARPI rate of change and the PLR rate of change.

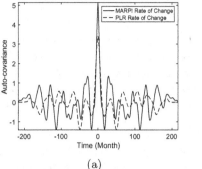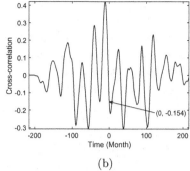

(a)   (b)

Fig. 2. (a) Auto-covariance of MARPI rate of change and PLR rate of change. (b) Cross-correlation (normalized version of cross-covariance) of MARPI rate of change and PLR rate of change.

The periodicity of the rate of change observed from the auto-covariance is around 30 months, as confirmed by the result of the auto-covariance spectrum in Fig. 3.

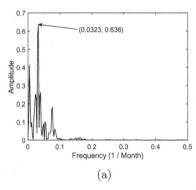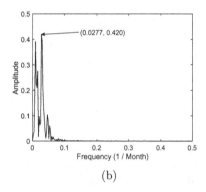

(a)   (b)

Fig. 3. Single-sided auto-covariance spectrum of (a) MARPI rate of change and (b) PLR rate of change. The components with the largest amplitude in (a) and (b) have periodicity $1/0.0323 = 31.0$ Month and $1/0.0277 = 36.1$ Month respectively.

It is interesting to consider why housing price time series is quasi-periodic. We found evidence that the housing price is correlated with PLR, which happened to have a quasi-periodic variation during this period of study. While the underlying reason is beyond the scope of this study, it is natural to expect the occurrence of cycles in the world economy. When consumption heats up, PLR will be raised to prevent overheating, and when the economy is weak, PLR will be reduced to stimulate consumption. As shown in Figs. 1, 2, 3, there is a close correlation between the periods of two time series. Furthermore, the cross-correlation (normalized version of cross-covariance) is negative near 0 months showing that housing prices drop (rise) when the mortgage rate rises (drops).

5   Experiments

In this section, we show that the HOPE kernel can be applied to the real time series to improve the predictive power of the GP. At the same time, to contrast the prediction results using the new kernel, other popular kernels, such as Spectral Mixture (SM) with number of components $Q = 12$, Squared Exponential (SE), and Rational Quadratic (RQ) kernel (as shown in Table 1 [1,2]) are also applied to predict the data. Since the size of our data is small (around 200 data points), Broyden-Fletcher-Goldfarb-Shanno (BFGS) algorithm [13] with cubic interpolation is applied to optimize the log-likelihood function. The `fitrgp.m` and `fminunc.m` functions from MATLAB [14] are used to implement the GP and optimization. The mean function μ of the GP model is set to be a constant function $\mu(x) = c$. For the optimization of hpyperparameters, we randomly sample 40 initializations of hyperparameters for each model and optimizations are repeated. The final optimal hyperparameters are the ones which yield the highest log-likelihood $\mathcal{L}$ for the model. During the sampling, the order n in HOPE is sampled between 1 and 12.

Table 1. Expressions of one-dimensional popular kernels, where $\tau = |x_i - x_j|$.

Kernel	Expression	Hyperparameters
SM	$k_{SM}(x_i, x_j) = \sum_{q=1}^{Q} \omega_q \exp\left(-2\pi^2\tau^2\nu_q\right)\cos\left(2\pi\tau\mu_q\right)$	ω_q, ν_q, μ_q
SE	$k_{SE}(x_i, x_j) = \sigma^2 \exp\left(-\frac{\tau^2}{2\ell^2}\right)$	σ, ℓ
RQ	$k_{RQ}(x_i, x_j) = \sigma^2 \left(1 + \frac{\tau^2}{2\alpha\ell^2}\right)^{-\alpha}$	σ, α, ℓ

5.1   Predictions on MARPI Rate of Change

To illustrate the predictive power of the GP model using the HOPE, we choose two subsets from the time series of the MARPI rate of change as the training set, one containing the first 152 observations in the time series and the other constituting the first 171 observations, and the remaining data as testing sets. The predictions using different kernels can be found in Fig. 4.

GP models using all kernels perform well in reconstructing the training inputs, since the predictions on training data given by the models almost overlap with the real data in the training set. However, in future predictions, models using SE and RQ have little predictive power, since the estimated mean observations in the future (outside the training set) diverges largely and quickly from the testing data, and their values tend to fall back to the mean function c. The predictions using SM are unsatisfactory, since there are many huge gaps between the estimated mean observations and real data. On the other hand, the predictions given by the model using HOPE do not deviate from the testing data dramatically and can capture the trend of the future data up to a considerable number of steps. At the same time, Fig. 4 indicates that generally, the errors may increase when we predict the values to a more distant future.

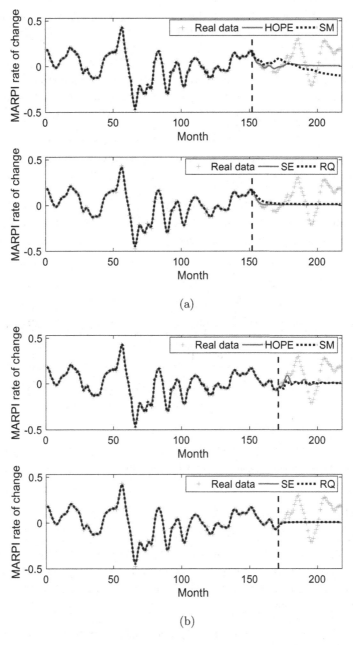

Fig. 4. Predictions on MARPI rate of change using different kernels. The training data are on the left hand side of the vertical dashed line. (a) The first 152 observations as the training inputs. (b) The first 171 observations as the training inputs.

Another observation is that for the training set with 152 (171) observations, the optimized log-likelihood $\mathcal{L}$ of HOPE and SM are 270.9 (311.7) and 300.4 (329.9) respectively. Although the SM has higher log-likelihood, Fig. 4 shows that it has poorer performance in predicting the unobserved data. One suggested reason for this is that using SM, possibly overparameterized, leads to the overfitting so that the predictive power of the GP is suppressed. Meanwhile, this observation can be considered as the evidence that compared to SM, HOPE is able to effectively prevent the GP from overfitting and give better predictions on the future data.

Moreover, we use another example to illustrate the problem of overfitting induced by the SM. For one set of training inputs (first 200 observations of MARPI rate of change), the predictions given by two GP models using SM ($Q = 12$) are shown in Fig. 5. Compared to the model with log-likelihood $\mathcal{L} = 419.5$, the model with the higher log-likelihood $\mathcal{L} = 428.1$ has rather poorer performance on predicting the future data.

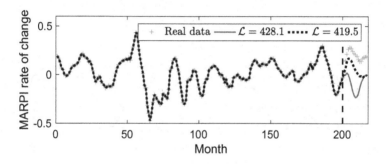

Fig. 5. Predictions on MARPI rate of change with first 200 observations as training inputs, which are on the left hand side of the vertical dashed line. $\mathcal{L}$ is the log-likelihood of the model. Both models are using SM with $Q = 12$.

5.2   Performance and Stability

Since GPs can be really sensitive to the initialization of the hyperparameters as well as the input training data, we further compare the prediction stability of different kernels by measuring their mean squared error (MSE) and following the procedure of time series prediction performance assessment introduced by Hyndman [15]. Specifically, experiments are repeated using different training sets, each one having additional one observation than the previous one, and during each experiment and for each training set the MSE in h-step horizon (that is, from 1-step prediction up to h-step prediction) is computed to investigate the performance of prediction. Initially, the training set has 151 observations. And the average MSE of experiments of each kernel can be found in Table 2. The smaller the average MSE, the more stable the model, because small average MSE indicates that the model can maintain relatively good performance when the training inputs are changed. As a comparison, the variance of the whole data set is 0.0232. Table 2 shows that the stability of HOPE outperforms other kernels in several-step horizon prediction, which also justifies that HOPE has better performance in capturing the future trends of the time series. It also indicates that the predictive power fades away when the GP model is trying to estimate the data in distant future, since the average MSE increases dramatically when the prediction horizon increases. And we can also see that the performances of SM are relatively poor, which might be the consequences of overfitting.

Table 2. Average MSE using different kernels

Horizon	HOPE	SM	SE	RQ
6-step	**0.0078**	0.0088	0.0092	0.0086
9-step	**0.0118**	0.0149	0.0130	0.0130
12-step	**0.0148**	0.0171	0.0153	0.0159

6 Application to Sunspot Time Series

To further illustrate the power of HOPE, we apply it to another well known time series, yearly mean total sunspot number, which can be obtain from http://www.sidc.be/silso/datafiles. To make comparisons, the popular kernels in Table 1 (SM with $Q = 20$) are also applied. The data from year 1700 to year 1899 (totally 190 data points) are used as training inputs, and the data from year 1900 to year 1941 (totally 42 data points) are considered as testing set. The mean function of the GP is set to be constant function $\mu(x) = c$. BFGS algorithm and 100 random initializations of hyperparameters are applied to search for optimal estimations of hyperparameters for each model, and the order n in HOPE is sampled between 1 and 20.

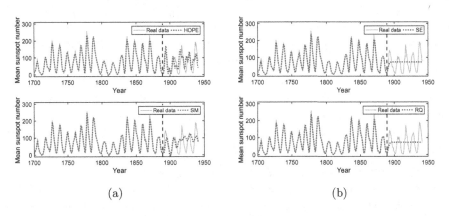

(a)   (b)

Fig. 6. Predictions on yearly mean total sunspot number with the first 190 observations as training inputs, which are on the left hand side of the vertical dashed line. (a) Predictions given by HOPE and SM with $Q = 20$. (b) Predictions given by SE and RQ.

Predictions on the yearly mean total sunspot number given by different kernels are illustrated in Fig. 6. Similar to the cases of predictions on MARPI rate of change, for sunspot time series, SM ($\mathcal{L} = -857.5$) has a larger log-likelihood than HOPE ($\mathcal{L} = -885.0$). However, for the performance on predicting the testing data, compared to HOPE (MSE = 1947.5), SM (MSE = 2272.3) gives poorer predictions, which indicates the problem of overfitting caused by SM. Also, predictions on future data given by both SE (MSE = 2588.4) and RQ (MSE = 2599.7) are quite poor, which can be seen in Fig. 6(b), though their log-likelihood values ($\mathcal{L}_{SE} = -887.7$ and $\mathcal{L}_{RQ} = -887.5$) are comparable to that of HOPE.

7   Conclusion

In this paper, we have discussed the Gaussian process (GP), which is a powerful regression model to predict the time series. Using a new kernel (HOPE) based on the information extracted from the auto-covariance of the training time series, we improved the predictive power of the GP model and it compares favourably with models using other popular kernels. Our work shows that the choice of the kernel is a very essential factor to the performance of GP. Choosing a kernel from some popular choices may not work for noisy quasi-periodic data, in which case extracting the kernel from the covariance functions provides an alternative.

The Gaussian process can be a powerful predictor in financial time series analysis. Since we have only used Gaussian processes to analyze the single time series, we believe that by incorporating multiple time series, such as Prime Lending Rate and trading volume, Gaussian processes may capture more information underlying the series, and give more accurate predictions. However, the model cannot predict the value at arbitrary time steps ahead, so we may further investigate the applicable range that the predicted values are still close to the real data. On the other hand, we may also focus on predicting a few steps ahead so that we can determine its accuracy and stability in short-term prediction.

Acknowledgements. We thank Prof. Chihiro Shimizu for forwarding the data to us. This work is supported by the Research Grants Council of Hong Kong (grant numbers 16322616 and 16306817).

References

1. Rasmussen, C.E., Williams, C.K.: Gaussian Process for Machine Learning. MIT Press, Cambridge (2006)
2. Wilson, A.G., Adams, R.P.: Gaussian process kernels for pattern discovery and extrapolation. In: Dasgupta, S., McAllester, D. (eds.) Proceedings of the 30th International Conference on Machine Learning. Proceedings of Machine Learning Research, vol. 28, pp. 1067–1075. PMLR, Atlanta (2013)
3. Brahim-Belhouari, S., Bermak, A.: Gaussian process for nonstationary time series prediction. Comput. Stat. Data Anal. **47**(4), 705–712 (2004)

4. Lee, C.H.L., Liu, A., Chen, W.S.: Pattern discovery of fuzzy time series for financial prediction. IEEE Trans. Knowl. Data Eng. **18**(5), 613–625 (2006)
5. Kim, K.J.: Financial time series forecasting using support vector machines. Neurocomputing **55**(1–2), 307–319 (2003)
6. Cao, L.J., Tay, F.E.H.: Support vector machine with adaptive parameters in financial time series forecasting. IEEE Trans. Neural Netw. **14**(6), 1506–1518 (2003)
7. Kim, T.Y., Oh, K.J., Kim, C., Do, J.D.: Artificial neural networks for nonstationary time series. Neurocomputing **61**, 439–447 (2004)
8. Hamada, K., Kashyap, A.K., Weinstein, D.E.: Japan's Bubble, Deflation, and Long-Term Stagnation. MIT Press, Cambridge (2011)
9. Tsay, R.S.: Analysis of Financial Time Series, vol. 543. Wiley, Hoboken (2005)
10. Hui, E.C., Yue, S.: Housing price bubbles in Hong Kong, Beijing and Shanghai: a comparative study. J. Real Estate Finance Econ. **33**(4), 299–327 (2006)
11. Phillips, P.C., Shi, S., Yu, J.: Testing for multiple bubbles: historical episodes of exuberance and collapse in the S&P 500. Int. Econ. Rev. **56**(4), 1043–1078 (2015)
12. Yiu, M.S., Yu, J., Jin, L.: Detecting bubbles in Hong Kong residential property market. J. Asian Econ. **28**, 115–124 (2013)
13. Fletcher, R.: Practical Methods of Optimization. Wiley (2013)
14. MATLAB: Statistics and Machine Learning Toolbox Release 2017b. The MathWorks Inc., Natick (2017)
15. Hyndman, R.J.: Measuring forecast accuracy. In: Gilliland, M., Sglavo, U., Tashman, L. (eds.) Business Forecasting: Practical Problems and Solutions, pp. 177–184. Wiley (2015)

Social Systems

Least Cost Rumor Influence Minimization in Multiplex Social Networks

Adil Imad Eddine Hosni[1,2], Kan Li[1(✉)], Cangfeng Ding[1,3], and Sadique Ahmed[1]

[1] School of Computer Science, Beijing Institute of Technology, Beijing 100081, China
{hosni.adil.emp,likan,dcf}@bit.edu.cn, ahmad01.shah@gmail.com
[2] Ecole Militaire Polytechnique, BP 17, 16046 Algiers, Algeria
[3] School of Mathematics and Computer Science, Yanan university, Yan'an, China

Abstract. This paper deals with the issue of rumors propagation in online social networks (OSNs) that are connected through overlapping users, named multiplex OSNs. We consider a strategy to initiate an anti-rumor campaign to raise the awareness of individuals and prevent the adoption of the rumor for further limiting its influence. Therefore, we introduce the Least Cost Anti-rumor Campaign (LCAC) problem to minimize the influence of the rumor. The proposed problem defines the minimum number of users to initiate this campaign, which reaches a large number of overlapping users to increase the awareness of individuals across networks. Due to the NP-hardness of LCAC problem, we prove that its objective function is submodular and monotone. Then, we introduce a greedy algorithm for LCAC problem that guarantees an approximation within $(1 - 1/e)$ of the optimal solution. Finally, experiments on real-world and synthetics multiplex networks are conducted to investigate the effect of the number of the overlapping users as well as the networks structure topology. The results provide evidence about the efficacy of the proposed algorithm to limit the spread of a rumor.

Keywords: Rumor propagation · Multiplex online social networks Optimization · Rumor influence minimization

1 Introduction

Online social networks (OSNs), such as Twitter, Facebook, and Weibo[1], are becoming more and more popular. These OSNs are playing a crucial role in our life, and they have infiltrated every aspect of it. Recently, statistics[2] show that three-quarters of Facebook users and the half users of Instagram use that sites daily. Additionally, two-thirds of U.S. adults receive at least one of their news on OSNs. Moreover, a phenomenon has emerged from OSN called "the viral propagation," which refers to a spread of information in a broader way in small laps of

---

[1] www.weibo.com.

[2] www.pewresearch.org.

© Springer Nature Switzerland AG 2018
L. Cheng et al. (Eds.): ICONIP 2018, LNCS 11306, pp. 93–105, 2018.
https://doi.org/10.1007/978-3-030-04224-0_9

time. This phenomenon has a positive effect such as the spread of innovation and sharing new ideas [17]. However, its adverse effect could not be avoided which can spread misinformation or disinformation in form of a rumor [4]. The spread of such information can shape public opinion [5] and create political issues [8] leading to damage the reputation of individuals or companies. By apprehending the threat posed by the spread of rumors on OSNs, it is mandatory to keep them as a trustworthy source of information by minimizing the influence of rumors and avoid any repercussions.

Nowadays, individuals often join several OSNs in which it is possible for users to connect their accounts across multiple OSNs at the same time [14]. Consequently, the overlapping users connect OSNs into a multiplex structure, where the rumors can propagate across multiple networks simultaneously through these users. These overlapping users create bridges for rumors to spread across multiple OSNs with. The most of the approaches [3,4,7,13,19,20,23] dealing with rumor influence minimization (RIM) work under a closed world assumption. They postulate that the rumor can only propagate from a node to another in a network and they cannot be influenced by external sources; The role of the overlapping nodes is ignored and failed to capture the multiplex structure of OSNs. Thus, theses RIM approaches cannot be applied directly in a multiplex OSNs, and to the best of our knowledge, there are no previous works which investigate the RIM in multiplex OSNs.

Therefore, this paper attempts to address the rumor influence minimization problem in multiplex OSNs. We consider the strategy to lunch an anti-rumor campaign to raise the awareness of individuals to prevent the adoption of the rumor and limit its influence. This strategy aims to confine the rumor in one network and minimize its influence across multiple OSNs. Accordingly, we introduce the Least Cost Anti-rumor Campaign (LCAC) problem. We propose to select the least number of users to initiate this campaign, which reaches a large number of overlapping users so as raise the awareness of individuals across networks. Since this problem is NP-hard, we have proved that its objective function is submodular and monotone. Consequently, we introduce a greedy algorithm for the LCAC problem that guarantees an approximation of within 63% of the optimal solution. Finally, experiments are conducted in a real-world and synthetics multiplex OSNs to evaluate the performance of the proposed algorithm.

The paper is organized as follows. Section 2 introduces the related work. In Sect. 3, we formulate the problem in multiplex OSNs and introduce several related definitions. In Sect. 4, we prove the submodularity of the objective function for LCAC problem and present the greedy algorithm. In Sect. 5, we display the results of experiments. Finally, we conclude this paper in Sect. 6.

2   Related Work

Last few years, there has been a growing interest in the RIM problem. The RIM problem is dual to the influence maximization (IM) problem which was formulated for the first time into an optimization problem by [12]. They studied the IM on the widely-used information diffusion models: the Independent Cascade (IC) and the Linear Threshold (LT) models which are the milestone for different IM problems in general and RIM in particular. The IM aims to boost the spread of information in an OSN, while the goal of RIM is to limit the influence of undesirable rumors. Considering the topology structure of the networks, human behaviors, or social interactions, many works have proposed strategies to diminish the influence of rumors. Works investigated the blocking nodes or link [13,20,23] strategies to limit the spread of undesirable information. While, other researchers have proposed to initiate a truth campaign that fights the false campaign [3,4,7,19] to limit the influence of rumors. Accordingly, researchers have improved the IC [4] and LT [3] models to a Multi-Campaign Model.

However, few there are the works which investigate the IM in multiplex OSNs. The works of [6,22] are the first researchers to explore the IM problem on multiplex networks where they focus on the understanding of the impact of these networks on the information diffusion. Then, attention has been pointed to the multiplex IM problems. Some researchers propose to combine all networks into one network by representing an overlapping user as one unique super node [18,24]. However, these works consider homogeneous diffusion process across all the networks in which [14] argue that these works cannot preserve the heterogeneity of the layers. Thus, [14] formulated this problem by considering heterogeneous propagation models. Nevertheless, scholars [15,25,26] have been attracted by the study of the impact of the multiplex network on the propagation of rumors under the epidemic models. However, to the best of our knowledge, there are no previous works which investigate the RIM in multiplex OSNs.

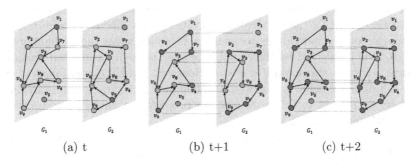

Fig. 1. An example of rumor propagation in multiplex OSNs with two layers from time t to $t+1$. Nodes are represented in green while the red nodes are infected individuals. Nodes in blue indicate a nodes added to the layers in order that each network has the same number of nodes. (Color figure online)

3   Problem Definition

3.1   Multiplex Online Social Networks

We consider an OSN as a directed graph $G = (V, E)$ where the set of nodes V represents the users and the set of edges E can be seen as relationships among individuals. Thus, we define a multiplex OSNs.

Definition 1. *A multiplex OSNs with k networks is a set $G^k = \{G_1 = (V_1, E_1), G_2 = (V_2, E_2), ..., G_k = (V_k, E_k)\}$ where $G_i = (V_i, E_i)$ is a directed graph representing an OSN. If a node exists in more than one OSN, then this node is added to set the overlapping users P of the multiplex G^k. Without loss of generality, we consider each network of the multiplex has a same number of nodes. Therefore, if a node $v \in G_i$ does not belong to G_j we add this node to G_j as an isolated node. Then for each node, interlayer edges are added to connect its adjacent interlayer copies across all the multiplex networks. Finally, we consider the set of all users of the multiplex OSNs as $V = \bigcup_{i=1}^{k} V_i$ where $|V| = N$.*

Figure 1 shows an example of multiplex OSNs with two layers $G^2 = \{G_1 = (V_1, E_1), G_2 = (V_2, E_2)\}$. Each graph G_i is referred to a layer network of the multiplex G^2. Interlayer edges are represented in dash lines to connect same nodes in each layer. The blue nodes represent the added nodes to each layer so that each network has the same number of nodes. Since they are isolated nodes, the blue nodes do not have any impact on the propagation process in the layer.

3.2   Rumor Influence Propagation Model

Generally, the rumor diffusion mechanism is seen similar to the spread of epidemics [5]. Thus, when the rumor reaches a node, we consider that this node is infected by a rumor. The rumor influence propagation model describes the process of the dissemination of a rumor through the network. The studies have been widely exploiting the classical LT and IC models. The LT model describes the behavior of the individuals to adopt a behavior when their neighbors do. Hence, a node is infected by the rumor when the ratio of its infected neighbors surpasses a certain threshold $0 < \theta < 1$. However, the LT model failed to consider the individual behaviors as well as the impact of the rumor on individuals. Therefore, we present in this work a rumor propagation model based on the IC model in the multiplex OSNs.

Initially, a set of individuals are infected called the rumor originator set S_R. Then, the whole process proceeds in discrete time steps where the rumor will propagate in the form of cascade from one node to another. Each time step t, an infected node u at time $t-1$ will have a single chance at time t to infected one of its neighbors v with probability $p_{u,v}(t)$ at each layer of the multiplex. When overlapping node v is infected in one graph G_i, its adjacent interlayer copies become infected in all OSNs. Finally, the propagation process ends when there are no more nodes to be infected in the layers of G^k. Inspired by the work of [11,20], we evaluate the rumor transmission probability $p_{u,v}(t)$ from a node u to v in two steps: the sending probability and acceptance probability $p_{u,v}(t) = p_u^{send}(t)p_{v,u}^{acc}$. Firstly, the sending probability estimates the chances of a user to send a rumor to his neighbors. Considering the work of [11], they postulate that the individual's attraction to the rumor is initially large and then exhibits a gradual downtrend. Thus, the sending probability is giving as follows

$$p_u^{send}(t) = \frac{P_0}{Log(10 + t)}, \tag{1}$$

where P_0 is the initial sending probability at time step t_0. Then, the acceptance probability evaluates the chances of an individual to accept a rumor from his neighbor. We consider high-degree nodes have a higher chance to send the rumor and a greater ability (authority) to influence other nodes, but they cannot be easily influenced called the *"celebrity effect."* However, a high-degree degree node could easily affect another high-degree. Therefore, we defined balanced weighted probability after considering the impact of the influence of both the sender u and the receiver v, which is defined as follows

$$p_{v,u}^{acc} = \frac{1}{1 + d_v^i/d_u^i}, \tag{2}$$

where $p_{v,u}^{acc}$ is the acceptance probability of the node v from u, and d_u^i and d_v^i are the connection degrees of nodes v and u in the network G_i.

As show Fig. 1 multiplex G^2, we illustrate the propagation in multiplex OSNs with two layers. The nodes are represented in green, and infected nodes are illustrated in red. At time t the set of rumor originator is $S_R = \{v_1, v_5\}$ (see Fig. 1(a)). At time $t + 1$, the rumor propagates through the multiplex, where the nodes v_2 and v_7 are infected by v_1 in layer G_1. Simultaneously, v_4 and v_9 are infected by v_5 in layer G_2 (see Fig. 1(b)). These newly infected nodes are infected at the same moment across all the networks. The reason behind, in real life a rumor infects the individuals and not the accounts in OSN in which she/he may spread the rumor in any OSN. Moreover, if we observe the layers of G^2 separately; we can see the network G_2 as an example, the nodes v_2 and v_7 are infected by an external source of this layer. These users will eventually infect other individuals as shown in Fig. 1(c).

3.3   Problem Formulation

To overcome the adverse effect of rumors, the researchers have proposed various nodes blocking strategies to limit the spread of the rumor. However, As [2] have highlighted that if the blocking period exceeds a certain threshold, the satisfaction of an individual toward an OSN is reduced. We distinguish few works which have investigated the time blocking of users in these strategies as [20]. Furthermore, the most of the works dealing with rumor detection problems in OSNs [9,10] are performed in a single network. Generally, Twitter has become the data source par excellence for collection and analysis of rumors. Then, since the rumor are detected after a period of propagation, it is very challenging to stop its spread after it has already infected a proportion of the individuals. Therefore, instead of excluding individuals from these strategies, we suggest involving the users in this process to be the main actors in minimizing the influence of the rumor. Against this backdrop, we consider the strategy to launch an anti-rumor campaign to raise the awareness of individuals to prevent the adoption of the rumor and further limiting its influence.

Considering that users join several OSNs, this last one can be connected through the overlapping users forming a multiplex structure of OSNs. These overlapping users create bridges for rumors to travel across multiple OSNs. In this study, we assume that the rumor and the anti-rumor campaign spread in the network by the same propagation model across all the networks. We note this two cascades respectively as C_R and C_{R_a}. Therefore, each node can be reached either by C_R and C_{R_a} in which it can be infected or aware of the rumor. However, we postulate that if a node is reached by the two cascade at the same time, this user will be aware of the rumor. We consider that a rumor is detected in an OSN, we select the least number of nodes to spread the anti-rumor campaign which reach a large number of overlapping nodes. This strategy aims to confine the rumor in one network and minimize its influence across multiple OSNs. However, as any problem, we have a limited number of users to be selected for this strategy noted k. Thus, we define our problem as follows.

Definition 2. *Least Cost Anti-rumor Campaign (LCAC) problem: Given a multiplex OSN $G^k = \{G_1 = (V, E_1), G_2 = (V, E_2), ..., G_k = (V, E_k)\}$, a set of overlapping nodes P, a network originator of the rumor G_R and positive constant k. The goal is to find a least number of nodes less or equal than k to start an anti-rumor campaign such that they can reach a large number of overlapping users so as to raise the awareness of individuals across networks.*

4 Proposed Solution

In this section, we present our solution for the LCAC problem in which we prove that its influence function is submodular and monotone. Then, we introduce a greedy algorithm which guarantees efficiently approximated to within a factor of $(1 - 1/e)$.

4.1 Submodularity of the Influence Function of LCAC Problem

Considering $\sigma(.)$ is the influence function of the LCAC problem where $\sigma(A)$ represents the number of the overlapping nodes reached by the set A. Since it is NP-hard to compute a minimum number of nodes to maximize $\sigma(.)$, we need to find approximation algorithms for it. The submodularity and monotonicity of the functions $\sigma(.)$ present an excellent way to obtain an approximation algorithm for our problem. $\sigma(.)$ is submodular if its meets the following condition

$$\sigma(A \cup \{v\}) - \sigma(A)) \geqslant \sigma(B \cup \{v\}) - \sigma(B), \tag{3}$$

where $A, B \subset V$, $A \subseteq B$ and $v \notin B$. In another word, σ is submodular if it has the diminishing marginal return property. Moreover, we can say that $\sigma(.)$ is monotone, if for all sets $A \subset B \subset V$, $\sigma(A) \leq \sigma(B) \leq \sigma(V)$.

To prove the submodularity of σ, we follow the approach of [7] of the footprint assignment method which follows these rules: at each time step, when a node v is infected by $u \in S_{R_a}$ or S_R at time t a footprint of the propagation is left on the node containing the source of infection and the step time (u, t). Since once a node is infected by the C_R or C_{R_a}, it will never change its status, we keep only the footprint with the smallest time from each source of infection. Given a multiplex $G_n = \{G_1, G_2...G_n\}$, a network G_i where the rumor appears, a set of infected individual S_R and a set anti-rumor originator S_{R_a}. We contract the propagation cascade of the rumor C_R and the anti-rumor C_{R_a}. We note $OP(S_{R_a})$ is the set of the overlapping nodes reached by this anti-rumor campaign before the rumor. Then, the expected influence of the anti-rumor campaign is defined as $\sigma(S_{R_a}) = E_{(C_R, C_{R_a})}(|OP(S_{R_a})|)$. By exploiting the following lemmas, we prove the submodularity and the monotonicity the $OP(.)$, and consequentially we prove the same proprieties for σ.

Lemma 1. *Given any nodes u and v, u is in the cascade C_R if it exists a footprint (w, t) in u where $w \in S_R$. v is in the cascade C_{R_a} if it exists a footprint (w', t') in v where $w' \in S_{R_a}$.*

Then, given a set S and C_S propagation cascade of the set S, the following lemmas show the conditions for any node $v \in OP(S)$ and $v \in OP(S), OP(\{u\})$.

Lemma 2. *A node $v \in OP(S)$ only under the following conditions: (1) v belongs to C_R and C_S. (2) the existence of footprint (t, w) where t is the smallest time in all footprints of the node v in C_R and $w \in S$.*

Lemma 3. *A node $v \in OP(S)$ and $v \in OP(\{u\})$ where $u \notin S$ only under the following conditions: (1) v belongs to C_R and C_S. (2) the existence of footprint (t, w) in which $w \in S$ where t is the smallest time in all footprints of the node v in C_R. (3) the existence of footprint (t', u) where t' is the smallest time in all footprints of the node v in C_R. Accordingly, if we consider the case of $OP(\{u\}) = \{v\}$ where $|OP(\{u\})| = 1$ in this situation then, $|OP(S \cup \{u\})| - |OP(\{u\})| = 0$, since $v \in OP(S)$ and $OP(\{u\})$.*

Theorem 1. *The influence function of the LCAC problem $\sigma(A)$ is submodular and monotone.*

Proof. To prove this theorem, we need to demonstrate that the function $|OP(.)|$ is submodular and monotone. Given a sets $A, B \subset V$ and $A \subset B$ and for any node $v \notin B$. For any node node $w \in OP(B), OP(\{v\})$ as shown in Lemma 3 this node may belong or not to $w \notin OP(A)$. Therefore, the marginal gain brought by v to the set A must be greater than equal to the marginal gain brought by v to the set B as $A \subset B$ than we have $|OP(B \cup \{v\})| - |OP(B)| \leq |OP(A \cup \{v\})| - |OP(A)|$ is always true. We conclude that $|OP(.)|$ is submodular. Then, $|OP(A)|$ is monotone since $\forall v \notin A$, then $|OP(A \cup \{v\})| \geq |OP(A)|$. $\square$

4.2   Greedy Algorithm

According to the previous analysis, we introduce a greedy algorithm for LCAC problem. This algorithm maximize the influence function by selecting each iteration the node with a largest marginal gain.

Algorithm 1. Greedy algorithm of the LCAC problem

Input: $G^k = \{G_1, G_2, ..., G_k\}$ multiplex OSN, network of the origin of the
rumor G_R, initial infected users S_R and positive integer k.
initialization;
$S_{R_a} = \emptyset$;
while $\sigma(S_{R_a}) < |P|$ **and** $|S_{R_a}| < k$ **do**
$\quad u = \underset{v \in V \backslash (S_R \cup S_{R_a})}{\arg\max} [\sigma(S_{R_a} \cup \{v\}) - \sigma(S_{R_a})];$
$\quad S_{R_a} = S_{R_a} \cup \{v\};$
Output: Set of anti-rumor originator S_{R_a}

---

5   Experiments

This section analyses the results of the conducted experiments presented in two parts. The first part aims to highlight the impact of the overlapping users as well as the topology structures of OSNs on the proposed strategy on synthetic networks. The second part illustrates the performance of the proposed algorithm on real-world networks.

5.1   Data Description

In this part, we exploited two multiplexes OSNs (synthetic and real-world) presented in Table 1. For the first multiplex networks G_1^4, We consider four synthesized OSNs based on scale-free (SF) networks and small-world (SW) networks. The SF networks G_1 and G_2 were generated according to Barabasi-Albert model [1] with 5000 nodes and average degree 20; the exponent in the power-law degree distribution generated by this method is 2. The SW networks G_3 and G_4 were generated according to Watts-Strogatz model [21] with 5000 nodes and average degree 40; the rewriting probability was set to $p = 0.3$. For the second multiplex network G_2^3, we exploited the data-set of [16] consisting of three real-world networks crawled from Facebook, Twitter, and YouTube.

5.2   The Role of the Overlapping Users and Network Topology

This part will spot the light on the impact of the overlapping nodes and the networks topology structure on the rumor propagation on multiplex OSNs in general and on the proposed strategy in particular. Therefore, we run experiments on the synthetic multiplex G_1^4 in which we variate the number of overlapping nodes $|P|$. Then, by changing the size of the anti-rumor set $|S_{R_a}| = 1\%N$,

Table 1. The employed data.

Multiplex networks	#Nodes	Network	#Nodes	Avg. deg.
G_1^4	10000	Scale-free	5000	40
			5000	40
		Small-world	5000	40
			5000	40
G_2^3	6407	Facebook	663	1.78
		Twitter	5540	11.52
		YouTube	5702	14.84

$5\%N$ and $10\%N$, we illustrate how the number of the overlapping nodes affect the propagation of the rumor, where $|S_R| = 2\%N$. Figure 2 shows the obtained results for $P = 10\%N$, $15\%N$, $20\%N$ and $25\%N$. Apparently, when the size of the anti-rumor set nodes S_{R_a} increases, the number of infected nodes decreases. However, when the number of overlapping nodes $|P|$ increases, the impact of the rumor is reduced illustrated by the number of the infected individuals. Moreover, it is observable that the propagation speed has an increasing dependency with $|P|$. This observation can be explained by the fact that the proposed algorithm selects the nodes that can reach a larger number of overlapping users to spread the anti-rumor campaign. Thus, the anti-rumor campaign will have a more significant impact across OSNs; as a result, it reduces the influence of the rumor. The evidence from these results shows the rationally and the correctness of our assumptions to minimizing the influence of the rumor.

Furthermore, we investigated the impact the topology structure of the networks on our strategy to minimize the influence of the rumor. Figure 3 illustrates the evolution of the number of infected nodes for different networks originators $G_R = G_2$, and $G_R = G_3$. Results are showing a wider propagation of the rumor in the four networks when it network originator is G_2 compared to G_3. However, when $G_R = G_3$, we observe a significantly higher number of infected node in G_3 compared to the others networks. Moreover, we can see that the rumor propagates faster in SW networks G_3 and G_4 in comparison with SF networks G_1 and G_2. Then, in Fig. 4 we see that the number of infected nodes is always higher when the rumor is originated from SF networks. This phenomenon is explained by the specific topology of the SW networks and SF networks. The SW networks are denser than the SF networks where each node has an equivalent authority in a network. Consequently, the information will spread faster in this networks compared to the FS networks which favor the propagation of the anti-rumor campaign as well; we can observe the influence of the rumor is reduced by the anti-rumor campaign. However, The SF networks favor the existence of the high degree nodes (hub nodes) which are highly conductive and have a greater impact as it was mentioned in Sect. 3.2. Hence, the impact of the rumor will have a more significant effect when the hub nodes are infected. Hence, the anti-rumor campaign has a relatively lower performance when the rumor is originated from SF networks.

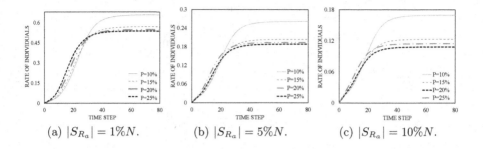

(a) $|S_{R_a}| = 1\%N$.   (b) $|S_{R_a}| = 5\%N$.   (c) $|S_{R_a}| = 10\%N$.

Fig. 2. Rate of the infected nodes for $G_R = G_2$ and $|S_R| = 2\%N$.

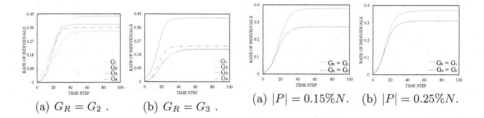

(a) $G_R = G_2$.   (b) $G_R = G_3$.   (a) $|P| = 0.15\%N$.   (b) $|P| = 0.25\%N$.

Fig. 3. Rate of the infected nodes for   **Fig. 4.** Rate of the infected nodes for
$|S_R| = |S_{R_a}| = 2\%N$ and $|P| = 0.15\%N$. $|S_R| = |S_{R_a}| = 2\%N$.

5.3 The Performance of the Proposed Strategy

This section aims to evaluate the performance of the proposed algorithm to minimize the influence of a rumor on a real-world multiplex OSNs G_2^3 shown in Table 1. Since there are no previous works to compare our results, we evaluate the performance of our algorithms in comparison with algorithms based on the following heuristics: (1) **Max degree heuristic**, select nodes according to the descendant order of their out degree; (2) the **Random heuristic** which is used as the baseline. We run experiments by varying the size the rumor originator set $|S_R| = 3\%N$, $5\%N$ and $10\%N$ as well as the size of the anit-rumor set $|S_{R_a}| = 2.5\%N$, $3\%N$, $5\%N$ and $10\%N$. We set Twitter network as the rumor originator and the results are illustrated in Figs. 5, 6 and 7. It is clearly seen that the number of the infected nodes decreases when $|S_R|$ decreases and $|S_{R_a}|$ increases. However, as an overall observation, it is seen that the proposed algorithm presents the best performance among all. Nevertheless, the proposed algorithm has a significantly better performance when $|S_{R_a}|$ is higher. The proposed algorithm selects nodes to spread the anti-rumor campaign that reaches a large number of overlapping users before the rumor. Consequently, the influence of the rumor is reduced in all the networks as it is displayed in the results. Moreover, we can see that the

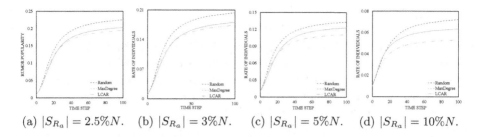

(a) $|S_{R_a}| = 2.5\%N$.   (b) $|S_{R_a}| = 3\%N$.   (c) $|S_{R_a}| = 5\%N$.   (d) $|S_{R_a}| = 10\%N$.

Fig. 5. Evolution of the rate of the infected nodes for $|S_R| = 3\%N$.

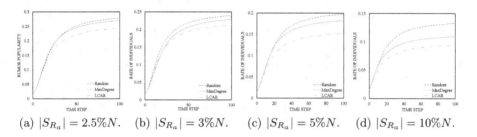

(a) $|S_{R_a}| = 2.5\%N$.   (b) $|S_{R_a}| = 3\%N$.   (c) $|S_{R_a}| = 5\%N$.   (d) $|S_{R_a}| = 10\%N$.

Fig. 6. Evolution of the rate of the infected nodes for $|S_R| = 5\%N$.

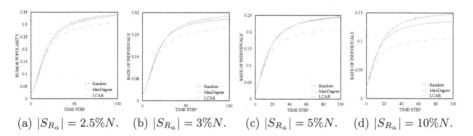

(a) $|S_{R_a}| = 2.5\%N$.   (b) $|S_{R_a}| = 3\%N$.   (c) $|S_{R_a}| = 5\%N$.   (d) $|S_{R_a}| = 10\%N$.

Fig. 7. Evolution of the rate of the infected nodes for $|S_R| = 10\%N$.

Max Degree algorithm has a similar performance compared to the baseline in some cases. The reason behind, this heuristic selects nodes with higher authority where even though these nodes will have a significant impact on the network nevertheless it will not reduce the impact of the rumor across the networks. The evidence from these results confirms the excellent performance of the proposed strategy in selecting target nodes accurately to minimize the influence of the rumor in multiplex OSNs.

6 Conclusions

This paper investigates rumor influence minimization problem in multiplex online social networks (OSNs) connected through overlapping users. We introduce in this study the Least Cost Anti-rumor Campaign (LCAC) problem. The strategy behind this problem is to select the least number of users to initiate an anti-rumor campaign and raise the awareness of a large number of overlapping users to spread it across the networks. We introduce a greedy algorithm for this problem that guarantees an approximation within $(1 - 1/e)$ of the optimal solution by proving that its objective function is submodular and monotone. Experiments performed in real-world multiplex OSNs provide the evidence about the performance of the proposed algorithm. Besides, the results in synthetic multiplex OSNs show that the overlapping nodes could contribute the diminishing of the influence of the rumor. Then, we illustrate the impact of the topology structures of OSNs on the rumor propagation in multiplex OSNs in general and on the proposed strategy in particular.

Acknowledgment. The Research was supported in part by National Basic Research Program of China (973 Program, No. 2013CB329605).

References

1. Barabási, A.L., Albert, R.: Emergence of scaling in random networks. Science **286**(5439), 509–512 (1999)
2. Bhatti, N., Bouch, A., Kuchinsky, A.: Integrating user-perceived quality into web server design. Comput. Netw. **33**(1–6), 1–16 (2000)
3. Borodin, A., Filmus, Y., Oren, J.: Threshold models for competitive influence in social networks. In: Saberi, A. (ed.) WINE 2010. LNCS, vol. 6484, pp. 539–550. Springer, Heidelberg (2010). https://doi.org/10.1007/978-3-642-17572-5_48
4. Budak, C., Abbadi, A.E.: Limiting the spread of misinformation in social networks. In: Distribution, pp. 665–674 (2011)
5. Daley, D., Kendall, D.: Epidemics and rumours. Nature **204**(4963), 1118 (1964)
6. Dinh, T.N., Nguyen, D.T., Thai, M.T.: Cheap, easy, and massively effective viral marketing in social networks: truth or fiction? In: ACM Conference on Hypertext and Social Media, pp. 165–174. ACM (2012)
7. Fan, L., Lu, Z., Wu, W., Bhavani, T., Ma, H., Bi, Y.: Least cost rumor blocking in social networks. In: IEEE 33rd International Conference on Distributed Computing Systems (2013)
8. Garrett, R.K.: Troubling consequences of online political rumoring. Hum. Commun. Res. **37**(2), 255–274 (2011)
9. Hamidian, S., Diab, M.T.: Rumor detection and classification for Twitter data. In: Proceedings of the Fifth International Conference on Social Media Technologies, Communication, and Informatics (SOTICS), pp. 71–77 (2015)
10. Hamidian, S., Diab, M.T.: Rumor identification and belief investigation on Twitter. In: ACL, pp. 3–8 (2016)
11. Han, S., Zhuang, F., He, Q., Shi, Z., Ao, X.: Energy model for rumor propagation on social networks. Phys. A Stat. Mech. Its Appl. **394**, 99–109 (2014)

12. Kempe, D., Kleinberg, J., Tardos, É.: Maximizing the spread of influence through a social network. In: KDD, p. 137 (2003)
13. Kimura, M., Saito, K., Motoda, H.: Minimizing the spread of contamination by blocking links in a network. In: AAAI, pp. 1175–1180 (2008)
14. Kuhnle, A., Alim, M.A., Li, X., Zhang, H., Thai, M.T.: Multiplex influence maximization in online social networks with heterogeneous diffusion models. IEEE Trans. Comput. Soc. Syst. **5**(2), 418–429 (2018)
15. Ma, J., Zhu, H.: Rumor diffusion in heterogeneous networks by considering the individuals' subjective judgment and diverse characteristics. Phys. A Stat. Mech. Its Appl. **499**, 276–287 (2018)
16. Magnani, M., Rossi, L.: The ML-Model for multi-layer social networks. In: ASONAM, pp. 5–12. IEEE Computer Society (2011)
17. Montanari, A., Saberi, A.: The spread of innovations in social networks. Proc. Natl. Acad. Sci. **107**(47), 20196–20201 (2010)
18. Nguyen, D.T., Zhang, H., Das, S., Thai, M.T., Dinh, T.N.: Least cost influence in multiplex social networks: model representation and analysis. In: 2013 IEEE 13th International Conference on Data Mining pp. 567–576 (2013)
19. Tong, G., et al.: An efficient randomized algorithm for rumor blocking in online social networks. IEEE Trans. Netw. Sci. Eng. (2017)
20. Wang, B., Chen, G., Fu, L., Song, L., Wang, X.: DRIMUX: dynamic rumor influence minimization with user experience in social networks. IEEE Trans. Knowl. Data Eng **29**(10), 2168–2181 (2017)
21. Watts, D.J., Strogatz, S.H.: Collective dynamics of 'small-world' networks. Nature **393**(6684), 440 (1998)
22. Yagan, O., Qian, D., Zhang, J., Cochran, D.: Information diffusion in overlaying social-physical networks. In: 46th Annual Conference on information Sciences and Systems (CISS), pp. 1–6. IEEE (2012)
23. Yao, Q., Shi, R., Zhou, C., Wang, P., Guo, L.: Topic-aware social influence minimization. In: Proceedings of the 24th International Conference on World Wide Web - WWW 2015 Companion, vol. 1, pp. 139–140 (2015)
24. Zhang, H., Nguyen, D.T., Zhang, H., Thai, M.T.: Least cost influence maximization across multiple social networks. IEEE/ACM Trans. Netw. (TON) **24**(2), 929–939 (2016)
25. Zhang, L., Su, C., Jin, Y., Goh, M., Wu, Z.: Cross-network dissemination model of public opinion in coupled networks. Inf. Sci. **451**, 240–252 (2018)
26. Zheng, C., Xia, C., Guo, Q., Dehmer, M.: Interplay between SIR-based disease spreading and awareness diffusion on multiplex networks. J. Parallel Distrib. Comput. **115**, 20–28 (2018)

Estimation of Student Classroom Attention Using a Novel Measure of Head Motion Coherence

Naoyuki Sato[1]($\boxtimes$) and Atsuko Tominaga[2]

[1] Department of Complex and Intelligent Systems, School of Systems Information
Science, Future University Hakodate, 116-2 Kamedanakano,
Hakodate, Hokkaido 041-8655, Japan
satonao@fun.ac.jp
[2] Center for Meta Learning, School of Systems Information Science,
Future University Hakodate, 116-2 Kamedanakano,
Hakodate, Hokkaido 041-8655, Japan

Abstract. Video-based head motion analysis has often been used to estimate student attention in the classroom. However, individual head motions variously depend on semantic events in the classroom (e.g., lecture slides), making it difficult to stably estimate student attention. In this article, we propose an index of students' attention in the classroom based on head motion coherence among students. We evaluated this index using 40 students' data recorded during a series of four classes. Results indicated that both head motion coherence and amplitude depended on the type of classroom activity the students were engaged in (e.g., lecture, individual, or group work) while motion coherence at an individual level was stable across the series of classes. These results suggest that head motion coherence captures elements of students' attention and it may also reflect the role of long-term, individual features (e.g., personality and motivation) in attention.

Keywords: Educational technology · Video motion analysis
Interpersonal synchronization · Neuroeducation

1 Introduction

Video-based head motion analysis [1] has often been used to estimate student attention in the classroom, information that is useful for teachers in their construction of efficient and engaging lectures [2]. Use of this method has demonstrated that head motion amplitude in the classroom inversely correlates with students' attention defined by their subjective evaluation [3]. In addition to the head motion, head orientation information was combined to estimate student attention [3]. These studies successfully demonstrated the link between head motion and students' attention. However, in the classroom, semantic events which can include, for example, audio or visual stimuli (e.g., lecture slides), are

© Springer Nature Switzerland AG 2018
L. Cheng et al. (Eds.): ICONIP 2018, LNCS 11306, pp. 106–117, 2018.
https://doi.org/10.1007/978-3-030-04224-0_10

highly complicated and individual response of head motions to the events are thought to depend on various factors, such as the cognitive state (e.g., motivation, interests in the class) and preexisting knowledge of the students. Therefore, it is important to evaluate the stability of the evaluation of student attention estimation against various semantic events in the classroom.

In this article, we propose an index of students' attention in the classroom based on head motion coherence among students. The notion of motion coherence stems from the hypothesis proposed by Raca [4]. Raca proposed that student motion synchronization reflected students' cognitive states, which can range from active listening to the teacher to absent-mindedness. Given this, the present article sought to quantify students' attentional states across a time series and correlate this with group behavior, which is thought to reflect the many classroom semantic contexts (Fig. 1). We assessed this index using 40 students' data recorded during a series of four classes and evaluated whether this measure (1) depended on the type of classroom activity students were engaged in and (2) reliably characterized students' cognitive status across a series of classes.

Fig. 1. Student head motion coherence in the classroom. (a) During lecture, students who are listening to the teacher more often move their heads in a synchronous manner. (b) During group work, students in the same group more often move their heads synchronously. (c) During individual work, students' heads move independently. The coherence of these head motions was hypothesized to serve as a measure of student attention and reactivity to the class material.

2   Method

2.1   Participants

Forty students, during a series of four classes (90 min each, first-year undergraduate students [29 males, age ranged from 18 to 19], class title: 'Basic strategies of the network economy', ~240 attendees, period: 4 weeks from Nov. 11, 2017), were recruited by an announcement at the first day of the class. They were explained that their participation was completely voluntary and independent of the grades of the class, and participants were not compensated. Prior to the participation, all participants provided informed consent. This study was approved by the Ethics Committee of the Future University Hakodate (ID:201703, approval date: July 10, 2017).

2.2 Video Recording

The class consisted of three phases: a lecture (students listened to a teacher talk), group work (groups of four students discussed a specific topic assigned by the teacher), and individual work (students solved several questions on a computer). The classroom seating arrangement is depicted in Fig. 2. Students in the same group were also seated in the same row. Seating was changed after day two of the experiment. Three video cameras (Sony, HDR-CX485) were located at the front of the classroom and recorded participant behaviors. All cameras were located >2 m away from participants and were carefully angled so as to avoid capturing non-participants. In order to preserve participants' privacy, the recorded video data were removed from the camera after copying to a hard disk which was stored in a key locker after extracting motion data with anonymization.

2.3 Video Data Analysis

Video preprocessing was performed using the following multi-step procedure: First, a video (1920 × 1080 pixels, 60 Hz sampling rate) was manually segmented to create 40 movies of individual, differentially-seated participants. Second, for each individual movie, a mask image to include the participant's face was created as faces were sometimes occluded by other participants or by the computer. Individual participant postures were categorized among 25 classes using the hierarchical clustering of grayscaled images sampled across 1 s intervals. Masks were then manually created for each individual posture. Third, optical flow in the face area was calculated using the Lucas-Kanade derivative of Gaussian (implemented by the MATLAB opticalFlowLKDoG.m function). Finally, head motion was calculated as the amplitude of average optical flow in the face area, as determined by mask area skin color (hue color ranged from 0.04 to 0.1). Motion amplitude was normalized by seat width in the video.

Head motion coherence was defined as follows: amplitude of the i-th participant at time t is given by $a_i(t)$. The correlation coefficient between head motion of the i th and j-th participants, $a_i(t)$ and $a_j(t)$, during time period t to $t + T$, is given by $c_{ij}(t)$. The head motion coherence of the i-th participant at time t, $s_i(t)$, was defined by the following equation:

$$s_i(t) = \sum_{j \neq i} c_{ij}. \tag{1}$$

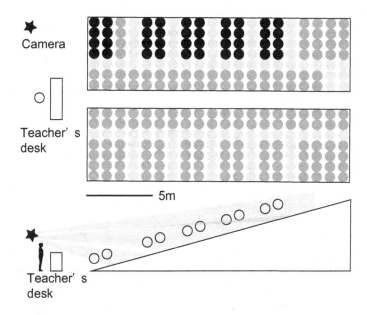

Fig. 2. Student classroom seating. Black, dark, and light gray circles indicate participating students, other students, and empty seats, respectively. The star and rectangle at the bottom of the panel indicate locations of the camera and teacher's desk, respectively. Bottom figure indicates the side view of the lecture theater while gray triangles indicate angles of the three cameras.

While it depends on the number of participants who simultaneously move with the target participant, head motion coherence is independent of motion amplitude. We hypothesized that a participant's attentional state would reflect his/her reactivity to the classroom context in multiple semantic contexts.

In the analysis, the dependency of head motion coherence on type of classroom activity was evaluated by two-way ANOVA (40 participants × three activity types). Averaged head motion coherence for individual participants was then assessed in terms of its stability across the series of four classes. Values across days were compared using regression analyses ($N = 40$) and Z-scored correlation coefficients integrated for each pair of days were evaluated. Finally, the influence of the number of participants on the stability of individual head motion coherence was assessed using a random sampling test.

3   Results

3.1   Calculation of Head Motion Coherence

Figure 3a shows the temporal evolution of head motion signals collected from all 40 participants on Day 2 across 90 min. Grayscale coloring indicates the amplitude of head motion and horizontal and vertical axes indicate time and

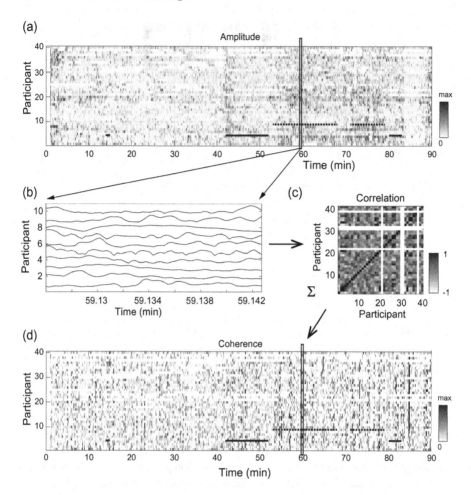

Fig. 3. Head motion across a 90 min class session (Day 2). (a) Head motion amplitude. Horizontal and vertical axes denote time and participant ID, respectively. Solid and dashed bold horizontal lines in the plot indicate periods of individual and group work, respectively. Other time periods were considered to be lecture time (during which the teacher was talking/teaching). (b) Example 1 s segment of motion amplitude. (c) A matrix of correlation coefficients for the time series of motion amplitudes (b). White strips denote no associated value. (d) Head motion coherence. Values were calculated from the correlation matrix at each time segment. See text for additional details.

participant ID, respectively. Solid and dashed horizontal lines indicate periods of individual and group work. The motion amplitudes appeared to associate with these lecture events. However, except for these periods, no obvious tendency of head motion signals, related to some postulated classroom events or individual difference were observed.

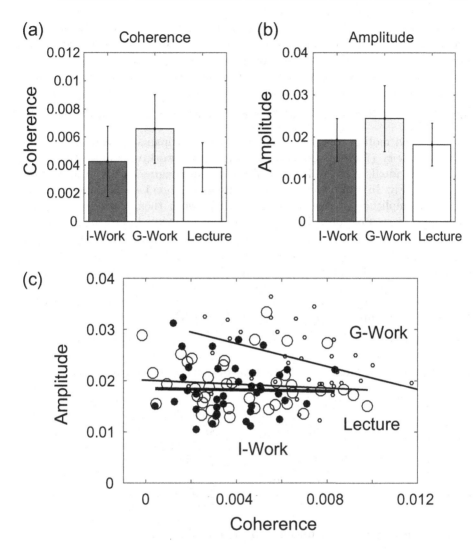

Fig. 4. Relationship of head motion signals to classroom activity type. (a) Average head motion coherence during each class activity. 'I-Work' and 'G-Work' indicate individual and group work, respectively. Error-bars indicate standard error ($N = 40$). (b) Average head motion amplitude during each type of class activity. (c) Correlation between head motion coherence and the amplitude during each type of class activity.

Head motion coherence was calculated from 1 s segments of head motion data (Fig. 3b). Using correlation coefficients for the motion signal at each time segment (Fig. 3c), the temporal evolution of head motion coherence was calculated (Fig. 3d). In the correlation analysis (Fig. 3c), motion coherence between each pair of participants was plotted where increased coherence would be associated with simultaneous response to teacher's instruction during lecture or cooperative behavior during group work. In the coherence map, individual tendency of the coherence motion in response to other students was quantified using Eq. (1). Vertical patterns in the coherence map, thought to be associated with teacher instruction, were more obvious here than in the amplitude map.

3.2 Relation Between Head Motion Coherence and Amplitude

Average head motion coherence for individual participants was related to type of class activity (Fig. 4a). Coherence during group work was larger than that during individual work or lecture. These findings agreed with head motion amplitude (Fig. 4b). To investigate individual association between motion coherence and amplitude, a correlation analysis between them during each classroom activity was performed. It was found that individual head motion coherence and amplitude values were inversely correlated during group work ($r = -0.35, t(38) = -2.26, p < 0.05$) (Fig. 4c), and non-significantly correlated during other class activities (individual work, $r = 0.08, t(38) = 0.51$, n.s.; lecture, $r = -0.12, t(38) = -0.76$, n.s.). This result suggested that the motion coherence and amplitude were differently quantified participants' behavior except for the period of group work in which participants with a large head motion were suggested to be less correlated with head motions of other students.

3.3 Stability of Head Motion Coherence Across Days

The stability of individual head motion coherence values was evaluated via regression analyses across days (Fig. 5a). This revealed that head motion coherence was stable across the four classes assessed here, despite a seating arrangement change after the second day. This suggests that these values reflect individual student characteristics during class time. Individual head motion amplitudes (Fig. 5b) were not as stable as motion coherence and there was a significant difference between their correlation coefficients (paired t-test, $t(5) = 3.54, p < 0.05$).

3.4 Influence of the Number of Participants

Head motion coherence was defined here by the average of head motion signal correlation coefficients between a target participant and others (Eq. (1)). Furthermore, the number of participants was a fundamental parameter in the determination of head motion coherence. Therefore, the influence of the number of participants was estimated using a random sampling test. Results of this are shown in Fig. 6 and demonstrate that head motion coherence was stable if more than 20 participants was used. Given this, the stability of head motion amplitude was independent of the number of participants used here (40).

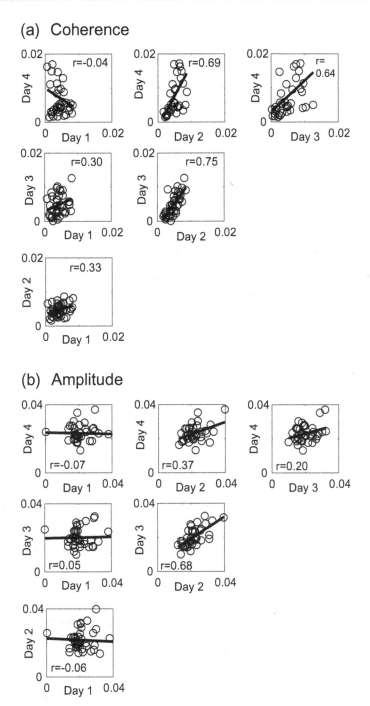

Fig. 5. Stability of average head motion coherence (a) and amplitude (b) over four successive classes. Note that the seating arrangement was changed after the second day.

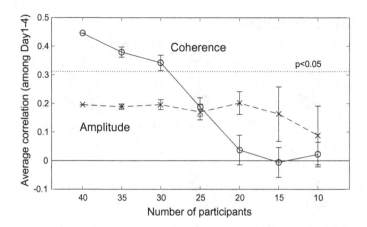

Fig. 6. Influence of the number of participants on the stability of calculated head motion coherence and amplitude. The stability of head motion coherence or amplitude was defined by the average correlation coefficient between individual head motion coherence or amplitude across multiple days. In this analysis, the number of participants was simulated by a random sampling of participants. This process was performed 10 times per condition. Error bars indicate standard error ($N = 10$).

4   Discussion

In the present article, we proposed a measure of student attention in the classroom based on head motion coherence among students. The results of our analysis of 40 participants' data across four days of successive 90-min classes revealed that head motion coherence reflected the type of class activity students were engaged in. Additionally, head motion coherence profiles mirrored those of head motion amplitude (Fig. 4a, b). However, individual head motion coherence was not significantly correlated with head motion amplitude (Fig. 4c) and appeared to be more stable than individual head motion amplitude across the four successive classes assessed (Fig. 5). These results suggest that student attention estimated by head motion coherence captured different aspects of attention than those captured by head motion amplitude (Fig. 7). This point is discussed in the following section.

4.1   Relationship Between Head Motion Coherence and Attention

The current study did not show any direct association between motion coherence and attention, in contrast to the findings of Raca [3] who clearly demonstrated the link between head motion amplitude and a questionnaire on attention. However, the head motion signals were also thought to capture students' attention according to the following considerations. First, the head motion coherence shared a common aspect of attention; as a definition, "attention" describes an internal state which enables selective processing of a part of complicated stimuli

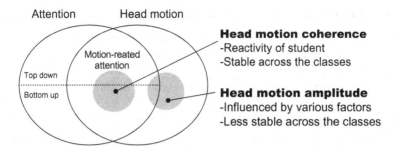

Fig. 7. Schematic illustration showing the relationship between attention and head motion signal. Head motion coherence was found to capture participant attentional state more directly than head motion amplitude.

and it can be independent of either head or eye movements. However, in natural environment, some type of attention would associate with the motions [5] that were quantified using head motion coherence and amplitude. Second, the head motion coherence associates with reactivity of students to the classroom event, and decreases in head motion associates with an increase in attention [3]. In contrast, as shown in Fig. 3a, the motion coherence was thought to reflect detailed classroom events, which may be the teacher's key sentences, such as "In summary . . ." or "The answer is . . .". This type of head motion was also thought to associate with the personality of each student, where interpersonal synchronization is known to largely depend on participant personality [6]. Therefore, the head motion coherence was thought to capture students' attention associated with individual participant characteristics (e.g., personality, learning motivation, and basic interest in the topic being taught). As a future study, it would be worthwhile to evaluate head motion coherence using educational measures, such as exam performance or peer-based feedback.

4.2   Possible Improvement of Calculation of Head Motion Coherence

The calculation involved in the assessment of participants' attention via head motion coherence required >20 participants (Fig. 6). This may be improved by modifying the definition of head motion coherence (Eq. (1)) to focus on a particular participant subgroup in which participants move synchronously during a particular class activity. Improvement of this parameter may allow for more reliably detecting head motion coherence and increasing the signal to noise ratio. In a related work by Raca [3], the spatial arrangement of seats in a classroom was found to impact motion coherence, with a ∼2 s delay in the motion synchronization of sleeping students. Thus, inclusion of these sleeping students in the calculation of head motion coherence may also improve the stability of head motion coherence.

4.3 Possible Contribution to Neuroeducation Research

Neuroeducation [7–9] is a growing field of research that aims to apply insights from neuroscience to educational settings. For example, a recent study used electroencephalogram (EEG) signals, simultaneously measured from multiple participants [10,11], to estimate students' cognitive states in the classroom. Information from studies such as this may be useful to teachers for their design of curricula and efficient, engaging classes. As neuroscience techniques have traditionally involved laboratory experimentation, in which complicating or confounding variables are carefully excluded from experimental tasks, applying these techniques to the classroom can be challenging. Student participants often vary in parameters including attention, attitude, preexisting knowledge, social and emotional characteristics, and other individual differences. Thus, one of important factors in bridging neuroscience and education research is the quantification of students' cognitive states in the classroom context.

As a technology that might be used in conjunction with classroom EEG measurement, video-based head motion coherence was found here to accurately quantify the timing of semantic events (as seen in the vertical patterns in Fig. 3d). These events, which can include, for example, audio or visual stimuli (e.g., lecture slides), are usually highly complicated and thus difficult to define. Furthermore, with a duration that can be <1 s, defining motion-artifact-free segments without semantic disturbance during these events, which is required for neuroscientific analysis, is made possible by using combined methodologies. The current study proposes and substantiates the use of a key method that might be used to bridge neuroscience and educational technology and research in future.

Acknowledgements. This work was supported by JSPS KAKENHI Grant Number 26540069.

References

1. Murphy-Chutorian, E., Trivedi, M.M.: Head pose estimation in computer vision: a survey. IEEE Trans. Pattern Anal. Mach. Intell. **31**(4), 607–626 (2009)
2. Raca, M., Dillenbourg, P.: System for assessing classroom attention. In: The Proceedings of the Third International Conference on Learning Analytics and Knowledge, pp. 265–269 (2013)
3. Raca, M., Kidzinski, L., Dillenbourg, P.: Translating head motion into attention-towards processing of student's body-language. In: The Proceedings of the 8th International Conference on Educational Data Mining, No. EPFL-CONF-207803 (2015)
4. Raca, M.: Camera-based estimation of student's attention in class. Thesis, EPFL (2015)
5. Itti, L., Koch, C.: Computational modelling of visual attention. Nat. Rev. Neurosci. **2**(3), 194–203 (2001)
6. Rennung, M., Göritz, A.S.: Prosocial consequences of interpersonal synchrony: a meta-analysis. Zeitschrift für Psychologie **224**(3), 168–189 (2016)
7. Ansari, D., De Smedt, B., Grabner, R.H.: Neuroeducation-a critical overview of an emerging field. Neuroethics **5**(2), 105–117 (2011)

8. Ansari, D., Coch, D.: Bridges over troubled waters: education and cognitive neuroscience. Trends Cogn. Sci. **10**(4), 146–151 (2006)
9. Goswami, U.: Neuroscience and education: from research to practice? Nat. Rev. Neurosci. **7**(5), 406–413 (2006)
10. Poulsen, A.T., Kamronn, S., Dmochowski, J., Parra, L.C., Hansen, L.K.: EEG in the classroom: synchronised neural recordings during video presentation. Sci. Rep. **7**(43916) (2017). https://doi.org/10.1038/srep43916
11. Sato, N., Sato, T., Okazaki, T., Takami, M.: Electroencephalogram dynamics during social communication among multiple persons. In: Lee, M., Hirose, A., Hou, Z.-G., Kil, R.M. (eds.) ICONIP 2013. LNCS, vol. 8226, pp. 145–152. Springer, Heidelberg (2013). https://doi.org/10.1007/978-3-642-42054-2_19

Structure, Attribute and Homophily Preserved Social Network Embedding

Le Zhang[1,2,3], Xiang Li[1,2,3], Jiahui Shen[1,2,3], and Xin Wang[3(✉)]

[1] School of Cyber Security, University of Chinese
Academy of Sciences, Beijing, China
{zhangle,lixiang9015,shenjiahui,wangxin}@iie.ac.cn
[2] State Key Laboratory of Information Security,
Chinese Academy of Sciences, Beijing, China
[3] Institute of Information Engineering, Chinese Academy of Sciences,
Beijing, China

Abstract. Network embedding is to map nodes in a network into low-dimensional vector representations such that the information conveyed by the original network can be effectively captured. We hold that a social network mainly contains three types of information: network structure, node attributes, and their correlation called homophily. All of these information could be potentially helpful in learning an informative network representation. However, most existing network embedding methods only consider one or two types of these information, which are possibly leading to generate unsatisfactory representation. In this paper, we propose a novel algorithm called Structure, Attribute, and Homophily Preserved (SAHP), which jointly exploits the aforementioned three information for learning desirable network representation. And we design a joint optimization framework to embed the three information into a consistent subspace where the interplay between them is captured toward learning optimal network representations. Experiments conducted on three real-world social networks demonstrate that the proposed algorithm SAHP outperforms the state-of-the-art network embedding methods.

Keywords: Network embedding · Network representation learning
Social network

1 Introduction

Nowadays, an increasing growth of social networks, such as WeChat, Facebook and Twitter, produces massive amounts of networked data. Mining valuable information from such networked data is of great benefits for human beings [26, 27]. However, the complexity of these networked data is the great challenge for us. To deal with this problem, one promising strategy is network embedding. It aims to map each node in the network to a low-dimensional vector representation space while preserving the neighborhood relationship between the nodes. After

© Springer Nature Switzerland AG 2018
L. Cheng et al. (Eds.): ICONIP 2018, LNCS 11306, pp. 118–130, 2018.
https://doi.org/10.1007/978-3-030-04224-0_11

that, the learned representations could be directly applied to subsequent network analysis tasks, including node classification [11,12,18], community detection [2, 13,24] and anomaly detection [3,9,14].

Majority of network embedding methods are based on network structure [8, 19,21]. Nevertheless, these methods primarily focused on the network structural information but ignore attribute information associated with each node, which are resulting in generating suboptimal representation. It is well recognized that node attributes are also one of the most important features of social networks, which can measure the attribute-based similarity of nodes [11,27]. As a result, node proximities are enhanced by the attribute information. Then, nodes sharing similar attributes are likely to have similar vector representations. It is more reasonable to embed the network structure with node attributes considered.

Moreover, node attributes are inherently correlated to the network structure [10,11], which can be explained by the principle of homophily [15]. Homophily encourages nodes with common attributes to be densely connected [6,26]. When we re-examine in terms of overall the network, it's easy to find that homophily could divide the whole social network into densely-connected homogeneous parts that are gotten weak connection with each other [6]. For the nodes within the same homogeneous part, even they have a litter weak structural relationship in the original network, their similarities will be strengthened by the principle of homophily. Then, they may also have similar representation. Hence, the property of homophily should be considered in the network embedding process.

Recently, most related works either only utilize the network structure [8,19,21] or simply combine node attribute with network structure [11,27] for network representation, which are possible leading to unsatisfactory performance. It is intuitive to improve the performance of embedding algorithms with all the aforementioned three information considered. Whereas, this solution is challenging in three aspects as follows. First, due to privacy concerns, node attributes are often very sparse, incomplete and noisy. It is a tough job to extract the useful information from them. Second, it is very difficult to detect the homogeneous parts both in the original network and in the learned representation space at the same time. Third, in the embedding process, integration of these three types of information is a hard task due to the bewildering mutual interplay among them.

To overcome the above problems, in this paper, we propose a Structure, Attribute, and Homophily Preserved (SAHP) network embedding algorithm, which takes advantage of all the three information sources, including network structure, node attributes and homophily, to jointly generate the effective representations for nodes. Specifically, we adopt random walk to capture the network structural information at first. Second, we embed the attribute information via non-linear mapping. It can filter out the noisy information from node attributes and preserve the valuable information consistent with topological structure. Third, in order to reflect the principle of homophily both in the primary network structure and in the learned vector representation space, we model the homogeneous parts in the network as a multivariate Gaussian distribution, then perform the Gaussian mixture model (GMM) to enforce homophily constraints among the nodes in the learned vector space. Finally, we formulate a joint optimization framework to integrate three parties and capture the interplay between them. In summary, the following three major contributions are in our paper:

- Our proposed SAHP algorithm is simultaneously considered all three information sources (*i.e.*, network structure, node attribute and homophily) for learning an effective network representation.
- We introduce a joint optimization framework to integrate the aforementioned three information sources into a joint embedding space, in which captures their mutual interplay to generate an optimal representation for each node.
- We employ node classification experiments on three real social network datasets to validate the effectiveness of our proposed method SHAP.

2   Related Works

In this section, two groups of network embedding methods are briefly listed: pure network structure based methods and content augmented methods.

Network structure-based methods focused on considering that the learned representation vector space preserved the network structure only. Inspired by the Skip-Gram model [17], DeepWalk [19] adopted the truncated random walks to sample node sequences, and then applied them to the Skip-Gram model to train node representations. LINE [21] formatted two different objective functions to preserve the first-order proximity and the second-order proximity between nodes respectively. Node2Vec [8] modified the truncated random walk into biased random walk, and then fed DeepWalk [19] to learn node representations. Most recently, several deep learning models are introduced to capture the rich network structural information from nodes to learn the network representation [4,22,23]. Nevertheless, the above works only used network structure but ignored node attributes to learn the network embedding.

Content augmented methods took node attributes into consideration for network representation learning. At the first attempt, TADW [25] imported the node textural features into DeepWalk through the matrix factorization formulation. TriDNR [18] used three types, including network structure, node content and labels, to jointly train the informative node representations. LANE [11] explored the potential of incorporating labels with network structure and node attributes in the embedding process, while preserving their correlations. UPP-SNE [27] projected profile features via a non-linear mapping into network structure to learn a joint network representation. From their reported results, node attributes could provide helpful gains to enhance the performance of network embedding.

3   Methodology

In this section, firstly we give several notations that used in the paper. Secondly, we describe our problem. Besides, we introduce the proposed algorithm SAHP in detail. Finally, we present the optimization process for SHAP.

3.1   Problem Statement

Considering a social network is an undirected graph $G = (\mathcal{V}, \mathcal{E}, \mathcal{A})$, where $\mathcal{V}$ is the set of nodes, then $\mathcal{E} \subseteq (\mathcal{V} \times \mathcal{V})$ is the set of edges, and $\mathcal{A}$ is the node attribute matrix for all nodes in G, where row $a_i \in \mathbb{R}^d$ represents the attribute feature associated with the node $v_i \in \mathcal{V}$. Our problem **is to** embed each node $v_i \in \mathcal{V}$ in G to a low-dimensional vector representation $\Phi(v_i) \in \mathbb{R}^{2m}$, where $2m \ll |\mathcal{V}|$. And the mapped node representations are directly taken as the input to subsequent network analytical tasks, like node classification.

3.2   Preserving Network Structure Information

Inspired by the idea of Skip-Gram [16], DeepWalk [19] extends from learning word representation to learn the node representations of a network. DeepWalk performs random walk on the network to generate the context for nodes. Then, it maximizes the occurrence of a node v_i and its context nodes within t window size, $\{v_{i-t}, \cdots, v_{i+t}\} \setminus v_i$. And the representation for v_i, $\Phi'(v_i)$ is computed as:

$$P\left(\{v_{i-t}, \cdots, v_{i+t}\} \setminus v_i | \Phi'(v_i)\right) = \prod_{j=i-t, j \neq i}^{j=i+t} P\left(v_j | \Phi'(v_i)\right) \tag{1}$$

The probability $P\left(v_j | \Phi'(v_i)\right)$ in Eq. (1) is modeled by softmax function:

$$P\left(v_j | \Phi'(v_i)\right) = \frac{\exp\left(\Phi(v_j) \cdot \Phi'(v_i)\right)}{\sum_{v \in \mathcal{V}} \exp\left(\Phi(v) \cdot \Phi'(v_i)\right)} \tag{2}$$

In DeepWalk, a node v has two vectors: $\Phi'(v)$ is the node itself and $\Phi(v)$ is the context for other nodes. Obviously, directly computing the term $P(v_j | \Phi(v_i))$ in Eq. (2) is computationally expensive, because we want to iterate through all contexts. To reduce the computation complexity, we adopt the Skip-Gram with Negative Sampling method [16]. Hence, we speed up the optimization process by adopting the following objective function:

$$\mathcal{O}_1 = -\sum_{i=1}^{|\mathcal{V}|} \sum_{j=1}^{|\mathcal{V}|} n(v_i, v_j) \left[\log \sigma\left(\phi_j \cdot \phi'_i\right) + \sum_{l=1}^{N} \log \sigma\left(-\phi_{v_i^l} \cdot \phi'_i\right)\right] \tag{3}$$

For convenience, we set $\Phi(v_i)$ to ϕ_i, and $\Phi'(v_i)$ to ϕ'_i. $\sigma(x) = 1/(1 + \exp(-x))$ is the sigmoid function. And $n(v_i, v_j)$ is the number of times that the node v_j occurs in the node v_i's context in a set of generated random walk sequences. If v_j never occurs in v_i's context, the value of $n(v_i, v_j)$ is to 0. Here, v_i^l is the index lth sampled negative nodes for v_i according to the probability $r_i^{0.75}$ (r_i is the v_i's degree). In total, there are N negative nodes. By minimizing the Eq. (3), the connected nodes are projected closely together in the representation vector space. Consequently, the space can preserve the original network structure.

3.3   Preserving Node Attribute Information

Motivated by the fact that node attributes could be potentially helpful to enhance network representation learning [11,27], we incorporate node attributes into the DeepWalk [19] framework. Following the idea of [27], we map node attributes a_i to its node representation $\varphi(a_i)$ via a non-linear mapping:

$$\Phi'(v_i) = \varphi(a_i) \tag{4}$$

here $\varphi(\cdot)$ is a non-linear mapping function. And the kernel mapping is a common choice, where $\varphi(a_i)$ may be infinite dimensional. Nevertheless, according to [20], the infinite dimensional kernel space could be mapped to some low-dimensional feature space. And thus, it could provide the feasible for $\varphi(a_i)$ to be low-dimensional. Then, the low-dimensional feature mapping $\varphi(\cdot)$ is calculated as follows:

$$a_i \to \varphi(a_i) = \frac{1}{\sqrt{m}} \left[\cos\left(\mu_1^T a_i\right), \cdots, \cos\left(\mu_m^T a_i\right), \right.$$
$$\left. \sin\left(\mu_1^T a_i\right), \cdots, \sin\left(\mu_m^T a_i\right) \right]^T \tag{5}$$

following the idea of the distribution from the Fourier transform of the kernel function, $\{\mu_1, \cdots, \mu_m\}$ are the m projection sampled directions. As mention above, node attribute are inherently correlated to the network structure [11,26, 27]. In the purpose to exploit this correlation, we plan to incorporate the feature mapping $\varphi(\cdot)$ with network topological structure in the DeepWalk framework. Therefore, we substitute $\varphi(a_i)$ for ϕ'_i in Eq. (3). Then the objective function in Eq. (3) is rewrote as:

$$\mathcal{O}_1 = -\sum_{i=1}^{|\mathcal{V}|} \sum_{j=1}^{|\mathcal{V}|} n(v_i, v_j) \left[\log \sigma\left(\phi_j \cdot \varphi(a_i)\right) + \sum_{l=1}^{N} \log \sigma\left(-\phi_{v_i^l} \cdot \varphi(a_i)\right) \right] \tag{6}$$

Here, the representation ϕ_j is a $2m$-dimensional vector. By solving the Eq. (6), we embed network structure and node attributes into a consistent subspace, and the interplay between them is able to promote each other to generate a consistent embedding representation space that contains the information both from topological structure and node attributes.

3.4   Preserving Homophily Information

As the mentioned above, the principle of homophily divides a network into some densely-connected homogeneous parts that are weakly-connected between them. To reflect the property of homophily among the nodes, we should enforce the nodes within the homogeneous parts to be close enough to each other. Therefore, we combine the above-mentioned network embedding process with Gaussian mixture model (GMM). That is, we consider a node v_i's out representation $\Phi(v_i)$ is generated by a multivariate Gaussian distribution $\mathcal{N}(\psi_k, \Sigma_k)$ from the homogeneous part $z_i = k$. In this paper, we use class labels for representing the homogeneous part. Then, for any node $v \in \mathcal{V}$ in G, we can formulate the likelihood as:

$$\prod_{i=1}^{|\mathcal{V}|} \sum_{k=1}^{K} p(z_i = k) \, p(v_i | z_i = k; \phi_i, \psi_k, \Sigma_k) \qquad (7)$$

where K is the number of homogeneous parts in G, $p(z_i = k)$ is the probability of node v_i belonging to the homogeneous part k, $\psi_k \in \mathbb{R}^{2m}$ is a mean vector and $\Sigma_k \in \mathbb{R}^{2m \times 2m}$ is a covariance matrix. To be simple, we set $p(z_i = k)$ as π_{ik}. And hence, we have $\pi_{ik} \in [0, 1]$ and $\sum_{k=1}^{K} \pi_{ik} = 1$. In this process, these π_{ik}'s indicate that the probability for each node v_i is derived from one of the homogeneous parts in the network G. Besides, $p(v_i | z_i = k; \phi_i, \psi_k, \Sigma_k)$ is defined as a multivariate Gaussian distribution:

$$p(v_i | z_i = k; \phi_i, \psi_k, \Sigma_k) = \mathcal{N}(\phi_i | \psi_k, \Sigma_k) \qquad (8)$$

After that, in order to encourage the homophily between nodes to be close enough in the representation space, we define the objective function as:

$$\mathcal{O}_2 = -\frac{\beta}{K} \sum_{i=1}^{|\mathcal{V}|} \log \sum_{k=1}^{K} \pi_{ik} \cdot \mathcal{N}(\phi_i | \psi_k, \Sigma_k) \qquad (9)$$

where $\beta \geq 0$ is a trade-off parameter. As claimed by [1], we can use the log-concavity equation to optimize the Eq. (9). Then, the objective function is rewrote as:

$$\mathcal{O}_2 = -\frac{\beta}{K} \sum_{i=1}^{|\mathcal{V}|} \sum_{k=1}^{K} \log \left[\pi_{ik} \cdot \mathcal{N}(\phi_i | \psi_k, \Sigma_k)\right] \qquad (10)$$

By optimizing the Eq. (10), the homophily information is not only reflected in the original network, but also retained in the embedding representation space.

3.5   Joint Optimization

In this paper, there are three information sources, including network structure, node attributes and homophily, considered for the node representation learned. Due to the heterogeneity and high dimensionality of these information, we cannot be directly concatenated them. In order to maximize the interplay among them toward learning the informative node representations, we adopt the unification approach. Then, the ultimate objective function for SAHP is defined as:

$$\mathcal{O} = \mathcal{O}_1 + \mathcal{O}_2 \qquad (11)$$

Since our objective in Eq. (11) consisted of node embedding and homophily detection, we decompose the joint optimization process into two sub-parts, then use an alternating way to optimize this problem.

Fixed (v, μ), Optimize (Π, Ψ, Σ). In this part, Eq. (11) is simplified as an inferring issue. Then we adopt *expectation maximization* (EM) method [5] to get (Π, Ψ, Σ). However, we notice that $\mathcal{O}_2$ might be negative infinity, when diag (Σ_k) becomes zero. To avoid this situation, we particularly introduce the constraints of diag $(\Sigma_k) > 0$ for each $k \in \{1, \cdots, K\}$. Here, we randomly reset $\Sigma_k > 0$ and ψ_k, whenever a diag (Σ_k) has zero at the beginning. Therefore, we can repetitive update the parameters (Π, Ψ, Σ) by:

$$\pi_{ik} = \frac{N_k}{|\mathcal{V}|} \tag{12}$$

$$\psi_k = \frac{1}{N_k} \cdot \sum_{i=1}^{|\mathcal{V}|} \gamma_{ik} \cdot \phi_i \tag{13}$$

$$\Sigma_k = \frac{1}{N_k} \cdot \sum_{i=1}^{|\mathcal{V}|} \gamma_{ik} \cdot (\phi_i - \psi_k)(\phi_i - \psi_k)^T \tag{14}$$

where $\gamma_{ik} = \frac{\pi_{ik} \mathcal{N}(\mathbf{v}_i | \psi_k, \Sigma_k)}{\sum_{k=1}^{K} \pi_{ik} \mathcal{N}(\mathbf{v}_i | \varphi_k, \Sigma_k)}$ and $N_k = \sum_{i=1}^{|\mathcal{V}|} \gamma_{ik}$. Inspired by [2], we also initialize $\mathbf{v}$ by DeepWalk results in our experiment, which may make the constraints of diag $(\Sigma_k) > 0$ be easily satisfied. And then, the inference of (Π, Ψ, Σ) could have converged at fast.

Fix (Π, Ψ, Σ), Optimize (v, μ). In this part, Eq. (11) is treated as a optimization issue. We optimize $\mathcal{O}_1$ and $\mathcal{O}_2$ via using the *stochastic gradient descent* (SGD) approach. And therefore, for each node $v_i \in \mathcal{V}$, we have:

$$\frac{\partial \mathcal{O}_1}{\partial v_i} = -\sum_{i=1}^{|\mathcal{V}|} \sum_{j=1}^{|\mathcal{V}|} n(v_i, v_j) [1_i(j) \cdot \sigma(-\phi_j \cdot \varphi(a_i)) \varphi(a_i) \tag{15}$$

$$-\sum_{l=1}^{N} 1_i(v_i^l) \cdot \sigma\left(\phi_{v_i^l} \cdot \varphi(a_i)\right) \varphi(a_i)\Bigg]$$

$$\frac{\partial \mathcal{O}_2}{\partial v_i} = \frac{\beta}{K} \cdot \sum_{k=1}^{K} \pi_{ik} \cdot \Sigma_k^{-1}(\phi_i - \psi_k) \tag{16}$$

where $1_i(\cdot)$ is an indicator function. For all $\mu_s \in \{\mu_s\}_{s=1}^m$, we also have:

$$\frac{\partial \mathcal{O}_1}{\partial \mu_s} = -\sum_{i=1}^{|\mathcal{V}|} \sum_{j=1}^{|\mathcal{V}|} n(v_i, v_j) \left[\sigma(-\phi_j \cdot \varphi(a_i)) \frac{\partial \varphi(a_i)}{\partial \mu_s} \varphi(a_i) \tag{17} \right.$$

$$\left. -\sum_{l=1}^{N} \sigma\left(\phi_{v_i^l} \cdot \varphi(a_i)\right) \frac{\partial \varphi(a_i)}{\partial \mu_s} \varphi(a_i) \right]$$

where $\frac{\partial \varphi(a_i)}{\partial \mu_s}$ is a $d \times 2m$ Jacobian matrix.

Algorithm 1. The proposed algorithm SAHP

Input: Graph $G = (\mathcal{V}, \mathcal{E}, \mathcal{A})$, walk length l, paths per node γ, embedding dimension m, homogeneous parts K, trade-off parameter β.

Output: The learned vector representations $\Phi'(v)$ for all node $v \in \mathcal{V}$

1: $\mathcal{P} \leftarrow \text{RandomWalk}(G, l, \gamma)$;
2: Count $n(u, v)$ in all random walk sequences $\mathcal{P}$;
3: Initialize $\{\mu_s\}_{s=1}^m$ with random distribution $[-1, 1]$;
4: Initialize $\Phi'(v_i)$ by Eq.(4) and $\Phi(v_i)$ by DeepWalk [19] with $\mathcal{P}$, respectively.
5: **for** $iter = 1 : T_1$ **do**
6: **for** $subiter = 1 : T_2$ **do**
7:   Update π_{ik}, ψ_k and Σ_k by Eq.(12), Eq.(13) and Eq.(14)
8:   **for** $k = 1, \cdots, K$ **do**
9:    **if** $diag(\Sigma_k)$ has zero item **then**
10:     Randomly reset $\Sigma_k > 0$ and $\psi_k \in \mathbb{R}^{2m}$
11: **for all** node $v_i \in \mathcal{V}$ **do**
12:   SGD on $\Phi(v_i)$ by Eq.(15);
13: **for all** node $v_i \in \mathcal{V}$ **do**
14:   SGD on $\Phi(v_i)$ by Eq.(16);
15: **for** $s = 1$ to m **do**
16:   SGD on μ_s by Eq.(17);
17: Construct the mapped image $\Phi'(v)$ for each $v \in \mathcal{V}$ with $\{\mu_s\}_{s=1}^m$.

Algorithm 1 shows the network representation generation process. In lines 1–4, we adopt random walk to generate the sequences and calculate the statistics $n(v_i, v_j)$, then initialize $\Phi(v_i)$ and μ. In lines 6–10, we fix (v, μ) and optimize (Π, Ψ, Σ) for detecting the homogeneous parts in the embedding generation space. In lines 15–14, we fix (Π, Ψ, Σ) and optimize (v, μ). Particularly, we update node embedding caused by GMM in lines 13–14, update the parameters $\{v_s\}_{s=1}^{|\mathcal{V}|}$ in lines 11–14, and the parameters $\{\mu_s\}_{s=1}^m$ in lines 15–16, respectively. Finally we construct the mapped image $\Phi'(v_i)$ as the representation for each node $v_i \in \mathcal{V}$ in line 17.

4 Experiments

In this section, we first present the datasets and experimental settings that will be used in our paper. Then, we perform node classification experiments and analyze the experimental results. Finally, we investigate the parameters sensitivity.

4.1 Experimental Setup

Datasets. Three real social network datasets are used in our experiments. **Ego-Facebook**[1] is a social friendship network for Facebook users, which contains

---

[1] https://snap.stanford.edu/data/.

4,039 nodes and 176,468 edges. Every node has a 477-dimensional vector that indicates its attribute information. We set user's education types as class labels. The other two datasets are **Hamilton** and **Rochester**[2], both of which are collected from 100 US university Facebook networks. The two networks consist of 2,314 nodes, 192,788 links, and 4,563 nodes, 322,808 links respectively. Every node uses a 144-dimensional, and a 235-dimensional feature vector as its attribute information separately. The student/faculty status flags are treated as class labels. The statistics of three datasets are listed in Table 1.

Table 1. The statistics of three real social network datasets

	Ego-Facebook	Hamilton	Rochester
# of nodes	4,039	2,314	4,563
# of edges	176,468	192,788	322,808
# of attributes	477	144	235
# of labels	4	6	6

Baselines. The following methods are comparing in our paper:

- **DeepWalk** [19] learns network representation using network structure only.
- **Node2Vec** [8] extends DeepWalk [19] to generate the network representation based on network structure only.
- **LINE** [21] only considers network structure to learn network representation.
- **LANE** [11] maps three kinds of information (*i.e.,* network structure, node attributes, labels) into a consistent latent space as the final node representation. Here, we only choose the version that ignores label information.
- **UPP-SNE** [27] takes user profile information as node attribute, and seamlessly projects it with network structure into a consistent embedding subspace to train the representation for nodes in a network.

Settings. We adopt node classification task to evaluate the quality of different methods. To be fair comparisons, we vary the training ratio from 1% to 6% by an increment of 1%. For each training ratio, we randomly pick up a small number of labeled nodes as training set, and the rest of the nodes removed their labels are for testing set at first. Then, we train a linear SVM implemented by Liblinear [7] based on the training set (nodes). And we use the learned SVM classifiers to predict the test set (nodes). After that, we use the Accuracy metric as the evaluation criteria to measure the classification performance. We repeat this process 10 times and report the averaged results in Tables 2, 3 and 4.

We set all baselines as default parameters according to they reported. For all datasets, the learned dimension is set to $m = 128$. In SAHP algorithm, the number of negative samplings N is 5, trade-off parameter β is 0.1, walks per node γ is 10, learning rate η is 0.1, walk length l is 40, window-size t is 10.

---

[2] https://escience.rpi.edu/data/DA/fb100/.

4.2 Node Classification

From the results shown in Tables 2, 3 and 4, we find that the proposed SAHP consistently outperforms all baselines. On Ego-Facebook, SAHP gains nearly 2% improvement over the best baselines. And on Hamilton and Rochester, SAHP achieves around 2.4% and 3% improvement over the best baselines, respectively.

To be specific, we observe that our proposed method SAHP consistently outperforms the pure structural embedding methods, such as DeepWalk [19],

Table 2. Accuracy (%) of the classification experiment on Ego-Facebook.

Dataset	Ego-Facebook					
Training ratio	1%	2%	3%	4%	5%	6%
DeepWalk	58.61	59.62	60.79	60.91	61.44	62.23
Node2Vec	58.95	59.34	61.39	60.68	62.11	63.08
LINE	66.56	67.65	68.66	69.29	69.97	69.47
LANE	73.22	74.44	74.97	75.49	69.95	76.25
UPP-SNE	83.81	84.13	84.96	84.69	85.03	85.58
SAHP	**85.70**	**86.40**	**86.73**	**87.29**	**87.93**	**88.42**

Table 3. Accuracy (%) of the classification experiment on Hamilton.

Dataset	Hamilton					
Training ratio	1%	2%	3%	4%	5%	6%
DeepWalk	84.77	86.16	87.63	89.44	89.75	90.44
Node2Vec	84.85	87.46	88.42	88.78	89.68	90.39
LINE	80.40	81.82	83.43	86.78	86.79	87.87
LANE	79.42	80.04	80.54	81.43	81.16	82.25
UPP-SNE	85.35	86.89	89.14	90.68	90.76	91.71
SAHP	**87.69**	**89.04**	**91.82**	**92.12**	**93.01**	**93.88**

Table 4. Accuracy (%) of the classification experiment on Rochester.

Dataset	Rochester					
Training ratio	1%	2%	3%	4%	5%	6%
DeepWalk	82.29	83.12	87.63	84.16	84.55	90.44
Node2Vec	83.01	83.45	83.84	88.78	84.37	85.39
LINE	81.29	81.80	82.43	83.09	84.99	83.78
LANE	80.89	80.98	80.78	80.88	81.11	81.03
UPP-SNE	83.29	84.89	85.89	86.14	87.20	87.83
SAHP	**85.99**	**86.59**	**88.16**	**89.15**	**89.38**	**90.66**

Node2Vec [8] and LINE [21]. This fact demonstrates that incorporating node attributes makes the node embedding representation obtain better performance. In addition, SAHP also achieves a better performance than the content-augmented network embedding baselines, including LANE [11] and UPP-SNE [27]. This fact can be explained that, by benefiting from incorporating the homophily information, SAHP get more informative representations than LANE and UPP-SNE. It demonstrates that homophily could provide an important property for network embedding, especially when node attributes are exploited.

4.3 Parameter Sensitivity

SAHP has three major parameters: dimension m, walk length per node l and the trade-off parameter β. To investigate the effect of these parameters, we perform the experiments on node classification to evaluate the effectiveness. In this process, there are any two of the three parameters fixed, and the training ratio is set at 10%, then we study the impact of the third one in the measure of classification accuracy. As shown in Fig. 1, we can observe that on Ego-Facebook, Hamilton and Rochester datasets, the performance of SAHP is less sensitive to these three parameters. This fact shows that the SAHP model is stable.

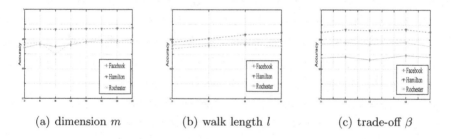

(a) dimension m (b) walk length l (c) trade-off β

Fig. 1. The effect of parameters m, l, and β

5 Conclusion

In this paper, we propose a Structure, Attribute and Homophily Preserved Social network embedding method that considers three information sources, including network structure, node attributes, homophily, to jointly train an effective vector representation for a social network. A joint optimization framework is also designed to map the network into a consistent embedding representation space that could enhance the node proximity by taking advantage of these three information sources as much as possible. As a result, the learned vector representations not only preserve the topological structure proximity between nodes and their attribute similarity, but also explicitly reflect the principle of homophily among the nodes in this new space. Experimental results indicate that the proposed algorithms SAHP achieves improvements than the state-of-the-art network embedding methods.

As nodes in a real-world network are associated with abundant textual features, in the future, we plan to apply the SAHP model to effectively incorporate these features of nodes with other information in the network embedding process.

Acknowledgments. This work is supported by the National Key Research and Development Program of China with Grant No. 2016YFB0800504, and by the National Natural Science Foundation of China with Grant No. U163620068.

References

1. Boyd, S., Vandenberghe, L.: Convex Optimization. Cambridge University Press, Cambridge (2004)
2. Cavallari, S., Zheng, V.W., Cai, H., Chang, K.C.C., Cambria, E.: Learning community embedding with community detection and node embedding on graphs. In: CIKM, pp. 377–386 (2017)
3. Chen, T., Tang, L.A., Sun, Y., Chen, Z., Zhang, K.: Entity embedding-based anomaly detection for heterogeneous categorical events. In: IJCAI, pp. 1396–1403 (2016)
4. Dai, Q., Li, Q., Tang, J., Wang, D.: Adversarial network embedding. In: AAAI, pp. 2167–2174 (2018)
5. Dempster, A.P., Laird, N.M., Rubin, D.B.: Maximum likelihood from incomplete data via the EM algorithm. J. R. Stat. Soc. Ser. B (Methodol.) **39**, 1–38 (1977)
6. Easley, D., Kleinberg, J.: Networks, Crowds, and Markets: Reasoning About a Highly Connected World. Cambridge University Press, Cambridge (2010)
7. Fan, R.E., Chang, K.W., Hsieh, C.J., Wang, X.R., Lin, C.J.: LIBLINEAR: a library for large linear classification. J. Mach. Learn. Res. **9**, 1871–1874 (2008)
8. Grover, A., Leskovec, J.: node2vec: scalable feature learning for networks. In: SIGKDD, pp. 855–864 (2016)
9. Hu, R., Aggarwal, C.C., Ma, S., Huai, J.: An embedding approach to anomaly detection. In: ICDE, pp. 385–396 (2016)
10. Huang, X., Li, J., Hu, X.: Accelerated attributed network embedding. In: SIAM, pp. 633–641 (2017)
11. Huang, X., Li, J., Hu, X.: Label informed attributed network embedding. In: WSDM, pp. 731–739 (2017)
12. Kipf, T.N., Welling, M.: Semi-supervised classification with graph convolutional networks. In: ICLR, pp. 1–14 (2017)
13. Li, Y., Sha, C., Huang, X., Zhang, Y.: Community detection in attributed graphs: an embedding approach. In: AAAI, pp. 338–345 (2018)
14. Liang, J., Jacobs, P., Sun, J., Parthasarathy, S.: Semi-supervised embedding in attributed networks with outliers. In: SIAM, pp. 153–161 (2018)
15. McPherson, M., Smith-Lovin, L., Cook, J.M.: Birds of a feather: homophily in social networks. Annu. Rev. Sociol. **27**(1), 415–444 (2001)
16. Mikolov, T., Chen, K., Corrado, G., Dean, J.: Efficient estimation of word representations in vector space. arXiv preprint arXiv:1301.3781 (2013)
17. Mikolov, T., Sutskever, I., Chen, K., Corrado, G.S., Dean, J.: Distributed representations of words and phrases and their compositionality. In: NPIS, pp. 3111–3119 (2013)
18. Pan, S., Wu, J., Zhu, X., Zhang, C., Wang, Y.: Tri-party deep network representation. In: Proceedings of IJCAI, pp. 1895–1901 (2016)

19. Perozzi, B., Al-Rfou, R., Skiena, S.: DeepWalk: online learning of social representations. In: SIGKDD, pp. 701–710 (2014)
20. Rahimi, A., Recht, B.: Random features for large-scale kernel machines. In: Advances in Neural Information Processing Systems, pp. 1177–1184 (2008)
21. Tang, J., Qu, M., Wang, M., Zhang, M., Yan, J., Mei, Q.: LINE: large-scale information network embedding. In: WWW, pp. 1067–1077 (2015)
22. Tu, K., Cui, P., Wang, X., Wang, F., Zhu, W.: Structural deep embedding for hyper-networks. In: AAAI, pp. 426–433 (2018)
23. Wang, D., Cui, P., Zhu, W.: Structural deep network embedding. In: SIGKDD, pp. 1225–1234 (2016)
24. Wang, X., Cui, P., Wang, J., Pei, J., Zhu, W., Yang, S.: Community preserving network embedding. In: AAAI, pp. 203–209 (2017)
25. Yang, C., Liu, Z., Zhao, D., Sun, M., Chang, E.Y.: Network representation learning with rich text information. In: IJCAI, pp. 2111–2117 (2015)
26. Zhang, D., Yin, J., Zhu, X., Zhang, C.: Homophily, structure, and content augmented network representation learning. In: ICDM, pp. 609–618 (2016)
27. Zhang, D., Yin, J., Zhu, X., Zhang, C.: User profile preserving social network embedding. In: IJCAI, pp. 3378–3384 (2017)

Neural Network Collaborative Filtering for Group Recommendation

Wei Zhang, Yue Bai, Jun Zheng[(⊠)], and Jiaona Pang[(⊠)]

Computer Center, East China Normal University,
3663 Zhong Shan Rd. N., Shanghai, China
{jzheng,baiyue}@cc.ecnu.edu.cn

Abstract. In the group recommender system, most of methods through aggregating individual preferences of each member in the group to group preference, which neglect the correlation among the members of the group. In this paper, group recommendation based on neural collaborative filtering (GNCF) and convolutional neural collaborative filtering (GCNCF) frameworks are proposed, which simulate the interaction between the members of the group and make recommendations directly for the group. GNCF and GCNCF frameworks predict group ratings by learning user-item interaction matrices. They project sparse vectors to dense vectors by utilizing the full connection layer, and improve the nonlinear capability of the model by using the deep neural networks. Comparing with the traditional method, our method builds a new group recommendation model, and its effectiveness is well demonstrated through experiments.

Keywords: Group recommendation · Neural network
Context-aware · Collaborative filtering

1 Introduction

Collaborative filtering is the most widely used technique in recommender system, it generates recommendation information for the user by finding a similar group to the user. Matrix factorization (MF) is the most popular means of collaborative filtering, which decomposed user-item interaction matrices into latent features that simulate user-item interactions, and provides users with more accurate recommendations. Lee *et al.* [12] proposed the prototype of MF algorithm and decomposed user rating matrix into users and items latent vectors. Based on it, a large number of researchers had improved MF. For example, Koren *et al.* [10] proposed a new method which added users and items biases to the original factorization function. Rendle *et al.* [16] further improved the performance of the recommender system by simulating the interaction between contextual information through the factorization machines (FMs).

The majority of recommender system makes recommendations for individual users, however, many of the activities in life (such as restaurant dining, watching movies and group travel) require group participation. When recommending

© Springer Nature Switzerland AG 2018
L. Cheng et al. (Eds.): ICONIP 2018, LNCS 11306, pp. 131–143, 2018.
https://doi.org/10.1007/978-3-030-04224-0_12

information for a group, the system must consider the preferences of each group member. Seko et al. [18] represented the group preference model as a multidimensional feature space, in which the group recommendation information can be obtained by calculating the distance between the candidate item and the high-density area. In recent years, with the introduction of context-aware, the performance of the recommender system has been further improved. For example, Li et al. [13] proposed a group-coupon recommender system which consider user preference, geographic convenience, and friends influence. Liu et al. [14] provided recommendation information for tour groups through location context and seasonal context. Bok et al. [5] proposed a method of recommending groups for users based on user profile and social networks. And Lu et al. [15] proposed a hierarchical Bayesian model based on location. They first utilized the group members, the group flow area and the group preference to generate the group geography topic model, then recommended activity venues for groups by combining a geographic model with a collaborative filtering framework.

Neural networks have made great progress in the fields of natural language processing, image processing, pattern recognition and so on. The combination of traditional recommendation algorithm and neural network is the development trend of recommender system. In the recommender system for individual users, Devooght et al. [6] proposed a recurrent neural network (RNN) framework. They regarded consumer behavior as a time-sequence problem, and utilized RNN to predict the user's next consumption behavior. He et al. [8] proposed a neural collaborative filtering framework to predict user's next interactive item by learning user-item interaction functions through the multi-layer perceptrons (MLP) and MF. Dieleman and Wang et al. [7,24] proposed to use neural networks to describe texts, music, and images, and further assists the recommender system. Comparing with recommender system for individual users, the known research seldom involves the application of neural network for group recommendation. Hu and Jian et al. [9] constructed a general depth model by collective deep belief networks and dual-wing restricted Boltzmann machines, it is applicable to many other areas that study the group behavior with coupled interactions among members. Wang et al. [23] proposed a group recommendation algorithm based on the probability matrix factorization of dynamic convolution. In this algorithm, they first used CNN and matrix factorization model to analyze user preferences comprehensively through combining with service description document, profile file and time factors, then used the average strategy to merge the individual preference into group preference.

Traditional group recommendation technologies mainly utilizes simple linear fusion to extend personal preference to group preference, which is too much dependent on personal recommendation technology, and magnify the error recommended by the group. In this paper, we take recommending movies for groups as an example, and utilize contextual information, such as the location, time and weather of the members of the group, to directly recommend the information that meets the interests of the group. The main contributions of this paper are as follows:

1. We proposed group recommendation based on neural collaborative filtering (GNCF) and convolutional neural collaborative filtering (GCNCF) frameworks, which simulate interaction between groups and items by mapping sparse vectors into dense vectors.
2. We showed that GNCF and GCNCF is a generalization of FMs, and it is proved by experiments that the use of neural network can better describe the nonlinear relationship between the groups and the items.
3. We performed experiments on Movielens and Netflix datasets, and compared the results with the traditional methods to verify the effectiveness and generality of the proposed algorithm.

2   Preliminaries

2.1   Problem Statement

For predicting individual rating questions, we first define users $U = \{u_1, u_2, ..., u_m\}$, and items $I = \{i_1, i_2, ..., i_n\}$. The rating R is predicted by user U and item I, the target function $y : U \times I \rightarrow R$. $y(u, i)$ is the rating of user u for item i. The task of the rating prediction is to approximate prediction rating $\hat{y}(u, i)$ to the original rating $y(u, i)$. Similarly, for predicting group rating questions, we can also define the groups $G = \{g_1, g_2, ..., g_z\}$, items $I = \{i_1, i_2, ..., i_m\}$, and the target function $y : G \times I \rightarrow R_g$, in which R_g is the rating of group g for item i. The aim of the rating prediction is to approximate prediction rating $\hat{y}(g, i)$ to the original rating $y(g, i)$.

2.2   Context-Aware

In context-aware recommender system, recommendation information is influenced by user context-aware. For example, people are more likely to watch comedy movies when they are sad. Defining the user's context variable $c \in C$, the user's mood can be expressed as C (C = happy, sad, etc.). FMs model utilized context-aware information to predict the ratings, it can be expressed as:

$$y : U \times I \times C_3 \times ... \times C_s \rightarrow R, \tag{1}$$

where U denotes users, I denotes items, C_i denotes contextual information, and subscript i denotes the number of contextual information, and starts counting from 3. The method proposed in this paper also uses this encoding.

In Fig. 1, we shows an example for context-aware data. In which C_3 denotes mood and C_4 denotes the weather:

$U = \{Abby, Bob, Cathy\}$
$I = \{Titanic, Spider - man, ForrestGump, ResidentEvil\}$
$C_3 = \{Sad, Normal, Happy\}$
$C_4 = \{Sunny, Cloud, Rain\}$

The left side of Fig. 1 shows the user's viewing record, for example, the first line means Abby rated Titanic with 4 stars, and she had watched this movie with Bob while she was Happy and the weather was sunny.

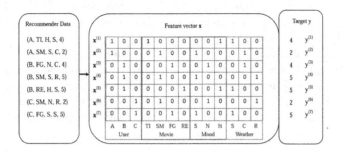

Fig. 1. Binary representation of the user viewing record

2.3   Rating Prediction with Factorization Machines

FMs [16] is the generalization of generalized matrix factorization, and the contextual information is added to the traditional matrix factorization. FMs models all interactions between pairs of variables with the target, including nested ones, by using factorized interaction parameters:

$$\hat{y}(\boldsymbol{x}) = w_0 + \sum_{i=0}^{s} w_i x_i + \sum_{i=1}^{s} \sum_{j=i+1}^{s} \hat{w}_{ij} x_i x_j, \tag{2}$$

where $\hat{w}_{ij}$ are the factorized interaction parameter between pairs:

$$\hat{w}_{ij} := <\boldsymbol{v}_i, \boldsymbol{v}_j> = \sum_{f=1}^{k} v_{if} v_{jf}, \tag{3}$$

where w_0 is the global bias, w_i models the interaction of the target with the i-th variable, and $\hat{w}_{ij}$ models the factorized interaction a pair of variables the target.

Comparing the FMs model with the generalized matrix factorization model [21], FMs not only decomposed interaction matrix between user and item, but also decomposed all pairwise interactions with all context variables.

In the group recommendation, we first get each user rating of the item in the group by using the FMs method, and then utilize the preference aggregation strategy to obtain the score of the group.

3   General Framework

3.1   Group Neural Collaborative Filtering Framework

Group neural collaborative filtering uses MLP learning user-item interaction matrix. As shown in Fig. 2, u_j denotes the j-th user, and the input layer includes

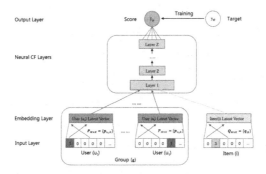

Fig. 2. Group neural collaborative filtering framework

multiple users. These users belong to the same group $u_j \in g_l^t$, and the number of users in the group is t. The user information is encoded by the binary code format like Fig. 1, and y_{gi} denotes the group's real ratings for item i, $\hat{y}_{gi}$ denotes the model's output.

Above the input layer is the embedding layer, which is used to represent the sparse information of the user into dense binary vectors. The generated vectors can be regarded as the model used to describe the user and item latent vectors. The obtained n dense vectors are connected in series, and the connection method is widely used in multi-modal depth learning [22,25]. The linked vectors are fed into NCF layers, which is used to learn the interaction between the user and the item latent vectors. The output layer can be regarded as a multi-class problem. Comparing with the FMs method, the framework can give the model a higher non-linear modeling ability. Since the principle of GNCF is similar to those of GCNCF mentioned below, it is only different from CF Layers. The GNCF framework is no longer detailed in this section, the principle of GNCF can be referred to the following GCNCF.

3.2 Group Convolutional Neural Collaborative Filtering Framework

In GNCF, latent vectors are connected to a group-item vector by serial connection and sent to NCF Layers for training. If we change the serial link method to a parallel link method, we can get the group feature matrix. Each row of this matrix denotes a vector. The matrix obtained by the attended mode can further model the recommendation process using convolutional neural networks. The model structure is shown in Fig. 3.

The GCNCF framework adds a convolutional pooling layer to the GNCF framework. The group feature matrix consists of latent vectors learned by each group member. In the convolution process, multiple convolution kernel is used to extract different local features of the group feature matrix. The pooling operation is used after the convolution is completed. The purpose of pooling is to reduce the number of parameters and extract the most useful eigenvalues in the matrix. When we recommend information for a large scale group, the

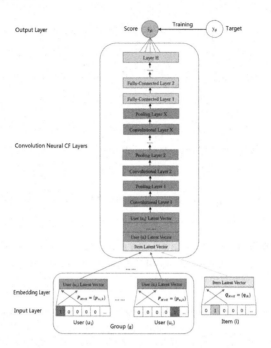

Fig. 3. Group convolutional collaborative neural collaborative filtering framework

recommendation matrix obtained by the embedding layers may be very large. After being processed by the pooling layer, the parameters of the model are reduced, which effectively prevents the model overfitting. After some convolution pooling operations, the extracted local features are synthesized into global features in the fully connected layer, and the global features are further learned. Finally, we output the classification results at the output layer.

The GCNCF model is defined as follows:

$$\hat{y}_{gi} = f(\boldsymbol{P}_1^T \boldsymbol{v}_{u_1}^U, \boldsymbol{P}_2^T \boldsymbol{v}_{u_2}^U, ..., \boldsymbol{P}_t^T \boldsymbol{v}_{u_t}^U, \boldsymbol{Q}^T \boldsymbol{v}_i^I | \boldsymbol{P}_1, \boldsymbol{P}_2, ..., \boldsymbol{P}_t, \boldsymbol{Q}, \Theta_f), \qquad (4)$$

where $u_1, u_2, ..., u_t \in g$, $\boldsymbol{P}_j \in R^{M \times K}$ and $\boldsymbol{Q} \in R^{N \times K}$ denote the latent factor matrix of users and items, and Θ_f denotes the model's parameters, respectively. The function f can be expressed as follows:

$$
\begin{aligned}
&f(\boldsymbol{P}_1^T \boldsymbol{v}_{u_1}^U, \boldsymbol{P}_2^T \boldsymbol{v}_{u_2}^U, ..., \boldsymbol{P}_t^T \boldsymbol{v}_{u_t}^U, \boldsymbol{Q}^T \boldsymbol{v}_i^I | \boldsymbol{P}_1, \boldsymbol{P}_2, ..., \boldsymbol{P}_t, \boldsymbol{Q}, \Theta_f) \\
&= \phi_{out}(\phi_Z(...(\phi_X(...\phi_2(\phi_1(\boldsymbol{P}_1^T \boldsymbol{v}_{u_1}^U, \boldsymbol{P}_2^T \boldsymbol{v}_{u_2}^U, ..., \boldsymbol{P}_t^T \boldsymbol{v}_{u_t}^U, \boldsymbol{Q}^T \boldsymbol{v}_i^I)...))...)),
\end{aligned}
\qquad (5)
$$

where ϕ_{out} denotes output layer function, ϕ_X and ϕ_Z denote the mapping function for X-th CF layer and Z-th fully-connected layer, respectively.

3.3   Learning GNCF and GCNCF

The group rating prediction can be transformed into a multi-objective classification problem, we use the softmax function at the output layer to output the probability of each category. The cross-entropy loss function is defined as follows:

$$J(\Theta) = -\sum_{gi\in y}\sum_{n=1}^{N} sign(y_{gi} = n)\log p(\hat{y}_{gi} = n|P, Q.\Theta_f), \tag{6}$$

where N denotes the number of categories. It represents the level of rating in this article.

Although cross-information entropy rarely investigated papers about the recommendation algorithm, literature [8] proved the effectiveness of this method through experiments.

3.4 GNCF, GCNCF and FMs

Generalized MF decomposes the matrix into user and item latent vectors. As a comparison, FMs takes full advantage of contextual information to decompose interaction matrix between users and items. FMs can also be seen as a special form of GNCF and GCNCF. For G(C)NCF, a fully connected layer is first used to project users and items binary sparse vectors to latent feature vectors. Then (C)NCF layers are used to learn group feature matrix, and output layer predicts group rating for which category. Supposed that $\boldsymbol{p}_{u_j}$ is the user latent vector $\boldsymbol{P}_j^T \boldsymbol{v}_{u_j}^U$, $\boldsymbol{q}_i$ is the item latent vector $\boldsymbol{Q}^T \boldsymbol{v}_i^U$. We define the mapping function of the first (C)NCF layer as:

$$\phi_1(\boldsymbol{p}_{u_1}, \boldsymbol{p}_{u_2}, ..., \boldsymbol{p}_{u_t}, \boldsymbol{q}_i) = \sum_{a=1}^{t}\sum_{b=a+1}^{t} \boldsymbol{p}_{u_a} \odot \boldsymbol{p}_{u_b} + \sum_{a=1}^{t} \boldsymbol{p}_{u_a} \odot \boldsymbol{q}_i, \tag{7}$$

where $\odot$ denotes the element-wise product of vectors and the output of the model can be expressed as follows:

$$\hat{y}_{gi} = a_{out}(\boldsymbol{h}^T(\sum_{a=1}^{t}\sum_{b=a+1}^{t} \boldsymbol{p}_{u_a} \odot \boldsymbol{p}_{u_b} + \sum_{a=1}^{t} \boldsymbol{p}_{u_a} \odot \boldsymbol{q}_i)), \tag{8}$$

where a_{out} and $\boldsymbol{h}$ denote the output layer activation functions and connection weights, respectively. If we use an identity function for a_{out} and enhance $\boldsymbol{h}$ to uniform vector of 1, then the model can be regarded as the deformation of FMs. Comparing with FMs, the G(C)NCF framework takes advantage of the nonlinear modeling ability of neural network to further learn the feature matrix.

The input layer and the embedding layer of GNCF and GCNCF are the same, but the difference is that GCNCF has increased the feature extractor composed of Convolution layer and Pooling layer. The GNCF directly establishes the connection between each group member by the full-connection layer and directly considers the preferences of each group member. This approach is more difficult to converge to the optimal solution. As a comparison, GCNCF makes full use of the local perceptive ability of convolutional neural networks, and uses different convolution kernel to learn the different preferences of some members in the group, and then the information of the local group members is integrated into the global information at the high layer. So this approach can find group's favorite information more accurately.

4   Experiments

In order to verify the effectiveness of the proposed algorithm, we explore the GNCF and GCNCF method from three aspects.

1. Comparing with the state-of-the-art group recommendation algorithm, we verify the effectiveness of our algorithm.
2. Network structure exploration is also the focus of this article. By setting different number of embedding factors and predictive factors, the effect on the experimental results can be observed.
3. Setting up different group size training multiple models to verify the generality of the proposed algorithm.

4.1   Experimental Settings

Datasets. This paper uses the MovieLens [1] and Netflix [2] datasets, which contains 100 thousand ratings for 1682 movies by 943 users and 100 million ratings for 17 thousand movies by 480 thousand users, respectively. We construct a semi-simulated dataset by adding reasonable context generation rules. The constructed data set includes:

User: ID, age, sex, and occupation
Item: ID, release date and type
Rating: user rating for the item
Context-aware: time, mood, location and weather.

Baltrunas *et al.* [3] divides the group into 2 parts, random groups and high inner similar group. Considering the randomness of members in a group, we select 10 users as a group (the limitation of dataset size) and the rating records of 100,000 groups randomly. During the experiment, 80% of the data was selected as the training set and 20% of the data as the test set. Since the group's ratings is not given in the data set, the mean value strategy [4] is used here to obtain the group's ratings.

Evaluation Protocols. The evaluation indexes in this paper are RMSE [16] and nDCG [3]. RMSE measures the degree of proximity between prediction ratings and actual ratings, and it can accurately predict user's preference degree of recommendation information. nDCG is used to measure the satisfaction of users with a recommendation list, the main idea means users prefer items that appear in the front of the recommendations list.

Parameter Settings. The model parameters are initialized to a normal distribution random number with a mean of 0 and a standard deviation of 0.005. With a small batches of Adam to optimize this model, the batch size is set to 64, the learning rate is set to 0.005. The activation functions of the convolutional layer and the fully connected layer are selected as rectified linear units (ReLU). This experiment takes the GCNCF structure for two identical convolution pooling layers. Convolution kernel parameters are taken as follows, $size = 3 * 3$, $number = 16$, $strides = [1,1,1,1]$, the maximum pool method is used in the pooling layer, and there is $strides = [1,2,2,1]$. After the pooling layer, we use local response normalization (LRN) to create the competition mechanism of local neurons. In the GCNCF framework, both the number of neurons in the embedding layers and the last fully-connected layer determine the performance of the model, which is referred to here as the embedding factors and the predictive factors. The model is evaluated by selecting different embedding factors [8, 16, 32, 64] and predictive factors [16, 32, 64, 128]. For example, when the embedding factors is 16, the predictive factors is 32 and the number of group members is 10, the GNCF structure of embedding layer and NCF layer are $11 \times 16 \rightarrow 256 \rightarrow 128 \rightarrow 64 \rightarrow 32$, the GCNCF structure of embedding layer and CNCF layer are $11 \times 16 \rightarrow 6 \times 8 \times 16 \rightarrow 3 \times 4 \times 16 \rightarrow 64 \rightarrow 32$ (here 16 is the number of convolution kernels). Without special mention, GNCF and GCNCF use four CF layer (a convolution pool layer in GCNCF as a CF layer).

4.2 Effectiveness of Group Recommendation

Our algorithm is based on group modeling, and we will compare GNCF and GCNCF with KNN [17], FMs methods [16], CARS2 [19] by setting different predictive factors. For FMs and CARS2, the number of predictive factors is equal to the number of latent factors, and that in KNN is equal to the number of nearest neighbors. When we set the embedding factor size to 32, the experimental results are shown in Fig. 4.

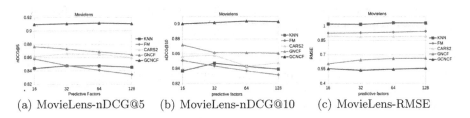

(a) MovieLens-nDCG@5   (b) MovieLens-nDCG@10   (c) MovieLens-RMSE

Fig. 4. Performance of nDCG@5, nDCG@10, and RMSE change with predictive factors using Movielens.

Figures 4 and 5 show the experimental results of GCNCF and GNCF in different datasets. The experimental results on the Netflix dataset are worse than those on Movielens dataset. In general, Netflix dataset is higher than Movielens dataset 4% in RMSE, and is lower than Movielens dataset 3% in nDCG, which is caused by the sparse data of Netflix. Comparing with the traditional method, GNCF and GCNCF have been significantly improved on both datasets, they have

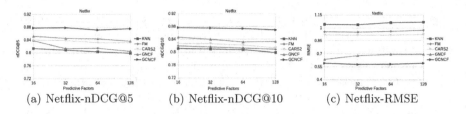

Fig. 5. Performance of nDCG@5, nDCG@10, and RMSE change with predictive factors using Netflix.

better non-linear modeling capabilities and can further describe the user-item matrix interaction by constructing a network structure. Moreover, the GCNCF has better performance than GNCF, it explains that the way of extracting local features by CNN is better than getting global features by using fully connection layers directly. The experimental results of GNCF and GCNCF are optimal when the predictive factors are 16 and 32, respectively. Here optimal values for these two methods are obtained for different size of predictive factors. The reason is that when GCNCF is extracted features from multiple convolution kernels, the dimension of mapping to full link layer will be relatively high, and few predictive factors can not learn enough features. While GNCF passes through multiple fully connected layers, the dimension has been reduced to a certain extent, the model can be well predicted without too many predictive factors.

4.3   The Structure of GCNCF

In the last section, we observed that the experimental results of GCNCF are better than that of GNCF. In this section, We further explore the different experiments by setting different the structure of GCNCF, the size of the embedding factors is set as 8, 16, 32, and 64, and the size of the predictive factors is 16, 32, 64, and 128. The experimental results are shown in Table 1.

When the number of predictive factors are 32 and embedding factors are 64, Table 1 shows the experimental results almost achieves an optimal value. With the increase of the number of embedding factors, the value of RMSE is reduced in general. For nDCG@5, the experimental results of embedding factors 32 are better than those of 64. But for nDCG@10, it is just the opposite. The reason is that the model can not fully learn all the features of the group when the embedding factors is 32. In other words, a small amount of predictive factors can provide high quality recommendation when the size of the recommendation list is small, but the quality of recommendation will decrease with the recommendation list increases.

The difference between the experimental results of Netflix and those of Movielens becomes smaller with the number of network nodes increases. It can be explained as follows, the complex network structure is more useful for mining the members of the group for the larger dataset (Netflix), but the effect on the experimental results are not obvious for the small dataset (Movielens).

Table 1. Performance of GCNCF with different network structures

E-Factors	P-Factors	16	32	64	128	16	32	64	128
		Movielens				Netflix			
8	nDCG@5	0.899	0.891	0.897	0.888	0.851	0.861	0.853	0.856
	nDCG@10	0.891	0.889	0.883	0.885	0.846	0.843	0.851	0.853
	RMSE	0.605	0.586	0.607	0.612	0.638	0.634	0.643	0.638
16	nDCG@5	0.905	0.906	0.904	0.895	0.863	0.865	0.863	0.853
	nDCG@10	0.889	0.893	0.893	0.884	0.867	0.862	0.871	0.866
	RMSE	0.582	0.577	0.585	0.598	0.623	0.617	0.618	0.619
32	nDCG@5	0.909	0.910	0.911	0.910	0.877	0.878	0.871	0.875
	nDCG@10	0.898	0.897	0.901	0.901	0.876	0.875	0.873	0.869
	RMSE	0.576	0.569	0.547	0.564	0.591	0.578	0.581	0.591
64	nDCG@5	0.901	0.910	0.905	0.904	0.869	0.871	0.864	0.867
	nDCG@10	0.900	0.903	0.903	0.902	0.879	0.873	0.881	0.884
	RMSE	0.564	0.536	0.551	0.557	0.579	0.557	0.567	0.563

4.4 Groups of Different Sizes

In theory, the proposed method can train any size of group model, but the size of the group cannot be set too large due to the reason of dataset. In this experiment, the group size is divided into 2, 4, 6, 8 and 10, and the experimental results are shown in Table 2.

Table 2. Performance of GNCF and GCNCF with different groups size

Group size	Movielens				Netflix			
	nDCG@10		RMSE		nDCG@10		RMSE	
	GNCF	GCNCF	GNCF	GCNCF	GNCF	GCNCF	GNCF	GCNCF
2	0.8243	0.8345	1.0796	0.964	0.8042	0.81	1.1189	1.0028
4	0.8417	0.8509	0.8338	0.7496	0.8171	0.8308	0.872	0.7874
6	0.8539	0.8801	0.754	0.6626	0.825	0.8485	0.7882	0.7008
8	0.8642	0.899	0.6645	0.5996	0.8366	0.8657	0.7019	0.6343
10	0.8719	0.9034	0.6041	0.5363	0.8462	0.8735	0.6355	0.5711

In the training set, it will be terminated when the loss is less 0.005. In the test set, the experimental results of GCNCF are better than those of CNCF, which shows that CNN has better nonlinear modeling capability than MLP. Due to the limitation of data set in this paper, the maximum size of the group is not more than 10. From the experimental results, we can see that the larger the size of the group, the better the results of the experiment. The reason is that the larger group provide more group characteristics, and the trained models provide better predictions. In particular, when the group size is set to 1, this framework is similar to the individual recommendation model proposed by [8]. The convolution pooling process of group feature matrix can be regarded as a feature learning process of a fully connected layer.

5 Conclusion

In this paper, group recommendation frameworks are proposed, and neural networks are applied to the group recommender system. Comparing with the traditional matrix factorization, the GNCF and GCNCF framework take advantage of the non-linear modeling ability of neural networks, which fully learn user-item interaction matrix, improve the accuracy of the recommended results. Meanwhile, the framework has good generality and opens up a new modeling method for group recommendation.

In the future, there is a lot of content to be explored. On the one hand, as the dataset used in this paper is small, the size of the group is not too large. In practical applications, the group membership can be divided into larger groups of 50, 100 and 200, the simple 4-layer networks structure can not meet the needs of certain recommender system, exploring for the new network structure becomes the focus of future work. On the other hand, the convolutional neural network is used for image processing, the size of training sets can be increased by flipping and cropping of images [11, 20]. In the GCNCF framework, data can be processed similarly. A group can consist of any number from 1 to N, and the dataset can be increased by changing the composition and number of members in the group. In the training process, we can first use all members in the group to pre-train the model, then use different number of group members to fine tune the model on this basis.

References

1. Movielens dataset. https://grouplens.org/datasets/movielens/1M/
2. Netflix dataset. https://www.netflixprize.com/
3. Baltrunas, L., Makcinskas, T., Ricci, F.: Group recommendations with rank aggregation and collaborative filtering. In: ACM Conference on Recommender Systems, pp. 119–126 (2010)
4. Berkovsky, S., Freyne, J.: Group-based recipe recommendations: analysis of data aggregation strategies. In: ACM Conference on Recommender Systems, pp. 111–118 (2010)
5. Bok, K., Lim, J., Yang, H., Yoo, J.: Social group recommendation based on dynamic profiles and collaborative filtering. Neurocomputing 209(C), 3–13 (2016)
6. Devooght, R., Bersini, H.: Collaborative filtering with recurrent neural networks (2016)
7. Dieleman, S., Schrauwen, B.: Deep content-based music recommendation. In: International Conference on Neural Information Processing Systems, pp. 2643–2651 (2013)

8. He, X., Liao, L., Zhang, H., Nie, L., Hu, X., Chua, T.S.: Neural collaborative filtering, pp. 173–182 (2017)
9. Hu, L., Cao, J., Xu, G., Cao, L., Gu, Z., Cao, W.: Deep modeling of group preferences for group-based recommendation. In: Proceedings of the Twenty-Eighth AAAI Conference on Artificial Intelligence, AAAI 2014, pp. 1861–1867. AAAI Press (2014). http://dl.acm.org/citation.cfm?id=2892753.2892811
10. Koren, Y., Bell, R., Volinsky, C.: Matrix factorization techniques for recommender systems. Computer **42**(8), 30–37 (2009)
11. Krizhevsky, A., Sutskever, I., Hinton, G.E.: ImageNet classification with deep convolutional neural networks. In: International Conference on Neural Information Processing Systems, pp. 1097–1105 (2012)
12. Lee, D.D., Seung, H.S.: Algorithms for non-negative matrix factorization, pp. 535–541 (2000)
13. Li, Y.M., Chou, C.L., Lin, L.F.: A social recommender mechanism for location-based group commerce. Inf. Sci. **274**(8), 125–142 (2014)
14. Liu, Q., Chen, E., Xiong, H., Ge, Y., Li, Z., Wu, X.: A cocktail approach for travel package recommendation. IEEE Trans. Knowl. Data Eng. **26**(2), 278–293 (2013)
15. Lu, Z., Li, H., Mamoulis, N., Cheung, D.W.: HBGG: a hierarchical Bayesian geographical model for group recommendation (2017)
16. Rendle, S., Gantner, Z., Freudenthaler, C., Schmidt-Thieme, L.: Fast context-aware recommendations with factorization machines. In: International ACM SIGIR Conference on Research and Development in Information Retrieval, pp. 635–644 (2011)
17. Sarwar, B., Karypis, G., Konstan, J., Riedl, J.: Item-based collaborative filtering recommendation algorithms. In: International Conference on World Wide Web, pp. 285–295 (2001)
18. Seko, S., Yagi, T., Motegi, M., Muto, S.: Group recommendation using feature space representing behavioral tendency and power balance among members. In: ACM Conference on Recommender Systems, pp. 101–108 (2011)
19. Shi, Y., Karatzoglou, A., Baltrunas, L., Larson, M., Hanjalic, A.: CARS2: learning context-aware representations for context-aware recommendations. In: ACM International Conference on Conference on Information and Knowledge Management, pp. 291–300 (2014)
20. Simonyan, K., Zisserman, A.: Very deep convolutional networks for large-scale image recognition. In: Computer Science (2014)
21. Srebro, N., Rennie, J.D.M., Jaakkola, T.: Maximum-margin matrix factorization. Advances in Neural Information Processing Systems, vol. 37, no. 2, pp. 1329–1336 (2005)
22. Srivastava, N., Salakhutdinov, R.: Multimodal learning with deep Boltzmann machines. In: International Conference on Neural Information Processing Systems, pp. 2222–2230 (2012)
23. Wang, H., Dong, M.: Latent group recommendation based on dynamic probabilistic matrix factorization model integrated with CNN. J. Comput. Res. Dev. 113–121 (2017)
24. Wang, H., Wang, N., Yeung, D.Y.: Collaborative deep learning for recommender systems, pp. 1235–1244 (2014)
25. Zhang, H., Yang, Y., Luan, H., Yang, S., Chua, T.S.: Start from scratch: towards automatically identifying, modeling, and naming visual attributes. In: ACM International Conference on Multimedia, pp. 187–196 (2014)

Influence of Clustering on the Opinion Formation Dynamics in Online Social Networks

Rajkumar Das[1,2(✉)], Joarder Kamruzzaman[1,2], and Gour Karmakar[1,2]

[1] Department of CSE, Bangladesh University of Engineering and Technology,
Dhaka, Bangladesh
rajkumardash05@yahoo.com,
{joarder.kamruzzaman,gour.karmakar}@federation.edu.au
[2] School of Science, Engineering and IT, Federation University Australia,
Ballarat, Australia

Abstract. With the advent of Online Social Networks (OSNs), opinion formation dynamics continuously evolves, mainly because of the widespread use of OSNs as a platform of social interactions and our growing exposure to others' opinions instantly. When presented with neighbours' opinions in OSNs, the natural clustering ability of human agents enables them to perceive the grouping of opinions formed in the neighbourhood. A group with similar opinions exhibits stronger influence on an agent than the individual group members. Distance-based opinion formation models only consider the influence of neighbours who are within a confidence bound threshold in the opinion space. However, a bigger group formed outside this distance threshold can exhibit stronger influence than a group within the bound, especially when that group contains influential or popular agents like leaders. To the knowledge of the authors, the proposed model is the first to consider the impact of clustering capability of agent and incorporates the influence of opinion clusters (groups) formed outside the confidence bound. Simulation results show that our model can capture several characteristics of real-world opinion dynamics.

Keywords: Opinion · Clustering · Centrality · Consistency

1 Introduction

The dynamics of forming public opinion in a society is still poorly understood due to the inherent complexity of human behaviour under the presence of various social influences. Over the last decade, the exposure of individuals to others' opinions, the characteristics of their interactions and the extent of the social influences have evolved, mainly because of the wide acceptance and prevalent use of Online Social Networks (OSNs) as the platform of expressing opinions and as the medium of social interactions. Opinion formation models need to embrace

© Springer Nature Switzerland AG 2018
L. Cheng et al. (Eds.): ICONIP 2018, LNCS 11306, pp. 144–155, 2018.
https://doi.org/10.1007/978-3-030-04224-0_13

this major shift in public opinion formation process by incorporating the factors originating from the OSNs that affect the nature and composition of the final opinion.

The seminal works of DeGroot [3], Clifford and Sudbury [4], Hegselmann and Krause [5], Deffuant and Weisbuch [6] and Sznajd [2] have established opinion formation modelling as an active area of research, which continues to attract researchers across multiple disciplines. The existing opinion formation models moslty measure the influence between two interacting agents as a function of their opinion distances [1, 5–7, 9–11]. In these models, an agent is influenced by another agent, only if their opinion distance is bounded by a confidence interval. However, the bounded confidence models and their derivatives fail to consider the impact of any agents beyond this confidence bound, especially when these neighbouring agents cluster together to form a strong group, which the agent cannot ignore while updating her opinion. When presented with neighbours' opinions in OSNs, the natural clustering ability of human beings enables them to find if there is any group formed in the neighbourhood. In the presence of multiple groups, the influence on agent depends on several factors such as, (i) the distance of the group from the agent's opinion, (ii) the size of the group and the number of neighbours in that group, (iii) the closeness/compactness of the group members' opinions, and (iv) the presence of influential members in the group. Please note that bounded confidence models consider all neighbours inside the confidence bound as the only group that has influence on an agent. However, a group can be formed outside the confidence bound of an agent, and still can exhibit stronger influence on the agent than a closer-distant group, specially if that distant group has influential agents. The existing models lack such consideration, thus creating a research gap that our proposed model addresses in this paper. To the knowledge of the authors, the proposed model is the first to incorporate the clustering capability of agent in opinion dynamics, and incorporates the influence of opinion clusters (groups) formed outside the agent's confidence bound.

In the proposed model, while updating opinion, an agent first determines the number of groups formed in her neighbours' opinions (i.e., opinion clusters) and computes the four influencing factors presented in the previous paragraph. The perceptual ability of human agents in clustering data can be modelled by existing clustering algorithms. We utilise the centroid based K-means clustering algorithm in the proposed model to emulate such ability, as K-means clustering is successfully used for semi-supervised learning in computer vision where computers automate the task of human visual system. It is also used in numerous applications requiring data analysis. The compactness of a group is represented by the consistency of the comprising opinions as proposed in [9]. On the other hand, the presence of any influential person in a group is determined by the centrality measure of the corresponding node in the underlying social network graphs.

The impact of number of groups present in neighbours' opinions on the steady state outcome of the dynamics is observed using simulation. Results shows that our model can exhibit the polarisation characteristics of bounded confidence models, even though we do not consider any confidence bound threshold. In bounded confidence models, the agents' opinions reach consensus in the steady state of the dynamics, for any confidence thresholds greater than a particular value. On the other hand, our model does not yield any such consensus as a steady state outcome. Please note that having no consensus is a natural phenomenon for opinion dynamics occurred in OSNs with a large number of people, where the participants hardy agree on any issue.

2   Proposed Opinion Formation Model

Consider that $G = (V, E)$ represents a social network, where V denotes the set of agents participating in the opinion formation process and E is the set of edges interconnecting the agents. For an agent $i \in V$, $N_i = \{j | j \in V \wedge (i, j) \in E\}$ designates the set of neighbours the agent is connected with in the underlying social network. Opinion is formed through an iterative process in discrete time steps $T = \{0, 1, 2, \ldots, t\}$. At each time step $T = t$, an agent i is associated with an opinion $O_i(t) \in [0, 1]$. Here, the range $[0, 1]$ is referred as the Opinion Space (OS). At $T = t$, agent i is also exposed to her neighbours opinions, represented as $O_{N_i}(t) = \{O_j(t) | j \in N_i\}$.

2.1   Conceptual Model

Figure 1 illustrates a conceptual diagram of the proposed model. In the figure, an agent opinion is placed at 0.3 at time t. While updating opinion, the agent observes five different groups in her neighbours' opinions placed at different distances from the agent. The number of neighbours in each group is measured by the agent along with the closeness of their opinions in the group. Moreover, the neighbouring agents also have social power, which is determined by the number of connections they have in their social networks. This factor is not shown in the diagram.

When neighbours are clustered into opinion groups, the agent is not influenced by any individual neighbour, instead individual groups separately exhibit their influences on the agent. Under these circumstances, the distance of the agent with the group, the number of agents in each group, their closeness in the opinion space and the influential power of the group members contribute to the overall impact of the group on the opinion forming agent. The agent chooses one from the groups to interact with and to update her opinion. In the following section, the model is described elaborately with the ways of computing the corresponding model parameters.

2.2   Computing Model Parameters

When an agent i is presented with her neighbours' opinions, the perceptual ability of the agent enables her to determine whether there is any formation of opinion clusters in the neighbourhood. Here, a opinion cluster represents a group of

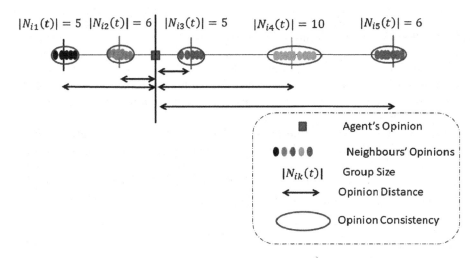

Fig. 1. Conceptual realisation of the proposed model showing important model parameters.

neighbours sharing similar thoughts (opinions) on a topic, which is clearly distinguishable from any other group of opinions on that topic. Human agent is capable of manually distinguishing those groups and counting the number of groups that the neighbours' opinions are divided into. Consider that neighbours' opinions of agent i are divided into ζ number of groups $\{N_{i1}(t), N_{i2}(t), \ldots N_{i\zeta}(t)\}$. Here, $N_{ik}(t) \subset N_i$ and $\cup_{k=1}^{k=\zeta}(N_{ik}(t)) = N_i$, denotes the kth group formed in i's neighbourhood, and $O_{N_{ik}}(t)$ denotes the set of opinions that belong to the neighbour group $N_{ik}(t)$. It is worth noting that the groups are identified by an agent based on the closeness of their opinions. A group $N_{ik}(t)$ is recognised by a representative opinion $\overline{O}_{N_{ik}}(t)$, which can be the average of the constituting opinions of that group,i.e., the average of opinions in the set $O_{N_{ik}}(t)$.

The influence of an opinion group $N_{ik}(t)$ on an agent i depends on (i) the opinion distance between the agent and the group $d_{ik}(t)$ and (ii) the inner-group distance S_{ik}, where S_{ik} represents the distance between the agents in that group $N_{ik}(t)$. The closer the neighbour group is from the agent (smaller $d_{ik}(t)$), the higher the impact of the group on the agent's opinion update. Similarly, the closeness of the group members in the opinion space increases its impact an any opinion forming agent. Equations (1) and (2) formally define $d_{ik}(t)$ and S_{ik} respectively.

$$d_{ik} = \| (O_i(t) - \overline{O}_{N_{ik}}(t)) \|, \tag{1}$$

$$S_{ik} = \frac{\sum_{j \in N_{ik}(t)}(O_j(t) - \overline{O}_{N_{ik}}(t))^2}{|N_{ik}(t)|}, \tag{2}$$

where, $\overline{O}_{N_{ik}}(t) = \frac{\sum_{j \in N_{ik}(t)} O_j(t)}{|N_{ik}(t)|}$ represents the centroid of the corresponding cluster and $|N_{ik}(t)|$ is the number of agents belonged to k^{th} cluster in agent i's neighbourhood. Here, the number of group members also strengthen or weaken the overall impact of the group on the agent.

However, the extent of influence also depends on several other factors. One of them is the distribution of the group's opinions in the opinion space. A densely packed group of opinions certainly exhibits more influence than a loosely coupled group. The closeness of a group of opinions can be measured by it's consistency as proposed in [9]. It utilises Shannon's entropy from information theory to measure the consistency of a set of values. The reason behind this is that entropy of a set with similar values is smaller that of a diverged valued set, thus having a reciprocal relationship with the notion of consistency. Thus, the consistency $\xi_{O_{N_{ik}}}(t)$ of a group $N_{ik}(t)$ of opinions in the neighbourhood of agent i can be measured by the following equation.

$$\xi_{O_{N_{ik}}}(t) = \frac{e_{max} - e_{O_{N_{ik}}}(t)}{e_{max}} \tag{3}$$

Here, $e_{O_{N_{ik}}}(t)$ is the Shannon's entropy which is defined as per Eq. (4), where p represents the probability.

$$e_{O_{N_{ik}}}(t) = \text{entropy}(O_{N_{ik}}(t)) = -\sum\nolimits_{O(t) \in O_{N_{ik}}(t)} p(O(t)) \times \log(p(O(t)) \tag{4}$$

While computing probability $p(O(t)$ of an opinion, we consider that the opinion space comprises of 10 equal length bins as motivated by the work presented in [9].

In Eq. (3), e_{max} is the maximum possible entropy value. Equation (3) ensures the reciprocal relationship between consistency and entropy, and the values to be normalised between $[0, 1]$. It is worth to mention here that, although S_{ik} and $\xi_{O_{N_{ik}}}(t)$ measure similar property of a set of data (here quality of an opinion group in terms of their closeness), the first one depends on the centroid of the group and can be highly impacted by outliers, whereas the latter one measures the quality of the distribution and suffers less from outliers' impact.

Finally, an opinion group has greater impact on an agent if the group members are influential persons in the society. An influential person in OSNs is usually followed by a large number of people, and her opinion is also shared and disseminated by those followers. Degree centrality of a node in the underlying social network graph captures the importance of a person in her opinion group. Degree centrality (C_i) of agent i is defined as the number of neighbouring agents i is connected with, i.e., $C_i = |N_i|$. The overall centrality $(C_{N_{ik}(t)})$ of a group $N_{ik}(t)$ is defined as per Eq. (5).

$$C_{N_{ik}(t)} = \frac{1}{\left(\frac{\sum_{j \in N_{ik}(t)} |C_{j*} - C_j|}{(|N_{ik}(t)| - 1) \times (|N_{ik}(t)| - 2)}\right)} \times \left(\frac{\sum_{j \in N_{ik}(t)} C_j}{|N_{ik}(t)|}\right) \tag{5}$$

where, $j* \in N_{ik}(t)$ is the node with highest centrality in the group $N_{ik}(t)$. Here, the first term in Eq. (5), as defined in [12], captures the difference between the centrality scores of all nodes with the highest central nodes. A group with similar centrality measures yield a higher value for the first term. On the other hand, the second term represents the average centrality scores of the set. The combination of these two factors ensures that both high and similar centrality scores of a group exhibits larger influence on an agent.

2.3   Opinion Update Process

While updating opinion at time step $T = t$, an agent measures the overall influence of each group in her neighbours' opinions using the parameters discussed in the above section, and find the group with higher overall influence than others. The influence of a group is measured using the following equation.

$$\mathcal{I}_{N_{ik}(t)} = \frac{\xi_{O_{N_{ik}}}(t) \times C_{N_{ik}(t)}}{d_{ik} \times S_{ik}} \tag{6}$$

The rationale for defining the influence using Eq. (6) is already explained in Sect. 2.2. In a nutshell, the smaller distance between an agent and a group of opinions, the larger is the influence on that agent. On the other hand, the other three factors has positive impact on the agent. The more the number of neighbours in a group, the higher the influence. Similarly, higher opinion consistency of the group reflects a close-knit group, which has stronger influence on the agent. Finally, the centrality measures the presence of influential members in a group, with higher values reflecting more influence on the agent.

Thus, the group with the highest influence, i.e., $N_{ik}^*(t)$ is determined using Eq. (7).

$$N_{ik}^*(t) = \operatorname*{argmax}_{k}(\mathcal{I}_{N_{ik}(t)}) = \operatorname*{argmax}_{k} \frac{\xi_{O_{N_{ik}}}(t) \times C_{N_{ik}(t)}}{d_{ik} \times S_{ik}} \tag{7}$$

After selecting the group to interact with, agent i updates her opinion using the weighted average of her own opinion and the representative opinion of the selected group, as defined in Eq. (8).

$$O_i(t+1) = \frac{1}{2} \times O_i(t) + \frac{1}{2} \times \overline{O}_{N_{ik}^*}(t) \tag{8}$$

The reason for using the weighted averaging in Eq. (8) is that it is the most predominant opinion update rule adopted in the literature for opinion formation models with a continuous opinion space $[0, 1]$. Here, the agent considers equal weights while combining her opinion with those of the neighbouring groups.

3   Simulation Results and Analysis

To emulate the social network graph, we adopted the scale-free graph with power-law degree-distribution for the social network connectivity. This is because social network resembles well with the scale-free graph. Our simulated social networks consisted of 1000 nodes to represent the opinion forming agents. The agents' initial opinions were sampled in the range $[0, 1]$ following an uniform random distribution. The most important parameter of the proposed model is the selection of clustering algorithm for determining the number of groups in an agent's neighbourhood. We adopted the K-means clustering algorithm to estimate the groups' location. The number of clusters ζ (groups in the context of our proposed model) is a configurable parameter for the K-means algorithm. Consequently, we simulated our model by varying the values of ζ. It is worth to mention here that ζ refers to the number of opinion groups that an agent is able to predict when she is presented with the neighbours' opinions. We varied ζ in the range of $[2, 10]$, where $\zeta = 2$ represents that an agent coarsely determines the group numbers, whereas $\zeta = 10$ denotes a fine-grained agent in determining the same. While measuring consistency of an opinion group using Eq. (3), we considered 10 equal length bins in the Opinion Space, as motivated by the opinion scale of 0 to 10 in a survey. Finally, degree centrality measured from the graph connectivity of the generated nodes in the simulation is considered to estimate the social influence of the participating agents, as per Eq. (5).

3.1   Varying the Number of Clusters ζ

The results obtained by varying the number of predictable clusters ζ are presented in Fig. 2. Figure 2(a) illustrate the results obtained for the value of $\zeta = 2$, whereas Figs. 2(b)–(c) depict the results for $\zeta = 5$. Finally, $\zeta = 10$ is chosen for Fig. 2(d).

The first rows of the figures show how the dynamics of opinion changes for the participating agents. Here, the X-axis stands for the time steps, while the Y-axis represents the opinion space $[0, 1]$. Initially, the opinions are distributed uniformly, which is also depicted in the second rows of the figures as a histogram of the distribution (labelled as "Initial Opinion"). The bar heights represent the fraction of agents that have opinions in the corresponding opinion bins. As discussed above, we use 10 equal length bins for generating the histogram distribution. As time progress, agents update their opinions by compromising with the opinion groups chosen based on their influence level as computed using Eq. (8), which is shown in the third row of the figures. As a result of this interactions, the groups become stronger by the joining of more agents in that groups.

At the steady state, the final opinions are merged together to form two strong polarisation groups for $\zeta = 2$ (Fig. 2(a)). When $\zeta = 5$, the number of polarisation groups, their strength (number of agents belonged to each polarisation group) and their position in the opinion space might vary. The final distribution of agents' opinions into multiple polarisation is illustrated with histogram distribution as labelled by "Final Opinion" (Fig. 2(b)–(c)). With $\zeta = 10$, the opinion formation process goes through an unstable dynamics, where agents are moving from one group to another (illustrated as jumps in the opinion space by the same agent), and ultimately results in the formation of a strong polarisation group along with some weak ones.

The outcome of the dynamics as presented in Fig. 2 resembles the outcome of a true opinion dynamics happened in OSNs and in societies in general. In

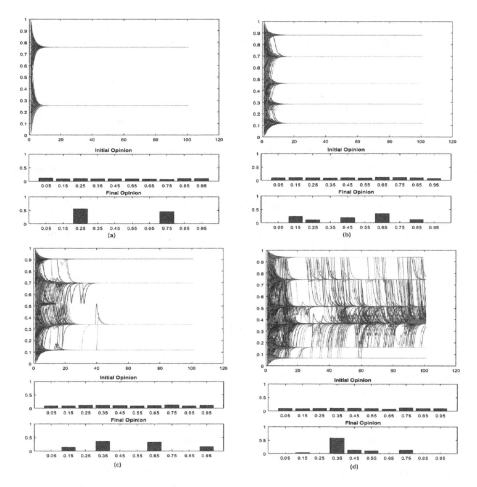

Fig. 2. Steady state outcome of the dynamics. (a) $\zeta = 2$, (b)–(c) $\zeta = 5$ and (d) $\zeta = 10$. In the figures, the first rows depict the traces of opinion changes for all participating agents, the second rows show the initial opinion distribution using histogram, whereas the third rows present the final opinion distribution

the presence of two strong schools of thought for any topics in the OSNs platform, people rarely converge to a global consensus value, as both opinions are supported by a large number of people. The debate may keep going for a long time. Even in the topic of discussion on the benefit and harm of the "Vaccination" debate, a large number of people in the OSNs argue against the necessity of child vaccination. In a democratic country with two major political parties, they have large number of supporters always opposing each other on most issue. With more than two political parties, or multiple competing opinions on a topic, each of them is also supported a portion of the society, as reflected by the results obtained for $\zeta = 5, 10$. Interestingly, some opinion groups might attract supporters in the early stages, however disappear in the long-run if it disseminate propaganda or untrue opinions. This is illustrated in Fig. 2 (c) (the opinion group formed in the middle).

3.2 Impact of Variable Clusters ζ_i

The results presented in this section were obtained for an agent population that have personalised clustering perception, i.e., the number of clusters present in the neighbouring opinion space perceived by each agent varies. For this, the number of predictable clusters ζ by agent i, here referred as ζ_i, are assigned randomly from a range $[\zeta_{min}, \zeta_{max}]$. It is different from the results presented in the prevision section, as for the earlier case, $\zeta_i = \zeta, \forall i$. For results presented in Fig. 3(a)–(c), the cluster range is $[2, 5]$, whereas for Fig. 3(d), the range is $[2, 10]$. The interpretation of this range is that, for range $[2, 5]$, an agent i can have a clustering ability between 2 to 5. Clustering ability of a particular agent remains the same throughout the simulation.

From the results, it is evident that the number of polarisation groups found in the final opinion distribution of the dynamics is less than ζ_{max}. For $\zeta_{max} = 5$ in Fig. 3(a)–(c), two very strong polarisations are formed as compared to results presented in Fig. 2(b)–(c) where more than two polarisation groups are formed, although their positions in the opinion space and the number of their constituting agents have been changed. Polarisation groups can be formed very quickly, with the agents merging to a consensus value for that polarization (both polarisation in Fig. 3(a), lower polarisation in Fig. 3(b) and upper polarisation in Fig. 3(c)). On the other hand, polarisation can be formed slowly, where the agents gradually merge towards a value (upper polarisation in Fig. 3(b) and lower polarisation in Fig. 3(c)). Sometimes a polarisation can start forming, but does not survive in the long-run. We have found another type of polarisation group that are small in size, but never be merged with a strong nearby group. The results for $\zeta_{max} = 10$ follow similar trend. The only difference is that the number of final polarisation groups are five instead of two found with $\zeta_{max} = 5$. The interesting finding for $\zeta_{max} = 10$ is that the dynamics are more unstable, as seen by agents moving between groups during the later stage of the dynamics.

The results are consistent with the outcome of real-world opinion formation dynamics. Although there might be multiple schools of thought at the start of any opinion formation dynamics, not all of them survive at the end. People start interacting and merging together to form strong group of opinions. Some opinions do not sustain in the long-run, if they cannot be substantiated by proper causes or evidences. On the other hand, a small group of opinions can survive if they are supported by an agent group who are reluctant to change their opinions at any cause.

3.3 Impact of Centrality

In this subsection, we present the impact of centrality on the proposed opinion formation dynamics. The number of predictable clusters ζ are kept same for all

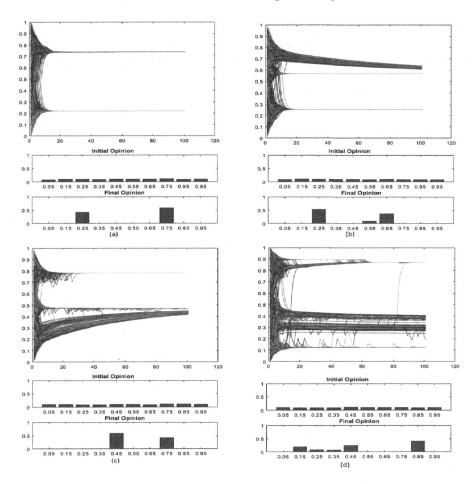

Fig. 3. Steady state outcome of the dynamics. (a)–(c) $[\zeta_{min}, \zeta_{max}] = [2, 5]$, (d) $[\zeta_{min}, \zeta_{max}] = [2, 10]$. In the figures, the first rows depict the traces of opinion changes for all participating agents, the second rows show the initial opinion distribution using histogram, whereas the third rows present the final opinion distribution

agents. To show the impact of centrality, the agents at one end of the opinion space (either 0 or 1 side) are assigned with higher centrality values, whereas the agents residing at the other end have lower centrality scores. Therefore, an opinion cluster formed with higher centrality agents has greater impact on an agent. The Figs. 4(a)–(b) present the simulation results with $\zeta = 2$, whereas Figs. 4(c)–(d) illustrate those for $\zeta = 5$. The left-side column of the figures depicts the results where the higher centrality nodes are placed at the '0' end of the opinion space, while for the right-side column central agents are assigned with opinions from the '1' end of the opinion space.

From the figures it is evident that the final opinions are divided into two polarisation groups. The size of the group formed near the high-centrality-agents end of the opinion space is larger than that formed at the other end. When, $\zeta = 2$, the dynamics becomes stable very quickly and two polarisation groups are well formed. On the contrary, for $\zeta = 5$, agents from weak polarisation group never stop moving towards the stronger group. Movements in the other direction also happen and the two polarisation groups never merge to a single opinion value. This is a consistent observation with the real world opinion dynamics. An opinion supported by people with higher degree centrality always gains popularity and adopted by a large number of followers. However, agents with less number of followers also survive with their opinions.

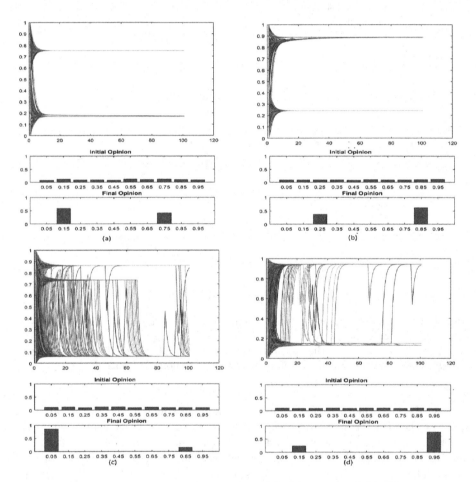

Fig. 4. Steady state outcome of the dynamics. (a)–(b) $\zeta = 2$ and (c)–(d) $\zeta = 5$. For (a) and (c) higher centrality nodes were placed at '0' side, whereas for (b) and (d) they are placed at '1' side of opinion space.

4   Conclusions

In this paper, we introduce an opinion formation model by considering the natural tendency of an agent to be influenced by the clusters formed in her neighbours' opinions. We emulate such natural tendency of a human being by adopting the well known K-means algorithm to determine the groups of opinions in an agent's neighbourhood. Instead of considering only the opinions within a predetermined distance threshold, an agent takes into account the influence of all opinion groups and chooses the most influencing one to interact with. Simulation results show that the clustering capability of agents has a considerable influence on the final opinion distribution. The number of polarisations depends on an agent's ability to perceive the number of opinion groups in her neighbourhood, the closeness of the group's opinions and the social power of the constituting agents. The persistent support of people on different schools of thought, the emergence and dissolution of opinions and the tendency of people to move between different opinions are some interesting natural phenomena captured by the proposed model.

References

1. Das, A., Gollapudi, S., Munagala, K.: Modelling opinion dynamics in social networks. In: Proceedings of the 7th ACM International Conference on Web Search and Data Mining, pp. 403–412 (2014)
2. Xia, H., Wang, H., Xuan, Z.: Opinion dynamics: a multidisciplinary review and perspective on future research. Int. J. Knowl. Syst. Sci. **2**(4), 72–91 (2011)
3. DeGroot, M.H.: Reaching a consensus. J. Am. Stat. Assoc. **69**(345), 118–121 (1974)
4. Clifford, P., Sudbury, A.: A model for spatial conflict. Biometrika **60**(3), 881 (1973)
5. Hegselmann, R., Krause, U.: Opinion dynamics and bounded confidence: models, analysis and simulation. J. Artif. Soc. Soc. Simul. **5**(3), 1–24 (2002)
6. Deffuant, G., Neau, N., Amblard, F., Weisbuch, G.: Mixing beliefs among interacting agents. Adv. Complex Syst. **3**(1), 87–98 (2000)
7. Moussaid, M., Kaemmer, J.E., Analytis, P.P., Neth, H.: Social influence and the collective dynamics of opinion formation. PLoS ONE **8**(11), e78433 (2013)
8. Hassan, R., Karmakar, G., Kamruzzaman, J.: Reputation and user requirement based price modelling for dynamic spectrum access. IEEE Trans. Mobile Comput. **13**(9), 2128–2140 (2014)
9. Das, R., Kamruzzaman, J., Karmakar, G.: Modelling majority and expert influences on opinion formation in online social networks. World Wide Web **21**(3), 663–685 (2018)
10. Etesami, S. R.: Hegselmann-Krause Opinion Dynamics in Finite Dimensions. Potential-Based Analysis of Social, Communication, and Distributed Networks, pp. 69–89 (2017)
11. Xie, Z., Song, X., Li, Q.: A review of opinion dynamics. In: Theory, Methodology, Tools and Applications for Modelling and Simulation of Complex Systems, pp. 349–357 (2016)
12. Freeman, L.C.: Centrality in social networks conceptual clarification. Soc. Netw. **1**(3), 215–239 (1978)

N2SkyC - User Friendly and Efficient Neural Network Simulation Fostering Cloud Containers

Aliaksandr Adamenko, Andrii Fedorenko, and Erich Schikuta(⊠)

University of Vienna, Währingerstr. 29, 1090 Vienna, Austria
alexadamenko@gmail.com, andriifedorenko@gmail.com,
erich.schikuta@univie.ac.at
http://www.cs.univie.ac.at

Abstract. Sky computing is a new computing paradigm leveraging resources of multiple Cloud providers to create a large scale distributed infrastructure. N2Sky is a research initiative promising a framework for the utilization of Neural Networks as services across many Clouds. This involves a number of challenges ranging from the provision, discovery and utilization of services to the management, monitoring, metering and accounting of the infrastructure.

Cloud Container technology offers fast deployment, good portability, and high resource efficiency to run large-scale and distributed systems. In recent years, container-based virtualization for applications has gained immense popularity.

This paper presents the new N2SkyC system, a framework for the utilization of Neural Networks as services, aiming for higher flexibility, portability, dynamic orchestration, and performance by fostering microservices and Cloud container technology.

Keywords: Problem solving environment · Neural networks
Cloud computing · Containers · Microservices

1 Introduction

The Cloud computing paradigm provides access to a set of computational power by aggregating heterogeneous resources and software and offering them as a single system composition. It hides the details of implementation and management of software and hardware from the end user. Cloud computing has been evolved from technologies like Cluster computing and Grid computing and has given rise to Sky computing [8].

Sky computing combines multiple Cloud-based infrastructures in such a way that the Sky providers aggregate the services scattered across several Clouds thus becoming the consumers of the Cloud providers. Sky computing in this way copes with the problem of vendor lock-in and extends the flexibility, transparency, and elasticity of the integrated infrastructure as compared to that of a single Cloud.

L. Cheng et al. (Eds.): ICONIP 2018, LNCS 11306, pp. 156–168, 2018.
https://doi.org/10.1007/978-3-030-04224-0_14

Sky computing has taken another step forward towards the realization of virtual collaborations, where resources are logical, and solutions are virtual. The exchange of information and resources among researchers is one driving stimulus for development. That is just as valid for the neural information processing community as for any other research community. As described by the UK e-Science initiative [2] several goals can be reached by the usage of new stimulating techniques, such as enabling more efficient and seamless collaboration of dispersed communities, both scientific and commercial.

In the course of our research we designed and developed N2SkyC, a microservice oriented Cloud container enabled neural network simulation environment, which is based on N2Sky [14], a virtual organization for the computational intelligence community. It provides access to neural network resources and enables infrastructures fostering multi Cloud resources. On the one hand, neural network resources can be generic neural network objects trained by a specific learning paradigm and training data for given problems whereas on the other hand these objects can also represent already trained neural networks, which can be used for given application problems. The vision of N2SkyC is the provisioning of a neural network problem solving virtual organization where any member of the community can access or contribute neural network objects all over the Internet.

In this paper, we present the new N2SkyC. The original system was developed as a monolithic application. That proved feasible for image-based virtualization mechanisms as in classical Cloud infrastructures. However, by the actual advent of containers and microservice architecture aiming for higher flexibility, portability, dynamic orchestration, and performance there was a need for a redesign of N2SKy to foster this new technology. By interpreting N2SkyC as a highly-distributed, service-based organizational body, a break-up of the monolithic system into various, self-contained N2SkyC microservices was done.

The layout of this paper is as follows: The state of the art of neural network simulators and the baseline research is given in Sect. 2. The architecture of N2SkyC is presented in Sect. 3 followed by a description of the use cases of the new N2SkyC platform in Sect. 4.2. The paper is closed by a summary and a short lookout for current and future research issues.

2   Related Work and Baseline Research

A virtual organization (VO) [4] is a temporary or permanent alliances for cooperation of human beings, organizations or enterprises that come together to share a wide variety of geographically distributed computational resources (such as workstations, clusters and supercomputers), storage systems, databases, libraries and particular purpose scientific instruments to present them as a unified integrated resource transparently [12].

Large number of neural network simulation environments have been developed during last years [11]. It started with systems for specific network families, as Aspirin/MIGRAINES [10], SOM-PAK [9]. Some systems aimed for a more comprehensive environment, as SNNS [17]. Resource management became a problem as well and numerous frameworks were developed to tackle this problem, such as JointDNN [3], SafetyNets [5], MCDNN [16]. With the advance of virtual resources by Grid and Cloud computing new collaborative environments motivated the authors of this paper to aim for an "everything about sharing" approach leading the way towards virtual collaborative organizations, as N2Grid [15] and N2Cloud [7]. In the course of our research, we designed and developed N2Sky [14], a virtual organization for the computational intelligence (CI) community, providing access to neural network resources and enabling infrastructures to foster federated Cloud resources. N2Sky is an artificial neural network provisioning environment facilitating the users to create, train, evaluate neural networks fostering different types of resources from Clouds of different affinity, e.g., computational power, disk space, networks, etc. N2Sky aroused strong interest even beyond the CI community[1].

To describe and identify neural network objects in N2Sky we developed ViNNSL (Vienna Neural Network Specification Language) [1]. ViNNSL allows for easy sharing of resources between the paradigm provider and the customers. ViNNSL is an XML-based domain specific language providing mechanisms to specify neural network objects in a standardized way by attributing them with semantic information. Initially, it was developed as a communication framework to support service-oriented architecture based neural network environments. Thus, ViNNSL is capable of describing the static structure, the training and execution phase of neural network objects in a distributed infrastructure, as Grids and Clouds.

3   N2Sky Architecture

Technically speaking, N2Sky is an artificial neural network simulator, which purpose is to provide different stakeholders with access to robust and efficient computing resource [14]. It was designed to provide natural support for Cloud deployment with distributed computational resources. However, the current N2Sky implementation is based on the Java programming language and deployed as a single monolithic application, which is not well aligned with the chosen paradigm. Main bottlenecks of such architecture are the inability to split the infrastructure workload according to the available computational resources within the "shared-nothing" infrastructure which leads to the inefficient usability of the system. That led us to a redesign of the N2Sky platform using microservices approach and the new technology stack for the Cloud infrastructure, which will allow us to utilize the benefits of Cloud computing to its full extent. This approach was not only used on the backend services but also on the frontend and its services. Microservices application architecture is a considerably new approach in software architecture, and like any new idea, it is evaluated in the variety of projects. The conclusions from practice are:

---

[1] http://cacm.acm.org/news/171642-neural-nets-now-available-in-the-Cloud/.

- It is simpler to develop and maintain small applications. However the complexity of such an approach does not disappear, it just transfers on the more abstract level: to the relationships between components.
- Microservices are easier to split between developers or other parties - as each part is not strongly bound to others, it is possible to work on the same project without a huge amount of collisions.
- Technological stack flexibility - each component can be implemented in a language that suits better the concrete task.
- Horizontal scalability - as services are small, it is cheaper and faster to deploy them horizontally. With a Cloud infrastructure, this feature becomes very important.

On the other hand, there are some drawbacks, which have to be considered:

- The complexity of the distributed system is higher than with a monolithic approach.
- Maintenance costs are also higher, due to the constant redeployment, monitoring, and logging.

There is no definitive answer about the feasibility of microservices architecture; it depends on the domain. But in the case of a Virtual Organization for computational science, where resources and scientists work together in parallel there is a necessity to have control over the executing environment. Thus, new components integration and deployment costs are lower, and researchers can allocate computational resources according to their demands without additional difficulties.

From the general point of view, the environment system components are divided into three main layers:

- Frontend UI,
- Orchestration tool layer, and
- Cloud platform layer.

In such architecture all communication with the Cloud infrastructure level is handled through the orchestration tool layer: spawning and stopping containers and virtual machines, uploading new containers, and updating them. That approach allows to focus on the deployment and maintenance process. In cases, when there is a need to edit and access internal components of the Cloud infrastructure directly, it can be done by the Cloud API. The Cloud platform provides infrastructure and environment for container deployment only, and it is not related to the current software application design. It provides great support for microservices paradigm implementation but implies no boundaries on the software developers.

The original N2Sky Java implementation is well designed: Java is a powerful programming language, which allows splitting the functionality between classes and maps them to the corresponding services of the platform. Source code and interaction between components are part of the same monolithic application.

Even that the application is well designed it creates the additional level of bindings between components of the software.

To redesigning the original architecture according to the microservices paradigm, it was necessary to perform additional steps:

- Decomposition of the business logic into microservices,
- Design storage according to data, stored by services,
- Neural network data archive: database with data, which is used to train and test neural networks, and
- Define endpoints and API for each service.

As a result, apart from technological and infrastructure improvements, the following changes to the main components of the system were performed:

- Neural network training and Neural network evaluation components are now separate microservices, which are run on demand. Resource allocation in such case can be performed on the orchestration layer, according to the computational cost of the task,
- Neural network data archive is a separate service on top of the database. It is not associated with a single database, as was in the previous implementation. It provides a specific API to add any external data sources,
- N2Sky Database has its own database storage,
- Business model module is split into two different parts: one part is responsible for the internal business logic of the application, and the second part is responsible for infrastructure management and handled by the orchestration component. Internal business logic functionality is also encapsulated into microservices.

The current working version of the new architecture with the described changes is presented in Fig. 1.

By implementing architectural changes to both infrastructure and software levels of the N2Sky platform, we achieve higher scalability and make the platform more suitable for usage in a Cloud environment. The shift to new technologies, as container base execution environment and orchestration tools for load and deployment management, increase efficiency and convenience of N2Sky neural network platform rigorously.

4   User-Centered Interface Based on Modular Frontend

The basis of N2SkyC is the user-centered design. Moving from the complex monolithic design to a more comprehensible one we fostered past user experiences. Thus, a fundamental requirement was to gain the functionality and increase the performance of the application.

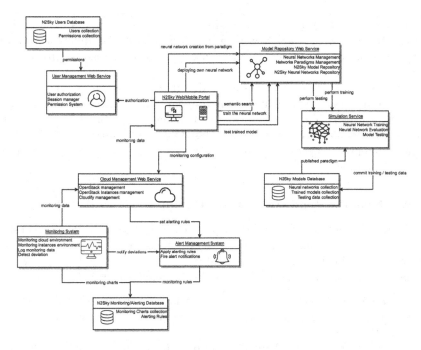

Fig. 1. New N2Sky architecture

4.1 User Role as an Independent User Interface

The N2SkyC web interface, as well as the mobile portal, provide an intuitive user interface. Since N2SkyC supports different knowledge level users it was decided to differentiate between various user roles. So, every user role has its own user interface, which deeply reflects the specific user needs:

- *Arbitrary User.* The Arbitrary User must not have deep knowledge of the neural network field or programming language skills. Her main goal is to find a suitable neural network for a given problem. Thus she uses already existing, trained neural networks and evaluates them.
- *Neural Network Engineer.* The Neural Network Engineer is allowed to create new neural network resources based on existing paradigms. She has access to her own dashboard and publicly available resources on the main application module. She can perform a semantic search for available neural network paradigms and use them. This user can create own neural network instances from existing neural network paradigms and can also train the running neural network instances and evaluate the trained models. She can share her trained neural networks by making them public.
- *Contributor.* The Contributor is an expert user, which has enough knowledge and experience to develop his own new neural network paradigm by using the ViNNSL template schema [1] and publish them on N2SkyC. This user can deploy neural networks on the N2SkyC environment as well as on her

own environment by providing training and testing endpoints. The goal of the contributor is to study how networks will behave with different network structures, input parameters and training data that can also be provided by other users.

– *System Administrator.* The System Administrator is a user who has full access to the application including environment management, monitoring and alerting features. The administrator can manage OpenStack and Cloudify instances. She also can shadow any N2SkyC user to observe the application from different perspectives. The administrator has access to all dashboards in every module.

4.2 Modular Frontend Application Design

Since N2SkyC supports microservices approach in the backend it was a design decision to apply the same approach on the frontend too.

Microservices in the frontend are small independent web applications, which are consolidated into one application. The main benefits of this approach are:

– *Maintainability.* It is possible to divide the application between different teams. Developers do not even need to have some knowledge about other parts of the application.
– *Diversity of technologies.* Monolithic approach makes the whole application stick to one framework. Microservices allow using any technology without the need to rewriting the application.
– *Independent deployment.* Every application has some releases periods, every release is accompanied by redeployment procedure. There is no need to redeploy the whole application, but just only the required components.

The microservices approach in frontend application is shown in "Fig. 2". It breaks the whole application into the following small micro applications:

– *Shell.* Is a top-level component, which wraps Components picker and Container for the component. It contains application specific configuration.
– *Component picker.* Is a router, which manages the micro applications.
– *Container for Component.* Container, where the component will be injected.
– *Micro Application.* Independent application, which can be written in any programming language, but has to use one of the shards.
– *Shards.* Is a code base, which is shared between Micro Applications. Shards can have multiple levels.

5   A Sample Workflow

To support Software as a Service (SaaS) distribution every web service can work independently. The different user roles can use N2SkyC both via the web portal and the N2SkyC API directly. The N2SkyC API allows to:

– Authorise users in the System.
– Create new neural network objects from existing paradigm.

- Deploy own neural network objects on N2SkyC environment.
- Perform training against own as well published neural network objects.
- Perform testing and evaluating trained models.

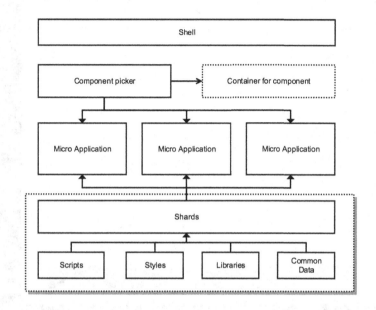

Fig. 2. Microservices approach in frontend application

Sample workflow overviews for different user roles show the microservices architecture in action, see Fig. 1:

1. *Contributor*
   (a) The Contributor authorizes via the N2SkyC portal using a browser on desktop PC or mobile device. She is redirected to her own dashboard according to permissions, which will be received from User Management Web Service.
   (b) The user described her own neural network paradigm using the ViNNSL template and deploying it in the N2SkyC Cloud as it shown in Fig. 3.
   (c) The Contributor performs training of her neural networks. Since she is an expert she can perform this operation using Simulation Service via Model Repository Web Service API.
   (d) The user publishes her paradigm via N2SkyC UI or available API.
   (e) The Contributor awaits that other N2SkyC users will use her neural network paradigm in order to monitor the behavior of the neural network.
   (f) The user modifies, redeploys and retrains her neural network after first results.

Fig. 3. The user uploads the neural network paradigm description in ViNNSL format

2. *Neural Network Engineer*
   (a) The Neural Network Engineer authorizes via the N2SkyC portal using a browser on a desktop PC or mobile device. She is redirected to her own dashboard according to permissions, which will be received from User Management Web Service.
   (b) From the dashboard, the user creates a neural network from existing paradigms using Model Repository Web Service. The user specifies the neural network structure as shown in Fig. 4.
   (c) The user performs training against her newly created neural network using the N2SkyC platform as shown in Fig. 5.
   (d) If the user is satisfied with a trained model, she can perform data analysis using N2SkyC.
   (e) The neural network engineer user publish her neural network and trained model in order to make it available to other N2SkyC users.
3. *Arbitrary User*
   (a) The Arbitrary User authorizes via the N2SkyC portal using a browser on a desktop PC or mobile device. She will be redirected to her own dashboard

according to permissions, which will be received from User Management Web Service.

(b) She performs a semantic search in order to find possible neural network as well as trained models according to her needs, see Fig. 6.

(c) The user copies existing neural networks and trained models into her project.

(d) The user performs training of copied neural networks with default input parameters data.

(e) The user evaluates trained neural network models with default parameters.

4. *System Administrator*

(a) The System Administrator authorizes via the N2SkyC portal using her browser on a desktop PC or mobile device. She will be redirected to the administration dashboard.

(b) The user observes the Cloud environment.

(c) She creates new monitoring charts showing specific metrics and adds it to the administration dashboard.

(d) She creates alerts for newly created monitoring

(e) The user is notified by the Alert Management System, if some events occur.

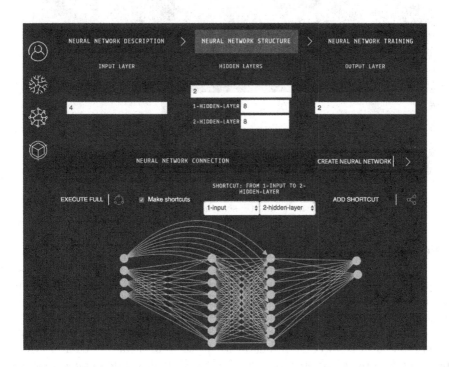

Fig. 4. The user specifes the neural network structure

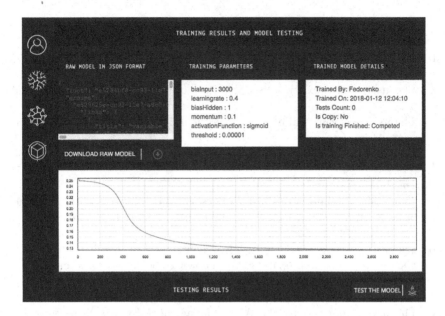

Fig. 5. The user performs neural network training

Fig. 6. The user performs semantic search in Neural Network Repository

6 Conclusion and Future Work

We presented the novel microservice based N2SkyC architecture, which aims for increased extensibility, portability, dynamic orchestration and performance fostering Cloud container technology.

Tied to the presented new architectural design is also a shift of the application domain of N2SkyC from neural networks to arbitrary machine learning operators. Hereby a new specification of ViNNSL is under development, which aims for compatibility with PMML (Predictive Model Markup Language) [6].

A further new development of N2SkyC is N2Query [13], which allows the semantic discovery of N2SkyC services through a natural language querying mechanism using ontology alignment mechanisms.

The novel N2SkyC system and its documentation can be accessed and downloaded from http://www.wst.univie.ac.at/projects/n2skyc.

References

1. Beran, P.P., Vinek, E., Schikuta, E., Weishaupl, T.: Vinnsl - the Vienna neural network specification language. In: IEEE International Joint Conference on Neural Networks IJCNN 2008, IEEE World Congress on Computational Intelligence, pp. 1872–1879. IEEE (2008)
2. e-Science: UK e-science programme (2016). http://www.escience-grid.org.uk. Accessed Jan 2018
3. Erfan Eshratifar, A., Saeed Abrishami, M., Pedram, M.: JointDNN: An Efficient Training and Inference Engine for Intelligent Mobile Cloud Computing Services, University of Southern California (2018)
4. Foster, I., Kesselman, C., Tuecke, S.: The anatomy of the grid: enabling scalable virtual organizations. Int. J. High Perform. Comput. Appl. $15(3)$, 200–222 (2001)
5. Ghodsi, Z., Gu, T., Garg, S.: SafetyNets: Verifiable Execution of Deep Neural Networks on an Untrusted Cloud, New York University (2017)
6. Guazzelli, A., Zeller, M., Lin, W.C., Williams, G., et al.: PMML: an open standard for sharing models. R J. $1(1)$, 60–65 (2009)
7. Huqqani, A.A., Li, X., Beran, P.P., Schikuta, E.: N2Cloud: cloud based neural network simulation application. In: The 2010 International Joint Conference on Neural Networks (IJCNN), pp. 1–5. IEEE (2010)
8. Keahey, K., Tsugawa, M., Matsunaga, A., Fortes, J.: Sky computing. IEEE Internet Comput. $13(5)$, 43–51 (2009)
9. Kohonen, T., Hynninen, J., Kangas, J., Laaksonen, J.: Som pak: The self-organizing map program package. Report A31, Helsinki University of Technology, Laboratory of Computer and Information Science (1996)
10. Leighton, R.R., Wieland, A.: The aspirin/migraines software tools, user's manual. Technical Report MP-91W00050 (1991)
11. Prieto, A., et al.: Neural networks: an overview of early research, current frameworks and new challenges. Neurocomputing 214, 242–268 (2016)
12. Schikuta, E., Fuerle, T., Wanek, H.: ViPIOS: the vienna parallel input/output system. In: Pritchard, D., Reeve, J. (eds.) Euro-Par 1998. LNCS, vol. 1470, pp. 953–958. Springer, Heidelberg (1998). https://doi.org/10.1007/BFb0057953
13. Schikuta, E., Magdy, A., Mohamed, A.B.: A framework for ontology based management of neural network as a service. In: Hirose, A., Ozawa, S., Doya, K., Ikeda, K., Lee, M., Liu, D. (eds.) ICONIP 2016. LNCS, vol. 9950, pp. 236–243. Springer, Cham (2016). https://doi.org/10.1007/978-3-319-46681-1_29
14. Schikuta, E., Mann, E.: N2Sky - neural networks as services in the clouds. In: The 2013 International Joint Conference on Neural Networks (IJCNN), pp. 1–8. IEEE (2013)
15. Schikuta, E., Weishaupl, T.: N2Grid: neural networks in the grid. In: Proceedings 2004 IEEE International Joint Conference on Neural Networks, vol. 2, pp. 1409–1414. IEEE (2004)

16. Shen, H., Philipose, M., Agarwal, S., Wolman, A.: MCDNN: An Execution Framework for Deep Neural Networks on Resource-Constrained Devices, University of Washington and Microsoft Research (2014)
17. Zell, A., et al.: SNNS (stuttgart neural network simulator). In: Neural Network Simulation Environments, pp. 165–186. Springer, Boston (1994). https://doi.org/10.1007/978-1-4615-2736-7_9

Question Rewrite Based Dialogue Response Generation

Hengrui Liu[1], Wenge Rong[1(✉)], Libin Shi[2], Yuanxin Ouyang[1],
and Zhang Xiong[1]

[1] School of Computer Science and Engineering, Beihang University, Beijing, China
[2] Sino-French Engineer School, Beihang University, Beijing, China
{14061024,w.rong,libin.shi,oyyx,xiongz}@buaa.edu.cn

Abstract. Dialogue response generation is a fundamental technique in natural language processing, which can be used in human-computer interaction. As the quick development in neural networks, the sequence to sequence (seq2seq) model which employed recurrent neural networks (RNN) encoder-decoder has archived great success in machine translation. Many researchers began to apply this model in dialogue response generation. However, the conventional seq2seq model counters several problems, e.g., grammatical mistake, safe response and etc. In this paper, motivated by the great success of generative adversarial networks (GANs) in generating images, we propose an improved seq2seq framework by employing GANs to rewrite questions in order to retrieve more information from the question. Afterwards we combine the original question and the rewritten question together to generate responses. The experiments on the public Yahoo! Answers dataset demonstrated the proposed framework's potential in dialogue response generation.

Keywords: Dialogue generation · Generative adversarial networks
Question rewriting

1 Introduction

Dialogue response generation can be used in many fields of human-computer interaction, e.g. order a task or offer tutoring. The goal of dialogue response generation is to generate reasonable responses to a question. Earlier widely used dialogue models are normally based on a large corpus of templates and a large database called knowledge base. However, these models might not generate suitable responses if questions are not included in the templates. Also, they can not generate reasonable responses for open-ended questions [2]. Another kind of models called retrieve based models was proposed to learn responses from historical data [23]. These models' common method is to calculate the similarity of question and responses. However, these models are also not capable of generating responses when questions are not shown in historical data.

© Springer Nature Switzerland AG 2018
L. Cheng et al. (Eds.): ICONIP 2018, LNCS 11306, pp. 169–180, 2018.
https://doi.org/10.1007/978-3-030-04224-0_15

Nowadays, with the development of idea of deep learning, many methods based on artificial neural networks (ANNs) have been proposed, among which sequence to sequence (seq2seq) model [22] with attention mechanism [1,13] has achieved great improvement in neural machine translation (NMT). Seq2seq model generally contains an encoder and a decoder based on recurrent neural networks (RNN) to build a general end-to-end language model with fewer restrictions. As such researchers begin to use these models in dialogue response generation. However, it is widely found that these dialogue models tend to generate simple and safe responses, which means these responses may be weakly correlated to the question.

There are several possible reasons for these phenomena and one of them is that the information in the question may be not sufficient [25]. Many questions contain the same structure such as "What is ...", "How can ..." etc, which makes language model not able to obtain enough information to generate responses. Furthermore, though RNN is a powerful structure to model sentence, it is more suitable for predicting because of its recurrent compute mode [15]. However, the meaning of sentences can be changed by a single word, which is hard for RNN to capture the critical information in the question.

To relieve this problem, many methods have been proposed. For example, Li et al. [11] tried to avoid generating simple and safe responses by penalizing these responses' generation probabilities. Xing et al. [25] tried to bring more information by using topics which related to the question. In this work, we also want to bring more information when generating responses. Specifically, we proposed question rewriting method to retrieve more information from questions by using generative adversarial networks (GANs) [4] to rewrite questions, which is an unsupervised learning framework. It contains a generator and a discriminator. The generator is responsible to generate rewritten questions with the input of original question. A discriminator is responsible to distinguish whether the rewritten question is semantically close to the original question. There are no target sequences and the direction of training generator is led by the discriminator, which give feedback to the generator by distinguishing the original questions (real sample) and rewritten questions (fake sample). Afterwards we combine the original question and rewritten question as the response generating input to generate responses.

Convolution neural network (CNN) is able to capture local features of data [14]. This structure can enhance the local features by pooling operation. Motivated by the different advantages and limitations in RNN and CNN, we further propose to improve seq2seq framework in GAN's generator by using both the RNN and CNN in the encoding phase. In our encoder, RNN is responsible to generate the initial state of the decoder and the encoded context by timesteps. CNN is responsible to generate a global encoded context. We expect that encoder can capture more comprehensive information from the input questions so that the rewritten question can contain more useful information. Our experimental study demonstrates that the rewriting method incorporated with the improved seq2seq model has the potential to generate more reasonable responses.

The rest of this paper is organized as follows. Section 2 introduces related work about dialogue generation. Section 3 describes the details of our proposed framework, including the network structure of rewriting framework and the improved seq2seq framework. The experimental study is presented in Sects. 4 and 5 concludes this paper and indicates some possible future research directions.

2   Background

When we treat dialogue response generation as a seq2seq task, it is similar to a machine translation task where we give the model a sentence and the model generate a translated sentence. But dialogue response generation is a more difficult task because the mapping relationship is more complicated [19]. In general, seq2seq model has several challenges in terms of safe response and weakly related responses [26]. There are many ideas proposed to solve these problems which can be roughly divided into two category. The first one is to optimize the seq2seq structure. For example, Li et al. [11] proposed a model to maximize the mutual information between the response and the question which use a term in lost function to penalize the safe responses but this model is difficult to train because penalization term is difficult to tract. Similarly Shao et al. [21] employed the attention mechanism in the decoder to generate longer sentences.

Another direction is to bring more information into the response generation process. Seq2BF model [16] used two seq2seq models in forward decoding and backward decoding. The model will choose a keyword when generating response as such the safe response problem will be relieved. A similar work by Xing et al. [25] used the twitter-LDA model to find topic words and then use a joint attention to generate response with topic words. Also, alternative information like emotions [8] can be used in generating responses.

Question rewrite is a popular field in web search which can bring more information by rewriting the query. In web search, people can use question rewriting to improve search ranking performance and an example of such methods is statistical machine translation (SMT) [18]. There are some other rewrite models relying on deep learning methods, such as encoder-decoder architecture [5]. These methods can generate a list of questions which will be selected by the learning-to-rank algorithm. However, these methods are mainly based on the frequency of word terms, therefore semantic information should also be considered.

Deep generative models have the ability to learning from large unlabeled data. The semantic information can be caught when using these model to generate sentences [12]. But many popular generative models such as deep belief nets (DBN) [6], variational autoencoder (VAE) [10] are suffered from the intractable probabilistic computations. Goodfellow et al. proposed GAN to overcome this challenge [4], which plays a minmax game between a generator and a discriminator and this framework has archived great success in computer vision field [3]. However GAN is hard to apply in generating text because of the non-differentiability of natural language words. There are two kinds of methods proposed to solve this problem, one method is to change the non-differentiable function to an approximate differentiable function [28], another is to adopt reinforcement learning method [27]. In this paper, we adopt policy gradient given by discriminator to guide to training process of the generator.

3   Methodology

3.1   Question Rewriting Framework

The proposed question rewrite based dialogue response generation framework is depicted in Fig. 1, where the left part in dotted rectangle is the detail architecture of the generator.

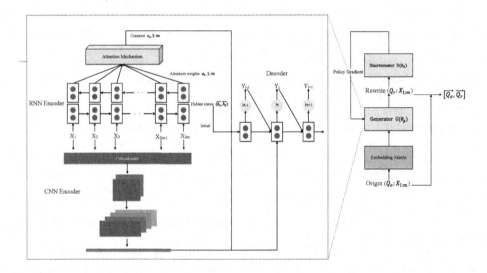

Fig. 1. Improved response generating framework

Generator. Generator is used in this paper to improve seq2seq model, whose task is to estimate the probability of Y by given X:

$$
\begin{aligned}
p(Y|X) &= \prod_{t=1}^{n} p(y_t|x_1, x_2, ..., x_n, y_1, y_2, ..., y_{t-1}) \\
&= \prod_{t=1}^{n} \frac{exp(f(h_t - 1, c_{yt}))}{\sum_y exp(f(h_{t-1}, c_y))}.
\end{aligned}
\tag{1}
$$

In the forward pass, Generator G encodes the input question $Q_o(x_1, x_2, ..., x_m)$ into m hidden states, where m indicates the length of input question. We improved the basic seq2seq model by employing RNN and CNN to joint encode input question which will be used both in question rewriting and response generating. Its architecture is shown in the right side of Fig. 1. The RNN encoder of the generator adopted bidirectional RNNs [20] which include a forward RNN and a backward RNN. The forward RNN reads input question from begin to end $(X_1, X_2, ..., X_m)$, then outputs hidden states $(\overrightarrow{h_1}, \overrightarrow{h_2}, ..., \overrightarrow{h_m})$. The backward RNN reads input question in reverse order $(X_m, X_{m-1}, ..., X_1)$ and outputs hidden states $(\overleftarrow{h_1}, \overleftarrow{h_2}, ..., \overleftarrow{h_m})$. Finally, we concatenate the forward hidden states and the backwards hidden states as our final hidden states of encoder, i.e., $h_t = [\overrightarrow{h_t}, \overleftarrow{h_t}]$. The hidden states of encoder is wrapped by attention mechanism to get context vector c_t. In order to relieve the gradient vanishing/exploding problem, we change the RNN units to Long Short-Term Memory (LSTM) [7] units.

CNN is known for its ability to capture local feature and has been widely used in image processing. Also, it is widely used in text classification [9], NMT [14]. Here we also adopt CNN as part of the encoder. For instance, suppose the input X's length is L, after word embedding operation, the input sequence is $S \in \mathbb{R}^{L \times D}$, where D is the dimension of word vectors. We adopt S as the input of CNN encoder:

$$S = [s_1 \oplus s_2 \oplus ... \oplus s_L], \tag{2}$$

where s_i is word vector. The filter size is $l \times D$. l represents that choosing l words to do convolution operation:

$$c_i = \sigma(wx_{i:i+l-1} + b), \tag{3}$$

where w, b are weight parameters, c_i is feature vector get from convolution operation. Suppose stride length is 1, we can get feature map $C = [c_1, c_2, ..., c_{L-l+1}] \in \mathbb{R}^{L-k+1 \times D}$. We can get the different number of feature map by adjusting the number of filters. Also, we adopt max-pooling operation between the two convolution operation. Different from RNN encoder, CNN encoder outputs a global context vector $c \in R^{1 \times N}$ where N is the feature dimension. This context vector will participate in every timestep in decode phase. The input of the decoder's LSTM units will be:

$$x_t = [y_{t-1}, c_t, c_g], \tag{4}$$

where c_t is RNN encoder's context vector in timestep t wrapped by attention mechanism, similar to the encoder in question rewriting framework's generator. c_g is the global context vector produced by CNN encoder.

When generating text at decoder stage, we use LSTM as the basic units. The decoder is responsible to estimate the joint probability over each predicted output token:

$$G(y|x) = \prod_{t=1}^{m} p(y_t|y_{<t}, x). \tag{5}$$

For instance, at every timestep, the decoder will output the probability distribution of the vocabulary. Then we use *argmax* function to get the most likely words:

$$h_t^{dec} = LSTM_{dec}(y_{t-1}, h_{t-1}^{dec}, c_t),$$
$$p(y_t|y_{<t}, x) = softmax(W s_t + b), \qquad (6)$$
$$y_t = argmax(p(y_t|y_{<t}, x)),$$

where h_t^{dec} are hidden states of decoder cells.

Discriminator. In this research we employed CNN as discriminator. The discriminator's task is to distinguish whether an input question is original question or generated question. We can consider it as a classification task. The structure is shown in Fig. 2.

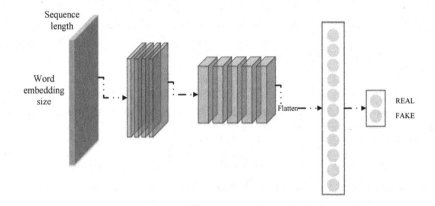

Fig. 2. Discriminator structure

3.2   Policy Gradients Training

After softmax operation, y_t is not differentiable, we can leverage the policy gradients to update the parameters of generator θ_g. Policy gradients are widely used in reinforcement learning training [24]. Here we consider generator as a policy network and the loss (reward) function can be defined as:

$$\sum_{y \in P_{\theta_g}} Q(y)G(y|x), \qquad (7)$$

where $Q(y)$ is action-value function which is the rewards of rewritten question generated by G. We define it as:

$$Q(y) = log(1 - D(y)). \qquad (8)$$

In GANs, generator needs to fool discriminator D, in other words, maximize $D(y)$. So we can change it to minimize L. The gradients of L is calculated as follows:

$$\nabla_{\theta_g} L = \sum_{y \in P_{\theta_g}} [\nabla_{\theta_g} Q(y) G(y|x) + Q(y) \nabla_{\theta_g} G(y|x)]$$

$$= \sum_{y \in P_{\theta_g}} Q(y) \nabla_{\theta_g} log(G(y|x)) \qquad (9)$$

$$= \mathbb{E}_{y \sim P_{\theta_g}} log(1 - D(y)) \nabla_{\theta_g} log(G(y|x)).$$

Then we can update the θ_g by:

$$\theta_g = \theta_g - \alpha \cdot \nabla_{\theta_g} L. \qquad (10)$$

4 Experimental Study

4.1 Experiment Configuration

We conduct our experiment on a subset of Yahoo! Answers corpus[1]. This dataset includes 4,483,032 questions and their candidate answers. Because of the limitation of computing resources, we randomly choose one category as our experiment dataset which contains 9,984 questions and answers. We use the best answer of each question as our expected answer. When processing this dataset, we use following rules: 1. The maximum length of questions and answers are limited to 50 and 20, which cover 85.35% questions and 73.95% answers. 2. Remove all special characters such as "!#$%&()*+,-./:;<=>?@^_'{—}~". 3. All alphabetical characters in the dataset have been converted to lowercase. We randomly shuffle the whole dataset and divided the dataset into training set, dev set and test set by 8:1:1.

The word embeddings used in this experiment is randomly initialized with truncated normal distribution. The dimension is 512. All LSTM cells used in the experiment with size 512. As for CNN encoder, we have 12 kinds of filters and the maximum filter numbers for a layer is 200. Furthermore, we employed Adam optimization in gradient descent, which is provided with the advantages of RMSProp and momentum. The learning rate is set to 0.0005.

We conduct the experiment with several baselines for comparison against our framework and also test whether their performance will be improved by rewriting input questions[2].

4.2 Evaluation Metric

The evaluation of response generation is an open problem so we follow the existing work and adopt the following metrics:

---

[1] https://webscope.sandbox.yahoo.com/catalog.php?datatype=l&guccounter=1.

[2] The source code is available at https://github.com/HenryL-study/GAN-for-Question-Rewrite.

BLEU. BLEU [17] is widely used in NMT and this score evaluates the similarity between the generate sentence and target sentence. In our experiment, we calculate BLEU-1, BLEU-2, BLEU-3, BLEU-4 for each model. It can be calculated by:

$$BLEU = BP \cdot exp(\sum_{n=1}^{N} w_n log p_n),$$

$$BP = \begin{cases} 1 & if\ l_c > l_s \\ exp(1 - \frac{l_s}{l_c})) & if\ l_c \le l_s \end{cases}. \tag{11}$$

Distinct-1 & Distinct-2. These metrics calculate the total distinct unigrams and bigrams in the generated response. Following the previous work [11], we divided the numbers by the total unigrams and bigrams. This score can evaluate how diverse the generate response are. The higher number indicates that the generated responses may have more information.

Perplexity. Perplexity is normally used to measure how well the model predicts a response. It is defined as Eq. 12. The lower the perplexity score are, the better generating performance the model will has.

$$PPL = exp(-\frac{1}{N} \sum_{i=1}^{N} log(p(Y_i))). \tag{12}$$

Human Evaluation. Besides these automatic metrics mentioned above, we also employed human to judge the quality of generating responses. We randomly shuffled all generated responses by experiment models and send them to human labelers. The score range is from 1 to 5. 1 indicates that the generated response is totally irrelevant to the question or have severe grammar mistake. 5 indicates that the generated response is not only relevant to the question, but also informative.

4.3 Results and Discussion

Tables 1 and 2 shows the comparison results of the proposed framework against different baselines on the dataset.

Table 1 shows all BLEU scores from different models. It is clear that rewritten questions will improve the performance of all models. Also our proposed framework obtains the highest score in this evaluation metric. Table 2 shows other evaluation metrics result. Our framework with question rewrite gets the highest human evaluation score. Also, our framework without question rewriting archives best scores in distinct-1 & distinct-2. If we adopt question rewriting methods, it will decrease the perplexity score and increase the human evaluation score.

Table 1. BLEU score

Model	BLEU1	BLEU2	BLEU3	BLEU4
RNN encoder	2.45	0.87	0.33	0.16
RNN encoder + question rewrite	4.09	1.43	0.73	0.46
CNN encoder	3.16	0.92	0.36	0.17
CNN encoder + question rewrite	3.73	1.21	0.66	0.48
RNN+CNN encoder	4.17	1.51	0.51	0.19
RNN+CNN encoder + question rewrite	4.93	1.93	0.98	0.61

Table 2. Results on other metrics

Model	distinct-1	distinct-2	Perplexity	Human
RNN encoder	1.51%	2.27%	346.78	1.42
RNN encoder + question rewrite	1.17%	1.97%	325.37	1.68
CNN encoder	1.42%	2.45%	355.06	1.41
CNN encoder + question rewrite	1.32%	2.29%	356.12	1.57
RNN+CNN encoder	3.46%	7.42%	372.28	1.64
RNN+CNN encoder + question rewrite	2.00%	4.07%	351.72	1.70

When we look at the generated responses, it is found that when we use the rewritten questions and the original questions to generate responses, the responses will be longer than responses generated by only using with original questions. Also, responses generated with original questions and rewritten questions have more nouns rather than "OK", "It is a question", which may cause higher BLEU scores and human evaluation scores. As responses are longer, the distinct-1 & distinct-2 scores will be lower, a possible reason is that when responses are longer, the frequency of common words will be higher.

But longer responses will be more natural and have less grammar mistakes, that is also part of reason why human evaluation score is higher.

We also visualize the attention mechanism to understand the how the rewritten question will help the generating process. As we can see in Fig. 3, the words in rewritten question like "implement" and "attachment" got more attention in generating response.

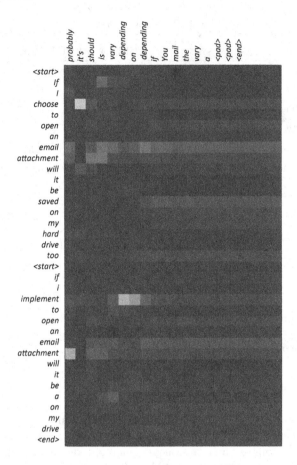

Fig. 3. Attention mechanism visualization

5   Conclusion and Future Work

In this research we propose a framework which use GAN to rewrite questions and use the joint encoder to encode questions which can get a better performance in response generating compared with the existed baselines. To relieve the non-differentiable problem in GAN's training process, we adopt policy gradient method. We also find the rewritten questions can improve the performance of baseline models which indicates rewritten questions can bring more information when generating responses. In response generating, except RNN encoder, we use CNN encoder to get a global context in order to fetch more local information.

Concerning the future work, we prepare to use other methods to build the rewriting framework. For example, we could apply the Bi-LSTM and CNN into the generator of the GAN. Furthermore, we will project to design more powerful and flexible generator model in our generating framework. We can modify the attention mechanism in response generating phase. It is also interest to consider using CNN in decoder part.

Acknowledgments. This work was partially supported by the National Natural Science Foundation of China (No. 61332018).

References

1. Bahdanau, D., Cho, K., Bengio, Y.: Neural machine translation by jointly learning to align and translate. CoRR abs/1409.0473 (2014)
2. Chen, H., Liu, X., Yin, D., Tang, J.: A survey on dialogue systems: recent advances and new frontiers. SIGKDD Explorations **19**(2), 25–35 (2017)
3. Denton, E.L., Chintala, S., Fergus, R., et al.: Deep generative image models using a laplacian pyramid of adversarial networks. In: Proceedings of 2015 Annual Conference on Neural Information Processing Systems, pp. 1486–1494 (2015)
4. Goodfellow, I., et al.: Generative adversarial nets. In: Proceedings of 2014 Annual Conference on Neural Information Processing Systems, pp. 2672–2680 (2014)
5. He, Y., Tang, J., Ouyang, H., Kang, C., Yin, D., Chang, Y.: Learning to rewrite queries. In: Proceedings of the 25th ACM International Conference on Information and Knowledge Management, pp. 1443–1452 (2016)
6. Hinton, G.E., Salakhutdinov, R.R.: Reducing the dimensionality of data with neural networks. Science **313**(5786), 504–507 (2006)
7. Hochreiter, S., Schmidhuber, J.: Long short-term memory. Neural Comput. **9**(8), 1735–1780 (1997)
8. Huang, C., Zaïane, O.R., Trabelsi, A., Dziri, N.: Automatic dialogue generation with expressed emotions. In: Proceedings of the 2018 Conference of the North American Chapter of the Association for Computational Linguistics: Human Language Technologies, pp. 49–54 (2018)
9. Kim, Y.: Convolutional neural networks for sentence classification. In: Proceedings of the 2014 Conference on Empirical Methods in Natural Language Processing, pp. 1746–1751 (2014)
10. Kingma, D.P., Salimans, T., Józefowicz, R., Chen, X., Sutskever, I., Welling, M.: Improving variational autoencoders with inverse autoregressive flow. In: Proceedings of 2016 Annual Conference on Neural Information Processing Systems, pp. 4736–4744 (2016)
11. Li, J., Galley, M., Brockett, C., Gao, J., Dolan, B.: A diversity-promoting objective function for neural conversation models. In: Proceedings of the 2016 Conference of the North American Chapter of the Association for Computational Linguistics, pp. 110–119 (2016)
12. Li, J., Monroe, W., Shi, T., Jean, S., Ritter, A., Jurafsky, D.: Adversarial learning for neural dialogue generation. In: Proceedings of the 2017 Conference on Empirical Methods in Natural Language Processing, pp. 2157–2169 (2017)
13. Luong, T., Pham, H., Manning, C.D.: Effective approaches to attention-based neural machine translation. In: Proceedings of the 2015 Conference on Empirical Methods in Natural Language Processing, pp. 1412–1421 (2015)
14. Meng, F., Lu, Z., Wang, M., Li, H., Jiang, W., Liu, Q.: Encoding source language with convolutional neural network for machine translation. In: Proceedings of the 53rd Annual Meeting of the Association for Computational Linguistics and the 7th International Joint Conference on Natural Language Processing of the Asian Federation of Natural Language Processing, pp. 20–30 (2015)

15. Mikolov, T., Karafiát, M., Burget, L., Cernocký, J., Khudanpur, S.: Recurrent neural network based language model. In: Proceedings of 11th Annual Conference of the International Speech Communication Association, pp. 1045–1048 (2010)

16. Mou, L., Song, Y., Yan, R., Li, G., Zhang, L., Jin, Z.: Sequence to backward and forward sequences: a content-introducing approach to generative short-text conversation. In: Proceedings of 26th International Conference on Computational Linguistics, pp. 3349–3358 (2016)

17. Papineni, K., Roukos, S., Ward, T., Zhu, W.: BLEU: a method for automatic evaluation of machine translation. In: Proceedings of the 40th Annual Meeting of the Association for Computational Linguistics, pp. 311–318 (2002)

18. Riezler, S., Liu, Y.: Query rewriting using monolingual statistical machine translation. Comput. Linguist. **36**(3), 569–582 (2010)

19. Ritter, A., Cherry, C., Dolan, W.B.: Data-driven response generation in social media. In: Proceedings of the 2011 Conference on Empirical Methods in Natural Language Processing, pp. 583–593 (2011)

20. Schuster, M., Paliwal, K.K.: Bidirectional recurrent neural networks. IEEE Trans. Sig. Process. **45**(11), 2673–2681 (1997)

21. Shao, L., Gouws, S., Britz, D., Goldie, A., Strope, B., Kurzweil, R.: Generating long and diverse responses with neural conversation models. CoRR abs/1701.03185 (2017)

22. Sutskever, I., Vinyals, O., Le, Q.V.: Sequence to sequence learning with neural networks. In: Proceedings of 2014 Annual Conference on Neural Information Processing Systems, pp. 3104–3112 (2014)

23. Wang, H., Lu, Z., Li, H., Chen, E.: A dataset for research on short-text conversations. In: Proceedings of the 2013 Conference on Empirical Methods in Natural Language Processing, pp. 935–945 (2013)

24. Williams, R.J.: Simple statistical gradient-following algorithms for connectionist reinforcement learning. Mach. Learn. **8**, 229–256 (1992)

25. Xing, C., et al.: Topic aware neural response generation. In: Proceedings of the 31st AAAI Conference on Artificial Intelligence, pp. 3351–3357 (2017)

26. Xu, J., Sun, X., Ren, X., Lin, J., Wei, B., Li, W.: DP-GAN: diversity-promoting generative adversarial network for generating informative and diversified text. CoRR abs/1802.01345 (2018)

27. Yu, L., Zhang, W., Wang, J., Yu, Y.: SeqGAN: sequence generative adversarial nets with policy gradient. In: Proceedings of the 31st AAAI Conference on Artificial Intelligence, pp. 2852–2858 (2017)

28. Zhu, J.-Y., Krähenbühl, P., Shechtman, E., Efros, A.A.: Generative visual manipulation on the natural image manifold. In: Leibe, B., Matas, J., Sebe, N., Welling, M. (eds.) ECCV 2016. LNCS, vol. 9909, pp. 597–613. Springer, Cham (2016). https://doi.org/10.1007/978-3-319-46454-1_36

A Lightweight Cloud Execution Stack for Neural Network Simulation

Benjamin Nussbaum and Erich Schikuta$^{(\boxtimes)}$

University of Vienna, Währingerstr. 29, 1090 Vienna, Austria
benjamin.nussbaum@gmail.com, erich.schikuta@univie.ac.at
http://www.cs.univie.ac.at

Abstract. This paper presents an execution stack for neural network simulation using Cloud container orchestration and microservices. User (or other systems) can employ it by simple RESTful service calls. This service oriented approach allows easy and user-friendly importing, training and evaluating of arbitrary neural network models. This work is influenced by N2Sky, a framework for the exchange of neural network specific knowledge and is based on ViNNSL, the Vienna Neural Network Specification Language, a domain specific neural network modelling language. The presented execution stack runs on many common cloud platforms. It is scalable and each component is extensible and interchangeable.

Keywords: Neural network simulation · Neural network modelling
Cloud computing · Container technology · Microservices
Service orientation

1 Introduction

Getting started with machine learning and in particular with neural networks is not a trivial task. It is a complex field with a high entry barrier and most often requires programming skills and expertise in neural network frameworks. In most cases a complex setup is needed to train and evaluate networks, which is both a processor- and memory-intense job. With Cloud computing getting more and more affordable and powerful, it makes sense to shift these tasks into the Cloud. There are existing platforms for neural network simulation (see [11] for a survey), however, lacking at least one of the following criteria:

- Simple, user friendly but effective interface (e.g. by RESTful calls)
- Neural network are described by a domain specific modelling language,
- No programming skills required to define and train neural network models,
- Platform is open-source,
- Can be deployed on-site and/or in the Cloud (of your choice), and
- Components are extensible and replaceable by developers.

Thus we developed a neural network execution stack, that achieves all of that.

© Springer Nature Switzerland AG 2018
L. Cheng et al. (Eds.): ICONIP 2018, LNCS 11306, pp. 181–192, 2018.
https://doi.org/10.1007/978-3-030-04224-0_16

The proposed system architecture fosters the *Kubernetes*[1] container orchestration and a Java based microservice architecture, which is exposed to users and other systems via RESTful web services and/or a web frontend. The whole workflow including importing, training and evaluating a neural network model is purely service oriented. The presented stack runs on popular Cloud platforms, like *Google Cloud Platform*[2], *Amazon AWS*[3] and *Microsoft Azure*[4]. Furthermore it is scalable and each component is extensible and interchangeable. This work is influenced by N2Sky [14], a framework to exchange neural network specific knowledge and aims to support *ViNNSL*, the Vienna Neural Network Specification Language [4,13]. ViNNSL is a domain specific modelling language that does not require programming skills to define, train and evaluate neural networks.

The presented project combines these techniques and demonstrates a prototype that is open-source and supported by common Cloud providers. Developers can integrate their own solutions into the platform or exchange components ad libitum.

The layout of this paper is as follows: The state of the art of neural network simulators and the baseline research is given in Sect. 2. In the following Sect. 3 we present the requirements and specification of our approach. The technology stack the developed system is based on is specified in Sect. 4. A use case demonstrating and evaluating our prototype is shown in Sect. 5. The paper is closed by a summary and a short look on future research.

2   Related Work and Baseline Research

Over the last decades, a very large number of artificial neural network simulation environments have been developed, which aim to mimic the behaviour biological neural networks [11]. It started with systems which were developed for specific network families, as Aspirin/MIGRAINES [10], SOM-PAK [9]. Some systems aimed for a more comprehensive environment, as SNNS [16]. With the advance of virtual resources by Grid and Cloud computing new collaborative environments motivated the authors of this paper to aim for an "everything about sharing" approach leading the way towards virtual collaborative organisations, as N2Grid [15] and N2Cloud [7]. In the course of our research, we designed and developed N2Sky [14], a virtual organisation for the computational intelligence (CI) community, providing access to neural network resources and enabling infrastructures to foster federated Cloud resources. N2Sky is an artificial neural network provisioning environment facilitating the users to create, train, evaluate neural networks fostering different types of resources from Clouds of different affinity, e.g., computational power, disk space, networks, etc [12]. N2Sky aroused strong interest even beyond the CI community[5].

---

[1] https://kubernetes.io.

[2] https://cloud.google.com/kubernetes-engine.

[3] https://aws.amazon.com/eks.

[4] https://azure.microsoft.com/services/container-service.

[5] http://cacm.acm.org/news/171642-neural-nets-now-available-in-the-Cloud/.

To describe and identify neural network objects in N2Sky we developed ViNNSL (Vienna Neural Network Specification Language) [4,13]. ViNNSL allows for easy sharing of resources between the paradigm provider and the customers. ViNNSL is an XML-based domain specific language providing mechanisms to specify neural network objects in a standardized way by attributing them with semantic information. Initially, it was developed as communication framework to support service-oriented architecture based neural network environments. Thus, ViNNSL is capable of describing the static structure, the training and execution phase of neural network objects in a service-oriented infrastructure, as Grids and Clouds.

3 Requirements and Specification

As starting point for a novel system adhering to our envisioned goals we defined functional and non-functional requirements for the system to be developed.

3.1 Functional Requirements

Due to the fact that neural network training requires a lot of computing power, the main requirement is to design an architecture that can be executed in the Cloud or on-site cluster hardware.

To enable developers to extend the application, it is designed as a platform that is open-sourced and documented. An easy setup on a local computer and small micro-services with a clear structure and manageable code base make it easier to get acquainted with the architecture.

The neural network platform should also offer a way to be extended or used by external applications and services. Therefore, a well documented RESTful webservice is provided, that can be consumed by various clients.

3.2 Non-Functional Requirements

The execution stack shall comply with the following quality features:

- Standard RESTful API
- The user interface works on all common browsers and devices (responsive design)
- Loading time of the user interface should be less than three seconds

3.3 User Groups

The neural network execution stack focuses on two main user groups: data scientists and developers:

- *Data scientists* use the provided services in a deployed environment (Cloud or own computer) to develop and train their neural networks. The system should be easy to setup and no programming knowledge should be needed to get started.
- *Developers* can extend the neural network stack with features or use the provided web services to implement their own custom solution.

3.4   User Interface

The system has to provide two different user interfaces:

- a brittle but powerful RESTful API interface based on a pure and simple command line, which aims for high flexibility and reusability in a service oriented environment, and
- a web-based graphical user interface (GUI). The GUI is a web application that gives a quick overview of all neural networks and their training status. The frontend uses the RESTful API as backend source. However, it does not cover the whole function range of the API.

RESTful API. The functionality of the whole neural network simulation framework is defined by eight use cases realized by respective HTTP service calls.

1. **Import Neural Network:** An existing ViNNSL XML file with a neural network description is imported via the ViNNSL web service into the database. Actor: Data Scientist
   Call: The actor sends a *POST* request to the ViNNSL web service including a XML body.
2. **Train Neural Network:** An imported neural network is trained by passing the configuration over to the worker service.
   Actor: Data Scientist
   Call: The actor sends a *POST* request to the working service including the identifier of the neural network that should be trained.
3. **Monitor Training Status:** The Data Scientist monitors the training status to evaluate the trained network afterwards.
   Actor: Data Scientist
   Call: The actor sends a *GET* request to the status endpoint of the ViNNSL service including the identifier of the neural network that is in progress.
4. **Evaluate Neural Network:** The Data Scientist evaluates the accuracy of the network after its training.
   Actor: Data Scientist
   Call: The actor sends a *GET* request to the status endpoint of the ViNNSL service including the identifier of the neural network that is finished.
5. **Upload Files:** The Data Scientist uploads files, that are usable as datasets (e.g. CSV files or pictures) to the storage service.
   Actor: Data Scientist
   Call: The actor sends a *POST* request to the storage service endpoint containing a multipart file.
6. **List Neural Networks:** Imported neural networks are listed.
   Actor: Data Scientist
   Call: The actor sends a *GET* request to the ViNNSL web service optionally including a neural network identifier.
7. **Extend Service:** An existing micro service can be extended by developers.
   Actor: Developer
   Call: The developer downloads the source code and extends functionality of a micro service.

8. **Replace Service:** An existing micro service can be replaced by developers.
   Actor: Developer
   Call: The developer writes a new implementation of an existing service
   respecting the API definition.

Graphical User Interface. Figure 1 shows the user interface design for the
frontend web service. This GUI just uses the API calls presented above.

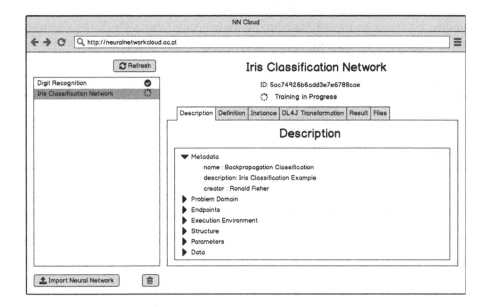

Fig. 1. User interface design for ViNNSL-NN-GUI

3.5 Neural Network Life Cycle

New neural networks are created by sending a POST request including a XML
ViNNSL network description in the request body. The *vinnsl service* creates a
new neural network based on the definition and answers with the HTTP status
code 201 (CREATED). The location header points to the URL where the created
network can be retrieved. The URL contains the unique identifier. Using this
identifier the next step is to add the ViNNSL definition XML file to the network.
This is done via a POST request appending the id and the /definition endpoint.
The XML file is placed in the request body. Resources that are required for the
training (like the training set) need to be uploaded to the storage service, which
returns a unique file id. Before the training can start, the training set needs to
be linked to the neural network. This is possible with the /addfile endpoint.

Next the network is marked for training by calling the worker service with
its identifier. The worker service confirms that the training is queued. As soon

as the training is finished, the worker service updates the neural network object with the result schema and uploads the trained binary model to the storage service for retraining.

A simple GET request to the ViNNSL service along with the identifier returns the current trained neural network model.

4 Technology Stack

To fulfil the requirements presented above for the envisioned system we chose a number of proven and widely accepted paradigms and tools. Guiding principles was acceptance and support by the software developer community and open-source availability:

RESTful services. Representational State Transfer (REST) is an architectural style for the development of webservices. A *REST client* is an application program interface that uses RESTful API HTTP requests to GET, PUT, POST and DELETE data.

Microservices. The micoservice architecture pattern is a variant of a service-oriented architecture (SOA). Monolithic applications bundle user interface, data access layer and business logic together in a single unit. In the microservice architecture each task has its own service. The user interface puts information together from multiple services. This leads in our approach to separation of the specific services in own components, as *database, storage, worker, GUI services etc.*

Docker Containers. Containers enable software developers to deploy applications that are portable and consistent across different environments and providers by running isolated on top of the operating system's kernel. As an organisation, Docker[6] has seen an increase of popularity very quickly, mainly because of its advantages, which are speed, portability, scalability, rapid delivery, and density [3] compared to other solutions. Building a Docker container is fast, because images do not include a guest operating system. The container format itself is standardized, which means that developers only have to ensure that their application runs inside the container, which is then bundled into a single unit. The unit can be deployed on any Linux system as well as on various Cloud environments and therefore easily be scaled. Not using a full operating system makes containers use less resources than virtual machines, which ensures higher workloads with greater density [8]. *Docker containers build the execution framework* for all our ViNNSL services.

Kubernetes Container Orchestration Technologies. As every single microservice runs as a container, we need a tool to manage, organise and replace these containers. Services should also be able to speak to each other and restarted if they fail. Services under heavy load should be scaled for better performance. To deal with these challenges container orchestration technologies come into place. According to a study from 2017 published by Portworx,

---

[6] https://docker.com.

Kubernetes [2] is the most frequently used container orchestration tool in organizations. *Kubernetes* realizes the *orchestration of all Docker services.*

Neural Network Execution. We chose Deeplearning4J, which is a Neural Network Execution Frameworks providing a deep learning programming library written for Java and the Java virtual machine (JVM) and a computing framework with wide support for deep learning algorithms. However, any other simulation framework can be easily integrated. *Deeplearning4J* implements the *neural network worker service.*

Putting all these components together leads to the following architecture of our envisioned system (see Fig. 2):

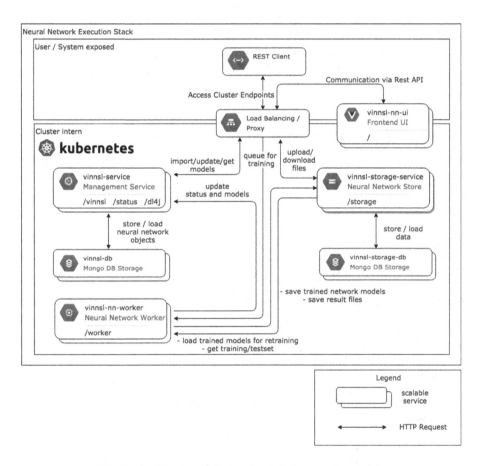

Fig. 2. Architecture of neural network execution stack

5   Use Cases

For demonstration of the simplicity of the usage of the presented neural network execution stack we present a practical use case where we create a neural network, train and evaluate it. This whole process consists of three RESTful POST calls only: one for the creation, one for the training specification, and one for the execution of the neural network. This is described in detail in the following.

5.1   The Iris Flower Data Set

Ronald A. Fisher published 1936 [6] a dataset that is known as the *Iris flower data set*. The data set [6] features 50 examples of three Iris species: Iris setosa, Iris virginica and Iris versicolor. A table lists four measured features from each sample: the length and the width of the sepals and petals. The dataset used for training is available from the UCI Machine Learning Repository [5] as a *CSV* (comma separated value) file[7]. The first example has a sepal length/width of 5.1 cm/3.5 cm, a petal length/width of 1.4 cm/0.2 cm and is an Iris setosa.

The first lines of the dataset explain the structure of the dataset. The columns are formatted for better readability. The species column is an enumerated value (0 = Iris setosa, 1 = Iris virginica, and 2 = Iris versicolor).

```
Sepal length, Sepal width, Petal length, Peta width, Iris species
5.1          , 3.5        , 1.4         , 0.2       , 0
4.9          , 3.0        , 1.4         , 0.2       , 0
...
```

5.2   Creation of the Neural Network

This step is done by a simple HTTP POST call

```
POST https://cluster.local/vinnsl
```

The body of the call contains the description of the 4-3-3-3 multi-layer back-propagation network in the domain specific ViNNSL modelling language:

```
<vinnsl>
  <description>
    <identifier><!-- will be generated --></identifier>
    <metadata>
      <paradigm>classification</paradigm>
      <name>Backpropagation Classification</name>
      <description>Iris Classification Example</description>
    </metadata>
    <creator> Nussbaum </creator>
    <problemDomain>
      <propagationType type="feedforward">
```

----

[7] https://archive.ics.uci.edu/ml/datasets/iris.

```
      <learningType>supervised</learningType>
    </propagationType>
    <applicationField>Classification</applicationField>
    <networkType>Backpropagation</networkType>
    <problemType>Classifiers</problemType>
  </problemDomain>
  <endpoints>
    <train>true</train>
    <retrain>true</retrain>
    <evaluate>true</evaluate>
  </endpoints>
  <structure>
    <input>
      <ID>Input1</ID>
      <size>4</size>
    </input>
    <hidden>
      <ID>Hidden1</ID>
      <size>3</size>
    </hidden>
    <hidden>
      <ID>Hidden2</ID>
      <size>3</size>
    </hidden>
    <output>
      <ID>Output1</ID>
      <size>3</size>
    </output>
    <connections>
      <fullconnected>
        <fromblock>Input1</fromblock>
        <toblock>Hidden1</toblock>
        <fromblock>Hidden1</fromblock>
        <toblock>Output1</toblock>
      </fullconnected>
    </connections>
  </structure>
  </description>
</vinnsl>
```

As successful response we get a 201 **CREATED** message. Aside from the HTTP Status Code, we also get HTTP headers in the response. The one needed for further requests is named `location`. The value of this field is the URL of the network that was created and can be used to get and update fields on the dataset.

5.3   Training of the Neural Network

For the training we have to specify additional parameters: The activation function is set to hyperbolic tangent, the learning rate is 0.1 and the training is limited to 500 iterations. A seed, set to 6, allows reproducible training score.

Also this step is done by a simple HTTP POST call. The {id} of the new network is returned by the creation step of the neural network.

```
POST https://cluster.local/vinnsl/{id}/definition
```

The body contains the additional parameter information:

```
<definition>
<identifier><!-- will be generated --></identifier>
  ...
 <resultSchema>
   <instance>true</instance>
   <training>true</training>
 </resultSchema>
 <parameters>
   <valueparameter name="learningrate">0.1</valueparameter>
   <comboparameter name="activationfunction">tanh</comboparameter>
   <valueparameter name="iterations">500</valueparameter>
   <valueparameter name="seed">6</valueparameter>
 </parameters>
 <data>
   <description>iris file, 3 classes, 4 inputs</description>
   <dataSchemaID>name/iris.txt</dataSchemaID>
 </data>
</definition>
```

The created neural network is queued for training by

```
POST https://cluster.local/worker/queue/{id}
```

which is acknowledge after success by 200 OK.

5.4 Execute Training and Evaluate Result

During the training it is possible to open the graphical user interface called *DL4J Training UI* in a browser, that is provided with the *Deeplearning4J* package, to see the learning progress of the neural network.

```
https://cluster.local/train/overview
```

Figure 3 shows the network training of the Iris Classification. The overview tab provides general information about network and training.

Testing takes place automatically after training and evaluates the accuracy of the trained neural network. In this case 65% of the dataset is used for training and 35% for testing.

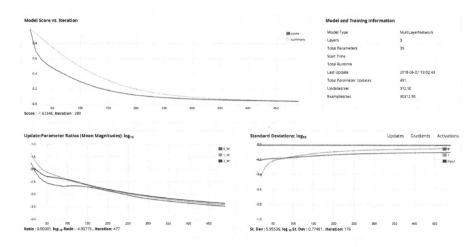

Fig. 3. *DL4J Training UI* shows training progress of Iris Classification network

```
Examples labeled as 0 classified by model as 0: 19 times
Examples labeled as 1 classified by model as 1: 17 times
Examples labeled as 1 classified by model as 2: 2 times
Examples labeled as 2 classified by model as 2: 15 times


==========================Scores==========================
# of classes:    3
Accuracy:        0.9623
Precision:       0.9608
Recall:          0.9649
F1 Score:        0.9606
Precision, recall & F1: macro-averaged (equally weighted
avg. of 3 classes)
==========================================================
```

By examining the result file it can be noticed that the accuracy of the network was 96 percent. All *Iris setosa* and *Iris versicolor* from the testset were recognized correctly, two *Iris virginica* were incorrectly recognized as *Iris versicolor*.

6 Conclusion and Future Work

In this paper we presented an effective and efficient way for the neural network simulation using simple RESTful webservices fostering Cloud container orchestration and microservices. A respective prototype system was demonstrated and evaluated on the Iris flower data set.

In the near future we will integrate our approach with other systems on the market. A first step is the use of Tensorflow [1] as alternative neural network execution engine. A second step will be the embedding of our presented system into the N2Sky simulation framework [14]. The system and its documentation can be accessed and downloaded from http://www.wst.univie.ac.at/projects/n2container.

References

1. Abadi, M., et al.: TensorFlow: a system for large-scale machine learning. In: OSDI, vol. 16, pp. 265–283 (2016)
2. Baier, J.: Getting Started with Kubernetes. Packt Publishing (2015)
3. Bashari Rad, B., Bhatti, H., Ahmadi, M.: An introduction to docker and analysis of its performance. IJCSNS Int. J. Comput. Sci. Netw. Secur. **17**(3), 228–235 (2017)
4. Beran, P.P., Vinek, E., Schikuta, E., Weishaupl, T.: ViNNSL - the Vienna Neural Network Specification Language. In: 2008 IEEE International Joint Conference on Neural Networks (IEEE World Congress on Computational Intelligence), pp. 1872–1879, June 2008
5. Dheeru, D., Karra Taniskidou, E.: UCI machine learning repository (2017). http://archive.ics.uci.edu/ml
6. Fisher, R.A.: The use of multiple measurements in taxonomic problems. Ann. Hum. Genet. **7**(2), 179–188 (1936)
7. Huqqani, A.A., Li, X., Beran, P.P., Schikuta, E.: N2Cloud: cloud based neural network simulation application. In: The 2010 International Joint Conference on Neural Networks (IJCNN), pp. 1–5. IEEE (2010)
8. Joy, A.M.: Performance comparison between linux containers and virtual machines. In: 2015 International Conference on Advances in Computer Engineering and Applications, pp. 342–346, March 2015
9. Kohonen, T., Hynninen, J., Kangas, J., Laaksonen, J.: SOM PAK: the self-organizing map program package. Report A31, Helsinki University of Technology, Laboratory of Computer and Information Science (1996)
10. Leighton, R.R., Wieland, A.: The Aspirin/MIGRAINES software tools, user's manual. Technical report MP-91W00050 (1991)
11. Prieto, A., et al.: Neural networks: an overview of early research, current frameworks and new challenges. Neurocomputing **214**, 242–268 (2016)
12. Schikuta, E., Fuerle, T., Wanek, H.: ViPIOS: the vienna parallel input/output system. In: Pritchard, D., Reeve, J. (eds.) Euro-Par 1998. LNCS, vol. 1470, pp. 953–958. Springer, Heidelberg (1998). https://doi.org/10.1007/BFb0057953
13. Schikuta, E., Huqqani, A., Kopica, T.: Semantic extensions to the Vienna neural network specification language. In: 2015 International Joint Conference on Neural Networks (IJCNN), pp. 1–8. IEEE, July 2015
14. Schikuta, E., Mann, E.: N2Sky - neural networks as services in the clouds. In: The 2013 International Joint Conference on Neural Networks (IJCNN), pp. 1–8. IEEE, August 2013
15. Schikuta, E., Weishaupl, T.: N2Grid: neural networks in the grid. In: 2004 Proceedings of IEEE International Joint Conference on Neural Networks, vol. 2, pp. 1409–1414. IEEE (2004)
16. Zell, A., et al.: SNNS (Stuttgart Neural Network Simulator). In: Skrzypek, J. (eds.) Neural Network Simulation Environments. The Kluwer International Series in Engineering and Computer Science, vol. 254, pp. 165–186. Springer, Boston (1994). https://doi.org/10.1007/978-1-4615-2736-7_9

Intelligent University Library Information Systems to Support Students Efficient Learning

Laszlo Barna Iantovics[1]($\boxtimes$), Corina Rotar[2], and Elena Nechita[3]

[1] University of Medicine, Pharmacy, Sciences and Technology of Targu Mures,
Tirgu Mures, Romania
ibarna@science.upm.ro
[2] "1 Decembrie 1918" University, Alba Iulia, Romania
[3] "Vasile Alecsandri" University, Bacau, Romania

Abstract. The extension of the actual university library information systems (ULISs), in order to offer intelligent support for the registered students, was identified - both by users and specialists - as a necessity. During the documentation process, performed by the students in the library, the need to formulate various issues for which they wish to find rapid answers frequently emerges. The students' requests could be very diverse, ranging from simple questions to complex issues. Our paper introduces a type of complex, intelligent university library information system that we refer to as *NextGenULIS*. A *NextGenULIS* is able to intelligently support students in finding answers to the issues formulated within a network of students, teachers and other possible intelligent agents. Several novel related paradigms, such as hybridization of a library information system, specialization in providing support, and complexity hiding, are also proposed. *NextGenULIS* is briefly compared with a recently introduced ULIS called *IntelligUnivLibSys*.

Keywords: Intelligent library information system
Computational intelligence in a library information system · Intelligent agent
Cooperative hybrid search

1 Introduction

The present paper approaches the *computational system's intelligence* in terms of advanced problem solving abilities (especially efficiency and flexibility) of difficult problems. The intelligent behavior is implemented at the system's creation, but many intelligent agent-based systems (IABSs) are designed to be able to improve it by autonomous learning during their life cycle. IABSs include intelligent agents (IAs) and intelligent cooperative multiagent systems (ICMASs). The learning process enhances the possibilities of a system, leading to new features [1, 2]: improved problem solving ability, self-adaptation, evolution, which can be achieved during several generations.

In [3] the authors outline the importance of computing technology for library and information science research. An integrated *Library Information System*, hereinafter referred as LIS (also known as *Library Management System*) [4, 5], is a planning system used to track the resources owned, orders made, and students who access the

© Springer Nature Switzerland AG 2018
L. Cheng et al. (Eds.): ICONIP 2018, LNCS 11306, pp. 193–204, 2018.
https://doi.org/10.1007/978-3-030-04224-0_17

resources. These may be books, scientific articles, book chapters, reference work entries, book series, journals etc. In what follows, a LIS used by a university library is called *University Library Information System* (ULIS).

Recent LISs offer numerous functionalities for librarians and for the registered students. In [6] the concept of universal library is introduced. In [7] an intelligent library and tutoring system is presented. In [8] an intelligent library management system based on an RFID technology is proposed. *IntelligUnivLibSys*, a complex ULIS based on a hybrid cooperative learning, is proposed in [4]. A study on the implementation of smart library systems using IoT is presented in [9], while [10] displays an interesting study related to the students' perception on the e-library.

From a librarian's points of view, a good ULIS should provide all the necessary functionalities. Studying a large number of ULISs, we concluded that, in fact, the assistance offered to the students is very poor. Therefore, we identified the enhancement of the ULISs with intelligent features as a necessity. We define the intelligence of such a system as the capacity to intelligently support the students along their documentation and learning processes that take place in the library.

Our approach on the intelligence of an ULIS considers its ability to solve problems intelligently. Computational intelligence (CI) covers the domains of evolutionary computation [11], neural networks [11], and fuzzy systems [12]. An ULIS could use one or more algorithms based on computational intelligence for different problem-solving, but this does not necessarily makes it an intelligent system. We further provide an illustrative example in this regard. Meanwhile, an intelligent ULIS could use computational algorithms for solving specific problems.

We propose a new type of intelligent ULIS that we called *NextGenULIS*. *NextGenULIS* is able to intelligently support students in finding answers to the issues formulated when they carry on documentation, during the learning sessions in the library. Students' queries are supposed to have different degrees of difficulty. Searching for a book that describes a specific topic, when the student cannot identify the book, is an example of a simple issue. The case of a student who reads an algorithm but he is not able to test it on some sample data, therefore he needs supplementary texts to accomplish this task, is a kind of difficult task for an ULIS. We will discuss our proposal and compare it with the *IntelligUnivLibSys* ULIS [4].

The upcoming part of the paper is structured as follows: state-of-the-art results related to IABSs and CI algorithms used by LISs are presented in Sect. 2. Section 3 introduces our ULIS proposal: Sect. 3.1 describes the *NextGenULIS* architecture, in Sect. 3.2 the cooperative search process is presented, while Sect. 3.3 introduces a novel paradigm called partial merging of LIS. In Sect. 4 the proposal is discussed. Section 5 presents the conclusions of the research.

2 Applicability of Intelligent Agent-Based Systems

In [13, 14] a comprehensive review of the literature related to the utilization of IABSs agents in modern libraries environment is performed. According to [13, 14] the IABSs have applications in two main areas: digital library and services in traditional libraries.

The performed research shows that the agents have a great potential in many areas related to the library context.

Many of the Artificial Intelligence (AI) researches are focused on topics of CI and IABSs. Among others, AI has many applications in education and library information science, like: estimating individualized treatment effects in students' studies [15] and intelligent tutoring systems [16]. Many difficult real-life problems solving requires CI. On another side, numerous situations require the use of IABSs [1, 2, 17]. As an illustrative scenario, the case of an early carrier physician, who may need a decision support system when he/she should take difficult decisions, can be mentioned.

An intelligent system [1, 2] can offer advantages in solving problems, versus a system that does not possess intelligence. Researches in IABSs are focused on developing embodied computing systems, which can be considered intelligent based on the nature of the problems that they can solve.

Very often, an agent should be able to communicate and cooperate with other artificial agents and/or humans [1, 2]. These characteristics are demanded in order to provide the agents with other capacities such as autonomous learning [1, 2]. Usually, autonomous learning allows the agents to adapt, for an efficient approach of the problems to be solved. The capacities to self-adapt and to self-evolve are frequently associated with the intelligent behavior, as well. Frequently, the self-adaptation of an agent is a consequence of autonomous learning.

In a cooperative multiagent system (CMAS), the intelligence can be considered at the multiagent system's level [1, 2]. Many researches prove that if the agents cooperate efficiently, they could solve difficult problems more properly, in an intelligent manner. The intelligence of an efficiently cooperating multiagent system could be superior to that of its member agents. It is the case of many cognitive multiagent systems. Some of them are composed of relatively simple agents, whose collective intelligence manifests at the multiagent system's level.

3   NextGenULIS - Intelligent Library Information System

3.1   NextGenULIS Architecture

This paragraph introduces the architecture of the *LIS* that we propose. *NextGenULIS* consists of a set denoted $IntCom = Stud \cup Edu \cup AS$. $Stud = \{Stud_1, Stud_2, ..., Stud_n\}$ is the set of the students who access the *LIS* services. $Edu = \{Edu_1, Edu_2, ..., Edu_m\}$ is the set of the educators; by "educators" we understand those persons who are in position to offer consultancy to the students, during the time they spend in the library. Educators could be teachers with different specializations at the university or other persons (such as librarians), able to offer help in finding answers to the issues for-mulated by the students. $AS = Ass \cup Asu$ represent the set of all the intelligent assistant agents of *NextGenULIS*. Each assistant agent holds information about the owner, which could be student or educator: $Ass = \{Ass_1, Ass_2, ..., Ass_n\}$ is the set of students' intelligent assistant agents. Each student enrolled in the library owns an assistant agent: $own(Stud_1) = Ass_1, own(Stud_2) = Ass_2, ..., own(Stud_n) = Ass_n. Asu = \{Asu_1, Asu_2, ...,$

Asu_m} is the set of educators' intelligent assistant agents. Each member of *Edu* owns an assistant agent: $own(Edu_1) = Asu_1,..., own(Edu_m) = Asu_m$.

As well, *NextGenULIS* includes a set of knowledge-based agents able to perform intelligent retrieval for students' queries: $Ia = \{Ia_1, Ia_2, ..., Ia_p\}$. These agents use computational intelligence algorithms. Moreover, each assistant agent also holds information about the other agents from $AS \cup Ia$. With the components and notations described above, the architecture of *NextGenULIS* is presented in Fig. 1.

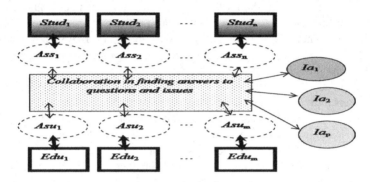

Fig. 1. The architecture of *NextGenULIS*

Each agent has two characteristics that we denote as *capability* and *capacity*. The *capacity* is determined by the computing resources that it can access. The *capability* is defined by the specializations of the agent: a set of problems/tasks that the agent can solve. Semantic analysis of a text (available in digital format) in the library is an example of specialization.

In the framework of *NextGenULIS*, we also assume that each educator owns several specializations based on which he can assist students. Students are also allowed to have various specializations, thus being able to help other colleagues.

Therefore, the system possesses two types of specializations: human (with the subtypes: educator and student) and artificial (with the subtypes: assistant agent specialization and intelligent agent specialization).

Let $IssType = \{IssType_1, IssType_2, ..., IssType_q\}$ represent the types of issues that could be solved in the frame of *IntCom*. An agent is capable of returning an answer if it has the adequate capability (problem-solving knowledge and the necessary data/information), as well as the proper capacity. The agents in *AS* have specializations that allow them to assist the owners (students and educators).

The agents in *Ia* have different sets of specializations. The concept of specialization of an intelligent agent has been extensively studied. Expert systems for assistance [14] are only an example, which has specializations detained in a knowledge base. However, specialization was rarely integrated into LISs, even if it has a great potential for university libraries. The field that benefited most of specialized intelligent agents is medicine, with several well-known medical expert systems.

3.2   NextGenULIS. Cooperative Search Process and Examples

The *Cooperative Hybrid Search* (*CHS*) algorithm presented below describes the cooperative search process that returns the answer to an issue denoted Pr_d, formulated by a student $Stud_j$, $1 \leq j \leq n$. Details and comments are given in what follows. The *CHS* algorithm uses the following notations: Pr_d is the formulation of the issue/question addressed by student $Stud_j$; St_d denotes the initial problem representation returned by Ass_j in a universal format, intelligible for humans and agents; an_d denotes an announcement, sol_d denotes the solution of St_d; $filtered(AS \cup Ia)$ is the selected set of agents to whom the announcement is transmitted (the selection is done based on the available information); "$\rightarrow$" denotes a communication process; "$\Rightarrow$" denotes an analysis followed by a processing. The search for finding the answer to an issue is sometimes a recursive process to which more entities (students, educators, and/or artificial agents) contribute consecutively. A complex issue can be passed from agent to agent until the answer to that issue is found or a failure in answering is reported (when the issue cannot be solved).

The series of transformations below illustrates a generic search process for finding the answer to a complex issue Pr_q (in a successful case):

$$Pr_q \Rightarrow St_a \Rightarrow St_a(S_q) \Rightarrow St_b(S_w) \Rightarrow \ldots \Rightarrow St_y(S_r) \Rightarrow St_g \qquad (1)$$

With S_q, S_w, ..., S_r we have denoted the specializations accessed during the search process. Processing St_a uses S_q and leads to St_b (a new issue to be solved), processing St_b uses S_w and provides St_c and so on, finally leading to St_g. St_g is the statement that contains the solution of issue Pr_q.

Let Ty_1, Ty_2, ..., Ty_n be the types of the parameters that describe the problem solving statements. All these types are known to all the agents; therefore they can interpret the parameters. Let N_1, N_2, ..., N_n be the specifications of the sub-questions/sub-issues generated from the initially formulated issue; id_1, id_2, ..., id_n the identifiers associated to the types Ty_1, Ty_2, ..., Ty_n; mi_1, mi_2, ..., mi_n the minimum number of values that should be retained in the parameters; ma_1, ma_2, ..., ma_n the maximum number of values that can be retained in the parameters.

With these notations, (2) presents the set of parameters used for the representation of the successive solving statements in a universally understandable form.

$$< [Ty_1|N_1|id_1|mi_1, ma_1]; \ [Ty_2|N_2|id_2|mi_2, ma_2]; \ \ldots; \ [Ty_n|N_n|id_n|mi_n, ma_n] > \qquad (2)$$

The following two examples are meant to illustrate the meaning of the parameters presented above.

Example "Melanoma". Melanoma is a type of cancer that develops from the pigment. We consider the scenario of a medical student interested by dermatology. He/she wants to know the early signs of malignant melanoma; therefore he/she searches for the information: "early signs of malignant melanoma". This issue formulation requires a single parameter (see (2), the parameters set). The form [Ty_1 = "signs of an illness"| N_1 = "early signs of melanoma"; $id_1 = 1$; $mi_1 = 1$; $ma_1 = \infty$] specifies that the parameter that could be associated to the type Ty_1 (type with number 1) may contain as values the specifications of the signs of an illness. During the search process, if the corresponding parameter is completed, it must contain the specification of at least one sign (mi_1 = the minimal number) and may specify an unlimited numbers of signs (ma_1 = the maximum number of admissible values).

CHS: Cooperative Hybrid Search Algorithm
```
IN: Pr_d;//A question formulated by a student.
OUT: Answ_d;//The answer to the formulated question.
Step 1. Formulation of an issue by a student.
@Stud_j formulates Pr_d;//Understandable to humans/agents.
@Ass_j helps Stud_j to formulate Pr_d in a universal format;
St_d:= Reprez(Pr_d);//St_d the established formulation.
Step 2. Initial analysis of the formulated Pr_d question.
@Ass_j checks if it is able to answer to the Pr_d;
//Based on its capacity and capability and communicating
with the agents. Let Answ_d be the answer to Pr_d;
If (Ass_j founds the answer to Pr_d) Then
    @report Answ_d to Stud_j; Endif
Step 3. Case if the answer to the question was not found.
While (the answer to the Pr_d issue has not been found) Do
//The cooperative hybrid search for finding the answer.
    @Ass_i is assigned to currently handle the issue.
    If (Ass_i jointly with the owner Stud_i is capable to
    processes St_d) Then
        [Stud_i,Ass_i](St_d)⟹St_h;//St_d is processed.
        If (St_h contains the solution of St_d) Then Goto Step4;
            Else Ass_i(St_h)⟹an_d;//The issue is announced.
            //an_d is transmitted to selected agents
                Ass_i(an_d)→filtered(AS∪Ia);
        EndIf
    Else Ass_i(St_d)⟹an_d;//St_d is announced in the system.
        //an_d is transmitted to selected agents.
        Ass_i(an_d)→filtered(AS∪Ia);
    EndIf
    While (the answering time has not expired) Do
        @Ass_i evaluates the bids to an_d;
    EndWhile.
    @Ass_i allocates the solving to a suitable agent As_b;
    from AS∪Ia who answered positively to the announcement.
EndWhile
Step 4. Formation of the problem solution.
if (Pr_d was solved) Then @sol_d is extracted from the last
    problem-solving statement;
    //Answ_d consist in sol_d and the last problem-solving.
    @report Answ_d to Stud_j;
        Else @Report Failure to Stud_j;
EndIf
EndCooperativeHybridSearch
```

Example "Wonders". "Wonders of the World" are the masterpiece of the skill and handwork of the people of ancient era. We consider the scenario of a student who develops a research that includes the question: "What is the list of the most significant Wonders of the World?". This issue formulation requires a single parameter: [Ty_9 = "Knowledge about culture"|N_9 = "Wonders of the World"; id_9 = 9; mi_9 = 1; ma_9 = ∞]. If it is completed, it should contain the specification of at least one (mi_9 = 1) wonder of the world and may specify an unlimited number of wonders (ma_9 = ∞). More students may answer consecutively to this question, each one being allowed to include at least one more new wonder, which was not previously included by other students.

The *CHS* algorithm presents the following search process. Initially, an issue-solving statement is formulated by a student in a very general form. At *Step 1*, the solving statement structured description is completed with some detailed data. On the path from the initial form of the issue to its answer, the problem-solving statement suffers consecutive transformations, if the answering process involves multiple, successive contributions. This is done by adding new data, the description is completed step-by-step. The aim of each added item is to lead to an exhaustive description of the solution, which is also completed step-by-step.

Formula (3) presents an issue-solving statement at a generic stage of the search process (after several consecutive improvements were performed).

$$([id_a; no_a; Kg_a], [id_b; no_b; Kg_b], \ldots, [id_t; no_t; Kg_t]) \tag{3}$$

For the example *"Melanoma"* presented above, let the answer be: [id_j; no_j; Kg_j] = [1; 5; "Asymmetry", "Borders (irregular)", "Color (variegated)", "Diameter (greater than 6 mm)", "Evolving over time"]. The interpretation of the answer items is as follows. id_j = 1: this is the identifier of the type "early signs of an illness"; no_j = 5: this is the number of the early signs retained; Kg_j = "Asymmetry", "Borders (irregular)", "Color (variegated)", "Diameter (greater than 6 mm)", "Evolving over time": these are the values of the five early signs.

For the example *"Wonders"*, let the answer be: [id_j; no_j; Kg_j] = [9, 10; "The Great Wall of China", "Petra", "Cristo Redentor Statue", "Taj Mahal of Agra", "Machu Picchu", "Hagia Sophia", "Chichen Itza", "Colosseum", "Leaning Tower of Pisa", "Roman Baths"]. In this case, id_j = 9 is the identifier type, and no_j = 10 denotes the number of retained Wonders of the World which are listed in Kg_j.

Step 3 of the *CHS* algorithm introduces the announcements. An announcement an_d is written as (4) where the parameters are as follows. St_j: the issue to be solved; *Emitted*$_j$: numerical value which holds the moment in time when an_d was generated; *Deadline*$_j$: numerical value which stores the maximum time allowed for searching for the answering to St_j.

$$an_d = \, <St_j; Deadline_j; Eligibility_j; Emitted_j> \tag{4}$$

Based on the values of *Deadline*$_j$ and *Emitted*$_j$, an agent who receives an_d specifies the remaining time for the processing of St_j. The value of *Eligibility*$_j$ specifies the eligibility criteria for the bid acceptance. For example, let us consider the specification of a very difficult issue to which an educator is expected to answer. In this case *Eligibility*$_j$ = "Educator": the agent who generates the problem announcement should specify that the problem is considered difficult and requires an educator to solve it.

We formally describe the response $R_{j,d}$ to the statement St_j announced by an_d as

$$R_{j,d} = \;< an_d; Of_i; Tm_i; Cb_i >$$ (5)

with the following parameters. Cb_i: list of the specializations that the announcement receiver can use when processing St_j; Tm_i: numerical value which specifies the estimated processing time needed for answering; Of_i: specifies the bid to the processing of St_j, with Of_i = "yes" (acceptance of processing) or Of_i = "no" (rejection of processing). When an agent receives the bids to an announcement, it can improve its decisions about how to respond to the announced problem using the information already embedded in the responses.

The algorithm *Cooperative Hybrid Search* describes - as its name points - the hybrid search for finding the answer to an issue formulated by a student. An elementary case holds when the issue formulated by a student $Stud_k$ is simply answered by another student $Stud_s$ or by an educator Edu_r. Obviously, a *NextGenULIS* must be able to solve multiple issues, which are simultaneously formulated by students. For this reason, the algorithm is presented in a general form, capturing the process of a hybrid search for answers to complex issues, eventually raised by many students and solved with the contribution of many students and/or educators and/or agents. The communication between them is efficiently and flexibly mediated by the assistant agents, whose role is to minimize the human effort.

3.3 Novel Paradigm Called Partial Merging of Library Information Systems

We now propose a novel paradigm that we call "partial merging of library information systems". Let us consider the case of a university library, at a given moment in time. If an issue is submitted and a human (student or educator) should answer to it, there is a certain probability to find a human available and qualified to answer. If the university has a larger number of students, the probability to find the answer easily is higher. Based on this observation, an approach we consider important for the future is the partial merging of a number of ULISs. This would contribute to an increase in the efficiency of searching answers to the issues formulated by the users, and to the internalization of the university libraries as a side effect. The partial merging of ULISs is very appropriate in case of universities with small number of students, but it is also beneficial in universities with large number of students, where these wish to find answers to more complex issues. In this case, the complexity and the variety of the required answers could require the existence of a larger number of human respondents.

In deciding upon the partial merging of library information systems, students' specializations are among the most important criteria to consider. For example, connecting all the medical university libraries in a country would emerge a highly complex global ULIS endowed with increased capacity and adaptability. These features come from the links between humans, generated by their collaboration, and improve as long as the collaboration increases.

We consider that, in time, *NextGenULIS* could improve the human interpersonal relations (student-student and student-teacher) that are formed during the interactions involved when searching answers to the formulated issues. New professional relationships and interactions are also facilitated.

4 Discussions

After performing an extensive study of the related scientific literature, we considered a fundamental limitation of actual university library information systems, namely the ability to offer intelligent assistance to the students. Based on our experience in the field of intelligent systems and computational intelligence, we considered as most appropriate the creation of ULISs based on the principles of intelligent hybrid multiagent systems. In this paper, the architecture of a highly complex ULIS called *NextGenULIS* was proposed. We consider that its features to be very appropriate for the next generation of intelligent library university information systems. Embedding the principles of a very complex system, *NextGenULIS* cannot be designed and developed in a single step. Its life-cycle requires generating more consecutive versions until a fully functional and efficient version is obtained.

As a novel paradigm, the *hybridization* of a LIS was introduced. The proposed hybrid system consists of: intelligent agents (able to make intelligent computations), supporting students (who can help their colleagues, therefore they can be considered part of the ULIS) and educators (part of the ULIS as well, due to their role in supporting students' information search process). In the context of our ULIS, the significance of the concept "educators" differs than the usual one, as it is accepted in classroom setting. Here we call "educators" the persons (be they teachers at the university or librarians) who are able to offer support in finding answers to various issues formulated by the students. It is obvious that acting as teacher in a library is different than teaching in the classroom. The notion of "hybrid system" was used in order to illustrate that both human and non-human (artificial) agents can contribute to the problem-solving process, but this behavior of the system is not necessarily transparent from the outside. The student who formulates the problem does not see when a human or an artificial agent contributes to its solving.

A novel hybrid search for answers to the formulated issues/questions, in the frame of a ULIS, was introduced. It is partially based on the *contract net task allocation protocol* described in the Artificial Intelligence literature in [18, 19]. During the documentation and/or learning process occurring in the library, a student formulates an issue to be solved. The search for a satisfactory, complete answer could involve students and/or intelligent agents and/or educators. A study performed on the complexity of the contract net task allocation protocol in [20] is presented. Contract net was approached even in recent studies and researches [21–23].

In the frame of the *NextGenULIS*, the search process is efficient and flexible. It allows finding answers to a wide variety of formulated issues (containing structured or unstructured data, information and knowledge) that require an intelligent approach.

NextGenULIS can find answers for simple questions and for more complex issues as well. A simple question may have an immediate response. Most often, a complex issue does not get an immediate response; the search for finding its answer could consist in several consecutive processing steps.

The hybrid search algorithm proposed in this paper allows students to find the answers to questions and issues formulated by them. Since other students participate in this process, the latter can also benefit from this non-formal learning process. This is a solution to handle the complexity of finding answers.

From the perspective of the cognitive systems theory, the intelligence of *NextGenULIS* can be considered to manifest at the system's level. Each issue formulated by a student is solved cooperatively. One or more intelligent agents and/or students and/or educators contribute at this cooperative search. The issues solved by *NextGenULIS* can have different types and complexities (ranging from very simple to extremely complex ones).

The system *IntelligUnivLibSys* [4] can be mentioned as another recent ULIS being able to offer intelligent help for personalized learning of students. In [4] there were defined some novel paradigms in the context of a novel kind of cooperative hybrid personalized learning, such as learning role, sub-role, and learning intelligence level. The main difference between *NextGenULIS* and *IntelligUnivLibSys* consists in the usability. *IntelligUnivLibSys* is proposed with the purpose of efficient learning of difficult problems solving in a library. *NextGenULIS* has the main advantage that offers more flexibility in learning, communication and collaboration between the students. Our future research will consider integrating the ideas of the two systems, in order to obtain a highly complex ULIS that is appropriate to a broad range of needs.

5   Conclusions

Some of the recent, modern University Library Information Systems (ULISs) are moderately complex, being able to offer different functionalities for librarians and for the students. A low degree of intelligent behavior was identified as one of their major drawbacks. Therefore, we analyzed various aspects related to intelligence in ULISs: what does ULIS intelligence consists of and what Computational Intelligence means for ULISs. An ULIS can use computational intelligence algorithms, but this does not mean that it can be considered an intelligent system.

The intelligence of an ULIS was defined as the ability to intelligently support the enrolled students during their documentation and learning processes in the library. This imposed the analysis of the learning paradigm in the context of an ULIS, where learning is different from that in a classroom-based setting. When learning in a library with the support of an ULIS, the learning process of the students is not coordinated by a teacher. Students can be interested in topics that are different from those approached in the classroom. A novel paradigm that we introduced consists in the ULIS hybridization and its *Cooperative Hybrid Search* algorithm that it uses in order to find answers to the students' demands.

The complexity of the proposed system *NextGenULIS* is hidden from the external point of view. An issue formulated by a student is treated autonomously by the system.

It develops an efficient search for the answer, by intelligently combining the human and computing systems capabilities and capacities. At the moment of an issue formulation, the student does not need to know any details. The internal complexity of the system is handled by the specific operations of the system. As a conclusion, the intelligent library information system that we proposed is very complex. It cannot be designed and developed in a single step: its life cycle is expected to have more consecutive versions, until a fully functional form is obtained. Some interesting studies related to complexity issues and to complex systems are available for the reader in the chapters of the book [24]. One of the main subjects approached in the book was the study of nonlinear dynamics in the context of complex systems. In the frame of *NextGenULIS*, the interactions are nonlinear and dynamic.

Our next researches will approach the study of different learning algorithms that can be implemented in the frame of the *NextGenULIS,* increasing the system's intelligence by more efficient and flexible help offered to the students in learning and documentation. The second research direction will consist in studying and measuring the ULIS intelligence. We already proposed a metric for measuring the machine intelligence of cooperative multiagent systems in [25].

Acknowledgment. The authors acknowledge the support of the project "Bacău and Lugano - Teaching Informatics for a Sustainable Society", co-financed by a grant from Switzerland through the Swiss Contribution to the enlarged European Union.

This publication does not necessarily reflect the position of the Swiss government. The responsibility for its content lies entirely with the authors.

References

1. Hajduk, M., Sukop, M., Haun, M.: Cognitive Multi-agent Systems. Structures, Strategies and Applications to Mobile Robotics and Robosoccer. Studies in Systems, Decision and Control Series, vol. 138. Springer, Berlin (2019). https://doi.org/10.1007/978-3-319-93687-1
2. Weiss, G. (ed.): Multiagent Systems: A Modern Approach to Distributed Artificial Intelligence. MIT Press, Cambridge (2000)
3. Thelwall, M., Maflahi, N.: How important is computing technology for library and information science research? Libr. Inf. Sci. Res. **37**(1), 42–50 (2015)
4. Iantovics, L.B., Kovacs, L., Fekete, G.L.: Next generation university library information systems based on cooperative learning. New Rev. Inf. Netw. **21**(2), 101–116 (2016)
5. Raju, J.: Information professional or IT professional?: The knowledge and skills required by academic librarians in the digital library environment. Portal Libr. Acad. **17**(4), 739–757 (2017)
6. Reddy, R.: The Universal Library: Intelligent agents and information on demand. In: Adam, N.R., Bhargava, B.K., Halem, M., Yesha, Y. (eds.) Digital Libraries Research andTechnology Advances. ADL 1995. LNCS, vol. 1082, pp. 27–34. Springer, Heidelberg (1996). https://doi.org/10.1007/BFb0024597
7. Kaklauskas, A., Zavadskas, E., Babenskas, E., Seniut, M., Vlasenko, A., Plakys, V.: Intelligent library and tutoring system for brita in the PuBs project. In: Luo, Y. (ed.) Cooperative Design, Visualization, and Engineering. CDVE 2007. LNCS, vol. 4674, pp. 157–166. Springer, Heidelberg (2007). https://doi.org/10.1007/978-3-540-74780-2

8. Younis, M.I.: SLMS: a smart library management system based on an RFID technology. Int. J. Reason. Based Intell. Syst. **4**(4), 186–191 (2012)

9. Pandey, J., Kazmi, S.I.A., Hayat, M.S., Ahmed, I.: A study on implementation of smart library systems using IoT. In: 2017 International Conference on Infocom Technologies and Unmanned Systems (Trends and Future Directions) (ICTUS), pp. 193–197. IEEE Press (2017)

10. AlHamad, A.Q.M., AlHammadi, R.A.: Students' perception of E-library system at Fujairah University. In: Auer, M., Langmann, R. (eds.) Smart Industry & Smart Education. REV 2018. LNNS, vol. 47, pp. 659–670. Springer, Cham (2019). https://doi.org/10.1007/978-3-319-95678-7

11. Mirjalili, S.: Evolutionary Algorithms and Neural Networks. Theory and Applications. Studies in Computational Intelligence, vol. 780. Springer, Cham (2019). https://doi.org/10.1007/978-3-319-93025-1

12. Mahmoud, Magdi S.: Fuzzy Control, Estimation and Diagnosis. Springer, Cham (2018). https://doi.org/10.1007/978-3-319-54954-5

13. Liu, G.: The application of intelligent agents in libraries: a survey. Progr. Electron. Libr. Inf. Syst. **45**(1), 78–97 (2011)

14. Herron, J.: Intelligent agents for the library. J. Electron. Resour. Med. Libr. **14**(3–4), 139–144 (2017)

15. Beemer, J., Spoon, K., He, L., Fan, J., Levine, R.A.: Ensemble learning for estimating individualized treatment effects in student success studies. Int. J. Artif. Intell. Educ. **28**(3), 315–335 (2018)

16. Dermeval, D., Paiva, R., Bittencourt, I.I., Vassileva, J., Borges, D.: Authoring tools for designing intelligent tutoring systems: a systematic review of the literature. Int. J. Artif. Intell. Educ. **28**(3), 336–384 (2018)

17. Dent, V.F.: Intelligent agent concepts in the modern library. Libr. Hi Tech. **25**(1), 108–125 (2007)

18. Sandholm, T.W.: An implementation of the contract net protocol based on marginal cost calculations. In: 1993 National Conference on Artificial Intelligence, pp. 295–308. AAAI Press, California (1993)

19. Smith, R.G.: The contract net protocol: high-level communication and control in a distributed problem solver. IEEE Trans. Comput. **29**(12), 1104–1113 (1980)

20. Wooldridge, M., Laurence, M.: The complexity of contract negotiation. Artif. Intell. **164**, 23–46 (2005)

21. Wang, Z., Wang, S.: Distributed collaborative control model based on improved contract net. In: Xhafa, F., Patnaik, S., Zomaya, A. (eds.) Advances in Intelligent Systems and Interactive Applications, IISA 2017. AISC, vol. 686, pp. 779–784. Springer, Cham (2018). https://doi.org/10.1007/978-3-319-69096-4

22. Gupta, A.K., Gallasch, G.E.: Equivalence class verification of the contract net protocol-extension. Int. J. Softw. Tools Technol. Transf. **18**(6), 685–706 (2016)

23. Xie, Y., Wang, H.: A group cooperative decision support system based on extended contract net. Group Decis. Negot. **23**(5), 1191–1217 (2014)

24. Macau, E.E.N. (ed.): A Mathematical Modeling Approach from Nonlinear Dynamics to Complex Systems. Nonlinear Systems and Complexity, vol. 22. Springer, Cham (2019). https://doi.org/10.1007/978-3-319-78512-7

25. Iantovics, L.B., Rotar, C., Niazi, M.A.: MetrIntPair—a novel accurate metric for the comparison of two cooperative multiagent systems intelligence based on paired intelligence measurements. Int. J. Intell. Syst. **33**(3), 463–486 (2018)

Neural Networks Assist Crowd Predictions in Discerning the Veracity of Emotional Expressions

Zhenyue Qin, Tom Gedeon[(⊠)], and Sabrina Caldwell

Research School of Computer Science,
Australian National University, Canberra, Australia
{zhenyue.qin,sabrina.caldwell}@anu.edu.au, tom@cs.anu.edu.au

Abstract. Crowd predictions have demonstrated powerful performance in predicting future events. We aim to understand crowd prediction efficacy in ascertaining the veracity of human emotional expressions. We discover that collective discernment can increase the accuracy of detecting emotion veracity from 63%, which is the average individual performance, to 80%. Constraining data to best-performers can further increase the result up to 92%. Neural networks can achieve an accuracy of 99.69% by aggregating participants' answers. That is, assigning positive and negative weights to high and low human predictors, respectively. Furthermore, neural networks that are trained with one emotion data can also produce high accuracies on discerning the veracity of other emotion types: our crowdsourced transfer of emotion learning is novel. We find that our neural networks do not require a large number of participants, particularly, 30 randomly selected, to achieve high accuracy predictions, better than any individual participant. Our proposed method of assembling peoples' predictions with neural networks can provide insights for applications such as fake news prevention and lie detection.

Keywords: Emotion veracity · Crowd prediction · Neural network
Fake news

1 Introduction

Acted emotions are facial expressions whose performers do not carry genuine feeling [2]. By using acted emotions, human beings attempt to convince others that they are experiencing the pretended mental state. For example, sales staff act smiles to their customers in order to express friendly attitudes. Commonly acted emotions in our daily life include anger, surprise, fear, and happiness. These four emotions are assumed to be recognizable regardless of cultural background [7]. The ability to distinguish between acted and genuine emotion expression can aid in lie detection, advertising effect assessment, and other applications [3,18].

Research indicates that in general, human beings perform poorly in verbally differentiating between genuine and acted emotion expressions [2]. Hossain et al.

© Springer Nature Switzerland AG 2018
L. Cheng et al. (Eds.): ICONIP 2018, LNCS 11306, pp. 205–216, 2018.
https://doi.org/10.1007/978-3-030-04224-0_18

reported that the accuracy of participants' verbal responses in respect of smile veracity was approximately 60% [2], which was better than random guessing (50%). However, their work did not investigate the accuracy of a collection of participants' responses to a particular emotion expression as a whole. Instead, they considered only the accuracy of an individual's reply. That is, an individual might demonstrate poor performance in discerning the emotion veracity. Nevertheless, utilizing a majority response from a group of individuals as their final collaborative answer may present a higher-accuracy response.

Moreover, researchers in social sciences have shown the promise of utilizing the crowd to predict future events [19]. Their work indicated that the prediction results from crowds, of five US presidential elections, were more accurate than the traditional pools 74% of the time [9,16]. Due to this higher accuracy of crowd forecasting, it has been adopted by a range of industries including healthcare companies, technology corporations and so on [14].

However, the previous applications of crowd forecasting weighted all the participants' responses with the same importance. That is, they did not acknowledge that people's predicting capabilities can vary. To supplement this defect, in 2015, researchers improved the traditional forecasting methodology by extracting top-performing predictors through an array of prediction tasks and assigning them into elite teams called *superforecasters* [12]. These teams composed of the selected top-performers demonstrated a 50% greater accuracy than traditionally assembled crowd forecasting teams [5].

Nonetheless, all of these previously conducted experiments on crowd forecasting provided financial rewards for correct predictions. As such, a natural question to ask is whether crowd forecasting can still give high-quality predictions without any payments. A study conducted in 2004 showed a positive result by comparing the prediction qualities from providing participants with real and play money [15]. That is, recompense is not essential for stimulating high-quality predictions. Furthermore, both crowd predictions exhibited more significant power than individual humans [15].

Inspired by the success of crowd predictions, we would like to know whether aggregating a group of people's responses will lead to better answers for discerning emotion veracity. Furthermore, considering Meller et al.'s results on the success of utilizing elite teams [12], we also found that some participants may also demonstrate better capability in discerning pretended emotion expressions than others. We would also like to know whether a team of elites detecting emotion veracity will present higher accuracy than a team consisting of average participants.

Moreover, it is also worthwhile investigating if a neural network can distinguish better emotion discerners and assign higher weights to their responses. In contrast, there may also exist people who are particularly poor in the emotion recognizing task and who tend to always give incorrect answers. We wish to know whether neural networks can learn to assign them negative weights to flip their responses so that even these poor performers can make contributions to a higher accuracy.

Furthermore, we would also like to know the minimum number of participants to achieve a highly accurate result, which can make contributions to reducing the data collection cost when utilizing this technique. Additionally, we also want to know whether elites who can discern one emotion expression accurately can still give a high-accuracy performance for detecting other emotions. Similarly, we wish to study the transfer learning ability [13] of our neural networks. That is, whether a neural network built for discerning one emotion can also work well on other emotions. In the future, we wish to discover the potential similarity between distinguishing genuine and acted emotion expressions and identifying fake news. Ideally, the methodologies in discerning emotion veracity can be applied to other forms of veracity detection and hence be useful in the prevention of fake news spread.

2   Methods

2.1   Stimuli

Previous work conducted by Hossain and Gedeon investigated people's ability to distinguish acted and genuine smiles, using both verbalized descriptions and pupil dilation responses from the participants [8]. Chen et al. extended this research to investigate the emotion of anger [2] with the same methodology. However, Hossain and Gedeon used minimal stimuli [8], where videos are cropped to only include the faces [11], while Avezier et al. suggested that the contextual backgrounds for displaying emotional expressions are essential to differentiating emotions [1]. We used the same stimuli as what Chen et al. have collected and utilized, which include contextual backgrounds [2].

The raw videos for the experiments were sourced from YouTube. Each emotion type consisted of 20 videos, which were selected considering balancing ethnicity, gender, and background context to reduce unnecessary noise in the experiment. Genuine emotion expressions were collected from live news reports as well as documentaries and acted ones were obtained from movies containing similar scenes.

2.2   Neural Networks

A simple feedforward neural network [17] is utilized to give participants different weights in order to value high/low-quality responses differently. Specifically, the neural network contains 117 input neurons, corresponding to 117 participants' answers to a stimulus video. In addition, we also provide the neural network with the correct label of that video. That is, whether the video is a genuine or acted emotional expression, 1/0 respectively. The number of inputs corresponds to the number of human predictors. For example, if there are 117 participants, then the input layer will have 117 nodes. Each input corresponds to a participant's prediction. Specifically, 1 for genuine and 0 for acted emotions. This structure guarantees that the network can learn from the behavior of specific individuals, as each input represents one person.

The network has a single hidden layer consisting of 10 neurons, which use the logistic activation function, and achieved excellent results. We attempted various numbers of hidden neurons, up to 20 which all produced similar results. The network classifies its inputs, corresponding to the likelihood of the emotion presented in the video as being genuine or acted emotion. We train our neural network with stochastic gradient descent and cross entropy. We trained for 5000 epochs, with a learning rate being 0.01. Additionally, we utilize leave-one-video-out cross-validation to test the performance of our neural networks. That is, we train a neural network to learn the reliabilities of each participant using the first 19 videos, and test with the last one video, and repeat 20 times, reporting average results. This is a highly reliable testing approach.

We are aware that the training with 5000 epochs may be likely to result in overfitting. Nonetheless, even overfitting still does not influence the performance of our neural network. This is because an overfit network means it has learned who are reliable predictors, which is the consequence that we expect. Moreover, We note that we used results for 80 videos in total, being 4 emotions × 20 videos per emotion. This is a considerable dataset for human data: 117 people × 80 videos × 2 mins (to watch a short video and decide on veracity, occasional short rests) = 312 h. Furthermore, we have also tested a neural network trained with significantly fewer epochs, namely 30. It showed very similar performance with 5000 epochs (98.75%).

2.3   Elite Detectors

Similar to *superforecasters* who are teamed up to give accurate predictions on future events [12], we also selected elites from the participants who demonstrated higher individual accuracy in discerning emotional veracity. Afterward, we utilized the majority response from those elites as the unified decision on a particular emotional expression. We also tested a range of elite sizes in order to find the best-performing elite ratio.

3   Results

3.1   A Collective Voting Approach Can Increase the Accuracy of Human Ability to Discern Emotion Veracity

Previous research indicates that the individual capability of discerning emotion veracity, including smile and anger, is little above randomly guessing [2,8]. It is also expected that a similar level of accuracy will be obtained for fear and happiness [4,6]. Although individuals tend to give low-accuracy answers for emotion veracity, our experiments revealed that by adopting the most common response among a crowd, the accuracy can be increased to 80% for determining the veracity of an emotional expression. For example, if the majority prediction from a group believes that an emotion is genuine, then the final prediction will be genuine.

3.2 Teaming up Elites Can Increase the Accuracy of Discerning Emotion Veracity

Preliminary analysis on participants' responses indicated that people's capability to discern emotion veracity varied. That is, there exist some people who demonstrated higher accuracies in detecting veracity of emotional expressions. Our results revealed that teaming up better emotion veracity detectors and averaging their responses can produce higher accuracies than considering all the participants' responses. We also found that an elite ratio of 5% would lead to the best performance in detecting emotional veracity, at an accuracy of approximately 92%, as Fig. 1 indicates.

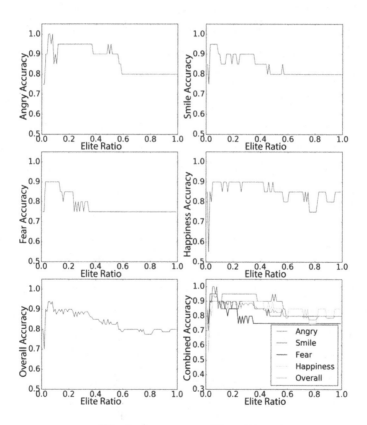

Fig. 1. Accuracy vs elite ratios

3.3   Neural Networks Assist in Synthesizing Participants' Responses to Present Higher Accuracy

The neural network, specified in Sect. 2.2, demonstrated high-level performance in discerning emotional veracity from participants' responses. Specifically, an overall 99.69% accuracy for all of four emotion types, tested by the leave-one-out cross-validation approach and take the average accuracies of 20 repeated runs. The quantity of training data matters to the performance of neural networks, such that more training data can result in higher accuracies. This is shown in our results: increasing the number of folds in cross-validation leads to corresponding performance growth, as Fig. 2 indicates. The shapes of the curves also tell us that, e.g, anger is most strongly recognized, and that of smiles, the least strongly. This is consonant with the Psychological literature of emotion recognition, see our Introduction, and a validation of our techniques that this just *falls out of the data.*

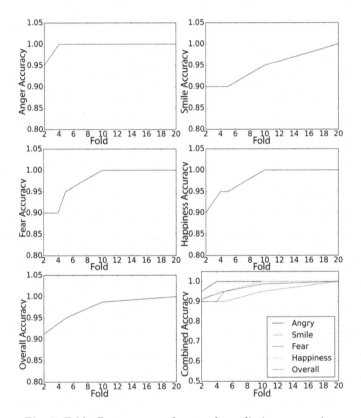

Fig. 2. Fold effects on neural network predicting accuracies

3.4 Neural Networks that Are Trained from One Emotion Perform Well on Other Emotions

In order to investigate the transfer learning ability of our neural networks, we trained them with one emotional expression data and tested their performance with the other three kinds of emotional data. For example, we trained a neural network with the anger data and tested using the smile, fear, and happiness data. As Table 1 indicates, our neural networks demonstrated high accuracy (90% on average) when identifying the veracities of other emotional expressions. This suggests that emotions may share similarities in the way they are acted or that human have generalized veracity detectors.

Table 1. Train one emotion and test on the other three

Anger	Smile	Fear	Happiness
Training	0.75	0.9	0.9
0.95	Training	0.9	0.9
0.95	0.85	Training	0.9
0.95	0.95	0.9	Training

3.5 Combining All the Emotion Data to Train Neural Networks Still Produces High-Accuracy Performance

In order to further investigate the similarities among predictability of veracity of the different emotional expressions, instead of dividing the data into four parts, corresponding to the four emotional expressions, and do the training as well as testing for each of them separately, we combined the data for all the four emotions. Sticking to the leave-one-out cross-validation approach, the accuracy increased to 100% with 20 reputational runs. Again, this implies that distinct emotions may share some likeness. It appears that people who are expert in distinguishing the veracity of one emotion are also excellent in discerning the

Table 2. Elites of one emotion to predict others

Anger	Smile	Fear	Happiness
Training	0.8	0.9	0.65
1.0	Training	0.9	0.85
1.0	1.0	Training	0.9
0.8	0.85	0.8	Training

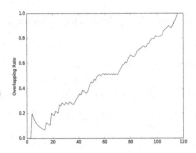 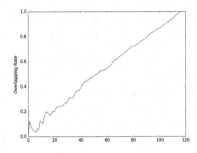

(a) Overlapping rate of elites for discerning different emotion veracities

(b) Overlapping rate of individual assigned weights and accuracies

Fig. 3. Overlapping rate analyses

veracity of other emotions. This can also be seen in Table 2, which shows the results of utilizing elites of one emotion to discern the veracities of other emotions.

Figure 3a demonstrates the overlapping of the top n elites for distinguishing the four emotional expressions. For example, we separately picked the top 5 elites in discerning each emotion type. Among these 20 people (5 elites times 4 emotions), 20% of them were the same.

3.6   The Effects of Participation Numbers on Accuracy

We wish to discover the relationship between accuracy levels and the numbers of participants. Finding a minimum number will minimize resource costs of using such crowd prediction techniques. To investigate, we randomly picked various numbers of participants from all the participants and repeated this process 20 times for each size of the group to ensure the reliability.

For example, we randomly chose 40 participants from all 117 and tested the accuracy using the 40 responses of those selected participants. In order to reduce noise resulting from the random selection, we repeated the action of randomly picking 40 participants and testing their accuracies 20 times, and averaged the 20 accuracy results as the final accuracy for this size of group.

As Fig. 4 indicates, when the number of participants increased, the accuracy also grew. Moreover, the growth became smoother when the total number of participants has reached approximately 20. This suggested that a high accuracy in discerning the emotion veracity did not require a very large size of total participants. In our case, specifically, 1/4 of the original participant size could lead to an overall over 99% accuracy of discerning all the four kinds of emotions.

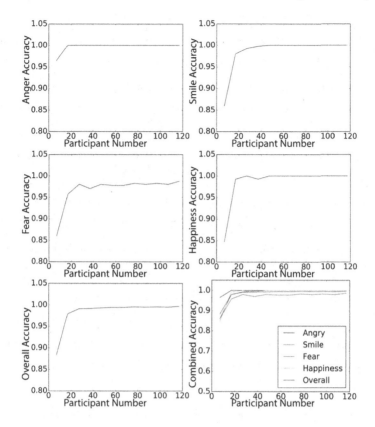

Fig. 4. Accuracies for different participant numbers

4   Discussion

4.1   An Explanation of Why Neural Networks Can Give High Accuracy Discerning Results

Our results indicated that neural networks can effectively aggregate various participants' responses and so identify the veracity of emotional expressions in order to give precise answers. We hypothesized that neural networks would assign positive weights to the responses of participants who give many correct answers and would assign negative weights to those participants who are more likely to give wrong responses. Thus, the wrong answers can still contribute to correct predictions. We suspected that this exploitation of incorrect responses explains why our neural networks performed better than teaming up elites.

In order to verify our hypothesis, we composed a dummy dataset in which the first three participants would always present correct answers, whereas the remaining seven would keep giving wrong answers.

After training with this dummy dataset, our neural networks would assign positive weights to the first three elements and negative weights to the remaining seven. Therefore, the hypothesis which states neural networks negated the responses of people who always present wrong answers in order to predict right results seemed to hold.

Furthermore, we would like to know whether the assigned weights to each participant correlated to the person's individual discerning accuracy. For example, if a subject's accuracy in detecting the emotional veracity is the highest, will the neural networks also assign him or her the highest weight? Neural networks fed by the dummy data above demonstrated a negative answer to this question. This is justified by the fact that although the individual accuracies of the first three participants were the same, their assigned weights varied significantly. In order to further understand this, we plotted the overlapping rate given the top n individuals. That is, among the top n participants sorted by their discerning accuracies and assigned weights, what is the proportion of overlapped individuals between these two? This is as Fig. 3b indicates. Specifically, given approximately the top half individuals, i.e., the top 60 participants, the overlapping rate is 54%. Moreover, the probability of maintaining the same or higher overlapping rate through random selection was 0.2610. This result was a little bit high to defend the null hypothesis, which stated that it was very unlikely to randomly generate the same overlapping rate without neural networks' operation.

Overall, this suggests that neural networks do give positive weights to people who are apt to give true answers and negative weights to those who present false responses, however, it is not necessarily perfectly correlated in assigning to levels of accuracy.

4.2 The Difficulty of Discerning the Different Kinds of Emotion Veracities Varies

As Figs. 2 and 4 indicate, anger converged faster than the other three emotional expressions. This implies that: 1. It does not require a lot of training data in order to discern the veracity of anger expressions; 2. It needs fewer participants in order to give accurate aggregated answers on distinguishing anger veracity. This means that it is easier to detect the veracity of anger expressions than others. This observation is consistent with the work of Mather and Knight [10], which indicated that anger is quicker to detect overall.

5   Future Work

In the future, we will extend our method of discerning the veracity of fake versus real news. First, we will conduct experiments on utilizing the wisdom of crowds collectively in order to recognize fake news. Second, if the crowd's accuracy on discerning fake news is satisfactory, privacy will be an issue to solve, as our neural networks need to be aware of the quality of specific participants' responses. However, people may resist the fact that they are found to often present incorrect answers or they may behave differently if they know that they are usually right.

6   Conclusion

In this paper, we aggregated peoples' verbalized responses to predict the veracity of emotional expressions comprising four universal emotions, namely anger, surprise, fear, and happiness. We discovered that by adopting collective voting instead of using individual responses, the accuracy of human discernment of emotion veracity overall could be increased from 63% to 80%. We also found that there exist people who demonstrate better abilities in ascertaining emotion veracity. By incorporating the responses from these *elite* predictors, the overall accuracy could be further increased to 92% in collective voting. Finally, we introduced neural networks to aggregate participants' responses and obtained an overall 99.7% accuracy. Additionally, we found that training with one emotion data leads to high accuracies when testing with the other three kinds of emotion data using our neural network approach, a novel emotion transfer result. Our neural networks did not require a large number of participants for high-accuracy performance. A closer look revealed neural networks achieved these high-level results by assigning positive and negative weights to the participants who tended to give consistently good and bad answers, respectively. However, the weightings of participants in the neural networks may not simply reflect the ranking of participants' accuracies on ascertaining emotion veracity. In the future, we will utilize the same methods to investigate the feasibility of utilizing crowds to discern genuine and fake news.

Acknowledgements. The authors are grateful to Aaron Manson for access to the data, and particularly acknowledge his efforts in data collection.

References

1. Aviezer, H., Hassin, R., Bentin, S., Trope, Y.: Putting facial expressions back in context. In: First Impressions, pp. 255–286 (2008)
2. Chen, L., Gedeon, T., Hossain, M.Z., Caldwell, S.: Are you really angry?: detecting emotion veracity as a proposed tool for interaction. In: Proceedings of the 29th Australian Conference on Computer-Human Interaction, pp. 412–416. ACM (2017)
3. Conroy, N.J., Rubin, V.L., Chen, Y.: Automatic deception detection: methods for finding fake news. Proc. Assoc. Inf. Sci. Technol. **52**(1), 1–4 (2015)
4. Du, S., Martinez, A.M.: The resolution of facial expressions of emotion. J. Vis. **11**(13), 24–24 (2011)
5. Frood, A.: Work the crowd. New Sci. **237**(3166), 32–35 (2018)
6. Gagnon, M., Gosselin, P., Hudon-ven Der Buhs, I., Larocque, K., Milliard, K.: Children's recognition and discrimination of fear and disgust facial expressions. J. Nonverbal Behav. **34**(1), 27–42 (2010)
7. Hassin, R.R., Aviezer, H., Bentin, S.: Inherently ambiguous: facial expressions of emotions, in context. Emot. Rev. **5**(1), 60–65 (2013)
8. Hossain, M.Z., Gedeon, T.: Classifying posed and real smiles from observers' peripheral physiology. In: 11th International Conference on Pervasive Computing Technologies for Healthcare, Barcelona, Spain, pp. 177–186, May 2017

9. Manski, C.F.: Interpreting the predictions of prediction markets. Econ. Lett. **91**(3), 425–429 (2006)
10. Mather, M., Knight, M.R.: Angry faces get noticed quickly: threat detection is not impaired among older adults. J. Gerontol. Ser. B Psychol. Sci. Soc. Sci. **61**(1), 54–57 (2006)
11. McLellan, T., Johnston, L., Dalrymple-Alford, J., Porter, R.: Sensitivity to genuine versus posed emotion specified in facial displays. Cogn. Emot. **24**(8), 1277–1292 (2010)
12. Mellers, B., et al.: Identifying and cultivating superforecasters as a method of improving probabilistic predictions. Perspect. Psychol. Sci. **10**(3), 267–281 (2015)
13. Pan, S.J., Yang, Q.: A survey on transfer learning. IEEE Trans. Knowl. Data Eng. **22**(10), 1345–1359 (2010)
14. Polgreen, P.M., Nelson, F.D., Neumann, G.R., Weinstein, R.A.: Use of prediction markets to forecast infectious disease activity. Clin. Infect. Dis. **44**(2), 272–279 (2007)
15. Servan-Schreiber, E., Wolfers, J., Pennock, D.M., Galebach, B.: Prediction markets: does money matter? Electr. Markets **14**(3), 243–251 (2004)
16. Surowiecki, J.: The Wisdom of Crowds. Anchor (2005)
17. Svozil, D., Kvasnicka, V., Pospichal, J.: Introduction to multi-layer feed-forward neural networks. Chemom. Intell. Lab. Syst. **39**(1), 43–62 (1997)
18. Teixeira, T., Wedel, M., Pieters, R.: Emotion-induced engagement in internet video advertisements. J. Mark. Res. **49**(2), 144–159 (2012)
19. Wolfers, J., Zitzewitz, E.: Prediction markets. J. Econ. Perspect. **18**(2), 107–126 (2004)

Large Scale Behavioral Analysis
of Ransomware Attacks

Timothy R. McIntosh[1($\boxtimes$)], Julian Jang-Jaccard[1], and Paul A. Watters[2]

[1] Massey University, Auckland, New Zealand
t.mcintosh@massey.ac.nz
[2] La Trobe University, Melbourne, Australia

Abstract. Ransomware is now the highest risk attack vector in cybersecurity. Reliable and accurate ransomware detection and removal solutions require a deep understanding of the techniques and strategies adopted by malicious code at the file system level. We conducted a large-scale analysis of more than 1.7 billion lines of I/O request packets (IRPs), and additional file system event logs, to gain deeper insights into malicious ransomware behaviors. Such behaviors include crypto-ransomware file system attacks achieved by either encrypting individual files or modifying the Master Boot Record (MBR). Our large-scale analysis shows that crypto-ransomware preferentially attacks certain file types; greedily performs file operations more frequently on more diverse types of files; randomizes novel filename generation for malicious executables; and exhibits a preference for alternating file access. We believe that these insights are vital to building the next generation of ransomware detection and removal solutions.

Keywords: Ransomware · Malware · Cybersecurity · File system

1 Introduction

In recent years, ransomware has developed to become one of the most significant cybersecurity threats on the Internet. According to a recent report [1], the global damage caused by ransomware is predicted to reach US$11.5 billion by 2019, up from US$315 million in 2015. This is a more than 35-time increase in just four years, and many cybersecurity experts have warned that cyber ransoms are the "fastest growing threats" since they first emerged in 2013 [2].

Ransomware executes on victims' computers by making important user documents and sensitive data inaccessible, and then demanding ransom payments from victims to release the restrictions. The ever-growing cases of high-profile ransomware attacks on hospitals, universities, government agencies and corporates have caused numerous disruptions in services offered by the affected entities, and indirect financial losses incurred due to the disruptions that are often more than ransom payments [3, 4]. In response to the rising ransomware threats, users are often advised to regularly backup their data, use security software, and be vigilant while opening files from unknown sources. However, ransomware developers can target unsophisticated users, e.g., those who often do not follow such recommendations, and continue to create new, evolved and more sophisticated attacks to evade detection.

© Springer Nature Switzerland AG 2018
L. Cheng et al. (Eds.): ICONIP 2018, LNCS 11306, pp. 217–229, 2018.
https://doi.org/10.1007/978-3-030-04224-0_19

Current defense solutions, often based on pure-detection approaches such as malware signature matching, are insufficient, since modern ransomware implements multiple techniques, e.g., polymorphism, to evade detection [5]. A forward-thinking solution to the issue could be to equip operating systems with a generic and practical self-defense system against ransomware intrusions. This would only be possible when the security developers have a large scale understanding of what ransomware attacks are, and how the code behaves under conditions that are likely to lead to successful detections.

The main purpose of this paper is to investigate and present characteristics of crypto-ransomware attacks based on our large-scale study of analyzing hundreds of ransomware samples, and the billions of system calls generated by them. We performed quantitative and qualitative analysis on the file system activities of benign applications and crypto-ransomware, and identified a range of hitherto unknown behavioral characterizations. These insights were only made possible by the large-scale analysis.

2   Related Work

A long-term ransomware study between 2006 and 2014 on 1,359 ransomware samples [5] was the first to analyze a significant quantity of ransomware samples, and contributed to the theory base of strategically monitoring, analyzing and protecting file systems integrity. Scaife et al. developed CryptoLock [6], which aimed to halt malicious processes as early as possible if they were found to be tampering with large amounts of user data. CryptoLock checked several pre-defined indicators during runtime, e.g., file type changes, similarity, file deletions, file type funneling. Their system managed to achieve a 100% detection rate against 492 real-world ransomware samples. However, the shortfall of their technique was with file loss. Even after successful detection, there was a medium loss of 10 files that were permanently damaged by ransomware and became irrecoverable.

In contrast, Kharraz et al. proposed UNVEIL [7], a dynamic analysis system that automatically created an artificial execution environment. UNVEIL was able to identify previously unknown evasive ransomware before most security vendors, such as discovering the new family of SilentCrypt. However, the consumption of CPU usage and RAM allocation was high, making it unsuitable for an endpoint solution.

At a lower level, Continella et al. implemented ShieldFS [3], a self-healing, ransomware-aware file system that checked both file system I/O activities and cryptographic primitives in processes in system memory, and updated a set of adaptive profiling models. ShieldFS made assumptions that the ransomware would use a known cryptographic library, would pre-compute the key schedule in predictable locations in the memory, and that the encryption key schedule could be scanned and located easily. The researchers acknowledged the estimated average runtime overhead was 26%, possibly due to employing copy-on-write technique to shadow protected file copies.

For real-time analysis, Kharraz and Kirda presented Redemption [8], a generic system that monitored file system I/O request patterns on a per-process basis. Redemption performed system-wide monitoring, and used content-based features (entropy, file overwrite, delete operations) and behavior-based features (directory traversal, output file types, access frequency), to calculate a malice score and estimate how malicious a process could be. While the implementation of Redemption achieved a low runtime overhead of 2.6%, its decision to calculate a malice score per-process could potentially mark legitimate Windows modules as malicious; this could occur when some ransomware inject themselves into legitimate Windows processes like svchost.exe, or when ransomware execute in the form of PowerShell scripts.

While previous approaches presented useful progress towards a more generalizable solution, these techniques were not based on a more nuanced understanding of how ransomware works across a wide range of samples. For example, does ransomware target system or user files? Or, by focusing on protecting a subset of all files, can scalability problems observed in previous solutions be remedied? These are the kinds of behavioral knowledge deficits that we have tried to remedy in this study.

3 Approach and Methodology

3.1 Ransomware Dataset

To achieve a comprehensive dataset of file system IRPs, we used two sources of data: the same IRP logs collected and studied by ShieldFS in [3] for our own quantitative analysis, and file system event logs collected using a custom-implementation of a file system monitoring utility we developed for qualitative analysis. The ShieldFS dataset contains 1.7 billion IRPs generated by 2,245 applications, either benign or infected by ransomware samples. The IRP log includes information of the IRP operation type, operation time, Process ID, IRP major operation type, IRP minor operation type, and file name. The file paths and the file names have been hashed except for the file extension names. This makes it impossible to discover any corresponding relationship between processes and any specifics about file operations.

To resolve such limitation, we further developed a file system monitoring utility to capture the information of which process performed what type of file operations to what files and when. This allowed us to thoroughly investigate and characterize file system activities by benign software applications and malicious ransomware.

3.2 Data Analysis

Data cleansing was performed on the IRP logs to exclude log entries that did not contain valid process names or target file names. Because the original IRP logs were in the format of compressed 7z files containing txt files, all files were first decompressed into different folders. Another utility was used to parse the text files, count the numbers of log entries grouped in different combinations of parameters (e.g., operation time, file extension names, Process ID etc.), and export the results into csv files. The csv files were then opened using Microsoft Excel to generate graphs.

File system event logs collected by our utility were displayed as unhashed plaintext output on the terminal after capture. This enabled us to associate file system events with software operations performed by us and develop an understanding of what happened at the file system level when testers performed different software operations.

4 Characteristics of Crypto-Ransomware Behaviors

In this section, how crypto-ransomware typically attack victims' systems is characterized based on the findings from our data analysis.

4.1 Crypto-Ransomware Must Attack the File Systems

Researchers [5] found that although different families of crypto-ransomware carried out attacks with different levels of sophistication, they shared similar characteristics from a file system perspective: there is a sudden and significant change in the file system activities involving either the MBR, the IRPs or a combination of both. To encrypt user file contents, ransomware must call Windows file system APIs, which will in turn generate IRPs and send them through the I/O stack. [5] proposed to protect the MBR because the Seftad ransomware family locked up the MBR of victim computers to prevent proper booting. [5] was conducted before the discovery of Petya, which overwrites MBR to gain privileged access to encrypt the Master File Table (MFT) of NTFS partitions [9]. MFT contains file metadata information, including how they are stored in different locations of the disk partition. Without the MFT, even if MBR is restored, the operating system cannot easily reconstruct the user files [9].

Crypto-ransomware encrypts user data using strong encryption algorithms and a key obtained from the remote criminal server [6]. Due to the nature and engineering of NTFS relying on a healthy MFT, it is possible to capture such data-centric behaviors of crypto-ransomware to develop effective detection and defense mechanisms [5, 6].

4.2 Two Entry Points of Targeted Attacks

Selected User Files based on File Types. Many crypto-ransomware samples, such as TeslaCrypt and WannaCry, have been found to selectively encrypt files, often based on file types. They quietly search and index victims' files in the background. They select files based on filename extensions, and target documents, photos and presentations, with pdf, odt, docx, pptx and txt file types being the most frequently attacked [5]. Selecting and encrypting certain types of files is more efficient, allowing the encryption to complete within the shortest time possible, before the intrusion is noticed. Operating system modules usually do not get encrypted; there is no valuable individual data in them and they can be recovered by reinstalling the operating system.

Master Boot Record. Since Microsoft Windows Vista, Windows offers two different disk-partitioning options: the MBR and a Globally Unique Identifier Partition Table (GPT) [10]. The MBR was introduced earlier and was designed for BIOS-based systems; it contains a piece of simple executable code, known as "bootstrap" or "bootloader", which will in turn select and load the actual operating system. GPT works with UEFI-based systems and contains a protective copy of MBR. UEFI provides a simple Boot Manager to select an operating system, and each operating system provides its own bootloader. When SecureBoot is enabled in UEFI settings, only properly signed and trusted modules can be loaded during the boot process.

A few variants of ransomware, such as Petya and Redboot, would instead attempt to attack the MBR first, in order to gain control during the next reboot to be able to encrypt the MFT, a database containing information about every file and directory on an NTFS volume. On a BIOS/MBR-based system, encrypting the MFT is carried out in a two-staged attack, which involves (1) modifying the MBR and rebooting to gain control and (2) encrypting the MFT after the reboot [9, 11]. Most unsophisticated users do not need to perform operations that require writing into MBR of the boot drive, unless they are installing a new operating system. Because detailed file information can be either stored in MFT entries or external spaces described by MFT entries, it would be difficult and sometimes impossible to rebuild MFT to regain access to files; the encryption of MFT in combination with a compromised bootloader leaves most victims no choice but to pay the ransom. On a UEFI/GPT-based system with SecureBoot enabled, Petya would still write its own bootloader into the MBR. However, when the system reboots, the Boot Manager inside UEFI cannot find a signed and trusted bootloader, and would fail to boot the operating system. As a result, Petya does not damage the actual MFT on a UEFI/GPT-based system but would still make it unusable until the GPT is repaired.

4.3   Preference to Access Non-system Files and Folders

Users are often likely to store their files in certain folders of their choice [12]. Users overwhelmingly preferred location-based file search, by going to the most likely storage folder and performing a file listing and browsing [12]. There are three special folders on Microsoft Windows platforms: "Windows" for modules of the operating system, installed drivers, system logs and services, "Program Files" containing installed applications, and "Documents and Settings" (later renamed as "Users" since Windows Vista) containing the folders for each user account on the computer, including domain profiles [10, 13]. A few documents and shell folders are created by Windows inside each user folder, such as "Documents", "Pictures" and "Desktop", where most computer users usually store their user files [13].

Crypto-ransomware demonstrates a preference to access non-system files and folders. Each IRP log collected by [3] contained information on where the target file was located, but if the files were not located in "Windows" or "Program Files" folders, the folder names in the path were hashed. The IRP logs were re-examined based on the location of the target files, to count what percentage of files accessed were in the "Windows" folder, in "Program Files" folders or in other folders. The overall algorithm to investigate the percentage of IRP access against different folders is demonstrated in the Algorithm 1. If the paths contained "\Windows\", the target files were considered to be in the "C:\Windows" folder. If the paths contained "\Program Files", the target files were in the "C:\ Program Files" or "C:\ Program Files (x86)" folders.

```
Algorithm 1 Calculating Percentage of IRP Access
procedure CalculateIRPAccess (LogFolder)
  define countWindows = 0, countProgram = 0, countOthers = 0
  foreach LogFile in LogFolder
    while ((Line = LogFile.ReadLine()) is not FileEnd)
    TargetName = Line.ParseFileName()
    if TargetName has "\Windows\" then
      ++countWindows
    else if TargetName has "\Program Files" then
      ++countProgram
    else
      ++countOthers
  print countWindows, countProgram, countOthers
```

Figure 1 shows that the average percentage of IRP access to other folders (the solid bars) is 47% on benign machines (left), but it rose significantly to the average of 87% on infected machines (right) during ransomware attacks, showing that crypto-ransomware attacks increased the percentage of IRP access to files and folders that were neither in the "Windows" folder nor in "Program Files" folders. Given that computer users do not usually store user files in those folders, there appeared to be a strong preference for crypto-ransomware to access user files.

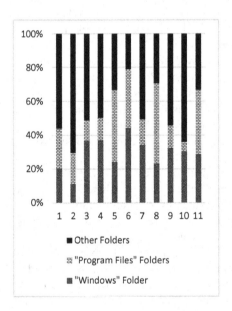 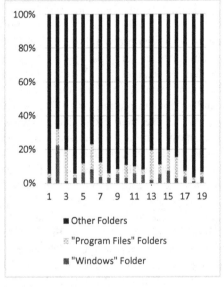

Fig. 1. Percentage of IRP access on benign machines (left) vs on ransomware-infected machines (right)

4.4   Aggressiveness Towards File Systems

Past research [5] noted the unusual and aggressive file system activities by crypto-ransomware, which generated a large amount of create, read, write and rename operations on many different types of non-system files. In [7], they discovered that statistically, crypto-ransomware performed more diverse types of file operations (create, read, modify, rename, delete, move etc.), on more types of files (all files within a wider set of file extensions), and much more frequently. Another study [8] concluded that "modify" and "move" were potentially more dangerous file operations that were more often involved in file encryption performed by crypto-ransomware.

We calculated the speed of IRP requests to modify file contents or file information of PDF files outside of different folders as shown in Algorithm 2. We chose the PDF files because PDF files were the most abundant file type of documents in user files in the IRP logs. The IRP operations were grouped and counted by operation time accurate to minutes, to calculate the speed of modifications to PDF files per minute.

```
Algorithm 2 Calculating Speed of IRP Requests to PDF Files
procedure Calculate_IRP_Speed_PDF (LogFolder)
    define dict as SortedDictionary <time, count>
    foreach LogFile in folder
        while ((line = LogFile.ReadLine()) is not FileEnd)
            OperationTime = line.ParseMinute()
            TargetFileExtname = line.ParseTargetFileExtname()
            if OperationTime is not valid then
                continue
            if TargetFileExtname is not PDF then
                continue
            if dict.ContainsKey (OperationTime) then
                ++dict[OperationTime]
            else
                dict.InsertKeyValuePair (OperationTime, 1)
        foreach KeyValuePair <time, count> in dict
            print time, count
```

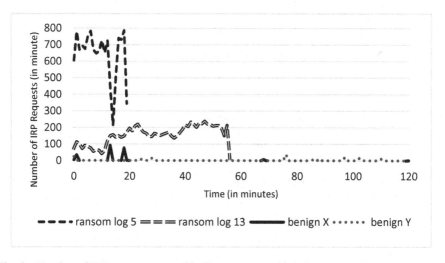

Fig. 2. Number of IRP requests to modify file contents or file information of PDF files outside "Windows" and "Program Files" directories in each log

As shown in Fig. 2, ransomware-infected systems demonstrated extremely high file operations within shorter time periods. For example, "the ransom log 5" had an average of 600–800 IRP requests per minute to modify files within less than 60 min. In contrast, both benign systems only generated a very small number of IRPs per minute (< 100) over a long period of time.

4.5   Greediness for Modifying More Diverse Types of Files

Both [3] and [5] noted that crypto-ransomware could modify the file contents of more diverse types of files (based on file extension names) than many other benign applications, because of the need to encrypt as many files as possible and because temporary files with random filenames during the encryption process can be generated.

We examined the IRP logs to count the number of different file types, based on file extension names, modified by the program which modified the most file types outside the "Windows" and the "Program Files" folders in each log. To modify files, the IRP request must contain the major function IRP_MJ_WRITE. A dictionary was created, using the program name as the key and a HashSet of file extension names as the value. For each new line of log, if the program name existed in the dictionary, the extension name of the target file was added to the HashSet linked to the program; otherwise a new dictionary record was created containing the program name and file extension name. After processing all records, the numbers of different extension names in each HashSet were counted, and the highest number of different file types modified by a single program would be revealed.

```
Algorithm 3 Calculating Diversity of File Modification Patterns
procedure Calculate_Diversity _Modifications (Logfolder)
    define AppList as SortedDictionary <string, HashSet<string>>
    foreach LogFile in Logfolder
        while ((Line = LogFile.ReadLine()) is not FileEnd)
            if Line does not contain "IRP_MJ_WRITE" then
                continue
            AppName = Line.ParseProgramName()
            FileExtname = Line.ParseTargetFileExtensionName()
            if ProgramName is not valid then
                continue
            if FileExtname is not valid then
                continue
            if AppList.ContainsKey(AppName) then
                AppList[AppName].Add(FileExtname)
            else
                define temp as new HashSet<string>
                temp.Add(FileExt)
                AppList.AddKeyValuePair(AppName, temp)
    foreach KeyValuePair<AppName, ExtHashSet> in AppList
        print AppName, ExtHashSet.Count
```

The key findings from Algorithm 3 are shown in Fig. 3, and can be summarized as:

- On the 11 benign systems not attacked by crypto-ransomware, with the label indicating benign 1–11 in the graph, the number of different file types that can be modified by a single program was between 5 and 51.
- On the 19 systems, with the label indicating ransom-log1–17 in the graph, during ransomware attacks, all but 2 systems had a single program in each system that was able to modify thousands of different file types. Upon the inspection of each log, many different file extension names were found to be associated with those applications.
- On systems infected by crypto-ransomware, ransomware executables could be involved in modifying file contents of more diverse file types – that was 500 times more on average!

Based on the findings, it is believed a "File Type Rule" on file system activities could be implemented to detect file modifications performed by crypto-ransomware.

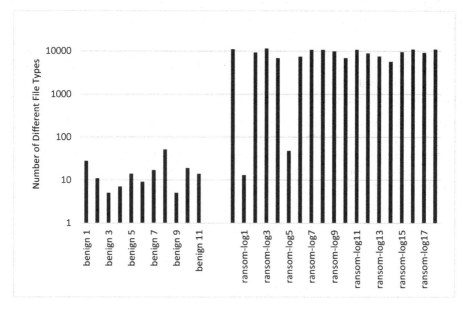

Fig. 3. Number of different file types modified by the program which modified most file types outside "Windows" and "Program Files" directories in each log

4.6   Change of Behaviors

Process Injection is defined as executing the code of one program in the memory of another process by forcing the carrier process to load a dynamic-linked library, and can cause the carrier process to exhibit unexpected behaviors [14]. Some Windows system modules, like rundll32.exe and svchost.exe, do not usually modify user files directly. The Windows File Manager Explorer.exe may perform "delete" and "rename" operations on user files, but only "modify" the content of "zip" files that are placed in user directories.

The logs of ransom-log1 and ransom-log5 were re-examined, due to their low number of file types of modified files (shown in Fig. 4). It was found that for ransom-log1, "svchost.exe" was responsible for creating or modifying 13 different types of files, including bmp files; "svchost.exe" is not a known viewer editor of bmp files, so it's likely that "svchost.exe" was injected with ransomware processes, or crypto-ransomware masqueraded as "svchost.exe" during the attacks. In the log of ransom-log5, a program named "lknwy-bc.exe" created or modified files of 48 different file types, including jpg, js, zip, docx, xlsx; it suggested that "lknwy-bc.exe" was either a sophisticated multi-purpose editor, or a ransomware executable. Searching the Internet for "lknwy-bc.exe" returned no results, suggesting the filename "lknwy-bc.exe" of the program appeared to be randomly generated and the program could be ransomware.

4.7 Randomness of File Information of Executables

According to the file event logs captured by our prototype utility, benign and legitimate software are usually designed and developed with pre-determined functional specifications. If the software are file editors, they usually only know how to open and decode certain types of files. Their executables usually have identifiable and meaningful filenames, and are installed to and executed from their installation paths; their filenames and installation paths usually do not change during the lifetime of the application. Some ransomware executables had the following characteristics:

Random File Names or Paths: Ransomware like CryptoLocker and TeslaCrypt generate executables in the "AppData\Roaming" folder of affected users with random file names. A few samples of GandCrab placed copies of executables into the "App-Data\Local\Temp" folder of affected users, renamed them into random file names like "poerlbm.exe" or "hociexd.exe", and executed from that location.

Fake Description: One WannaCry sample described itself as "DiskPart" in Fig. 4, while a Jigsaw sample carried the description of "Firefox".

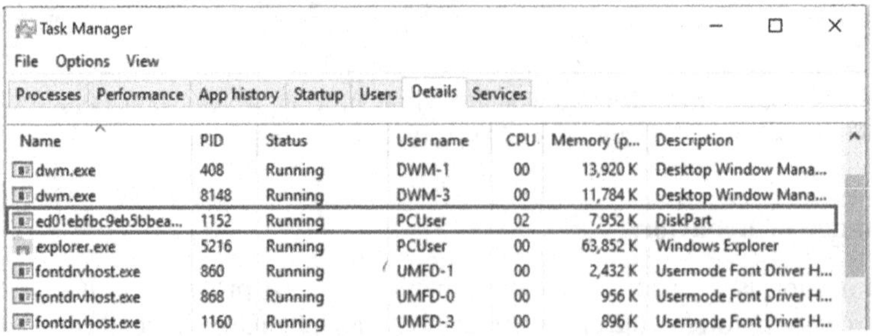

Fig. 4. WannaCry sample describing itself as "DiskPart" in windows task manager

4.8 Alternative File Access Patterns

On a system not infected by any ransomware, utilities like Disk Cleanup (cleanmgr. exe) may perform many file operations within a short period of time, but the file operations are usually of the same type, e.g. batch-deleting and batch-renaming. Most utilities do not modify file contents. The utilities used to compress files only "modify" files of archive formats. On a system infected by ransomware, there is often an alternating pattern of file operations. A file gets modified and often renamed or moved before the next file gets modified and so on.

5 Discussion

Based on the findings in Sect. 4, we believe a forward-thinking behavior-based detection system on crypto-ransomware activities should focus on protecting the file system, by applying physical control to reject unauthorized MBR modifications on the boot drive and statistical controls on monitoring and restricting file system activities on user files. Protecting the MBR and preventing malicious modifications can ensure ransomware or any malware would not take control over the system during the next reboot by writing malicious bootloader into MBR. Blocking MBR modification disrupts a critical step of Petya attack, causing Petya to error and exit.

Using statistical controls to monitor and restrict file system activities can help to distinguish file system activities by benign applications from those by crypto-ransomware and to terminate those malicious activities at the earliest time possible to minimize file damages we saw earlier in the state-of-art [3, 8]. In addition, such statistical control can also provide opportunities of using machine learning to make judgements on what activities may be malicious or unintended without human knowledge.

Because ransomware has a preference to access non-system user files, the efficiency of monitoring algorithm can be significantly improved, by not monitoring file system activities of all directories unnecessarily like previous researchers [3, 6–8]. Ransomware are generally more aggressive when modifying file contents, so additional security rules can be enforced, such as "speed rule" to detect excessive number of file content modifications within a given threshold time. Ransomware are found to be greedier for modifying more types of files, and an additional security rule can be applied to permit/deny file access depending on their file operation authorization or to detect applications that are too versatile by modifying different types of non-plaintext files with specialized encoding schemes. To prevent process injection by ransomware into known Windows operating system modules, it is possible to deny out-of-ordinary file system behaviors of those modules after profiling them. The randomness of file information of executables, either random file names or descriptions, can be used against those executables if the application file names are unknown or appear in unexpected directories. Finally, it is possible to implement a "File Access Pattern" rule to catch ransomware like WannaCry, which modifies, renames and moves one file, before continuing to modify the next file and so on.

6 Conclusion and Future Work

While previous studies on monitoring file system activities have provided useful insights on the nature of ransomware attacks, the depth and insights they provided were proven to have shortfalls. Due in large, the cause of the problem does not come from the lack of appropriate cyber algorithms [15] but from the quantity and quality of the data previous studies used which were incapable of driving more useful knowledge, especially given the sophistication of modern malware [16]. By running a large-scale behavioral analysis on hundreds of ransomware samples and billions of IRP calls generated by them, we were able to discover significant insights previously unknown. This is a significant step forward in the next generation of information processing that requires big data on ransomware behaviors.

References

1. Ransomware Damage Report 2017. https://cybersecurityventures.com/ransomware-damage-report-2017-part-2/. Accessed 24 June 2018
2. McAfee Labs Threats Report. https://www.mcafee.com/enterprise/en-us/assets/reports/rp-quarterly-threats-dec-2017.pdf. Accessed 24 June 2018
3. Continella, A., et al.: ShieldFS: a self-healing, ransomware-aware filesystem. In: Proceedings of the 32nd Annual Conference on Computer Security Applications, pp. 336–347. ACM (2016)
4. Symantec Internet Security Threat Report—April 2017. ISTR, vol. 22. https://www.symantec.com/content/dam/symantec/docs/reports/istr-22-2017-en.pdf. Accessed 27 Jan 2018
5. Kharraz, A., Robertson, W., Balzarotti, D., Bilge, L., Kirda, E.: Cutting the Gordian Knot: a look under the hood of ransomware attacks. In: Almgren, M., Gulisano, V., Maggi, F. (eds.) DIMVA 2015. LNCS, vol. 9148, pp. 3–24. Springer, Cham (2015). https://doi.org/10.1007/978-3-319-20550-2_1
6. Scaife, N., Carter, H., Traynor, P., Butler, K.R.: CryptoLock (and drop it): stopping ransomware attacks on user data. In: 2016 IEEE 36th International Conference on Distributed Computing Systems (ICDCS), pp. 303–312. IEEE (2016)
7. Kharraz, A., Arshad, S., Mulliner, C., Robertson, W.K., Kirda, E.: UNVEIL: a large-scale, automated approach to detecting ransomware. In: USENIX Security Symposium, pp. 757–772 (2016)
8. Kharraz, A., Kirda, E.: Redemption: real-time protection against ransomware at end-hosts. In: Dacier, M., Bailey, M., Polychronakis, M., Antonakakis, M. (eds.) RAID 2017. LNCS, vol. 10453, pp. 98–119. Springer, Cham (2017). https://doi.org/10.1007/978-3-319-66332-6_5
9. Fayi, S.Y.A.: What Petya/NotPetya ransomware is and what its remidiations are. In: Latifi, S. (ed.) Information Technology - New Generations. AISC, vol. 738, pp. 93–100. Springer, Cham (2018). https://doi.org/10.1007/978-3-319-77028-4_15
10. Halsey, M., Bettany, A.: Understanding windows file systems. In: Windows File System Troubleshooting, pp. 13–30. Apress, Berkeley (2015)
11. ESET vs Crypto-Ransomware: What, How and Why. https://cdn1.esetstatic.com/ESET/US/resources/white-papers/WhitePaper_ESET-vs-Crypto-Ransomware.pdf. Accessed 7 Mar 2018

12. Barreau, D., Nardi, B.A.: Finding and reminding: file organization from the desktop. ACM SigChi Bull. **27**(3), 39–43 (1995)
13. Agrawal, N., Bolosky, W.J., Douceur, J.R., Lorch, J.R.: A five-year study of file-system metadata. ACM Trans. Storage (TOS) **3**(3), 9 (2007)
14. Willems, C., Holz, T., Freiling, F.: Toward automated dynamic malware analysis using Cwsandbox. IEEE Secur. Priv. **5**(2) (2007)
15. Layton, R., Watters, P.: Determining provenance in phishing websites using automated conceptual analysis. In: eCrime Researchers Summit, 2009, pp. 1–7. IEEE (2009)
16. Alazab, M., Venkatraman, S., Watters, P., Alazab, M., Alazab, A.: Cybercrime: the case of obfuscated malware. In: Georgiadis, C.K., Jahankhani, H., Pimenidis, E., Bashroush, R., Al-Nemrat, A. (eds.) Global Security, Safety and Sustainability e-Democracy, pp. 204–211. Springer, Heidelberg (2012). https://doi.org/10.1007/978-3-642-33448-1_28

Image and Signal Processing

Estimation of 3-D Pose with 2-D Vision Based on Shape Matching Method

Bin Chen[1,2], Jianhua Su[1(✉)], Kun Lv[3], and Donge Xue[4]

[1] The State Key Laboratory of Management and Control for Complex Systems, Institute of Automation, Chinese Academy of Sciences, Beijing 100190, China
{chenbin2017,jianhua.su}@ia.ac.cn
[2] University of Chinese Academy of Sciences, Beijing, China
[3] China International Engineering Consulting Corporation, Beijing, China
18600863353@163.com
[4] Jiangxi University of Science and Technology, Ganzhou, China
18870773089@163.com

Abstract. Pose estimation is an important step in the grasping of workpieces. However, most previous works aim to use the 3D vision system to locate the 3D pose of the object. This paper develops a pose estimation of 3D object with 2D vision system. The proposed method includes two steps: (a) a hierarchy model of 2D views of the object is firstly constructed off-line; (b) the pose of object is then estimated by measuring the similarity of the model and target image. The proposed method is inherently robust against noise and illumination changes, and also efficient in real applications.

Keywords: 3D object recognition · Shape context · Similarity measure

1 Introduction

Picking of workpieces randomly placed on a conveyor belt require robots to precisely estimate the pose of the object. To achieve a precise location has been widely studied in the last decades due to its strong impact in the productivity for manufacturer. Recognition and pose estimation of the parts are mainly based on 2D or 3D vision techniques, where pose estimation with 2D vision system is ideal for a planar part whose three dimensions are negligible, and for complex objects, the 3D representation approach is highly preferable. However, the 3D vision still has some limitations [1]: (a) the cost of such industrial sensors is still higher than a conventional high resolution industrial camera; (b) with a 3D sensor is usually impossible to recognize specific patterns drawn on the object surfaces that may identify the correct object side.

In order to reduce the cost of the vision system, some researchers aim to acquire the pose of 3D parts with 2D vision system. Hinterstoisser et al. [2] developed a real-time template recognition approach to match the target object with the template and then detect its pose. In their recent work [3], a template-based Line-Mod approach with a Kinect is given to detect the objects. Rios-Cabrera and Tuytelaars [4] detected multiple specific 3D objects based on the Line-Mod template-based method in which they learned the template online and speed up the detection based on cascades. Brachmann

© Springer Nature Switzerland AG 2018
L. Cheng et al. (Eds.): ICONIP 2018, LNCS 11306, pp. 233–242, 2018.
https://doi.org/10.1007/978-3-030-04224-0_20

et al. [5] discuss the estimate of the 3D Pose of specific objects from a single RGB-D image, where the key concept is an intermediate representation in form of a dense 3D object coordinate labelling paired with a dense class labelling. Bonde et al. [6] presented a learning-based instance recognition framework from single view point clouds. They used a soft label shape features of an object to classify both the location and the pose of object. Borgefors et al. [7] estimate the pose of a target object by comparing the detected image and its precomputed 2D views with a matching metric, which is calculated by the edges similarity of the models and the image, but it suffers from the occlusion. Steger et al. [8, 10] computed the similarity by the dot product of the pixel gradient vector, where they claim the method is robust to the light and occlusion. Belongie et al. [9] developed the shape context method to calculate the similarity, which could obtain high matching accuracy but is time-consuming.

This paper develops a pose estimation of 3D object with 2D vision system. We firstly establish the 2D view library of the object with the 3D model of the workpiece. And then, the shape context is employed to calculate the similarity of the target contour and the sample contours in the 2D view library. The template matching method usually suffers from the noise and changes in illumination, which will reduce the precision of the location. Our work is similar to reference [10], but the main difference is that we measure the similarity of the 2D view and target image using a few points sampled from the shape contours, while [10] calculate the dot product of the pixel gradient vector as a similarity. Compared with [10], our method would reduce the complexity of the computation of the similarity.

The rest of the paper is organized as follows: the construction of 2D view library with 3D CAD model is firstly described in Sect. 2. And, the description of shape context of the sampling point is presented in Sect. 3. Finally, several experiments are given in Sect. 4.

2   Construction of Hierarchical View Library

Generally, a 2D view of an object could be obtained by projecting the CAD model of the object on a planar; hence, it is able to establish the 2D view library by projecting on various planar. Inspired by work [10], we use the "view ball" to establish the view library of the object, where a virtual camera is employed to project the object on the ball. Thus, the 3D pose of an object, i.e., $X(\alpha, \beta, r)$, could be described by a point on the sphere, where α denotes the longitude of the point, β denotes the latitude of the point, and r denotes the distance of the point from the center of the sphere (Fig. 1).

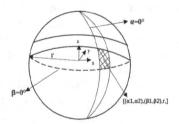

Fig. 1. The description of the "view ball".

By projecting the CAD model from different position, a vary view images of the model are obtained (Fig. 2).

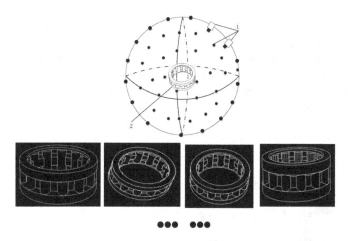

Fig. 2. Obtain 2-D projection images, where "1" represent virtual camera and "2" represent CAD model.

Once the 2D images of the object are captured, it is possible to establish the view library. In this work, we employ pyramid structure to layer the projection view, which includes three steps as shown in Fig. 3: (a) a view that achieves a degree of similarity is

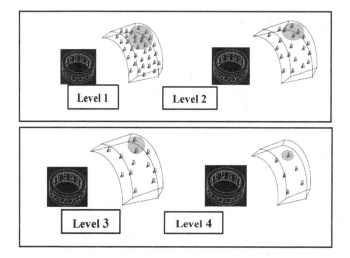

Fig. 3. Resulting aspects on pyramid levels 1 to 4 of the hierarchical model. The cameras of the aspects are visualized by small black pyramids. The blue region visualizes the set of aspects on the different levels that end up in a single aspect on the top pyramid level. (Color figure online)

firstly sorted as a 'aspect' class; (b) the 'aspects' are stored on the second floor of the pyramid store; (c) the similarities of the images on the second pyramid layer are calculated and clustered again until four layers pyramid are obtained.

As the pyramid model of the view library has been established, the matching search can be performed using of the hierarchical search method, which will improve the searching efficiency for an object needed to be identified [11–14]. A coarse scale searching is firstly performed at the bottom level, i.e., level 4. At each level, the similarities of the input image with all the nodes on the layer are calculated. And, the node with the largest similarity is selected, which would decide the child node. This process is repeated until all the nodes at the bottom of the pyramid are traversed. Repeat this process until we find the most similar model in first layer as shown in Fig. 4.

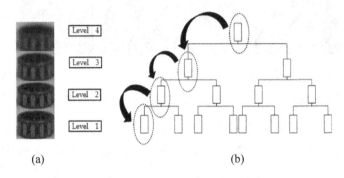

(a)   (b)

Fig. 4. (a) Object images; (b) object recognition using the hierarchy of views.

3   Computation of Similarity

This work aims to estimate the pose of a target object by calculating the similarity between the image of target and the template view in library.

It believes that there is a certain correspondence between the extracted feature and the geometric features of the 3D object [7–10]. Therefore, the position of the 3D object could be calculated from the corresponding feature set by matching the extracted feature with the geometric features of the 3D object. However, in the previous work [7–10], the calculation of similarity using the dot product of the pixel gradient vector is time consuming.

Borrowed the idea from the shape context [9], we introduce a new shape descriptor to match the extracted feature with the geometric features. The shape descriptor contains only coordinate information, through which we can calculate the Euclidean distance and cosine similarity between two shapes. Notice that the shape context [9] includes the gray-value, the location and the set of vectors originating from a point to all other sample points on the shape, but the proposed shape descriptor only contain the points (x_k, y_k) where k = 1, ..., n, on the contours to denote the shape context. For example, there are 50 points on the edge of the bearing to describe its shape as shown in Fig. 5.

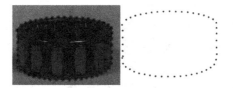

Fig. 5. Sample 50 points on the edge of the bearing.

Once the points on the shape are detected, the hierarchical search is performed by calculating the shape similarity of the object. So, for a new input image, the contour of the object is extracted by a simple image processing. For multi-object, the contours of the target object can be selected by it shape or area. In the following, a new algorithm is proposed for calculating the similarity between two contours.

Assuming that a point p_i is on a target shape (where i stands for the serial number, the same as bellow), our first task is to find a matching point on the sample shape. After finding all the matching points, we calculate the distance of the matching pairs, which is denoted by shape similarity. If the two shapes are identical, the distance between the corresponding points is zero. In other word, the shape similarity is 100%.

For example, if there are m points $P = \{p_1, p_2, p_3, \ldots, p_m\}$ on the target contour, we randomly select n sampling points from the set P. And then, we sort those points, where the point with the minimum x coordinate is denoted as the initial point, and another point closest to the first point is selected as the second point, and the point closest to the second point is selected as the third point, etc. After sorting the points, we denote p_i by the point on the target shape and q_i by matching point on the sample shape. We set a distance threshold d_{thd} to avoid the mistake matching. The mismatching points are also used as the penalty in calculating the similarity. That is, more mismatching points show the lower similarity of the two shapes.

The distance of the two shapes would be obtained by:

$$d_p = \frac{1}{n_{samp}} \sum_{i=1}^{n_{samp}} (p_i - q_i)^2 \quad \text{where } |p_i - q_i| \le d_{thd} \tag{1}$$

$$d_e = d_p / d_{thd} \tag{2}$$

where d_p represents the average distance between each matching point. Assuming that the maximum distance between two points is d_{thd}, which means that $0 < d_e \le 1$. Next, we define similarity between two point sets as $1 - d_e$.

The computations of similarity with Eqs. 1 and 2 suffer from the image noise. In order to improve the robustness of the matching, we introduce the vector sets to calculate the similarity of two shapes. Assuming that p_1 is the initial point, we connect

p_1 and other points to build vectors, and obtain a vector set $P_{vector} = \{\vec{p}_{12}, \vec{p}_{13}, \ldots \vec{p}_{1n_{samp}}\}$, as illustrated in Fig. 6.

Fig. 6. Vector sets from p_1 to the rest of points

We built vector set $Q_{vector} = \{\vec{q}_{12}, \vec{q}_{13}, \ldots \vec{q}_{1n_{samp}}\}$ for the sample shape. Therefore, the similarity of the vectors using the dot product could be defined as follows.

$$S_v = \left| \frac{1}{n_{samp} - 1} \sum_{i=2}^{n_{samp}} \frac{p_{1i} \cdot q_{1i}}{\|p_{1i}\| \cdot \|q_{1i}\|} \right| \tag{3}$$

It is expected that the higher similarity of the two shapes, the smaller angle between the two vector sets. We regard Sv as the cosine similarity between the two shapes.

Considering the Euclidean distance between the matching points and the cosine similarity between the matching vectors, we define the similarity function as:

$$S = \frac{1}{2}(1 + s_v - d_e) - \frac{n_{left}}{\alpha \cdot n_{samp}} \tag{4}$$

Where n_{left} is the number of points with wrong match.

Note that in this algorithm, the size of the sample point and the value of the threshold d_{thd} depend on the size of the actual target object and the size of the image. In the experiments, we set the number of selected points by $n_{samp} = m/20$, and the threshold by $d_{thd} = 0.00001$. The reason why n_{left}/n_{samp} is divided by α is that penalty term should not too big. Experiment shows that when $\alpha = 3$, the result is best.

4 Experiments

We firstly create a layered 2D view library by projecting the bearing CAD model from different pose, where the scope of the view library is shown in Table 1. And then, the three layers of the pyramid layer are shown in Figs. 7, 8 and 9, respectively. When the pyramid model is established, the number of layers is 3, and the number of 2D views from top to bottom is 27,208,1002.

Once we take a bearing picture shown in Fig. 10, we could find the model whose pose is most similar to the target object by the following steps: (a) we execute a rough

Table 1. The scope of the view library

Parameter	The meaning of the parameter	Value
$r_{min} - r_{max}$	Distance from object to camera	0.5 m–0.55 m
$\alpha_{min} - \alpha_{max}$	Longitude range	−90°–90°
$\beta_{min} - \beta_{max}$	Latitude range	0°–90°
d_{thd}	Points distance threshold	0.00001

search to find the model whose pose is similar with target object roughly in the third layer as shown in Fig. 7, where the result is marked by a red box. Thus, we could decide the child node. (b) we repeat the searching until all the nodes at the bottom of the pyramid are traversed. The model which is most similar with the target object has been selected in second layer and first layer, as shown in Figs. 8 and 9.

Fig. 7. The third layer of the view library

Fig. 8. The second layer of the view library

Fig. 9. The first layer of the view library

Fig. 10. The object's image

Since each 2D projection view corresponds to a position relationship between the camera and the actual object, it is easy to determine the position relationship between object and the camera by finding a 2D view that best matches the actual object (Fig. 11).

In order to test the experimental results, we place each object in two different places. In the same place, we take five pictures and use the mean value as the final position results. Table 2 shows the comparison of the results by the algorithm and the actual position of the target object.

To illustrate the difference between our approach and the approach proposed in [10], we compare the two different algorithms by calculating the accuracy and recognition times shown in Fig. 12, while [10] calculating the dot product of the pixel gradient vector as a similarity.

Table 2. Results of the accuracy evaluation

Object	Presented method			Dot_product-based method		
	Epos [mm]	Erot [°]	Time [s]	Epos [mm]	Erot [°]	Time [s]
Bearing	0.42	0.52	0.760	-	-	-
Metal polyhedron	0.51	0.67	0.794	-	-	-
Block	0.78	0.82	0.830	0.58	0.48	0.961

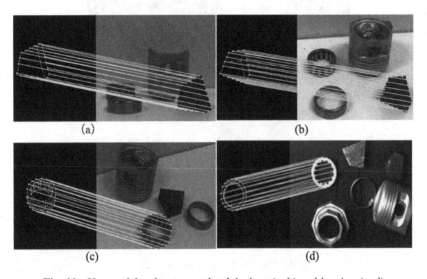

Fig. 11. Use model to locate metal polyhedron (a, b) and bearing (c, d)

It can be seen from Table 2, the position error is about 0.78 mm and the computation time is less than 1 s. The recognition of our approach is faster, whereas the accuracy is worse.

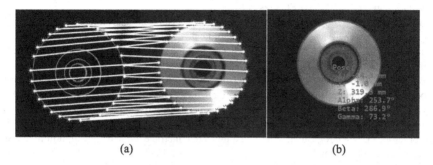

(a)   (b)

Fig. 12. Compare the two algorithms about the accuracy

5   Conclusion

This paper develops a pose estimation of 3D object with 2D vision system. We propose a new way of calculating the similarity between the model and the target image, which is based on the sampling point on the shape of the object. We firstly get the model and object contours and then sample the points on the contours and use the coordinate of the sampling points as the descriptor of the points. We sort those points on a certain way to determine the correspondence between two shapes. Through a distance threshold, we filter out the wrong match points. Finally, by calculating the Euclidean distance and cosine distance between these matching points, the similarity between the model and the target image is obtained, and the 3D position of the object is determined. The proposed approach has a significant advantage in terms of speed and anti-noise interference, relative to the way in which the similarity is calculated by the pixel value. However, the limitation of this algorithm is that it is difficult to guarantee the stability of the algorithm for objects with complex shapes, or objects without obvious contour features and difficult to extract, and the wrong results will be matched. Therefore, in the future work, we will improve the algorithm for this aspect.

References

1. Pretto, A., Tonello, S., Menegatti, E.: Flexible 3D localization of planar objects for industrial bin-picking with monocamera vision system. In: 9th IEEE International Conference on Automation Science and Engineering (CASE), Madison, pp. 168–175. IEEE Press (2013)
2. Hinterstoisser, S., et al.: Gradient response maps for real-time detection of textureless objects. In: 32nd IEEE Transactions on Pattern Analysis and Machine Intelligence, vol. 34, pp. 876–888. IEEE Press (2012)

3. Hinterstoisser, S., et al.: Model based training, detection and pose estimation of texture-less 3D objects in heavily cluttered scenes. In: Lee, K.M., Matsushita, Y., Rehg, J.M., Hu, Z. (eds.) ACCV 2012. LNCS, vol. 7724, pp. 548–562. Springer, Heidelberg (2013). https://doi.org/10.1007/978-3-642-37331-2_42

4. Rios-Cabrera, R., Tuytelaars, T.: Discriminatively trained templates for 3D object detection: a real time scalable approach. In: 14th IEEE International Conference on Computer Vision, Sydney, pp. 2048–2055. IEEE Press (2013)

5. Brachmann, E., Krull, A., Michel, F., Gumhold, S., Shotton, J., Rother, C.: Learning 6D object pose estimation using 3D object coordinates. In: Fleet, D., Pajdla, T., Schiele, B., Tuytelaars, T. (eds.) ECCV 2014. LNCS, vol. 8690, pp. 536–551. Springer, Cham (2014). https://doi.org/10.1007/978-3-319-10605-2_35

6. Bonde, U., Badrinarayanan, V., Cipolla, R.: Robust instance recognition in presence of occlusion and clutter. In: Fleet, D., Pajdla, T., Schiele, B., Tuytelaars, T. (eds.) ECCV 2014. LNCS, vol. 8690, pp. 520–535. Springer, Cham (2014). https://doi.org/10.1007/978-3-319-10605-2_34

7. Borgefors, G.: Hierarchical chamfer matching: a parametric edge matching algorithm. In: 8th IEEE Transactions on Pattern Analysis and Machine Intelligence, vol. 10, pp. 849–865. IEEE Press (1988)

8. Steger, C.: Similarity measures for occlusion, clutter, and illumination invariant object recognition. In: Radig, B., Florczyk, S. (eds.) DAGM 2001. LNCS, vol. 2191, pp. 148–154. Springer, Heidelberg (2001). https://doi.org/10.1007/3-540-45404-7_20

9. Belongie, S., Malik, J., Puzicha, J.: Shape matching and object recognition using shape contexts. In: 22nd IEEE Transactions on Pattern Analysis and Machine Intelligence, vol. 24, pp. 509–522. IEEE Press (2002)

10. Ulrich, M., Wiedemann, C., Steger, C.: Combining scale-space and similarity-based aspect graphs for fast 3D object recognition. In: 32nd IEEE Transactions on Pattern Analysis and Machine Intelligence, vol. 34, pp. 1902–1914. IEEE Press (2012)

11. Lowe, D.G.: Three-dimensional object recognition from single two dimensional images. Artif. Intell. **31**, 355–395 (1987)

12. Lowe, D.G.: Fitting parametrized 3-D models to images. In: 11th IEEE Transactions on Pattern Analysis and Machine Intelligence, vol. 13, pp. 441–450. IEEE Press (1991)

13. Costa, M.S., Shapiro, L.G.: 3D object recognition and pose with relational indexing. Comput. Vis. Image Underst. **79**, 364–407 (2000)

14. Wiedemann, C., Ulrich, M., Steger, C.: Recognition and tracking of 3D objects. In: Symposium on Pattern Recognition, vol. 5096, pp. 132–141 (2008)

Hybrid Iterative Reconstruction for Low Radiation Dose Computed Tomography

Jinghua Sheng[1(✉)], Bin Chen[1], Bocheng Wang[1,2], Qingqiang Liu[1],
Yangjie Ma[1], and Weixiang Liu[1]

[1] College of Computer Science, Hangzhou Dianzi University,
Hangzhou, Zhejiang, China
j.sheng@ieee.org, 1275695021@qq.com,
wangboc@zjicm.edu.cn, 172050037@hdu.edu.cn,
2468320459@qq.com, 619986206@qq.com
[2] Zhejiang University of Media and Communication,
Hangzhou, Zhejiang, China

Abstract. The purpose of this paper is to study the use of hybrid iterative reconstruction (HIR) technique for radiation dose reduction and its effect on low-contrast resolution. This method is designed to create prior information for improving image quality from low dose CT scanners. We compare the performance of lower radiation dose with the HIR and standard dose with the filtered back projection (FBP) using catphan®504 phantom, which is used to measure various image quality parameters. Results show that there are continuous linear reduction of noise and linear increase of CNR with increasing HIR levels compared to FBP for any given scanning protocol. It is possible to provide equivalent diagnostic image quality at low dose. In this paper, we use a quantitative method to evaluate the noise characteristics. Evidence from phantom tests demonstrates that the shape of NPS_{HIR} is shifted continuously to low frequency with increasing HIR levels compared to FBP for any given scanning protocol. Our study confirms that even if there are continuous reduction of noise and increase of CNR with increasing HIR levels, the performance of human observers did not seem to be improved simultaneously because coarser noise could appear. Our finding that the low-frequency components (HIR) are greater than one of FBP (previously believed) may result in the discrepancy between the performance of human observers and that of the ideal low-contrast objects.

Keywords: Computed tomography · Low contrast · Noise power spectrum
HIR · Image quality · Iterative reconstruction

1 Introduction

A major task in CT today is to minimize the radiation dose while preserve image quality. A limiting factor for radiation dose reduction has been the current CT reconstruction algorithm, filtered back projection (FBP). The main reason for the usage of FBP lies within its straight forward mathematical rational and relative low demand on computation. FBP has a major inherent obstacle for dose reduction: increased quantum noise with decreased radiation dose. Iterative reconstruction (IR) has been

© Springer Nature Switzerland AG 2018
L. Cheng et al. (Eds.): ICONIP 2018, LNCS 11306, pp. 243–256, 2018.
https://doi.org/10.1007/978-3-030-04224-0_21

widely used in SPECT/PET to improve the image noise and was introduced to CT some time ago [1]. A limiting factor of using iterative reconstruction, however, is the long computing time. A modified and computationally faster IR technique is recently developed. Examples include Adaptive Statistical Iterative Reconstruction (ARIS) by GE [2–5], iDose by Philips [6, 7], IRIS [8, 9] by Siemens and AIDR [10] by Toshiba (Adaptive Iterative Dose Reduction). These different iterative reconstruction algorithms were compared [11]. Several clinical and phantom studies consistently showed that image noise decreased and contrast-to-noise ratio improved with IR reconstructed images thus indicating great potential for dose reduction. To fully utilize the dose reduction potential with IR, its performance needs to be studied in correlation with diagnostic outcome. In general, image quality revolves around the ability to accurately resolve anatomy with two main features: high-contrast resolution (spatial resolution) and low-contrast resolution. It has been recognized, that image noise, measured as a simple pixel standard deviation, only reflects the noise magnitude and omits the spatial variation pattern, which could affect image visualization in particularly of low-contrast objects. Noise power spectrum (NPS) has been shown to provide better characterization of noise behavior on image systems [12]. The purpose of this study is to find out the effect of HIR compared with standard FBP, on image noise and NPS, and their correlation with the diagnostic performance in terms of spatial resolution and low contrast detectability. Different radiation doses were tested to understand the dose reduction limit using IR while providing sufficient spatial and low-contrast resolution.

2   Materials and Methods

2.1   Iterative Reconstruction Method in CT

Iterative image reconstruction is becoming popular for the following two reasons: (1) it is easy to model and handle projection noise, especially when the counts are low; and (2) it is easy to model the imaging physics, such as geometry, non-uniform attenuation, scatter, and so on. The reconstruction of an image from the acquired data is an inverse problem. Often, it is not possible to exactly solve the inverse problem directly. In this case, a direct algorithm has to approximate the solution, which might cause visible reconstruction artifacts in the image. Iterative algorithms approach the correct solution using multiple iteration steps, which is allowed to obtain a better reconstruction at the cost of a higher computation time.

In computed tomography, this approach was the one first used by Hounsfield. There are a large variety of algorithms, but each starts with an assumed image, computes projections from the image, compares the original projection data and updates the image based upon the difference between the calculated and the actual projections.

There are typically five components to iterative image reconstruction algorithms:

1. An object model that expresses the unknown continuous-space function $X(r)$ that is to be reconstructed in terms of a finite series with unknown coefficients that must be estimated from the data.

2. A system model that relates the unknown object to the "ideal" measurements that would be recorded in the absence of measurement noise. Often this is a linear model.
3. A statistical model that describes how the noisy measurements vary around their ideal values. Often Gaussian noise or Poisson statistics are assumed.
4. A cost function that is to be minimized to estimate the image coefficient vector. Often this cost function includes some form of regularization.
5. An iterative algorithm for minimizing the cost function, including some initial estimate of the image and some stopping criterion for terminating the iterations.

The basic process of iterative reconstruction is to discretize the image into pixels and treat each pixel value as an unknown. Then a system of linear equations can be set up according to the imaging geometry and physics. Finally, the system of equations is solved by an iterative algorithm. The system of linear equations can be represented in the matrix form as:

$$P = AX + R_N, \tag{1}$$

where $\mathbf{X} = [x_1, x_2, ..., x_j, ..., x_n]^T$ is pixel values of the image, $\mathbf{P} = [p_1, p_2, ..., p_i, ..., p_m]^T$ is projection measurements, a_{ij} in the M × N response matrix $\mathbf{A}$ is a coefficient that is the contribution from pixel j to the projection bin i, and R_N is an M × 1 random noise vector.

The diagram in Fig. 1 shows the basic procedure for using an iterative algorithm. Each loop in Fig. 1 represents one iteration.

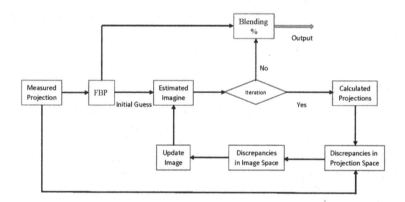

Fig. 1. Flow chart of iterative image reconstruction scheme

In iterative algorithm, a pixel is an area, which is used to form the projections of the current estimate of the image. It tries to solve a system of linear equations instead of trying to solve an integral equation. Iterative algorithms are used to minimize an objective function. This objective function can effectively incorporate the noise in the measurement. The desired image function $\hat{X}$ can be estimated by finding a penalized solution:

$$\hat{X} = arg \min_{X} \left\{ \|P - AX\|_2^2 + \lambda\|EX - D\alpha\|_2^2 \right\} \tag{2}$$

where $\| \cdot \|_2$ is L_2 norm, and the regularization parameter λ is a positive constant chosen to balance the data inconsistency(first term) and the penalized function(second term) from the prior information, which is typically structural feature. E is the matrix to extract prior information from the image; D is predetermined from a training set, which should contain representative structures in the image; α is intermediate variables.

The reconstruction method solves the low dose CT problem iteratively, which is designed to create prior information on target images for improving quality of images. It is an innovative statistical iterative reconstruction technique in which iterations are performed both in the projection and image domain, using sophisticated system and noise models in the domain that is best suited for the correction of the relevant noise/artifact characteristics while keeping structure intact. Primary role is noise reduction.

The HIR method works on image data area, reducing the time-consuming loops on raw data and noise removal is obtained in subsequent iterative steps with a smoothing process. It consists of the following 2 denoising components: (a) An iterative maximum likelihood-type sinogram restoration method based on Poisson noise distribution; and (b) A local structure model fitting on image data that iteratively decreases the uncorrelated noise. While the FBP method directly calculates the image in a single step, IR method repeatedly update the image for a more precise result with some prior information such as typically structural features. When performed, HIR allows users to adjust the image noise level by inputting a parameter called HIR level. The larger the HIR level is, the larger the noise reduction is, allowing users to prospectively decrease the dose at the time of the scan (expecting that HIR will cancel the associated increased noise level during the reconstruction process).

2.2   Phantom Study

In this study, catphan®504 phantom (The phantom Laboratory, Salem, New York) [13] was used for the acceptance testing and commissioning for clinical use of 64 slice CT. It is composed of several modules that can be used to measure various image quality indices. A commercial designed semi-iterative HIR algorithm was used in the reconstructions of the images from the raw data for different iterative percentages (HIR 1 to HIR 6). The image characteristics were found using clinical protocols and relevant phantom with and without HIR. Image Quality (IQ) metrics were assessed in same location to analyze noise, noise power spectrum (NPS) low contrast detectability and spatial resolution.

2.3   Data Acquisition

A Catphan®504 phantom was scanned using a 64 slice CT(Ingenuity, Philips Medical Systems) with two different tube voltages (100 kVp, and 120 kVp) at three different dose (CT Dose Index Volume) levels, a reference dose of 42 mGy and two reduced doses of about 50% and 75% (see Table 1). A collimation of 0.64 mm and 0.8 pitch were used. Data sets were reconstructed with FOV of 350 mm and 512 × 512 pixel matrix using FBP and different HIR levels at 4 mm thickness, and standard filter.

Table 1. Scanning protocols.

Voltage (KVp)	Current (mAs)	CTDI_vol (mGY)	Reduced dose	Reconstruction algorithm
120	650	42.36	0%	FBP, HIR 1–6
120	325	21.20	49.95%	Same as above
120	163	10.66	74.83%	Same as above
100	550	21.63	48.94%	Same as above
100	275	10.83	74.43%	Same as above

The low-contrast module (CTP 515) was used to study the effect of HIR on the visibility, which contains 3 groups of supra-slice disks with diameters ranging from 2-15 mm and subject contrasts of 1%, 0.5%, and 0.3%. The uniformity module (CTP 486) was used to calculate the spatial uniformity in the circular ROIs located in the center and four peripheral regions. The CTP486 module was used to calculate image noise and noise power spectrum (NPS).

2.4   Performance Evaluations

Image quality may be objectively measured in terms of physical measurements. These measures include: noise characteristics, and low contrast resolution.

2.4.1   Spatial Resolution

The resolution properties of an imaging system are commonly described by its modulation transfer function (MTF). The MTF of a radiographic system can be calculated using Fourier Transformation by either evaluating the response of the system to periodic patterns, or more commonly measuring the line spread function (LSP). The CTP528 module of the CatPhan®504 phantom contains a bead point object with a diameter of 0.28 mm. The point object can be used to calculate the modulation transfer function (MTF) which characterizes the spatial resolution of images. The images reconstructed using HIR algorithms are comparable with FBP in terms of spatial resolution by observing the maximum spatial resolution (lp/mm).

2.4.2   Image Noise and Noise Characteristics

The CTP 486 module in the Catphan®504 phantom contains a uniform material that is designed to have HU values within 2% (20HU) of the CT number of water. Image noise measurements were performed using the CTP 486 module. Standard deviation of pixel intensities over a region of interest (ROI) is indicative of the image noise. We used circular regions of interest (ROIs) with 150 mm² for evaluating noise and HU value (see Fig. 2). We report the mean of the noise evaluated in five ROIs that are located at the center and four peripheral regions of the phantom. The mean of the five mean HU values in ROIs was calculated to evaluate the HU stability.

As described elsewhere [12], to capture the noise characteristics, the noise power spectrum (NPS) is often used to characterize CT reconstruction algorithms. NPS measurements are best achieved using a large homogenous region. This ensures that measured frequency variations are due to noise properties from the system acquisition and the impact of the reconstruction algorithm without influence from variations due to physical object texture arising from anatomy or pathology. The homogeneous section of the CTP 486 module in the Catphan®504 phantom was reconstructed. A noise image was produced by subtracting two neighbor slices. The NPS was calculated by the following equation:

$$NPS(u, v) = \frac{1}{A} \left[\left| \iint n(x, y) e^{-2\pi(xu + yv)} dx dy \right|^2 \right] \tag{3}$$

where $n(x, y)$ is the reconstructed noise image, u and v are frequency variables, and A represents the area over which n is defined.

The NPS was measured using a centered box with M × M pixels of the homogeneous phantom region.

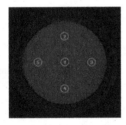

Fig. 2. ROIs for evaluating noise

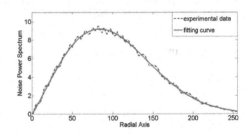

Fig. 3. Experimental data vs fitting curve

To examine the variance and spatial frequency characteristics of image noise, we measured the noise power spectrum (NPS) on reconstructed images obtained with the FBP and HIR algorithm. The NPS was calculated from the uniformity module of the Catphan®504 phantom images using the following procedure:

(a) The M × M matrix subimage located at the center was isolated from each phantom image;
(b) The mean of the isolated subimage was subtracted from the subimage to avoid having a direct current (DC) component in the Fourier transform.

The noise image was then zero-padded to a 512 × 512 image and was calculated using the 2D Fourier Transform averaged radially in the frequency domain to provide a 1-dimensional representation of the frequency distribution. To get a better idea of the power spectrum for one-dimensional NPS, we can do a rotational average of the two-dimensional NPS.

An analytical curve was suggested that fits the NPS accurately [14]. The curve we fit based on part data (not including low-frequency portion) is given by Eq. (2) that is similar to Weibull distribution.

$$NPS(x) = \begin{cases} \frac{A}{\alpha} x^m e^{-\frac{x^k}{\alpha}} & x \geq 0 \\ 0 & x < 0 \end{cases} \tag{4}$$

where x denotes the location of spatial frequency, A, m, k and α are unknown parameters to be solved.

The one-dimensional NPS can be calculated from a rotational average of the two dimensional NPS. Experimental data vs fitting curve are shown in Fig. 3.

The similarity in texture to FBP was defined as the ratio of the NPS for the low dose HIR technique (NPS_{HIR}) divided by the NPS for routine dose FBP (NPS_{FBP}) as a function of frequency. The resulting NPS ratio would be a value of NPS_{HIR} matched NPS_{FBP}. The spectrum change was calculated by the following equation:

$$spectrum\ change\ (SC) = \frac{1}{N} \sum_{i=1}^{N} \left| \frac{NPS_{HIR}(i)}{NPS_{FBP}(i)} - 1 \right| \times 100\% \tag{5}$$

where the N samples of the NPS curves represent the noise spectrum range of interest.

For estimate the noise characteristics, a value of spatial frequency splitting (SFS value) was calculated to divide noise power spectrum into two parts: lower image noise power spectrum and higher image noise power spectrum, which have the same amount of integral spectrum.

2.4.3 Low Contrast Resolution

The Catphan®504 phantom was used for imaging. The low-contrast module, CTP 515 (Fig. 4) consists of supra-slice and sub-slice targets. The supra-slice targets are of three contrast levels: 0.3%, 0.5% and 1.0%. The three outer groups include nine supra-slice targets with various diameters: 15.0 mm, 9.0 mm, 8.0 mm, 7.0 mm, 6.0 mm, 5.0 mm, 4.0 mm, 3.0 mm, and 2.0 mm. The low contrast detectability for any given contrast is defined as the diameter of the smallest visible sphere within the group.

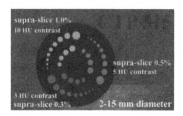

Fig. 4. Low-contrast module

Fig. 5. Signal (green) and background (blue) (Color figure online)

To find out the changes of image noise and possibly CT value associated with different scanning parameters and HIR algorithms, circular regions of interest (ROIs) with 50 mm^2 were placed on the image of the largest 1% contrast-filled sphere and on the background (BG) as indicated in the Fig. 5. Background ROIs were placed both at smaller and larger radii measured from the centre of the phantom. The two background values were averaged. The CNR is often used as a parameter for the image quality characterization to obtain practical measures of object detectability. CNRs were derived from mean CT numbers (CT) in the ROIs and the standard deviations (SD) in the background ROI as [15]:

$$CNR = \frac{|\mu_i - \mu_b|}{\sigma_b} \tag{6}$$

where μ_i is the mean CT number in HU of the tissue of interest, μ_b the mean HU of the background tissue and σ_b the standard deviation in HU of the background tissue.

3 Results

3.1 Spatial Resolution (MTF Estimation)

The acquisition and the image reconstruction parameters are presented blow: 150 mm field of view; 0.90 mm slice thickness; 768 × 768 matrix; standard resolution and standard reconstruction kernel (SRK) or high resolution and Y-sharp reconstruction kernel (YRK). Results of the estimation of MTF are summarized in Table 2.

Table 2. Modulation transfer function (MTF) of FBP and various HIR levels with SRK and YRK

Tube voltage/current	Reconstruction algorithm	MTF$_{50}$ (lp/cm) with SRK	MTF$_{50}$ (lp/cm) with YRK	MTF$_{10}$ (lp/cm) with SRK	MTF$_{10}$(lp/cm) with YRK
120 kVp, 325 mAs 21.20 mGy	FBP	3.398	6.533	6.007	8.945
	HIR 1–6	3.401–3.406	6.546–6.574	6.015–6.026	8.972–9.1777
120 kVp, 163 mAs 10.66 mGy	FBP	3.328	6.469	5.807	8.869
	HIR 1–6	3.331–3.354	6.475–6.499	5.914–5.953	8.905–9.059
100 kVp, 550 mAs 21.63 mGy	FBP	3.347	6.422	5.998	9.328
	HIR 1–6	3.354–3.374	6.450–6.598	6.006–6.014	9.369–9.512
100 kVp, 275 mAs 10.83 mGy	FBP	3.272	5.937	5.698	8.880
	HIR 1–6	3.289–3.371	5.995–6.082	5.793–6.001	8.955–9.156

The estimation of the MTF or spatial resolution are very similar for FBP and various HIR levels with any given scanning protocol in shape and position. There isn't obvious improvement in the spatial resolution of images reconstructed with different HIR levels compared with the FBP images at both 10% and 50% MTF. There is obvious change with different focal spot resolution and reconstruction kernels for MTF.

3.2 Image Noise

Reconstruction images with FBP and HIR levels are shown in Fig. 6. Figure 7 showed that there is continuous linear reduction of noise with increasing HIR levels for any given scanning protocol. With various HIR levels (HIR1–HIR6), image noise was improved from 0.40 to 1.86 at 325 mAs@120kVp, from 0.64 to 2.66 at 163 mAs@120kVp, from 0.42 to 1.92 at 550 mAs@100kVp, from 0.56 to 2.46 at 275 mAs@100kVp. For protocols with lower doses, every increasing HIR level resulted in relatively greater dose reduction. This indicates that HIR is effective to reduce noise with marked x-ray photon deprivation. However, as the HIR level increases, the images can have the appearance of being noise-free, which can present as an artifactual over smoothing of the images.

Fig. 6. Noise images: FBP (left); HIR4 (middle); HIR6 (right).

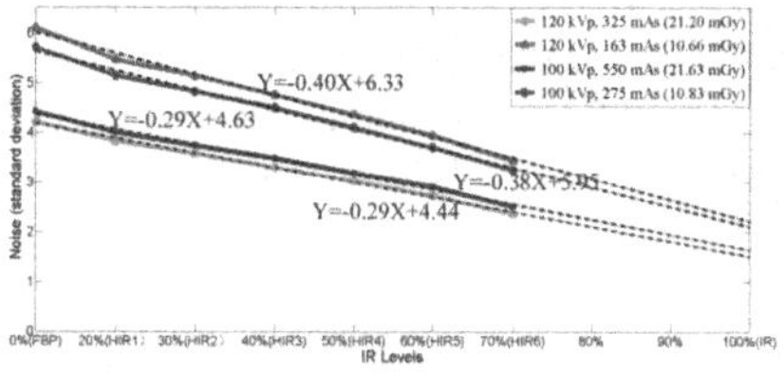

Fig. 7. Noise reduction with different HIR levels

3.3 Noise Power Spectrum

Noise or CNR isn't only parameter to evaluate the low contrast resolution of reconstructed image. HIR resulted in a noise texture described as coarser. Texture change was more evident as the increase of HIR levels.

To capture the noise characteristics, the noise power spectrum (NPS) is often used to characterize CT reconstruction algorithms. NPS measurements are best achieved using a large homogenous region. A 64×64 pixels noise image was produced by subtracting two neighbor slices in the homogeneous section of the CTP 486 module. The NPS was calculated by Eq. (3) and the curve was fitted by Eq. (4). The normalized NPS$_{HIR4}$ for any given scanning protocol were shown in 10, respectively.

In our study, the plots of the NPS (Figs. 8 and 9) show near identity of the frequency responses of images reconstructed with any given scanning protocol dose using FBP or different HIR level, respectively. This implies that the distribution of noise power remained the same for the entire spatial frequency. Consequently, the

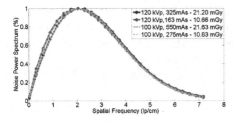

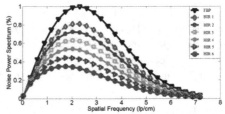

Fig. 8. Normalized NPS of HIR4 with different scanning protocols

Fig. 9. NPS$_{FBP}$ vs NPS$_{HIR}$ with different levels

noise texture of images reconstructed with FBP or HIR was not significantly different, although noise was significantly different for any given scanning protocol.

The NPS reflected improvements in the image noise with different HIR levels as compared with FBP at any given scanning protocol as shown in Fig. 9, which show the noise power spectrum normalized by the maximum NPS (NPS$_{FBP}$) at 163 mAs@120kVp. The results show different frequency responses of images reconstructed for FBP and various HIR levels, respectively. This implies that the distribution of noise power remained difference for the entire spatial frequency and noise frequency band is reduced with increasing HIR level for any given scanning protocol. Results demonstrate that HIR algorithm is effective to reduce noise based on the NPS analysis. Results of the estimation of the image noise, spectrum change (SC) for FBP and various HIR levels with different scanning protocols are summarized in Table 3.

Table 3. Image noise and spectrum change (SC).

Tube voltage/current	Parameters	FBP	HIR 1	HIR 2	HIR 3	HIR 4	HIR 5	HIR 6
120 kVp, 325 mAs	Noise	4.22	3.82	3.58	3.30	3.02	2.72	2.36
21.20 mGy	SC	0%	2.35%	3.90%	5.38%	8.64%	11.51%	15.63%
120 kVp, 163 mAs	Noise	6.10	5.46	5.14	4.76	4.36	3.94	3.44
10.66 mGy	SC	0%	2.48%	4.33%	5.95%	8.48%	10.80%	14.89%
100 kVp, 550 mAs	Noise	4.42	4.00	3.74	3.48	3.18	2.90	2.50
21.63 mGy	SC	0%	2.41%	3.86%	5.89%	8.63%	11.97%	15.38%
100 kVp, 275 mAs	Noise	5.70	5.14	4.82	4.50	4.10	3.70	3.24
10.83 mGy	SC	0%	2.42%	4.14%	6.31%	8.96%	12.36%	16.29%

The HIR NPS studies show an improvement for high-frequency noise compared with FBP NPS. Low contrast object is mainly determined by the low-frequency portion of the NPS. There are continuous increase of low-frequency components (reduction of SFS value) and larger spectrum change with increasing HIR level compared to FBP for any given scanning protocol. But even if there are continuous reduction of noise and increase of CNR with increasing HIR levels, the performance of human observers did not seem to depend on them absolutely. In some cases, coarser noise (low-frequency noise) could appear.

3.4   Low Contrast Resolution

Catphan phantom was employed as the imaging target: low contrast detectability and CNR were compared for routine dose CT with FBP and low dose CT with different HIR levels. Following figure show results calculated the CNR on images acquired at 120 and 100 kVp with the FBP and the different HIR levels using the CTP401 module of the Catphan® 504 phantom.

There are still very high CNR values using HIR method at the low-dose protocol. The plot in Fig. 10 is from a set of graphs that indicate the potential for dose reduction when using HIR. CNRs are generally decreased with decreasing radiation dose and increased with HIR levels, because noise decrease with increasing radiation dose and Hounsfield values are relatively contrast with varying radiation dose. We found CNR at a lower dose to be equivalent to that at a standard dose. For dose reduction of 50%, the CNRs of images reconstructed with HIR 5–6 at 325 mAs@120kVp, with HIR 4–6 at 550 mAs@100kVp are all higher than that of standard image. For dose reduction of 75%, the CNRs of images reconstructed with HIR 6 at 325 mAs@120kVp, and 550 mAs@100kVp are all higher than that of standard image.

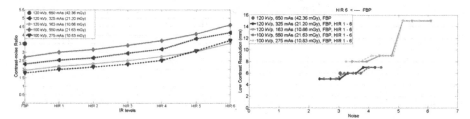

Fig. 10. CNR with different HIR levels for dose reduction

Fig. 11. Low-contrast resolution

3.5   Low Contrast Objects Detectability

A set of images with subject contrast of 0.5% were compared for FBP and different HIR levels (HIR 2, HIR 4 and HIR 6) at 163 mAs@120kVp with same reduced doses of 75%. The noise of Fig. 11(a)–(d) are 6.10, 5.14, 4.36 and 3.44, respectively. There is continuous reduction of noise with increasing HIR levels. Results show that there is not equally visibility evaluation. High HIR level may result in detail blurring.

Fig. 12. Images with subject contrast of 0.5%

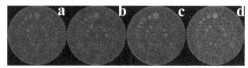

Fig. 13. Same noise with different scanning protocol

Figure 12 showed the subjective low-contrast resolution as a function of noise for different scanning parameters and HIR algorithms, which contains the supra-slice disks with diameters ranging from 2–15 mm and subject contrast of 0.5%. Each colored marker represent data from the same scanning protocol but seven varying HIR levels (FBP, HIR 1-HIR6); markers from right to left correspond to increased HIR levels. The low-contrast resolution in general improves with noise reduction.

Low contrast detectability and noise were compared for low dose CT with different HIR levels. Four images have similar noise with different HIR levels and radiation dose reduction in Fig. 13. The parameters of Fig. 13(a)–(d) are 325mAs@120 kVp and HIR 1; 163 mAs @ 120 kVp and HIR 6; 550mAs@100kVp, HIR 1; and 275 mAs@100kVp and HIR 6, respectively. The HIR NPS curves show an improvement for high-frequency noise compared with FBP NPS. Low contrast object is mainly determined by the low-frequency portion of the NPS. There is continuous increase of low-frequency components and larger spectrum change with increasing HIR level compared to FBP for any given scanning protocol. Even if there are continuous reduction of noise and increase of CNR with increasing HIR levels, the performance of human observers did not seem to be improved simultaneously because coarser noise (low-frequency noise) could appear. Our finding that the low-frequency components are greater than one of FBP (previously believed) may help to explain the discrepancy between the performance of human observers and that of the ideal low-contrast objects. Visually, it is quite apparent that HIR 4 stays much close the option, because HIR 6 has much greater low-frequency components.

Our results demonstrated that even at the same noise level, the low-contrast limit is different among different scanning protocols. There is no one-to-one correspondence between the noise and low-contrast resolution. Reducing scanning dose decreased low-contrast detectability at the same image noise level. Therefore it implicates that there is an examination specific limit on the dose reduction. As there are continuous reduction of noise and increase of CNR with increasing HIR levels, the performance of human observers did not seem to be improved simultaneously. According to Hanson's opinion [16], our finding that the low-frequency components (HIR) are greater than one of FBP (previously believed) may help to explain the discrepancy between the performance of human observers and that of the ideal low-contrast objects. Visually, it is quite apparent that HIR 4 stays much close the option selected in clinical medicine, although HIR 6 has much greater low-frequency components. These results agree well with clinical observation.

4   Conclusion

The phantom study provides objective assessment with HIR technique on image quality parameters. Compared with the conventional FBP, HIR results in Noise reduction with increase of HIR levels; improved contrast noise ratio; improved low contrast detectability and texture change or spectrum shift. Reducing scanning dose decreased low-contrast detectability at the same image noise level. Therefore, it implicates that there is a limit on the dose reduction when designated low-contrast structures are of clinical interest. There are continuous linear reduction of noise and linear increase of CNR with increasing HIR level for any given scanning protocol. Our study confirms that the MTF of the reconstructed image with HIR doesn't have decline along with the noise level decreasing. This indicates that high-contrast spatial resolutions were not lost while using HIR.

Our results demonstrated that even at the same noise level, the low-contrast limit is different among different scanning protocols. There is no one-to-one correspondence between the noise and low-contrast resolution. Reducing scanning dose decreased low-contrast detectability at the same image noise level. Therefore it implicates that there is an examination specific limit on the dose reduction when designated low-contrast structures are of clinical interest.

Noise or CNR isn't an only parameter to evaluate the low contrast resolution of reconstructed image. Even if there are continuous reduction of noise and increase of CNR with increasing HIR levels, the performance of human observers did not seem to be improved simultaneously because coarser noise (low-frequency noise) could appear. The HIR NPS curves show an improvement for high-frequency noise compared with FBP NPS. There is continuous increase of low-frequency components and larger spectrum change with increasing HIR levels compared to FBP for any given scanning protocol. HIR resulted in the change of noise texture. This change has the potential to affect spatial resolution and low contrast detectability. The shape of NPS is shifted continuously to low frequency with increasing HIR levels compared to FBP for any given scanning protocol.

Our finding that the low-frequency components (HIR) are greater than one of FBP (previously believed) may help to explain the discrepancy between the performance of human observers and that of the ideal low-contrast objects. Visually, it is quite apparent that HIR 4 stays much close the option selected in clinical medicine, although HIR 6 has much greater low-frequency components.

References

1. Hounsfield, G.N.: Computerized transverse axial scanning (tomography). Part 1. Description of system. Br. J. Radiol. **46**, 1016–1022 (1973)
2. Hara, A.K., Paden, R.G., Silva, A.C., Kujak, K.L., Lawder, H.J., Pavlicek, W.: Iterative reconstruction technique for reducing body radiation dose at CT: feasibility study. AJR **193**, 764–771 (2009)
3. Nuyts, J., De Man, B., Dupont, P., Defrise, M., Suetens, P., Mortelmans, L.: Iterative reconstruction for helical CT: a simulation study. Phys. Med. Biol. **43**, 729–737 (1998)
4. Liu, Y.J., Zhu, P.P., Chen, B., et al.: A new iterative algorithm to reconstruct the refractive index. Phys. Med. Biol. **52**, L5–L13 (2007)
5. Cheng, L., Fang, T., Tyan, J.: Fast iterative adaptive reconstruction in low-dose CT imaging. In: Proceedings of the IEEE International Conference on Image Processing, pp. 889–892. IEEE, New York, NY (2006)
6. Casey, B., Keen, C.: Philips Touts MRI Advances, CT dose reduction at RSNA. RSNA, Oak Brook (2009)
7. Noël, P.B., Fingerle, A.A., Renger, B., et al.: A clinical comparison study of a novel statistical iterative and filtered backprojection reconstruction. In: Physics of Medical Imaging Proceedings of SPIE, vol. 7961 (2011)

8. Division, Siemens Healthcare Imaging: Mathematical Approach Contributes to Lower Radiation Dose in Computed Tomography: Siemens Develops Innovative Method for Iterative Reconstruction of CT Images. Siemens, Erlangen (2009)
9. Bruder, H., Raupach, R., Sedlmair, M., Sunnegardh, J., Stierstorfer, K., Flohr, T.G.: Reduction of radiation dose in CT with an FBP-based iterative reconstruction technique (abstract). B-568, insight into imaging (ECR abstract book). S131 (2010)
10. Joemai, R.: Improved image quality in clinical CT by AIDR. Toshiba Med. Syst. J. Vis. **16**, 1–3 (2010)
11. Jensen, K., Catrine, A., Martinsen, T., Tingberg, A., et al.: Comparing five different iterative reconstruction algorithms for computed tomography in an ROC study. Eur. Radiol. **24**, 2989–3002 (2014)
12. Hsieh, J.: Computed Tomography Principles, Design, Artifacts, and Recent Advances, vol. 2. SPIE Press, Bellingham (2009)
13. Catphan@504 Phantom Manual (The phantom Laboratory, Salem, New York). http://www.phantomlab.com/library/pdf/catphan504manual.pdf
14. Benítez, R.B., Ning, R., Conover, D., Liu, S.H.: NPS characterization and evaluation of a cone beam CT breast imaging system. J. X-Ray Sci. Technol. **17**, 17–40 (2009)
15. Gupta, A.K., Nelson, R.C., Johnson, G.A., Paulson, E.K., Delong, D.M., Yoshizumi, T.T.: Optimization of eight-element multi-detector row helical CT technology for evaluation of the abdomen. Radiology **227**, 239–745 (2003)
16. Hanson, K.M.: Detectability in computed tomographic images. Med. Phys. **6**, 441–451 (1997)

Fast Single Image De-raining via a Weighted Residual Network

Ruibin Zhuge, Haiying Xia$^{(\boxtimes)}$, Haisheng Li, and Shuxiang Song

Guangxi Normal University, Guangxi 541000, China
653140685@qq.com, 573023049@qq.com, 187373363@qq.com,
songshuxiang@mailbox.gxnu.edu.cn

Abstract. Deep learning based methods for single image de-raining have shown great success in recent literatures. However, it is still a challenge to reduce the computation time while maintaining the de-raining performance. In this paper, we introduce a weighted residual network (WRN) to address above challenge. Inspired by the image processing knowledge that a rainy image can be decomposed into a base (low-pass) layer and a detail (high-pass) layer, we train the network on a weighted residual between the weighted detail layer of rainy image and the detail layer of clean image, which can significantly reduce the mapping range from input to output and easily employ the image enhancement operation on the base layer and the detail layer separately to handle the heavy rain with hazy looking. We also introduce a weighted convolution-deconvolution network structure to make the training easier. The first layer of network is a multi-scale convolution to expand the receptive field of the network. Our WRN requires less computation time for processing a test image because we set the stride of intermediate layers to 2 without zero-padding. Experiment results on both synthetic and real-world images demonstrate our WRN achieves high-quality recovery compared to several advanced methods of single image de-raining.

Keywords: Rain removal · Deep learning · WRN

1 Introduction

Many outdoor computer visual systems need clear and visible images, such as surveillance and navigation. The precision of most computer vision algorithms (e.g. object detection, location, tracking) depends on the quality of images. Under the rainy weather, objects are blurred by dense rain streaks, which severely degrades image quality. Thus, it is necessary to design effective approaches for removing rain streaks from images.

With the abundant temporal information from video sequences, great works have been made in video de-raining tasks [1,2,5,14,17,19]. Rain streaks can be easily detected and removed by using the frame difference information. However, it is challenging to remove rain streaks from a single image duo to the lack

© Springer Nature Switzerland AG 2018
L. Cheng et al. (Eds.): ICONIP 2018, LNCS 11306, pp. 257–268, 2018.
https://doi.org/10.1007/978-3-030-04224-0_22

temporal information. To handle this challenge, several algorithms [8,9,11,12] have been proposed. For example, the first single image de-raining algorithm was proposed by Kang et al. [8]. They utilized a bilateral filter to decompose the rain image into the low and high frequency parts. The rain streaks remained in the high frequency were removed by performing sparse coding and dictionary learning algorithms. In [9], a kernel regression algorithm was introduced for detecting the rain streaks in a single image while an adaptive nonlocal means filter was used to restore the detected rain streaks. Luo et al. [12] proposed a discriminative sparse coding algorithm to solve a non-linear screen blend model of rain images. In [11], Li et al. introduced simple patches based on the Gaussion mixture models (GMMs) to impose constraints on both the background and the rain layer for accommodating multiple orientations and scales of the rain streaks. These methods try to build models for rain streaks removal and employ optimizations to solve the mathematical models, which requires lots of manual parameters and computation resources.

Recently, the deep convolutional neural network (CNN) based methods have been proven usefully in the single image de-raining tasks [3,4,16,18]. In [3], Fu et al. separated the input rainy image into a base layer and a detail layer. They proposed a three-layer CNN to remove the rain streaks from the detail layer while a image enhancement was merged into the network to handle the heavy rain with hazy looking. To explore more effective information, Fu et al. [4] adopted the deep residual network (ResNet) structure [7] as the parameter layers and introduced a negative residual mapping to reduce the mapping range of the network. To deal with the environment with rain accumulation, Yang et al. [16] proposed a recurrent network structure to restore the rain accumulation images. A dehazing network was embedded in the base network to remove atmospheric veils. However, there still exist two challenges. One is the improvement for de-raining performance. As shown in Fig. 1, the results of methods [3,4,16] remain some rain streaks or artifacts. Another is the reduction for computation time. For example, method [16] requires approximately 126s for testing a 500 × 500 rainy image, which is hard for practical applications. To address aforementioned challenges, we propose a weighted residual network (WRN) to remove the rain streaks from a single image effectively. We first decompose a rainy image into a base layer and a detail layer by a low-pass filter. Then we directly use the clean image to subtract the base layer of rainy image to get the detail layer of clean image. Based on the decomposition operation, image enhancement can be used to enhance image visibility when the image suffers from heavy rain with hazy looking. After obtaining the detail layers, we introduce a weighted learning strategy to make the training easier. Different with methods in [3,4], it directly learns a weighted residual between the weighted rainy detail layer and the clean detail layer. Inspired by the encoder-decoder network with skip connections [13], we propose a weighted convolution-deconvolution network structure to adjust the contributions of parameter layers. We also use the multi-scale convolution to expand the receptive field and promote feature fusion. A key step of proposed network is the stride of intermediate layers is set to 2 without zero-padding, so

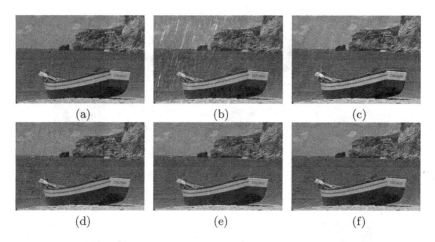

Fig. 1. De-raining results of different methods. (a) Ground truth (b) Rain image (c) The result of method [3] (d) The result of method [4] (e) The result of method [16] (f) The result of proposed WRN

that the computation time can be significantly cut down. We only require 1.9s to test a 500×500 rainy image on our CPU. The de-raining result of our WRN is shown in Fig. 1(f). It removes almost all of the rain streaks while preserving the image details well. Our contributions can be summarized as follows:

(1) We propose a weighted residual network (WRN) for single image rain removal. The designed network is composed by a weighted residual learning strategy and a weighted convolution-deconvolution network structure.
(2) Some strategies are used to boost the de-raining effectiveness: a multi-scale convolution; stride of intermediate convolution-deconvolution layers is set to 2 without zero-padding.
(3) Experiment results on both synthetic and real-world images prove the effectiveness of WRN.

2   Proposed Method

Our proposed WRN is illustrated in Fig. 2. It consists of a weighted residual learning strategy and a weighted convolution-deconvolution network.

2.1   Weighted Residual Learning Strategy

As described in [3], a rainy image X and the corresponding clean image Y can be separately decomposed into a base layer and a detail layer by a guider filter [6],

$$X = X_{base} + X_{detail}, Y = Y_{base} + Y_{detail} \qquad (1)$$

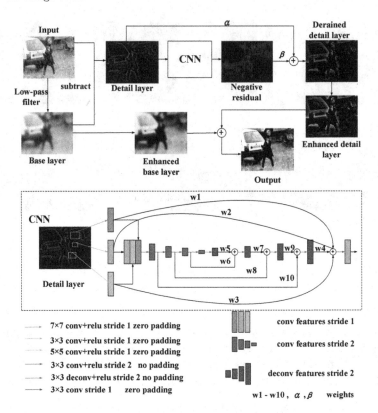

Fig. 2. The architecture of proposed WRN

where the subscript 'detail' denotes the detail layer, and 'base' denotes the base layer.

Detail layers are more sparse than the whole image, so training on the detail layers is more easier than directly training on the whole image. The detail layers are described in the following equations,

$$X_{detail} = X - X_{base}, Y_{detail} = Y - Y_{base} \qquad (2)$$

In [3], X_{base} and Y_{base} are considered as approximately equal, which are removed from the training process. Actually, X_{base} and Y_{base} can not be equal. As shown in Fig. 3, the pixel values of X_{base} are higher than the pixel values of Y_{base} duo to the influence of rain. But the network is trained to learn $X_{trained\ detail} \approx Y_{detail}$, which causes the following result,

$$(X_{trained\ detail} + X_{base}) > (Y_{detail} + Y_{base}) \qquad (3)$$

This will lead to the derained results to be brighter than the clean images. To address this limitation, we directly use the filtered result X_{base} as Y_{base} since the rainy image can be regarded as adding the rain streaks to the clean image,

$$Y_{base} = X_{base} \qquad (4)$$

| (a) | (b) | (c) | (d) |

Fig. 3. Examples of base layers. (a) Ground truth (b) Rainy image (c) The base layer of ground truth (d) The base layer of rainy image

According to Eq. 4, the negative residual mapping in [4] can be described as,

$$N_{neg} = Y - X = Y_{detail} + X_{base} - (X_{detail} + X_{base}) = Y_{detail} - X_{detail} \quad (5)$$

where N_{neg} refers to the negative residual mapping. To further reduce the mapping range of network, we introduce a weighted residual learning strategy (Fig. 2),

$$N_{neg} * \beta = Y_{detail} - X_{detail} * \alpha \quad (6)$$

where α and β indicate weights, which are adjusted in the training process.

The proposed learning strategy combines the advantages of method [3,4] and makes some improvements. Firstly, the prior of detail layer is used to reduce the mapping range. Different from the method [3], we just need a guider filter to get both the rainy detail layer and the clean detail layer, which reduces the training time. Secondly, the negative residual between the clean detail layer and rain detail layer becomes a special case of the proposed weighted residual since the parameters α and β are adjustable. Our proposed learning strategy is able to adjust contributions of the negative residual and the input detail while reducing the mapping range, which is has better learning than method in [4]. Thirdly, as shown in Fig. 2, we can easily employ the image enhancement operation on the base layer and the detail layer separately to handle the heavy rain looks hazy. Similar with method [3], we multiply the detail layer by two and use a non-liner function [10] to enhance the base layer,

$$Y_{enhanced} = (X_{base})_{enhanced} + 2(X_{detail}) \quad (7)$$

2.2 Weighted Convolution-Deconvolution Network

In [13], Mao et al. introduced a deep symmetrically link encoding-decoding network with skip connections for image restoration tasks, which tackles the problem of gradient vanishing and obtains performance gains. Based on the encoding-decoding network structure, we propose a weighted convolution-deconvolution network as parameter layers to make the training process easier, as shown in Fig. 2. Instead of using a symmetric encoding-decoding network, we first employ a multi-scale convolution to expand the receptive field and promote feature fusion.

Since the feature size needs to be same, we set the stride of convolution to 1 with zero-padding. Then we introduce a weighted module,

$$Y = w1 * Y_{deconv} + w2 * Y_{conv} \qquad (8)$$

where $w1$ and $w2$ refer to weights, Y_{conv} means the results of convolution layer. Y_{deconv} indicates the results of deconvolution layer. The result of each convolution is added to the result of corresponding deconvolution, including the multi-scale convolution layer and the intermediate convolution layer. Compared with the network in [13], the proposed weighted network can not only enjoy the gains of skip connection, but also adjust the contributions of convolution-deconvolution layers to make full use of the effective information. Since the computation time of network needs to be considered, we set the stride of intermediate convolution-deconvolution layers to 2 without zero-padding, which significantly cuts down the test time. Figure 4 shows the results of different learning strategies. We can see the proposed WRN with stride 2 convergences faster than the 10-layer network in [13] and the WRN with stride 1.

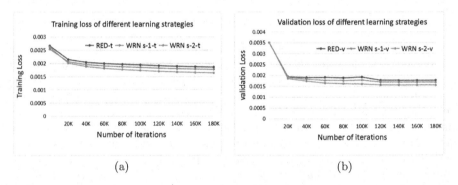

(a)   (b)

Fig. 4. Training loss and validation loss of different training strategies. 'RED' refers to the network in [13]. 's-1-t' means training loss with stride 1. 's-2-v' is the validation loss with stride 2.

2.3   WRN

Loss function. Combining the weighted residual learning strategy and the weighted convolution-deconvolution network, the objective function of WRN can be written as,

$$J = \frac{1}{N} \sum_{N}^{i=1} \| \beta * f(X_{i,detail}, W, b) + \alpha * X_{i,detail} - Y_{i,detail} \|_F^2 \qquad (9)$$

where N is the number of training images, $f(\cdot)$ is the proposed convolution-deconvolution network, W and b are the network parameters. α and β refer to weights, $\|\cdot\|_F$ means Frobenius norm.

Table 1. The detail setting of WRN

Layers	Filter sizes	Use ReLU	Features	Strides	Zero-padding
Convolution ×1	3×3/5×5/7×7	Yes	32/32/32	1	Yes
Convolution ×4	3×3	Yes	32	2	No
Deconvolution ×4	3×3	Yes	32	2	No
Convolution ×1	3×3	No	3	1	Yes

The proposed WRN has 10 layers to balance the trade-off between performance and computation efficiency. In the first layer, we use 3×3, 5×5 and 7×7 filter sizes to generate 32 features separately. The stride of each convolution is set to 1 with zero-padding to make sure the features are the same size. Then the features are concatenated to 96 followed by 4 convolution layers and 4 deconvolution layers. The stride of each layer is set to 2 without zero-padding and each layer generates 32 features. In the last layer, we use 3×3 convolution with 1 stride and zero-padding to fuse the different features. The detail design is shown in Table 1.

3 Experiments

In this section, the performance of our proposed WRN is compared with four advanced de-raining methods [3,4,11,16] on both synthetic and real-world images. All the networks are trained on a computer with Inter Xeon E3 CPU, 8GB RAM and NVIDIA Geforce GTX 750ti. We test the images on a PC with Inter Core i3 CPU. The implementation code of [11] is provided in Matlab version by authors. For methods of [3,4,16], the codes are exhibited on the websites[1,2,3].

3.1 Datasets

Since it is difficult to obtain the clean and rainy image pairs, we use three public synthetic datasets for comparison. **Dataset 1**: A large dataset is provided by [4], which contains 1000 clean images and 14000 synthetic rainy images. We random select 9100 image pairs to generate three million 64×64 rainy/clean image patch pairs for training. 700 pairs of rainy/clean images are used for validation. For testing, we random select 100 rainy/clean images from the remaining 4200 image pairs. **Dataset 2**: [16] provides 200 training image pairs and 100 testing image pairs. We use the 200 training image pairs to generate 800 thousand 64×64 image patch pairs for training. **Dataset 3**: 12 image pairs provided by [11]. We use the 100 testing image pairs from **Dataset 2** to generate 400 thousand 64×64 image patch pairs for training.

---

[1] https://xueyangfu.github.io/projects/cvpr2017.html.
[2] https://xueyangfu.github.io/projects/tip2017.html.
[3] http://www.icst.pku.edu.cn/struct/Projects/joint_rain_removal.html.

3.2 Parameter Setting

We use Adam with weight decay of 10^{-10} to train our WRN and set the mini-batch size to 20. The learning rate is initialized as 10^{-3} and divided by 10 at 100 K iterations for **Dataset 1**. We terminate training at 180 K iterations. For **Dataset 2** and **Dataset 3**, the learning rate is divided by 10 at 4 and 8 epochs separately. The training is terminated at 10 epochs. For the other methods, we use the optimal settings published in the literatures. The parameters α, β and 'w5-w10' of our WRN are initialized as 0.5 while the 'w1-w4' are initialized as 0.25.

Fig. 5. De-raining results of different methods on **Dataset 1**. (a) Ground truth (b) Rain image (c) The result of method [11] (d) The result of method [3] (e) The result of method [4] (f) The result of proposed WRN

Table 2. Average SSIM and PSNR on different datasets.

	Dataset 1	Dastaset 2	Dataset 3
Ground truth	1/Inf	1/Inf	1/Inf
Rain images	0.7046/21.34	0.8255/25.52	0.8371/28.82
Method [11]	0.7844/23.75	0.8712/28.36	0.8928/30.70
Method [3]	0.8333/21.97	0.9134/28.17	0.9033/29.42
Method [4]	0.8703/27.32	0.9161/31.39	0.8947/30.68
Method [16]	/	0.9696/35.21	0.9447/34.49
WRN	0.8888/28.31	0.9686/35.26	0.9542/35.05

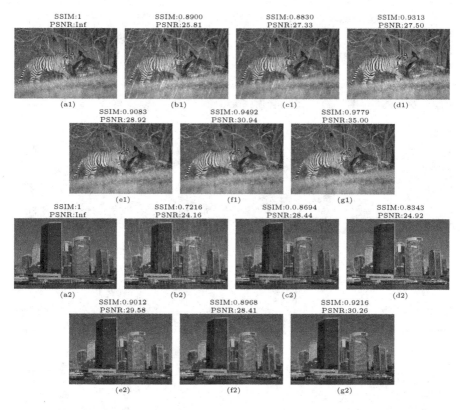

Fig. 6. De-raining results of different methods on **Dataset 2** and **Dataset 3**. (a1)-(a2) Ground truth (b1)-(b2) Rain images (c1)-(c2) Results of method [11] (d1)-(d2) Results of method [3] (e1)-(e2) Results of method [4] (f1)-(f2) Results of method [16] (g1)-(g2) Results of proposed WRN

3.3   Results on Synthetic Images

We compare the de-raining performance of different methods on synthetic images quantitatively and qualitatively. We use SSIM [15] and PSNR for quantitative evaluation. Since the training code of method [16] is not provided. We just compare it with other methods on **Dataset 2** and **Dataset 3**. The results of all synthetic images are shown in Table 2. As can be seen, the proposed WRN achieves the best quantitative performance compared to other methods. Our WRN is able to handle different types of the rainy images. It is more capable of learning by the weighted network structure, which can adjust the contribution between layers dynamically. We also show the qualitative comparison of different methods in Figs. 5 and 6. It can be clearly observed that the de-raining results of methods [3,4,11,16] can not remove all the rain streaks. Those methods contain rain artifacts or remove the image details. Our WRN can remove the rain streaks and preserve the image details effectively.

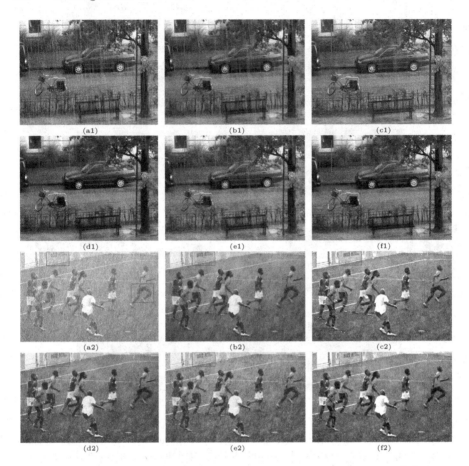

Fig. 7. De-raining results of different methods on real-world images. (a1)-(a2) Rain images (b1)-(b2) Results of method [11] (c1)-(c2) Results of method [3] (d1)-(d2) Results of method [4](e1)-(e2) Results of method [16] (f1)-(f2) Results of proposed WRN

3.4 Results on Real-World Images

We also compare the different methods on the real-world images. The qualitative results of real-world images are shown in the first and second row of Fig. 7. As we can seen, our proposed WRN provides better visual performance by preserving details, compared to other methods. In Fig. 7 (a2)-(f2), we use the image enhancement to handle the heavy rain with hazy looking. Methods [4,11,16] are based on the post-processed, which means directly employing the non-liner function [10] on the de-raining results. For method [3] and WRN, we employ the enhancement operation on the base layer and detail layer separately, as described in Sect. 2.1. We can observe the results based on decomposition are better than the post-processed results. The proposed WRN can remove rain streaks and enhance the visualization since it combines the effective de-raining network with image enhancement based decomposition.

Table 3. Average running time of different methods on three image sizes(seconds)

Image sizes	Method [11]	Method [3]	Method [4]	Method [16]	WRN
250 × 250	144.8	6.7	3.3	42.4	1.0
500 × 500	681.2	24.8	7.4	126.9	1.9
750 × 750	1557.9	54.2	14.6	234.6	3.5

3.5 Running Time

Table 3 shows the average running time of ten rainy images by using methods [3, 4, 11, 16] and our WRN. As observed, method [11] consumes lots of time because of the complex optimizations. The computation time of methods [3, 16] is limited by the redundant network parameters. Compared with the fast method [4], our proposed WRN takes about a quarter of the time for testing a 750×750 image since it has simple convolution-deconvolution structure with stride 2.

4 Conclusion

This paper has proposed a weighted residual network (WRN) for single image rain removal, including a weighted residual learning strategy and a weighted convolution-deconvolution network. The weighted residual learning strategy aims to reduce the mapping range of the network and make the training easier. Based on the encoder-decoder network, we proposed a weighted convolution-deconvolution network with a multi-scale convolution to explore more effective information and adjust the contributions of parameter layers. We set the stride of convolution-deconvolution to 2 without zero-padding to accelerate computing. Experiment results on different synthetic datasets and real-world images show our WRN achieves the state-of-the-art while less computation resources are required. The performance of proposed WRN can be improved by the fusion design of loss function, which is targeted for future work.

Acknowledgment. This work is supported by the National Natural Science Foundation of China (No.61762014, No.61462026 and No.61762012), the Opening Project of Guangxi Colleges and Universities Key Laboratory of robot & welding (Guilin University of Aerospace Technology), the Opening Project of Shaanxi Key Laboratory of Complex Control System and Intelligent Information Processing, and the Research Fund of Guangxi Key Lab of intelligent integrated automation.

References

1. Bossu, J., Hautiere, N., Tarel, J.P.: Rain or snow detection in image sequences through use of a histogram of orientation of streaks. Int. J. Comput. Vis. **93**(3), 348–367 (2011)
2. Eigen, D., Krishnan, D., Fergus, R.: Restoring an image taken through a window covered with dirt or rain. In: IEEE International Conference on Computer Vision, pp. 633–640 (2014)

3. Fu, X., Huang, J., Ding, X., Liao, Y., Paisley, J.: Clearing the skies: a deep network architecture for single-image rain removal. IEEE Trans. Image Process. **26**(6), 2944–2956 (2016)
4. Fu, X., Huang, J., Zeng, D., Huang, Y., Ding, X., Paisley, J.: Removing rain from single images via a deep detail network. In: IEEE Conference on Computer Vision and Pattern Recognition, pp. 1715–1723 (2017)
5. Garg, K., Nayar, S.K.: Detection and removal of rain from videos. In: Proceedings of the 2004 IEEE Computer Society Conference on Computer Vision and Pattern Recognition, CVPR 2004, vol. 1, pp. I-528-I-535 (2004)
6. He, K., Sun, J., Tang, X.: Guided image filtering. IEEE Trans. Pattern Anal. Mach. Intell. **35**(6), 1397–1409 (2013)
7. He, K., Zhang, X., Ren, S., Sun, J.: Deep residual learning for image recognition. In: IEEE Conference on Computer Vision and Pattern Recognition, pp. 770–778 (2016)
8. Kang, L.W., Lin, C.W., Fu, Y.H.: Automatic single-image-based rain streaks removal via image decomposition. IEEE Trans. Image Process. **21**(4), 1742–1755 (2012)
9. Kim, J.H., Lee, C., Sim, J.Y., Kim, C.S.: Single-image deraining using an adaptive nonlocal means filter. In: IEEE International Conference on Image Processing, pp. 914–917 (2014)
10. Li, Y., Guo, F., Tan, R.T., Brown, M.S.: A contrast enhancement framework with JPEG artifacts suppression. In: Fleet, D., Pajdla, T., Schiele, B., Tuytelaars, T. (eds.) ECCV 2014. LNCS, vol. 8690, pp. 174–188. Springer, Cham (2014). https://doi.org/10.1007/978-3-319-10605-2_12
11. Li, Y., Tan, R.T., Guo, X., Lu, J., Brown, M.S.: Rain streak removal using layer priors. In: Computer Vision and Pattern Recognition, pp. 2736–2744 (2016)
12. Luo, Y., Xu, Y., Ji, H.: Removing rain from a single image via discriminative sparse coding. In: IEEE International Conference on Computer Vision, pp. 3397–3405 (2015)
13. Mao, X.J., Shen, C., Yang, Y.B.: Image restoration using very deep convolutional encoder-decoder networks with symmetric skip connections. In: Advances in Neural Information Processing Systems, pp. 2802–2810 (2016)
14. Santhaseelan, V., Asari, V.K.: Utilizing local phase information to remove rain from video. Int. J. Comput. Vis. **112**(1), 71–89 (2015)
15. Wang, Z., Bovik, A.C., Sheikh, H.R., Simoncelli, E.P.: Image quality assessment: from error visibility to structural similarity. IEEE Trans. Image Process. **13**(4), 600–612 (2015)
16. Yang, W., Tan, R.T., Feng, J., Liu, J., Guo, Z., Yan, S.: Deep joint rain detection and removal from a single image. In: IEEE Conference on Computer Vision and Pattern Recognition, pp. 1357–1366 (2017)
17. You, S., Tan, R.T., Kawakami, R., Ikeuchi, K.: Adherent raindrop detection and removal in video. IEEE Trans. Pattern Anal. Mach. Intell. **38**(9), 1721–1733 (2016)
18. Zhang, H., Patel, V.M.: Density-aware single image de-raining using a multi-stream dense network. In: IEEE Conference on Computer Vision and Pattern Recognition (2018)
19. Zhang, X., Li, H., Qi, Y., Leow, W.K.: Rain removal in video by combining temporal and chromatic properties. In: IEEE International Conference on Multimedia and Expo, pp. 461–464 (2006)

SCE-MSPFS: A Novel Deep Convolutional Feature Selection Method for Image Retrieval

Dong-dong Niu[1,2], Yong Feng[1,2(✉)], Jia-xing Shang[1,2], and Bao-hua Qiang[3,4]

[1] College of Computer Science, Chongqing University, Chongqing 400030, China
[2] Key Laboratory of Dependable Service Computing in Cyber Physical Society,
Ministry of Education, Chongqing University, Chongqing 400030, China
{andyniu,fengyong,shangjx}@cqu.edu.cn
[3] Guangxi Key Laboratory of Trusted Software,
Guilin University of Electronic Technology, Guilin 541004, China
[4] Guangxi Cooperative Innovation Center of cloud computing and Big Data,
Guilin University of Electronic Technology, Guilin 541004, China
qiangbh@guet.edu.cn

Abstract. Effective image features are of vital importance for content-based image retrieval task. Recently, deep convolutional neural networks have been widely used in learning image features and have achieved promising results. However, there are still two questions need to be addressed. The first is the limitation of the image size in some works, and the second is the convolutional feature may not directly suitable for image retrieval. In our paper, we comprehensively solve these two problems by proposing a novel feature selection approach based on a pre-trained CNNs. Compared with others feature selection methods, our approach takes a two-stage strategy. The first stage is to select the effective feature sets using our proposed Median Sum Pooling Feature Selection method, and the second stage boosts the selected feature sets using the Space Channel Enhancement model. We evaluate our method on three benchmark datasets including Oxford5K, Paris6K, and Holiday. The experimental results show that our proposed method achieves competitive performance on both Oxford and Paris buildings benchmarks.

Keywords: Image retrieval · Feature selection
Convolutional neural networks

1 Introduction

Content-based image retrieval (CBIR) has been an active research topic in the computer vision society for decades due to its wide range of applications in both academic and industrial fields. Most existing approaches adopt low-level visual hand-crafted local feature, e.g. SIFT [1] and DoG, and then encode them with bag-of-words (BoW) model, vector locally aggregated descriptors (VLAD) or

© Springer Nature Switzerland AG 2018
L. Cheng et al. (Eds.): ICONIP 2018, LNCS 11306, pp. 269–281, 2018.
https://doi.org/10.1007/978-3-030-04224-0_23

Fisher vectors (FV) as the final representation. Since local features extracted by SIFT descriptors are good at capturing object characteristics well, such as edges and corners, it usually exhibits a better performance in instance image retrieval. However, three major issues still need to be addressed. One is that the SIFT-features lack discriminability [8] to tell the difference in images. Although this drawback can be relieved to some extent by embedding SIFT feature to higher dimensional space, there is still a huge semantic gap between SIFT-based image representation and human perception on image instances. The second issue is the strong burstiness effect [2], i.e. numerous descriptors are nearly similar in the same image and it largely degrades the quality of SIFT-based image representation for the image retrieval task. The last issue is that it is still a challenging problem for SIFT-feature to handle complex image change, such as shape deformation, illumination variation, and heavy occlusion.

Recently deep convolutional neural networks (CNNs) has demonstrated excellent performance on image classification task on the datasets such as PASCAL VOC and ImageNet Large Scale Visual Recognition Challenge (ILSVRC) [3,4], and it also has achieved great success in object detection [5], semantic segmentation [6], etc. The CNNs trained on a large annotated dataset like ImageNet automatically capture multilevel rich information at higher semantic levels and achieve superior performance compared to hand-crafted features. The middle layers of the CNNs preserve more specific information on edges, corners, patterns and structures and so they are more suitable for image retrieval compared with the fully connected layers. Some works [7–9] have shown that the convolutional feature maps extracted from the CNN can be viewed as a set of local descriptors, which can be aggregated into powerful global features for image retrieval. These local deep descriptors are learned by training a CNN on large annotated image dataset or fine-tuning a pre-trained CNN on task-specific datasets. The state-of-the-art results achieved by these works suggest that the local deep descriptors are more discriminative than the hand-crafted local feature. And some image retrieval algorithms using the off-the-shelf representation from CNNs have been proposed [10].

Although some approaches have applied pre-trained CNNs to extract general features for image retrieval and get promising outcomes, two questions still remain there. First, some works [8] have the limitation that the size of a test image must be the same as the training image. However different sizes of the input images may affect the behavior of convolutional layers as image pass through the network, and may have an unstable influence on the final retrieval results. Second, CNNs are trained for classification tasks by default, the feature from the final layer (or higher layers) are usually used for making decision because that this layer extract more semantic features for category-levels classification. So it is difficult to decide whether it directly extracts the feature from the final layers or higher layers for instance image retrieval.

In our paper, we comprehensively solve these two problems. For the first problem, we drop out the final fully connected layers so our CNN model have no limits on the image size. The key difference between our method and existing approaches is that we do not resize the input image as the work in [12] which means that our input image preserves more detailed information. The second problem is solved by our proposed Median Sum Pooling Feature Selection (MSPFS) method. The purpose of our method is to effectively select the discriminant feature sets which distinguish different types of images well and boost the final image representation. Inspired by the SEnet model [13], we then propose Space Channel Enhanced Median Sum Pooling Feature Selection (SCE-MSPFS). We add modeling channel part based on MSPFS method. So in this paper, we takes a two-stage strategy for the image retrieval. The first stage is to select the effective feature sets, and the second stage is to enhance the selection effect. More details will be given in the remaining sections. Our experiments show that applying our methods on the public image retrieval datasets significantly improves image retrieval accuracy.

2   The Proposed Approach

In this section, we first describe how to get the convolution features and explain it from two points of view based our understanding. We then propose some methods to boost the deep feature representation, namely MPFS, SPFS, and MSPFS. Then our proposed method MSPFS incorporates space channel enhancement model and we name it SCE-MSPFS method, which is inspired from SEnet model. At last, we describe the whole process of our image retrieval algorithm.

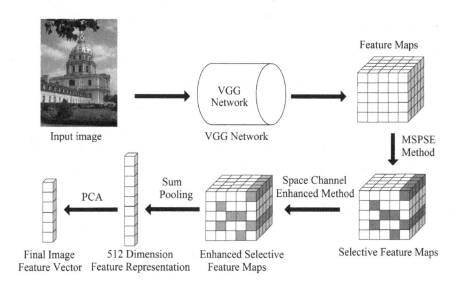

Fig. 1. Pipeline of our proposed retrieval method. (The figure best view in color)

2.1 Understanding Convolution Features

For an arbitrary-sized image, we put it through a pre-trained CNN (e.g., VGG-19 [4]) with all fully connected layers discarded. Then we extract a 3 dimension tensor from a layer in CNN. Now we donate $\mathcal{X} \in \mathbb{R}^{K \times H \times W}$ as the order-3 feature tensor from a layer(e.g., pool5), where W and H refer to the height and width of the feature map, and K is the total number of channels (or feature maps) in that layer.

We may view the deep feature from two aspects. One is that the feature tensor includes a set of 2 dimension feature map $\mathcal{F}_i(i = 1, ...K)$ and the Max-Pooling is in this way. From another point of view, $\mathcal{X}$ can also be considered as having $H \times W$ feature vectors and each vector is K-dimensional, and our methods are based on this perspective. The latter view is a popular perspective now and some works treat this kind of feature as the SIFT-based feature vector, and many works conduct experiments in this manner.

Let us give a specific example, We feed an image with 1024 in height and 768 in width into the pre-trained VGG-19 network and extract the pool5 feature, and we will obtain a $512 \times 32 \times 24$ activation tensor. So from the former view we get 512 feature maps and from the latter view we get 512-D feature vectors of size 32×24 from the latter view.

2.2 Selective Convolutional Features

We now propose some methods to boost the deep feature representation, namely MPFS (Max Pooling Feature Selection), SPFS (Sum Pooling Feature Selection), and MSPFS (Median Sum Pooling Feature Selection). The first two are straightforward methods while the last one is our proposed method. We get inspiration from the concept of RoI (Region of Interest) [21], which is just like a kind of attention mechanism and it can find the most interest response automatically. So compared with the traditional hand-craft features, our proposed methods may get the final feature representation more suitably.

A. MPFS. Max pooling [14] is first used to get the powerful feature vector representation and experiments show it does get effective image retrieval results. So we would like to try this method first and we formulate the MPSE method in the following. For every channel in $\mathcal{X}$, there are different activation responses and we choose the max activation value on that feature map to construct the final feature vector $\mathcal{F}_{max}$.

$$\mathcal{F}_{max} = \{\mathcal{F}_1, ..., \mathcal{F}_k, ..., \mathcal{F}_K\} \qquad \mathcal{F}_k = max\ x, \quad x \in \mathcal{X}_k \qquad (1)$$

B. SPFS. MPFS only use the channel max activation response, and the rest of activation values are deprived, which may contain more effective discriminate features than the max value. So based on that, someone propose SPSE method to

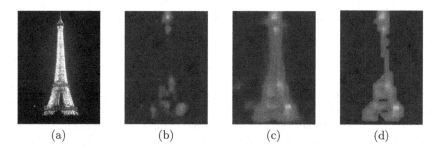

(a)   (b)   (c)   (d)

Fig. 2. Examples of visualization results of the methods MPFS, SPFS, MSPFS. The original image are showed on the first column. The second column displays the MPFS visualization result. The SPFS, MSPFS corresponding images are showed in the last two columns (The figure best view in color).

handle this situation. The core of this manner is including all activation values. Formally, we define the final feature vector as follows:

$$\mathcal{F}_{sum} = \{\mathcal{F}_1, ..., \mathcal{F}_k, ..., \mathcal{F}_K\} \qquad \mathcal{F}_k = \sum x, \quad x \in \mathcal{X}_k \qquad (2)$$

C. MSPFS. Compared with MPFS method, SPFS method uses all activation values to encode the feature vector, and the tensor values may include some noisy data which may effect the final image retrieval precises. These two methods are two extreme cases. One only uses the max value information and the other uses all of value activation. We select more effective elements to construct the final feature to make the results more discriminative.

Here we give a detailed description. In Fig. 2, Fig. 2(a) is the original input image. Figure 2(b) and (c) are the visualization results of MPFS and SPFS methods separately. The last image is the response visualization of our MSPFS method. We can see that all visualization results have a response to the main object in the input image. The difference between Fig. 2(b) and (c) is the strength of the response. Figure 2(b) have a dimmed light and there is some discontinuous light trace compared with Fig. 2(c). So it to some extent reflects the fact that the SPFS method is more discriminative than the MPFS method. However, we can also observe that there are some irrelevant light spots existing in Fig. 2(c). According to our former discussions, we call those irrelevant light spots as noise data. After the operation of our MSPFS method, the noise data are cut off in Fig. 2(d). The main object response is more clear than before. So our method exhibits its effectiveness. In the following part we will conduct a more comprehensive experiment to check whether our method is valid or not.

Our method works as follows, first, we get the sum along the channel, i.e., $S_{i,j} = \sum_1^K \mathcal{X}_{i,j}$, where $i \in (1, W)$ and $j \in (1, H)$. Now we will give the formulation of our method.

Here we refer to the concept of mask to simplify the whole process. let's denote

$$\mathcal{M}_{i,j} = \begin{cases} 1, & \text{if } S_{i,j} > median(S) \times \alpha \\ 0, & \text{otherwise} \end{cases} \tag{3}$$

The parameter α is a control factor for the purpose of selecting the useful feature map. We choose the median rather than max because in our experiments the median is more stable and exhibits better results. After getting the mask, we put the mask through the $\mathcal{X}$ along channel side, and finally come out the appropriate feature map, and then we do the same thing as the SPFS does on the masked feature map. The evaluation of the effectiveness of the control factor α will be tested in the experiment part.

2.3 Space Channel Enhancement Model

In this part, we discuss Space Channel Enhancement Model. We get the inspiration form the SEnet model. In the SEnet model, it focuses on the channels and proposes a novel architectural unit, i.e., "Squeeze-and-Excitation" (SE) block that adaptively recalibrates channel-wise feature responses by explicitly modeling interdependencies between channels. This method dynamically performs a weight on the channels so we also do the same thing based on our MSPFS method. The difference between our method and the SE block method is that our method is static, as a result, our network model does not need to be retrained, but the SE block is not. It needs plenty of training data to reshape the network weight to get better performance. So it is not possible to directly use their SE block model.

After operating MSPFS method, the feature maps are all zeros except the selective feature sets. So we can get every channel weighting according to the sparsity of the feature maps. Now we define the sparsity of every channel as Z_k:

$$Z_k = 1 - N_k \tag{4}$$

where $N_k = \frac{1}{W*H} \sum_{ij} \mathbb{1}[\lambda^{ij} > 0]$. And the sparsity of the channels may provide more distinguishability. So from this perspective, we get the same concept with inverse document frequency. Now we define the every channel weight C_k

$$C_k = \log \frac{\epsilon + \sum_i N_i}{\epsilon + N_k} \tag{5}$$

where ϵ is small constant to avoid divided by zero. Then we perform channel weights on MSPFS method.

2.4 Image Retrieval

For all database images and a query image, we first extract convolutional features as our proposed MSPFS method, and then use the SCE-MSPFS method to get the final 512 dimensional feature vector. Image retrieval is done by calculating the similarity measurements like L2 distance or cosine distance between the feature vector of the query image and database images. At the same time, we consider using PCA to compress the original feature vector to relatively low-dimensional vectors so that the computation of similarity measurements can be done efficiently. We test the effect of different final feature vector dimensions. More details will report in the experiments parts.

3   Experiments

In this section, we first describe the datasets and experimental environments. Then we discuss and analysis the impact of control factor α. At last we report the our image retrieval results and we also compare our results with other state-of-the-art works.

3.1   Datasets and Implementation Environments

We evaluate our methods on three publicly available image retrieval datasets: Holidays [15], Oxford5K [16] and Paris6k [17].

The Holidays contains 1491 vacation snapshots corresponding to 500 groups each having the same scene or object. One image from each group serves as a query. The performance is reported as average precision over 500 queries.

The Oxford5K contains 5063 photographs from Flickr associated with Oxford landmarks. 55 queries corresponding to 11 buildings(landmarks) are fixed, and the ground truth relevance of the remaining dataset w.r.t. these 11 classes is provided.

The Paris6k are composed of 6412 images of famous landmarks in Paris. Similar to Oxford5k, this dataset has 55 queries corresponding to 11 buildings/landmarks. The performance of all three datasets are reported as average precision.

In experiments, for the pre-trained deep model, the publicly available VGG-19 model is employed to extract deep convolutional descriptors using the Caffe deep learning framework. For all the retrieval datasets, the subtracted mean pixel values for zero-centering the input images are provided by the pre-trained VGG-19 model.

3.2   Effectiveness of Control Factor α

In this section, we investigate the impact of the control factor α. We experiment on the Paris6k and Oxford5K dataset to explore the effect of α. The final evaluation criteria is mAP. We compare our MSPFS method with the SPFS method in different dimensions.

As shown in Fig. 3, the first row is the results on the Paris6k dataset. Figure 3(a) is the comparison of our proposed MSPFS method and the SPFS method with the 512 dimensional feature vector, and Fig. 3(b) is the same as the Fig. 3(a) only the difference with 256 dimensional feature vector. The green curve is the result of our MSPFS method, and the blue curve is the SPFS method. From the first two images, we observe that when α is 0.8 we get the best performance. Although when α equals 0.2 the accuracy of our method is lower than the SPFS method, it performs better than the SPFS method in some situation. It is that the dynamic selection of α contributes to the better performance. The last two images in the first row are also the results on the Paris6k dataset. The result is consistent with the former results. With the difference in the former two images, our MSPFS method performs better than the MSPFS method in all range of α and it is attributed to the chosen of the median value as the mask selector rather than the max value as mask selector which is commonly used in other works. In the Fig. 2(d), there is almost 5 percent performance improvement of our MSPFS method over SPFS method.

The second row is results on the Oxford5K dataset, and the situation is almost the same as the Paris6K dataset. The performance of our method on the Oxford5K dataset is somehow inferior to the performance on the dataset, but our focuses are not on this. We concern more about whether our method is effective or not. From Fig. 3, we see that the control factor α does affect the final results, and when α belongs to 0.5-0.8 our method can get almost the best performance. The difference in the performance with different values of α may be huge, and the maximum performance gap is almost 5 percent. So we come to the conclusion that our method is effective compared with the SPFS method and the latter experiments also support our conclusion.

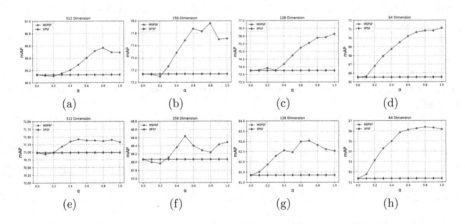

Fig. 3. Analysis of control factor α in different dimension and different datasets (The figure best view in zoom in).

3.3   Comparison of MSPFS and SCE-MSPFS

In this section, we will conduct experiments on our proposed MSPFS method and the SCE-MSPFS method to check whether our SCE-MSPFS is effective or not. We experiment on the Paris6k and Oxford5K datasets. The results are shown in the Fig. 4 and it does not include the result of SPFS because Fig. 3 has shown the same results so it is unnecessary to show the results in the Fig. 4.

Similar to Fig. 3, the first line is the results on the Paris6K dataset. Figure 4(a) is the comparison of our proposed MSPFS method and the SCE-MSPFS method with the 512 dimensional feature vector. The green curve is the result of our MSPFS method, and the blue curve is the SCE-MSPFS method. From Fig. 3(a), we can see that the SCE-MSPFS method is consistently outperforms the MSPFS method and when α is 0.8 it achieves get the best results which is the same as the results in Fig. 3(a). The largest performance gap shown in Fig. 3(a) between SCE-MSPFS method and the MSPFS method is one percent. The other images in the first line also get the similar results where the SCE-MSPFS method wins all the time. The second line are the results on the oxford dataset. The results in Fig. 4 (e) are just like Fig. 4(a), the difference is that the max gap on Fig. 4(e) is about two percents which is better than the gap on Fig. 4(a). The Fig. 4(h) is not like other image because there are some overlap, however our propose SCE-MSPFS method is still larger than the MSPFS method. From the results we can conclude that, our proposed SCE-MSPFS method is effective and we will set it as our final proposed method.

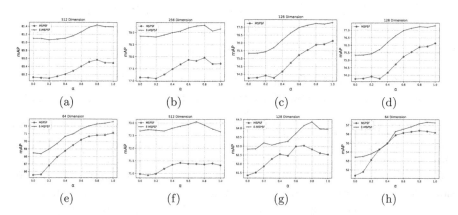

Fig. 4. The comparison of MSPFS and SCE-MSPFS in different dimension and on different datasets (The figure best view in zoom in).

3.4   Result Collection

The comparison results on Oxford5k, Paris6k, and Holidays datasets are reported in Table 1. First, we can observe that the SPFS method outperforms the MPFS method in all dimensions. As the dimension becoming lower, the mAP is also reduced at the same time. However in the MPFS results, the 256 dimension corresponds to the best performance in the Paris and Oxford datasets even surpass

Table 1. Comparison of different methods. The "Bold" values indicates the best performance in each method and the "Underline" values indicates best performance across all methods.

Method	Dimension	Paris	Oxford	Holiday
MPFS	512	71.1	59.8	-
	256	**71.5**	**59.9**	-
	128	68.1	52.2	-
SPFS	512	**80.1**	**71.0**	-
	256	77.1	68.1	-
	128	73.8	61.4	-
MSPFS	512	**80.6**	**71.4**	79.8
	256	78.0	68.6	77.1
	128	76.1	63.0	75.4
SCE-MSPFS	512	<u>**81.4**</u>	<u>**73.0**</u>	<u>**80.2**</u>
	256	79.3	71.0	77.9
	128	77.3	64.4	75.3
SCE-MSPFS with QE(10)	512	<u>**84.6**</u>	<u>**75.2**</u>	<u>**82.3**</u>
	256	81.7	74.1	80.7
	128	80.2	67.4	77.4

the 512 dimension performance. The reason may be that the MPFS method of 512 dimensions is not as compact and discriminative as the MPFS method of 256 dimensions. The MSPFS and SCE-MSPFS method also get the same result as the MPFS method. Secondly, our proposed three different selective feature methods have mutually superimposed effect. The MSPFS method enhance the SPFS method, and based on that, SCE-MSPFS method enhance the MSPFS method also. So compared with the MPFS method, the effect that SCE-MSPFS method boost the feature selection is obvious. Additionally, with the simply query expansion strategy, our proposed method get further performance improvement. In the next section, we will compare our method with the state-of-art methods.

3.5   Comparison to Other State-of-the-art Methods

In this section, we compare our proposed method with other state-of-the-art methods in image retrieval task. We report experimental results in Table 2. For the convenience of comparison, Table 2 is divided into three parts. The first part reports the retrieval results using SIFT local features or VLAD encoding. The second part and the third part report the retrieval results at dimensionality of 512 and 256.

From the first part in Table 2, we can see traditional SIFT features results is lower than the CNN-based feature results even in higher dimensions. From the second part in Table 2, our proposed method SCE-MSPFS achieve the highest

Table 2. Comparisons with STATE-OF-THE-ART image retrieval methods on the three public datasets

CNN-based Method	Dimension	Paris	Oxford	Holiday
Neural Codes [19]	4096	38.6	54.5	79.3
MOP [20]	2048	-	-	80.2
Tr. Embedding [18]	1024	-	56.0	72.0
Tr. Embedding [18]	512	-	-	70.0
Gong et al. [20]	512	-	-	78.3
Neural Codes [19]	512	-	43.5	78.3
R-MAC [11]	512	**83.0**	66.9	-
CroW [10]	512	79.6	68.2	**84.9**
MSPFS(Our method)	512	80.6	71.4	79.8
SCE-MSPFS(Our method)	512	81.4	**73.0**	80.2
Tr.Embedding [18]	256	-	-	65.7
Neural Codes [19]	256	-	43.5	75.9
SPoC [8]	256	-	53.1	80.2
R-MAC [11]	256	72.9	56.1	-
CroW [10]	256	76.5	68.4	**85.1**
MSPFS(Our method)	256	78.0	68.6	77.1
SCE-MSPFS(Our method)	256	**79.3**	**71.0**	77.9

performance on the Oxford5K dataset, and the result of MSPFS method is also surpass other methods on the Oxford5K dataset. On the Paris6K and Holiday datasets, our results are also higher than most of other methods. The R-MAC method achieves the best result on the Paris6K dataset and Crow method get the best performance on the Holiday. Because generating R-MAC features considers both multi-scale feature map and local spatial information and generating Crow feature also considers more complicated operations. From the third part, our proposed method achieve the best performance both on Paris6K and Oxford5K datasets. There are almost 3 percents improvements on both datasets. Thus our proposed MSPFS and SCE-MSPFS methods effectively improve the image retrieval results.

4   Conclusion

In this paper, we propose an efficient and straightforward MSPFS method based on the convolutional layer feature from the convolutional neural network. The MSPFS method provides a simple yet effective way to choose the most discriminative feature vector, which improves the retrieval accuracy significantly. Based on this, we then propose space channel enhanced MSPFS, namely SCE-MSPFS method, which largely boosts the final feature representation. Extensive experimental results show that the proposed methods exhibit competitive performance as compared with the state-of-the-art approaches.

Acknowledgments. This work was supported by National Nature Science Foundation of China (No. 61762025), Frontier and Application Foundation Research Program of CQ CSTC (No. cstc2017jcyjAX0340), The National Key Research and Development Program of China (No. 2017YFB1402400), Guangxi Key Laboratory of Trusted Software (No.kx201701), Guangxi Cooperative Innovation Center of Cloud Computing and Big Data (No.YD16E01), and Key Industries Common Key Technologies Innovation Projects of CQ CSTC (No. cstc2017zdcy-zdyxx0047), Chongqing Postdoctoral Science Foundation (No. Xm2017125), Social Undertakings and Livelihood Security Science and Technology Innovation Funds of CQ CSTC (No. cstc2017shmsA20013).

References

1. Lowe, D.G.: Object recognition from local scale-invariant features. In: ICCV, pp. 1150–1157 (1999)
2. Jegou, H., Douze, M., Schmid, C.: On the burstiness of visual elements. In: CVPR, pp. 1169–1176. IEEE Computer Society (2009)
3. Krizhevsky, A., Sutskever, I., Hinton, G.E.: Imagenet classification with deep convolutional neural networks. In: NIPS, pp. 1106–1114 (2012)
4. Simonyan, K., Zisserman, A.: Very deep convolutional networks for large-scale image recognition. CoRR abs/1409.1556 (2014)
5. Girshick, R.B., Donahue, J., Darrell, T., Malik, J.: Rich feature hierarchies for accurate object detection and semantic segmentation. CoRR abs/1311.2524 (2013)
6. Long, J., Shelhamer, E., Darrell, T.: Fully convolutional networks for semantic segmentation. In: CVPR, pp. 3431–3440. IEEE Computer Society (2015)
7. Arandjelovic, R., Gronát, P., Torii, A., Pajdla, T., Sivic, J.: Netvlad: CNN architecture for weakly supervised place recognition. In: CVPR, pp. 5297–5307. IEEE Computer Society (2016)
8. Babenko, A., Lempitsky, V.S.: Aggregating deep convolutional features for image retrieval. CoRR abs/1510.07493 (2015)
9. Gordo, A., Almazán, J., Revaud, J., Larlus, D.: Deep image retrieval: learning global representations for image search. CoRR abs/1604.01325 (2016)
10. Kalantidis, Y., Mellina, C., Osindero, S.: Cross-dimensional weighting for aggregated deep convolutional features. CoRR abs/1512.04065 (2015)
11. Tolias, G., Sicre, R., Jégou, H.: Particular object retrieval with integral maxpooling of CNN activations. CoRR abs/1511.05879 (2015)
12. Wei, X., Luo, J., Wu, J.: Selective convolutional descriptor aggregation for fine-grained image retrieval. CoRR abs/1604.04994 (2016)
13. Hu, J., Shen, L., Sun, G.: Squeeze-and-excitation networks. CoRR abs/1709.01507 (2017)
14. Zeiler, M.D., Fergus, R.: Visualizing and understanding convolutional networks. In: Fleet, D., Pajdla, T., Schiele, B., Tuytelaars, T. (eds.) ECCV 2014. LNCS, vol. 8689, pp. 818–833. Springer, Cham (2014). https://doi.org/10.1007/978-3-319-10590-1_53
15. Jegou, H., Douze, M., Schmid, C.: Improving bag-of-features for large scale image search. Int. J. Comput. Vis. **87**(3), 316–336 (2010)

16. Philbin, J., Chum, O., Isard, M., Sivic, J., Zisserman, A.: Object retrieval with large vocabularies and fast spatial matching. In: CVPR. IEEE Computer Society (2007)
17. Philbin, J., Chum, O., Isard, M., Sivic, J., Zisserman, A.: Lost in quantization: Improving particular object retrieval in large scale image databases. In: CVPR. IEEE Computer Society (2008)
18. Jégou, H., Zisserman, A.: Triangulation embedding and democratic aggregation for image search. In: CVPR, pp. 3310–3317. IEEE Computer Society (2014)
19. Babenko, A., Slesarev, A., Chigorin, A., Lempitsky, V.: Neural codes for image retrieval. In: Fleet, D., Pajdla, T., Schiele, B., Tuytelaars, T. (eds.) ECCV 2014. LNCS, vol. 8689, pp. 584–599. Springer, Cham (2014). https://doi.org/10.1007/978-3-319-10590-1_38
20. Gong, Y., Wang, L., Guo, R., Lazebnik, S.: Multi-scale orderless pooling of deep convolutional activation features. In: Fleet, D., Pajdla, T., Schiele, B., Tuytelaars, T. (eds.) ECCV 2014. LNCS, vol. 8695, pp. 392–407. Springer, Cham (2014). https://doi.org/10.1007/978-3-319-10584-0_26
21. Ren, S., He, K., Girshick, R.B., Sun, J.: Faster R-CNN: towards real-time object detection with region proposal networks. CoRR abs/1506.01497 (2015)

A Pathology Image Diagnosis Network with Visual Interpretability and Structured Diagnostic Report

Kai Ma, Kaijie Wu$^{(\boxtimes)}$, Hao Cheng, Chaochen Gu, Rui Xu, and Xinping Guan

Shanghai Jiao Tong University, 800 Dongchuan Rd, Shanghai 200240, China
{makaay, kaijiewu, jiaodachenghao, jacygu, xuruihaha, xpguan}@sjtu.edu.cn

Abstract. Despite recent advances in medical diagnosis domain, many challenges remain in obtaining more accurate conclusions and in presenting semantically and visually interpretable results during the diagnosis process. An interpretable diagnosis process is proposed through the implementation of a deep learning model. This consists of three interrelated models, an image model, an attention model and a conclusion model. The proposed image model extracts the semantic feature using convolutional neural networks (CNNs). The conclusion model, integrated with the semantic attributes attention model, aims to predict the conclusion label by long-short term memory (LSTM), which captures the discriminative relationship between semantic attributes. The network is trained in end-to-end way with different weight of each model. Based upon a cervical intraepithelial neoplasia images, diagnostic report and labels (CINDRAL) dataset, the approach demonstrates significant improvement when comparing the baseline in the conclusion result.

Keywords: Deep learning · Visual interpretability
Pathology diagnosis process

1 Introduction

In recent years, computer-aided medical diagnosis (CAD) has achieved remarkable progress with rapid development of deep learning. Traditional methods treat the process as a standard classification problem [1]. However, it is less efficient and of lower performance when diagnosing diseases [2]. The reason is the classification model simplifies the actual diagnosis process and lacks the discriminative information to support the conclusion. Doctors often find it difficult to understand how the model captures features and makes the diagnostic conclusion. Therefore, this challenge needs to be addressed through an interpretive method in order to support the decision-making process.

In clinical practice, process and output of the deep learning model, which is effective and important in the CAD process, must be evaluated by pathologists. In other words, the model should capture the discriminative features from the pathology image and generate the words, context and the visual attention regions to help pathologists

L. Cheng et al. (Eds.): ICONIP 2018, LNCS 11306, pp. 282–293, 2018.
https://doi.org/10.1007/978-3-030-04224-0_24

make a decision. Such a method is more efficient and convenient for medical doctors than models which only generate conclusion labels.

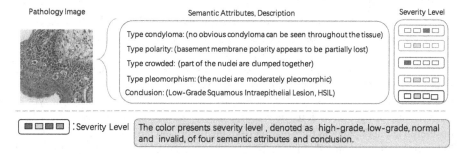

Fig. 1. One example in the CINDRAL dataset. It consists of a pathology image and a structured report (four semantic attributes, each with four state labels). No extra locations on images are needed in the dataset. (Best viewed in color).

This has prompted us for further research into the proposition of a model to automatically capture latent and discriminative features, then generates the report and attention region. Three key challenges remain to be addressed. The structured report has to contain different semantic attribute descriptions to support the conclusion. The output of the model must be clearly understood by medical professionals. The model needs to generate the discriminative attention region of semantic attribute label.

The above challenges are addressed through the implementation of an end-to-end network which consists of images, attention and conclusion models. The image model based upon CNNs extracts the discriminative features from the image. A new method is proposed for the attention model which generates the visible attention region and the structured report. The attention model is treated as a multi-label classification task so that the model generates full-structured context. In this approach, the conclusion model, combined with the attention model, possesses the ability to learn the contextual dependencies among the semantic attributes with the "memory" of LSTM for conclusion making.

Generally speaking, the main contributions of our method are:

- A new approach is proposed which can generate the structured report and support the conclusion on interpretable vision for pathology images.
- We build the pathology cervical intraepithelial neoplasia images, diagnostic report and attribute labels (CINDRAL) dataset (explained later in Sect. 3) with Shang Hai International Peace Maternity and Child Health Hospital (IPMCH).
- We perform extensive experiments and evaluations on the CINDRAL dataset, and demonstrate accuracy and effectiveness of the approach.

The rest of the paper is organized as follows. Section 2 reviews related works. Section 3 descripts the details of the dataset. The proposed model including the extracted features and how attention model works is described in Sect. 4. Then, Sect. 5 conducts detailed performance studies and analysis on CINDRAL dataset. Section 6 concludes the paper.

2 Related Work

2.1 Image and Conclusion Model

Recent advances in the performance of medical diagnostics [3, 5, 7, 14] have achieved rapid progress as a result of the development of deep CNNs [4] and the memory mechanism of recurrent neural networks (RNNs). Traditional methods treat the CAD as a classification problem, in cases such as skin lesions [6], lung squamous cell carcinoma [1], and conclusion of pathological images [8].

However, these methods for CAD typically aimed at finding one particular type of disease, concealing correlations between semantic attributes. Some researchers have considered latent dependent information from the report and pathology images by LSTM [11] in order to overcome this issue. The construction of CNN-RNN based framework model to predict the conclusion label of chest X-rays, is a prime example, Shin et al. [9]. This method implements CNN to extract the feature of a disease and RNN to describe the attributes of the disease. Another methods uses CNN to obtain visual and semantic features from chest X-rays while using hierarchical LSTM to get a more natural report and conclusion, Jing et al. [10]. The work most closely related to medical report and conclusion is recently contributed by Zhang et al. [12]. This group proposed the CNN-LSTM model to describe the semantic attributes report, which draws the conclusion. However, some words (e.g. "along", "the", and "is") present in their report (e.g. Polarity along the basement membrane is negligibly lost) contain no medical sematic information supporting the conclusion. In this case, the most important word in the sentence is "negligibly", which provides the discriminative features to help the LSTM draw the conclusion.

Following the work of Wang et al. [13], this problem can be addressed by changing the conclusion problem to multi-label classification which allows the conclusion model to obtain accurate sematic attribute description labels. Our methods can generate the attribute report including the conclusion, which has more accurate performance than the previously mentioned methods.

2.2 Attention Model

There has been a significant amount of research focused on the attention mechanism which achieves interpretability of context and vision in traditional natural image datasets, such as ImageNet [4], and so on [23, 30]. For the attention model, our works is similar to several previous works [15–20]. Xu et al. [18] proposed the sequence-to-sequence model and attention model in the image captioning task. In their work, the attention map was determined using the CNN features and the previously hidden states of LSTM. Pedersoli et al. [20] provided an association between the attention region and caption words.

These previous works have inspired researchers to improve in CAD domain, and several works have aimed to generate reports for medical images. Additionally, these works focus on interpretable visual diagnosis [8, 10, 12, 21, 22], which can support the pathologist's decision-making process.

Zhang et al. [12] introduced an attention model which focuses on the image region while every word is generated from the model. However, words, like the, a, along and so on, have no factual attention region in the image.

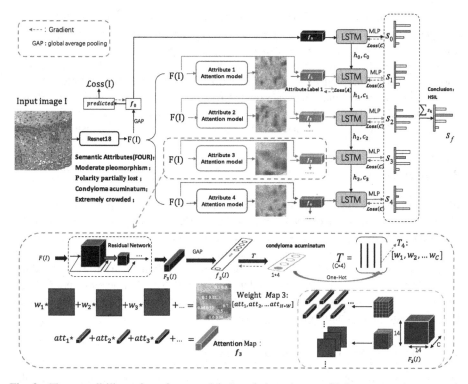

Fig. 2. The overall illustration of our model. A pathology image with its structured report and labels, presented as an example. Image model extracts the features, attention model demonstrates the attention region and the structured report, while conclusion model predicts the conclusion.

To address the problem, our method aimed at the visual interpretation of semantic attributes (e.g. in CINDRAL dataset, there are four type), rather than the single word from the report.

3   Dataset

The cervical intraepithelial neoplasia images, diagnostic report and labels (CINDRAL) dataset was collected in collaboration with Shanghai International Peace Maternity and Child Health Hospital (IPMCH). Whole-slide images (WSI) at $20\times$ magnification of stained tissue sections obtained from 50 patients at risk of cervical neoplasm. One thousand 600×600 RGB images were randomly selected from the dataset, close to cervical intraepithelial neoplasm.

The pathologists provided a paragraph describing four semantic attribute features (Fig. 1). These attributes included, the state of condyloma, cell polarity, cell crowding, and nuclear pleomorphism, followed by a diagnostic conclusion. The attributes and the conclusion are both comprised of four labels; normal, high-grade, low-grade, and insufficient information. Contained within different reports, each description of semantic attributes has two or three very similar description (i.e. "part of the nuclear are clumped together" and "extremely crowded nuclei can be seen"). Thus there are five sentences, four attributes and one conclusion, per image.

The dataset was pre-processed by cropping, rotating (90°, 180° and 270°), horizontal/vertical flipping for data augmentation. Then we randomly selected 20% of the images as testing data and the remaining 800 images for training and cross-validation. In the dataset, the four attributes and conclusion are treated as 5 separate tasks for the structured report and LSTM, trained to support the conclusion model.

4   Method

Figure 2 illustrates the architecture of the proposed method. The pathology image is first sent into a resnet18 ConvNet, which is fast to train and achieves similar or better performance than most commonly used VGG16 [24] or AlexNet [25] model. The network processes the image with convolutional (conv) layers to extract the conv feature map, denoted as F(I). Then the weight map is obtained from every attribute attention model and the conclusion model comprising an LSTM network predicts the last result from the attention map. Finally, the scores from four attention maps are fused to achieve the final conclusion label distribution.

4.1   Attention Model with Structured Report

As we all know, attention mechanism, the work in Xu et al. [18], is aimed to learn the attention region in the whole image so that it can support the final prediction. But in our work, we focus on the relationship between four semantic attributes, which implicitly provide the discriminate information for LSTM to draw the conclusion. So we propose attention models which present the four semantic attributes of the CIN images. The attention model dynamically computes a weight map for every attributes, which presents the visual attention region corresponding to input image.

For one attribute (Fig. 2), we duplicate the feature F(I) fourfold, since every residual network in the attention model can learn the accurate features for each attribute with different parameters. The residual network extracts features, denoted as $F_k(I)$ with a dimension of $256 \times (14 \cdot 14)$, and k = $\{1, 2, 3, 4\}$ represents the four attributes. After a global average pooling layer, feature denoted as $f_k(I)$ with dimension 1×256, is obtained from $F_k(I)$.

Specifically, following [27], the weight map can be computed as follows

$$O_k = \text{softmax}(f_k(I)T_k + b) \tag{1}$$

$$T_k^i = S_k T_k \tag{2}$$

$$W_k = (T_k^i)^T F_k(I) \tag{3}$$

where T_k is a learned fully connection layer parameter with dimension 256×4. $S_k, k = \{1, 2, 3, 4\}$ is the one-hot representation of the k-th image attribute generated by O_k.

Then we can get the one column of T_k, denoted as $T_k^i, i = \{1, 2, 3, 4\}$ with dimension 256×1, which contains the discriminate information of the k-th semantic attributes. Finally, the weight map, denoted as W_k with dimension 14×14, which corresponds to the attribute attentional regions in the input image is generated by bilinear interpolation.

The matrix W_k presents the visual interpretability for the pathologists. In other words, the W_k indicates which regions the attention model focuses on. As shown in Fig. 2, the weight map W_k and the feature map $F_k(I)$ need to work collaboratively in a fused manner so that the 256-dim feature f_k can contain the accurate k-th attribute feature selected by weight map.

$$f_k = W_k(F_k(I))^T \tag{4}$$

To better train the attention model, we use two loss functions: (1) we use the semantic attributes label for the feature $f_k(I)$, (2) and apply the severity level label of each attribute for the feature f_k. The motivation is two-fold. First, the feature generated by residual network can better extract the attribute information which is critical for next procedure. Second, a structured diagnostic report can be generated by the second loss function which can describe the symptom information of every attribute. The two loss functions serve as a supervision on the attention model, which can make sure the attention model training towards to accurate semantic features for the next conclusion model.

4.2 Conclusion Model

From our pathologists, we find that symptom descriptions of semantic attributes support the conclusion, and the latent relationship between attributes also provide support to lead the conclusion. Considering relevance and dependence, we adopt the LSTM network to draw the conclusion.

In generating diagnostic report of image captioning domain, there are a lot of works, like [8, 12, 18, 21, 22] and so on, treating the natural language word or sentence as the input for LSTM in training stage. However, the natural diagnostic report contains some words, like a, the, along, and so on, which present no medical information and provide less accurate feature to draw the conclusion. As shown in Fig. 2, our work proposed a new method, which treat the discriminate attribute features as the input of LSTM, to directly consider the critical feature into conclusion.

Following [11], LSTM is defined by the following equations

$$x_k = relu\left(W^{(x)}f_k + b_x\right), k \neq 0 \tag{5}$$

$$i_k = sigmoid\left(W^{(i)}x_k + U^{(i)}h_{k-1} + b_i\right) \tag{6}$$

$$f_k = sigmoid\left(W^{(f)}x_k + U^{(f)}h_{k-1} + b_f\right) \tag{7}$$

$$o_k = sigmoid\left(W^{(o)}x_k + U^{(o)}h_{k-1} + b_o\right) \tag{8}$$

$$\tilde{c}_k = tanh\left(W^{(c)}x_k + U^{(c)}h_{k-1} + b_c\right) \tag{9}$$

$$c_k = f_k * c_{k-1} + i_k * \tilde{c}_k \tag{10}$$

$$h_k = o_k * tanh(c_k) \tag{11}$$

$$z_k = relu\left(W^{(z)}h_k + b_z\right) \tag{12}$$

$$s_k = W^{(s)}z_k + b_s \tag{13}$$

where the computation process represents k-th (k $\neq$ 0) LSTM network, and f_k in Eq. (5) is a 256-dim vector containing the attribute information to support conclusion; h_{k-1} and c_{k-1} are the hidden state and memory cell of the previous LSTM; The k-th LSTM hidden state h_k is used to predict score distribution, denoted as s_k, of the conclusion by Eqs. (12) and (13). Note that we initialize the hidden state with the feature f_0 extracted from image, and the initial process for the LSTM network is set to be:

$$h_0 = f_0 \tag{14}$$

$$z_0 = relu\left(W^{(z)}h_0 + b_z\right) \tag{15}$$

$$s_0 = W^{(s)}z_0 + b_s \tag{16}$$

Then we can get the final predict conclusion, denoted as s_f, with softmax function and add five predicting score distributions, s_0, s_1, s_2, s_3, s_4:

$$s_f = s_0 + s_1 + s_2 + s_3 + s_4 \tag{17}$$

4.3   Network Optimization

The overall model has three sets of parameters: θ_I in the image model I, θ_A in the attention model A, θ_C in the conclusion model C. The overall optimization problem in our method is expressed as:

$$\max_{\theta_I, \theta_A, \theta_C} \mathcal{L}_I(l_c, I(I; \theta_1)) + \mathcal{L}_A(l_s, A[I(I; \theta_1); \theta_A]) + \mathcal{L}_C(l_c, C\{A[I(I; \theta_1); \theta_A]; \theta_C\}) \tag{18}$$

where (I, l_c, l_s) is a training tuple: I is a pathology image, l_c denotes the conclusion label, and l_s is the semantic attribute label. Modules I, A and C are supervised by three negative log-likelihood loss function $\mathcal{L}_I, \mathcal{L}_A$ and $\mathcal{L}_C$.

In the training stage, we adopt the method of Adam and standard back-propagation to optimize the joint model. For end-to-end training, we treat the loss function as two stages with different weights and learning rates. Thus, the training loss is computed through:

$$\mathcal{L}oss_{I,A,C} = \lambda(\mathcal{L}oss_I + \mathcal{L}oss_A) + (1 - \lambda)\mathcal{L}oss_C \qquad (19)$$

where $\mathcal{L}oss(I, A, C)$ is the joint loss function of the whole model. In the first 10 epoches, the parameter $\lambda(0 < \lambda < 1)$ is lager so that the accurate feature can be extracted from input image. With the training process, the λ is becoming smaller to support the better predicted conclusion.

Table 1. Diagnosis conclusion accuracy on CINDRAL

Method	Image classification							
Model	AlexNet		VGG16		Resnet18		Resnet34	
Pre-trained	✓	✗	✓	✗	✓	✗	✓	✗
DCA (%) ± std	71.2 ± 2.0	71.2 ± 2.5	66.0 ± 4.3	66.0 ± 5.8	78.4 ± 2.5	78.6 ± 1.5	77.2 ± 4.0	78.2 ± 3.5
Method	Image classification		Image captioning				Our	
Model	Densenst40		MDNET		Show and tell		Our Method	
Pre-trained	✓	✗	✓	✗	✓	✗	✓	✗
DCA (%) ± std	75.0 ± 2.4	75.4 ± 1.9	78.0 ± 2.3	78.2 ± 2.2	75.2 ± 4.5	74.9 ± 3.6	**85.0 ± 1.7**	**84.8 ± 2.4**

5   Experimental Results

In this section, we validate the proposed model on four aspects to demonstrate the significant improvement. The experiments are implemented on the CINDRAL dataset as follows: (1) we start by validating the diagnosis conclusion accuracy (DCA) with the purpose to show its superior performance against several other CNNs and image captioning methods; (2) then we conduct experiments on the semantic attributes prediction accuracy (SAPA) with medical diagnosis network [12] to prove our method have the ability to generate the same diagnosis report by different methods; (3) we conduct the experiment on the different sequence of semantic attributes with the purpose to validate the robustness of the conclusion model; (4) we demonstrate the attention region towards semantic attributes to support the conclusion on vision.

5.1   Diagnosis Conclusion Accuracy on CINDRAL

In the computer-aided medical diagnosis process, our pathologist usually expects to get more accurate conclusion so that the DCA of evaluation metrics is critical for our method. To validate the effectiveness of our model, we conduct the experiments by comparing with CNNs, AlexNet, VGG16, ResNet [28], DenseNet [26]. In this setting, the CNNs use the conclusion label and our model treats the semantic attributes and conclusion as the label to support that the semantic attributes do significant improvement for decision-making. We also consider the factor like pre-trained model on ImageNet. Moreover, we do the comparison with MDNET [12] in DCA, which our method is closest to.

From Table 1, we can find that pre-trained universal CNN model is not available in the medical image domain for the reason that there is a huge difference in features between the natural images from the ImageNet and pathology images. And with the increase of model depth, the CNNs models achieve the best accuracy,which is 78% in CINDRAL, as the same as MDNET.

Table 2. SAPA performance comparison with MDNET

Model	Condyloma		Cell polarity		Cell crowding		Pleomorphism	
	MDNET	Our	MDNET	Our	MDNET	Our	MDNET	Our
SAPA (%) ± std	72.4 ± 1.6	72.8 ± 1.0	76.0 ± 1.6	75.2 ± 2.3	76.0 ± 1.5	76.0 ± 1.5	78.0 ± 1.1	78.6 ± 1.7

Table 3. Influence of semantic attributes input sequence on model performance

Model	Sequence of attributes	DCA(%) ± std
CNN-RNN [29]	Previous order(ABCD)	76.6 ± 4.2
	Order 1 (BCAD)	65.4 ± 3.4
	Order 2 (DBCA)	70.2 ± 3.8
	Order 3 (CADB)	70.0 ± 3.1
Our method	Previous order (ABCD)	**85.0 ± 1.7**
	Order 1 (BCAD)	**85.2 ± 2.9**
	Order 2(DBCA)	**84.4 ± 1.3**
	Order 3(CADB)	**84.8 ± 1.2**

Our method achieves obviously better accuracy rate (84% in CINDRAL). The results demonstrate that our method, which treats the semantic attributes as label, substantially improves the performance of network.

5.2   Structured Diagnosis Report

In clinical practice, not only natural language but also visual interpretation is necessary for pathologists to understand the specific symptom towards every semantic attribute. Thus, we address the problem by generating the structured diagnosis report to help pathologists understand the rationale for the conclusion. In the experiment, we compare the discriminative information, semantic attributes prediction accuracy (SAPA), in the structured report with MDNET [12] in different manners.

Table 2 shows the mean scores over 5 folds. As can be observed, in the four semantic attributes, the accuracy of the result is almost as the same as MDNET. In our method, we achieve the prediction by treating the problem as multi-label classification problem, rather than treating the problem as image captioning with the natural language report in MDNET. Generally speaking, we propose a new way to generate the diagnosis report, which demonstrates the same performance as MDNET.

5.3   Different Sequence of Semantic Attributes

We conduct the experiment compared with CNN-RNN [29] on the purpose to investigate the influence on different sequences on LSTM. CNN-RNN model also considers the latent relationship between the attribute labels in natural images. Hence in our work, we change the attribute sequence. We denote the previous sequence, which is type condyloma, polarity, crowd and pleomorphism, as ABCD. Then we compare the DCA with the sequence of BCAD, DBCA and CADB.

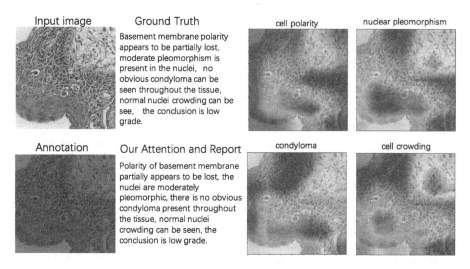

Fig. 3. The illustration of diagnosis report and four semantic attribute attention regions. Best viewed in color.

The results are shown in Table 3. Our proposed method outperforms the baseline model by demonstrating significantly improved DCA with the different sequence.

5.4   The Attention Model with Visually Interpretation

In this experiment, we show the attention region for each semantic attribute (an example in Fig. 3). The four attention models (as Fig. 2) computes and shows the attention region to interpret how the network support the diagnosis conclusion. Rather than the attention region for a single word in [12], we generate four attention maps to support four semantic attributes. Our pathologist draws the region of interest (ROI) which significantly support the decision-making process. we can observe the result which expresses strong correspondence between the pathologist annotations and our attention regions for four semantic attributes in the Fig. 3. Note that there are no regional annotations in the training stage. Our work demonstrates the model has learned the critical information to support its conclusion.

6 Conclusion

This paper presents a new approach to read the pathology image, generating structured diagnosis report with visual interpretation towards the attention region. Specifically, we propose the method to treat the four semantic feature as the input of LSTM, which learn the accurate information to support the conclusion. Practically, our pathologist also expresses appreciation to our work in the pathology image diagnosis. Experimental results on CINDRAL dataset demonstrate that our proposed deep model can significantly improve the performance in both accuracy and efficiency.

Acknowledgements. This work is supported by the National Key Scientific Instruments and Equipment Development Program of China (2013YQ03065101), the National Natural Science Foundation of China under Grant 61521063 and Grant 61503243.

References

1. Zhang, X., Su, H., Yang, L., Zhang, S.: Fine-grained histopathological image analysis via robust segmentation and large-scale retrieval. In: Computer Vision and Pattern Recognition, pp. 5361–5368 (2015)
2. Chang, H., Zhou, Y., Borowsky, A., Barner, K., Spellman, P., Parvin, B.: Stacked predictive sparse decomposition for classification of histology sections. Int. J. Comput. Vis. **113**(1), 3–18 (2015)
3. Cireşan, D.C., Giusti, A., Gambardella, L.M., Schmidhuber, J.: Mitosis detection in breast cancer histology images with deep neural networks. In: Mori, K., Sakuma, I., Sato, Y., Barillot, C., Navab, N. (eds.) MICCAI 2013. LNCS, vol. 8150, pp. 411–418. Springer, Heidelberg (2013). https://doi.org/10.1007/978-3-642-40763-5_51
4. Russakovsky, O., et al.: Imagenet large scale visual recognition challenge. Int. J. Comput. Vis. **115**(3), 211–252 (2014)
5. Kisilev, P., Walach, E., Hashoul, S., Barkan, E., Ophir, B., Alpert, S.: Semantic description of medical image findings: structured learning approach. In: British Machine Vision Conference, pp. 171.1–171.11 (2015)
6. Esteva, A., et al.: Corrigendum: dermatologist-level classification of skin cancer with deep neural networks. Nature **542**(7639), 115–118 (2017)
7. Chartrand, G., et al.: Deep learning: a primer for radiologists. Radiographics **37**(7), 2113–2131 (2017)
8. Zhang, Z., Chen, P., Sapkota, M., Yang, L.: TandemNet: distilling knowledge from medical images using diagnostic reports as optional semantic references. In: Descoteaux, M., Maier-Hein, L., Franz, A., Jannin, P., Collins, D.L., Duchesne, S. (eds.) MICCAI 2017. LNCS, vol. 10435, pp. 320–328. Springer, Cham (2017). https://doi.org/10.1007/978-3-319-66179-7_37
9. Kisilev, P., Walach, E., Barkan, E., Ophir, B.: From medical image to automatic medical report generation. IBM J. Res. Dev. **59**(2/3), 2:1–2:7 (2015)
10. Jing, B., Xie, P., Xing, E.: On the automatic generation of medical imaging reports (2017). arXiv:1711.08195
11. Surhone, L.M., Tennoe, M.T., Henssonow, S.F.: Long Short Term Memory. Beta Script Publishing (2010)
12. Zhang, Z., Xie, Y., Xing, F., Mcgough, M., Yang, L.: Mdnet: a semantically and visually interpretable medical image diagnosis network, pp. 3549–3557 (2017)

13. Wang, Z., Chen, T., Li, G., Xu, R., Lin, L.: Multi-label image recognition by recurrently discovering attentional regions. In: IEEE International Conference on Computer Vision, pp. 464–472 (2017)
14. Shi, X., Xing, F., Xie, Y., Su, H., Yang, L.: Cell encoding for histopathology image classification. In: Descoteaux, M., Maier-Hein, L., Franz, A., Jannin, P., Collins, D.L., Duchesne, S. (eds.) MICCAI 2017. LNCS, vol. 10434, pp. 30–38. Springer, Cham (2017). https://doi.org/10.1007/978-3-319-66185-8_4
15. Nam, H., Ha, J.W., Kim, J.: Dual attention networks for multimodal reasoning and matching, pp 2156–2164 (2016)
16. Pedersoli, M., Lucas, T., Schmid, C., Verbeek, J.: Areas of attention for image captioning, pp. 1251–1259 (2017)
17. Vinyals, O., Fortunato, M., Jaitly, N.: Pointer networks. In: Computer Science (2015)
18. Xu, K., Ba, J., Kiros, R., Cho, K., Courville, A., Salakhutdinov, R., Zemel, R., Bengio, Y.: Show, attend and tell: neural image caption generation with visual attention. In: Computer Science, pp. 2048–2057 (2015)
19. Yu, D., Fu, J., Mei, T., Rui, Y.: Multi-level attention networks for visual question answering. In: IEEE Conference on Computer Vision and Pattern Recognition, pp. 4187–4195 (2017)
20. Lu, J., Xiong, C., Parikh, D., Socher, R.: Knowing when to look: adaptive attention via a visual sentinel for image captioning, pp. 3242–3250 (2016)
21. Shin, H.C., Roberts, K., Lu, L., Demnerfushman, D., Yao, J., Summers, R.M.: Learning to read chest x-rays: recurrent neural cascade model for automated image annotation, pp. 2497–2506 (2016)
22. Wang, X., Peng, Y., Lu, L., Lu, Z., Summers, R.M.: Tienet: text-image embedding network for common thorax disease classification and reporting in chest x-rays (2018). arXiv:1801.04334
23. Everingham, M., Winn, J.: The pascal visual object classes challenge 2010 development kit contents. In: International Conference on Machine Learning Challenges: Evaluating Predictive Uncertainty Visual Object Classification, pp. 117–176 (2011)
24. Simonyan, K., Zisserman, A.: Very deep convolutional networks for large-scale image recognition. Computer Science (2014)
25. Krizhevsky, A., Sutskever, I., Hinton, G.E.: Imagenet classification with deep convolutional neural networks, pp 1097–1105 (2012)
26. Huang, G., Liu, Z., Maaten, L.V.D., Weinberger, K.Q.: Densely connected convolutional networks. In: CVPR (2017)
27. Zhou, B., Khosla, A., Lapedriza, A., Oliva, A., Torralba, A.: Learning deep features for discriminative localization. In: IEEE Conference on Computer Vision and Pattern Recognition, pp. 2921–2929 (2016)
28. He, K., Zhang, X., Ren, S., Sun, J.: Deep residual learning for image recognition, pp. 770–778 (2015)
29. Wang, J., Yang, Y., Mao, J., Huang, Z., Huang, C., Xu, W.: Cnn-rnn: a unified framework for multi-label image classification, pp. 2285–2294 (2016)
30. Lin, T.-Y., et al.: Microsoft COCO: common objects in context. In: Fleet, D., Pajdla, T., Schiele, B., Tuytelaars, T. (eds.) ECCV 2014. LNCS, vol. 8693, pp. 740–755. Springer, Cham (2014). https://doi.org/10.1007/978-3-319-10602-1_48

Sketch-Based Image Retrieval
via Compact Binary Codes Learning

Xinhui Wu$^{(\boxtimes)}$ and Shuangjiu Xiao

School of Software, Shanghai Jiao Tong University, Shanghai, China
kexuan@sjtu.edu.cn, xsjiu99@cs.sjtu.edu.cn

Abstract. With the exploding number of images on the Internet and the convenience of free-hand sketch drawing, sketch-based image retrieval (SBIR) has attracted much attention in recent years. Due to the ambiguity and sparsity of sketches, SBIR is more challenging to cope with than conventional content-based problem. Existing approaches usually adopt high-dimensional features which require high-computational cost. Furthermore, they often use edge detection and parameter-sharing networks which may lose important information in training. In this study, we propose a compact binary codes learning strategy using deep architecture. By leveraging well-designed prototype hash codes, we embed different domains input (sketch and photo) into a common comparable feature space. Besides, we present two separate networks specific to sketches and real photos which can learn very compact features in Hamming space. Our method achieves state-of-the-art results in accuracy, retrieval time and memory cost on two standard large-scale datasets.

Keywords: Deep learning · Hashing · Sketch-based image retrieval

1 Introduction

Sketches are highly abstract representations which express sufficient stories. Different from natural images, they are formed of a few hand-drawn strokes. Humans can draw simple sketches quickly without any reference, at the same time conveying information precisely. With such interesting characteristics, there exists much research dealing with sketch-based image retrieval [6,17,20], sketch-based 3D model retrieval [23] and sketch recognition [28].

In this paper, our research direction focuses on sketch-based image retrieval (SBIR). It aims at retrieving most similar results in image gallery collection by a query free-hand sketch. Figure 1 gives an example of retrieval flow. SBIR can solve the situation when it is hard to describe an object in words or query image is not available. In this situation, text-based image retrieval (TBIR) and content-based image retrieval (CBIR) [5,25,26] fail. Essentially, SBIR has two main advantages: (i) Science proves that people are sensitive to outlines [11,29]. Free-hand sketches can show enough key query points without noisy background. (ii) With the appearance of touch-screen mobile devices in recent years, drawing

© Springer Nature Switzerland AG 2018
L. Cheng et al. (Eds.): ICONIP 2018, LNCS 11306, pp. 294–306, 2018.
https://doi.org/10.1007/978-3-030-04224-0_25

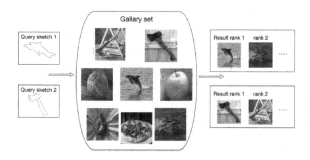

Fig. 1. An illustration of sketch-based image retrieval.

sketches becomes quite convenient. Even non-artists could draw sketches within few seconds.

However, SBIR confronts several challenges. First, sketches lack texture and color information. The feature of sparse lines is totally different with traditional images. Second, since people depict query sketch without reference as aforesaid, sketches usually exhibit large intra-class variations.

Over the past 25 years, the study on SBIR has developed rapidly [20]. Among them, the bulk of methods exploit traditional hand-crafted pipeline [6,18,19]. They usually first transform real images to detected edgemap photos in order to narrow the semantic gap between the sketch and the image. Then, hand-crafted features of both sketches and edgemap photos are extracted and fed into bag-of-words architecture. Whereas, their shallow features cannot handle large internal variations well. Recently, convolutional neural networks (CNN) [10] has shown great power on deep feature representation. By means of robust end-to-end deep frameworks, the deep methods [17,20] are superior to hand-crafted ones typically. Actually, the deep learning methods of SBIR come out much later than those in CBIR due to the lack of available fine-grained sketch dataset. Since 2016, the appearance of Sketchy [20] dataset boosts the development of SBIR. Though deep methods have achieved progress, they mainly calculate feature distances in Euclidean space with high complexity. It is not feasible when dealing with large-scale retrieval task. Hence, we introduce a deep hashing architecture to perform fast retrieval in Hamming space with low memory cost.

SBIR problem is a typical cross-domain retrieval case. To address the issue of aforementioned sketch challenges, previous works generally translate the real image to the approximate sketch in advance. However, salient structural information may be lost during edge extraction process. In addition, most of the existing methods adopt shared Siamese network. However, learning parameters individually will perform better if possible. Besides, previous works generally compare features in high-dimensional space requiring high-computational cost and long retrieval time.

In this paper, we present a novel deep hashing framework to solve sketch-based image retrieval. The main contributions of our work include: (i) Encoding supervised information into a semantic-preserving set of prototype hash codes to achieve better guide for deep training process; (ii) Presenting a deep hashing framework with two separate networks for sketches and photos, which is more suitable for cross-domain challenge, capturing the internal semantic relationship and cross-domain similarities at the same time; (iii) Achieving state-of-the-art results in accuracy, retrieval time and memory cost compared with existing methods on two large datasets.

2   Related Work

Sketch-Based Image Retrieval. Prior hand-crafted methods first use edge detectors like Canny [1] to generate edge or contour maps from real images. After that, they extract features of both sketches and generated edgemap photos, such as SIFT [15], HOG [2], SSIM [21], Gradient Field HOG [6] etc. Then, the bag-of-words framework is used to learn discriminative semantic representation. With the help of CNN, deep methods achieve better performance recently on category-level [13,17] and fine-grained SBIR [27]. Wang et al. [23] use a Siamese network to retrieve 3D models by a query sketch. Qi et al. [17] also adopt a similar Siamese strategy to solve category-level SBIR based on a small dataset Flickr15K [6], which is the first attempt in deep SBIR technique. Yu et al. [27] achieve a nice result in fine-grained search with triplet loss. To our best knowledge, DSH [13] is the only existing work that employs deep hash learning in SBIR. It uses a relatively complex semi-heterogeneous hashing framework, achieving a good performance in large-scale dataset. Despite that, our work surpasses DSH in evaluation with rather compact hash code learning.

Hashing Learning. Hashing is an effective method for fast image retrieval. It projects high-dimensional features to compact semantic-preserving binary codes, which are called hash codes. The mapping strategy is the crucial hashing function. Early unsupervised hashing methods include LSH [3], SH [24] and ITQ [5]. With the help of label information, supervised hashing can deal with more complicated semantics than unsupervised hashing. The representative ones are BRE [8], MLH [16] and KSH [14]. In recent years, deep hashing methods have shown promising power. CNNH [26], DPSH [12] and NINH [9] are representative methods. They leverage pair-wise or triplet-wise approaches to learn semantic similarity, while large storage of pair or triplet samples is required. Moreover, above-mentioned hashing methods are devoted to CBIR, which have not been specially designed for SBIR yet.

3   Methodology

3.1   Problem Formulation

We let $\mathcal{P} = \{p_i\}_{i=1}^{n_1}$ be the set of all real image photos and $\mathcal{S} = \{s_i\}_{i=1}^{n_2}$ be the set of all sketches, where n_1 and n_2 are the sample number of set $\mathcal{P}$ and

$\mathcal{S}$ respectively. The corresponding label set of real photos is denoted as $\mathcal{L}^{\mathcal{P}} = \{l_i^{\mathcal{P}}\}_{i=1}^{n_1}, l_i^{\mathcal{P}} \in \{1, 2, ..., C\}$. Each photo p_i is connected to one label tag $l_i^{\mathcal{P}}$ coming from total C classes. Similarly, we have the label set of sketches $\mathcal{L}^{\mathcal{S}} = \{l_i^{\mathcal{S}}\}_{i=1}^{n_2}$, $l_i^{\mathcal{S}} \in \{1, 2, ..., C\}$.

Our target is to learn a semantic-preserving hashing function mapping original photos $\mathcal{P}$ and sketches $\mathcal{S}$ to compact hash codes $\mathcal{B}^{\mathcal{P}} = \{b_i^{\mathcal{P}}\}_{i=1}^{n_1}$ and $\mathcal{B}^{\mathcal{S}} = \{b_i^{\mathcal{S}}\}_{i=1}^{n_2}$. These hash codes are d-bit binary representation $b_i^{\mathcal{P}}, b_i^{\mathcal{S}} \in \{0, 1\}^d$ that well preserving intrinsic semantics. For our particular task, we need to bridge the domain gap between real photos and sketches, in the meantime we should maintain the similarity relationship in original feature space for both domains themselves. More specifically, given two images, no matter which domain they belong to (perhaps a sketch or a natural photo): (i) If their corresponding labels are the same, they should be semantically similar all the way. In other words, the hamming distance of their hash codes has to be quite small. (ii) Otherwise, the distance between their codes should be pushed away as far as possible.

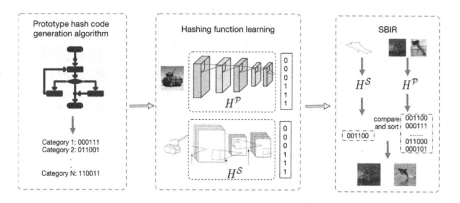

Fig. 2. Pipeline of our proposed idea. The first step is the prototype hash code generation algorithm. The next step is the hashing function learning procedure using prototype codes. The last step is the sketch-based image retrieval process.

To achieve our goal mentioned above, we design an efficient pipeline as shown in Fig. 2. It consists of three parts. Firstly, we encode a set of prototype hash codes fully utilizing the label semantic information. The next step is the hashing function learning procedure. We propose a novel deep hashing architecture that is specific to sketches and natural photos respectively with the help of generated prototype codes. Finally, we conduct sketch-based image retrieval process.

3.2 Prototype Hash Code

To mend the semantic gap in cross-domain situation such as our problem, we call for a comparable feature space applying to both sketches and real images. Hence, we specially design a common prototype binary encoding for both domains, which we call it prototype hash code. Since label information indicates the inherent semantic content for useful hashing learning aforesaid, a straight thinking is to generate a series of prototype hash codes based on label supervision. Given C label classes and d bit length of hash code, we denote the prototype set of C distinct hash codes as $\mathcal{B}^o = \{b_i^o\}_{i=1}^{C}$, $b_i^o \in \{0,1\}^d$. Among set $\mathcal{B}^o$, every code element b_i^o is matched with a class label. Then, during the subsequent hashing learning procedure, these prototype codes provide a firmly supervised support for efficient training. The target is to train a network which can output a hash code close to its nearest prototype code as much as possible.

To achieve our desired result, our generated prototype codes should meet some requirements. The hamming distance between every code pairs should be maximized, in order to capture more discriminative intrinstic structure and reduce the error rate in retrieval. That is to say, we need to enlarge the minimum hamming distance of this set to the greatest extent:

$$\max_f \left\{ d_{min} = \min_{i,j} \left\| b_i^o - b_j^o \right\|_H \right\}, f : \mathcal{L} \to \mathcal{B}^0 \tag{1}$$
$$s.t. \mathcal{B}^0 \in \{0,1\}^{d \times C}, b_i^o, b_j^o \in \mathcal{B}^0, i \neq j$$

where d_{min} is the crucial minimum hamming distance, $\|\cdot\|_H$ is the Hamming distance, $\mathcal{L}$ is the whole label set, f is our prototype encoding algorithm. In practice, our generation problem has no general sole solution in mathematics. Here, we search the feasible solution through controlling d_{min} to grow up increasingly. Starting with a relatively small d_{min}, we can easily find out a candidate set with more than C codewords satisfying the hamming distance between any codes is larger than d_{min}. Afterwards, we increase d_{min} by 1 repeatedly and follow the same searching strategy. Along with the increased d_{min}, the number of possible available codes of a candidate set will be reduced. We stop when the candidate set has less than the lower bound C codes. The last candidate set satisfying the C bound limitation is our final set resource. We randomly select C codes within it as optimal prototype code set $\mathcal{B}^o$. Consequently, we have our specially designed prototype result well maintaining discriminative essence. It is worth noting that with our generation algorithm, even very short hash codes can provide a large minimum hamming distance which is quite efficient for SBIR.

3.3   Deep Hashing Architecture

In this part, we propose a novel deep hashing architecture with two networks for sketches and photos individually as shown in Fig. 3. Such deep network can be seen as the hashing function which is a decisive factor in hashing learning. We denote hashing function as H^S for the sketch and H^P for the photo. As previously mentioned, sketches have a quite different appearance to real photos. It is unable to directly imitate mature network technique from CBIR here. If we let sketches and real images share the same network, the learned model will add extra noise on sketches and ignore detail structure of natural images, causing a bad effect on both domains. In addition, if we follow previous works to convert

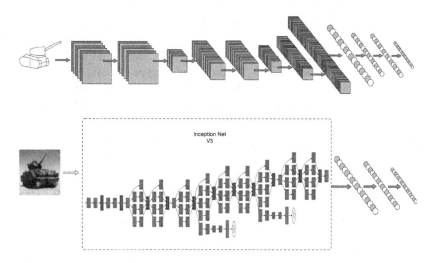

Fig. 3. An illustration of our proposed deep hashing retrieval architecture. We adopt two separate networks for two domains. The upper part is the network for sketch, and the lower part is the network for real photo.

natural images to approximate sketches by edge extraction in advance, it still has some defects. Actually, a sketch is different from a simple tracing of image boundary. People often draw sketches with geometry distortion and simply rely on their vague memory. We once test this edge extraction strategy and it turned out to be less effective.

Benefiting from prototype hash code introduced before, we adopt two separate networks for sketches and photos. The networks will learn a shared embedding in Hamming space. Prototype code set is a common binary embedding for sketches and photos both, guiding hashing learning at the last layer. Thus, two separate networks have the same mission that samples have to be grouped around the targeted prototype code.

For sketches, we adopt a carefully designed network containing 6 convolutional layers and 2 fully connected layers illustrated in the upper part of Fig. 3. The last layer is a fully connected layer with d nodes, relying on the hash code bit length. We also call it hash layer since it will encode the high-dimensional feature into binary-like one y_i via a following sigmoid activation. Sigmoid function has proved to be effective for hashing methods in that it could regulate features within a $(0, 1)$ real-valued range, becoming hash-like features. To train the sketch network end-to-end, we exploit Mean Squared Logarithmic Error (MSLE) with the supervision of prototype hash code:

$$\mathcal{L} = \frac{1}{n_1} \sum_{i=1}^{n_1} \|\log(b_i^o + 1) - \log(y_i + 1)\|^2 \tag{2}$$

where b_i^o is the prototype hash code that y_i referred to. As MSLE is more robust to overfitting than mean squared error, we choose MSLE as our learning objective. The detailed network configuration is illustrated in Table 1. The reason why we employ such kind of shallow network is based on sketch trait itself. The sketch is a grayscale image with rather sparse lines. Due to lack of abundant structure information and the unbalanced zero-one amount in sketches, a very deep network will fall into overfitting. Besides, we employ data augmentation procedure to further avoid overfitting situation and make limited training samples more robust. Each sketch sample runs through the augmentation preprocessing and then be fed into the network. A Random rotation, shear, zoom and translation will be applied.

Table 1. Detailed configuration of sketch network.

Layer	Filter Size	Filter Num	Stride	Pad	Activation	Output
Input	-	-	-	-	-	$1 \times 128 \times 128$
Conv	3×3	32	1	1	ReLU	$32 \times 128 \times 128$
Conv	3×3	32	1	1	ReLU	$32 \times 128 \times 128$
Dropout(0.25)	-	-	-	-	-	$32 \times 128 \times 128$
MaxPool	3×3	-	3	0	-	$32 \times 42 \times 42$
Conv	3×3	64	1	1	ReLU	$64 \times 42 \times 42$
Conv	3×3	64	1	1	ReLU	$64 \times 42 \times 42$
Dropout(0.25)	-	-	-	-	-	$64 \times 42 \times 42$
MaxPool	2×2	-	2	0	-	$64 \times 21 \times 21$
Conv	3×3	128	1	1	ReLU	$128 \times 21 \times 21$
Conv	3×3	2048	1	1	ReLU	$2048 \times 21 \times 21$
Dropout(0.25)	-	-	-	-	-	$2048 \times 21 \times 21$
GlobalAvgPool	21×21	-	21	0	-	2048
FC	1×1	256	-	-	ReLU	256
FC	1×1	256	-	-	ReLU	256
Dropout(0.5)	-	-	-	-	-	256
FC(hash)	1×1	d	-	-	Sigmoid	d

For real photos, we employ a revised successful deep network Inception-v3 [22] as shown in the lower figure of Fig. 3. Because the photo retrieval task is similar to traditional image-based task, we directly use a well-performing standard configuration from ILSVRC competition as our basic framework. Inception net is the winner of ILSVRC 2014 proposed by Google. Real photo input is downsampled to 140×140. Then we replace the last fully connected layers and softmax layer of original Inception net by a fully connected layer with 1024 nodes right after the global average pooling. At the end, a hash layer is applied exactly as sketch network works. We still use MSLE as our loss function.

3.4 Retrieval Process

The output of training network is real-valued feature y_i. To obtain the final hash code b_i, we binarize the activation output with a threshold:

$$b_i = sgn(y_i - 0.5),$$

$$sgn(x) = \begin{cases} 0 & x \leq 0, \\ 1 & x > 0 \end{cases} \tag{3}$$

Next, we carry out retrieval process. For SBIR, hashing retrieval process is a little more complicated than CBIR. (i) With our fine trained photo hashing function $H^{\mathcal{P}}$, all the real images in gallery set are transformed to compact d-bit hash codes. We denote the candidate hash pool as $\mathcal{P}$. (ii) Given a query sketch s^q, it will go through the trained sketch model $H^{\mathcal{S}}$ and output its sketch code b^q. (iii) We compare each candidate photo code in $\mathcal{P}$ with a query code b^q by calculating hamming distance. The hamming distance has a positive relation to similarity. Hence, it forms a rank of retrieval results in ascending order of distance.

4 Experiments

In this section, we demonstrate the effectiveness of our proposed method on sketch-based image retrieval. We conduct extensive experiments on two public datasets and our method is compared with several state-of-the-art methods. At last, we evaluate our method and verify its good performance.

4.1 Datasets

So far, the largest datasets in SBIR are *TU-Berlin Extension* and *Sketchy extension*. *TU-Berlin* benchmark [4] is aimed at sketch recognition and classification. It consists of 20,000 sketch images evenly belonging to 250 categories covering daily objects like teapot, car and horse. The extended TU-Berlin [30] dataset adds 204,489 real images in total as gallery set for sketch-based image retrieval. *Sketchy* [20] is the latest released dataset specifically collected for retrieval. It contains 75,471 sketches of 12,500 natural objects from 125 categories. The extended sketchy [13] provides another 60,502 natural images and merges original natural images into the retrieval gallery pool. Both two collections are convincing evaluation datasets for large-scale SBIR task.

For fair comparison to previous works, we follow the same experimental setting as DSH [13]. We randomly select 2,500 sketches (10 sketches per category) for TU-Berlin and 6,250 sketches (50 sketches per category) for Sketchy as test query sketches. And we use the remaining natural images and the rest of sketches for training.

4.2   Implementation Details

We implement our experiments on single GTX1060 GPU with 6GB memory. For sketch model, data augmentation is applied before entering the network. We perform random rotation in the range of 20 degrees, horizontal and vertical translation up to 25 pixels, zoom from 0.8 to 1.2 times and a 0.2 shear intensity to reduce overfitting. During training, batch size is set to 40 and the initial learning rate is 0.001. The model is trained for 200 epochs with Adam [7] optimizer. For photo model, we do training for 40 epochs with batch size of 128. And we still use Adam optimizer and the learning rate is 0.001.

Table 2. Performance comparison in SBIR with state-of-the-arts via mAP, Precision@200, Retrieval time per query and Memory load on TU-Berlin Extension.

Method	Dimension	TU-Berlin Extension			
		mAP	P@200	Retrieval time per query(s)	Memory load(MB)
HOG	1296	0.091	0.120	1.43	2.02×10^3
GF-HOG	3500	0.119	0.148	4.13	5.46×10^3
SHELO	1296	0.123	0.155	1.44	2.02×10^3
LKS	1350	0.157	0.204	1.51	2.11×19^3
Siamese CNN	64	0.322	0.447	7.70×10^{-2}	99.8
SaN	512	0.154	0.225	0.53	7.98×10^2
GN Triplet	1024	0.187	0.301	1.02	1.60×10^3
3D Shape	64	0.054	0.072	7.53×10^{-2}	99.8
Siamese-AlexNet	4096	0.367	0.476	5.35	6.39×10^3
Triplet-AlexNet	4096	0.448	0.552	5.35	6.39×10^3
DSH-32	32	0.358	0.486	5.57×10^{-4}	0.78
DSH-64	64	0.521	0.655	7.03×10^{-4}	1.56
DSH-128	128	0.570	0.694	1.05×10^{-3}	3.12
Our-12	12	0.550	0.622	3.04×10^{-4}	0.29
Our-24	24	0.561	0.634	4.48×10^{-4}	0.59
Our-32	32	0.573	0.650	5.43×10^{-4}	0.78
Our-64	64	0.591	0.668	6.99×10^{-4}	1.56
Our-128	128	0.613	0.693	9.72×10^{-4}	3.12

4.3   Results and Analysis

Our results are evaluated within the whole gallery set on extended TU-Berlin and Sketchy respectively. We compare our method with several state-of-the-art deep SBIR approaches including Siamese CNN [17], sketch-a-net (SaN) [28], GN

Table 3. Performance comparison in SBIR with state-of-the-arts via mAP, Precision@200, Retrieval time per query and Memory load on Sketchy Extension.

Method	Dimension	Sketchy Extension			
		mAP	P@200	Retrieval time per query(s)	Memory load(MB)
HOG	1296	0.115	0.159	0.53	7.22×10^2
GF-HOG	3500	0.157	0.177	1.41	1.95×10^3
SHELO	1296	0.161	0.182	0.50	7.22×10^2
LKS	1350	0.190	0.230	0.56	7.52×10^2
Siamese CNN	64	0.481	0.612	2.76×10^{-2}	35.4
SaN	512	0.208	0.292	0.21	2.85×10^2
GN Triplet	1024	0.529	0.716	0.41	5.70×10^2
3D Shape	64	0.084	0.079	2.64×10^{-2}	35.6
Siamese-AlexNet	4096	0.518	0.690	1.68	2.28×10^3
Triplet-AlexNet	4096	0.573	0.761	1.68	2.28×10^3
DSH-32	32	0.653	0.797	2.55×10^{-4}	0.28
DSH-64	64	0.711	0.858	2.82×10^{-4}	0.56
DSH-128	128	0.783	0.866	3.53×10^{-4}	1.11
Our-12	12	0.762	0.839	2.21×10^{-4}	0.11
Our-24	24	0.772	0.850	2.43×10^{-4}	0.21
Our-32	32	0.789	0.867	2.57×10^{-4}	0.28
Our-64	64	0.796	0.876	2.81×10^{-4}	0.56
Our-128	128	0.810	0.890	3.57×10^{-4}	1.11

Triplet [20], 3D Shape [23], as well as Siamese-AlexNet and Triplet-AlexNet [13]. Traditional hand-crafted methods HOG [2], GF-HOG [6]. SHELO [18] and LKS [19] are also included. To better demonstrate our outstanding performance, we conduct our experiments with 12, 24, 32, 64 and 128 bits hash code. This is the same setting compared with deep hashing method DSH [13]. During the comparison to SBIR baselines, we use a ranking based criterion mean Average Precision (mAP) and precision at top 200 (P@200) to evaluate the retrieval quality. Higher mAP and P@200 indicate a higher retrieval level in the ranking list. The memory load over the whole gallery images and retrieval time per query are also listed. Public results data of previous works are derived from DSH. The comparison results on two datasets are illustrated in Tables 2 and 3.

From our extensive comparison results, we have the following findings: (i) Our method outperforms all the baseline methods on both large-scale datasets. We increase around 4% and 3% mAP over TU-Berlin Extension and Sketchy Extension respectively. (ii) It is noteworthy that results on TU-Berlin are inferior to Sketchy due to much more categories and larger gallery size. (iii) Deep methods strongly beat traditional hand-crafted ones, proving the powerful ability of deep features. Additionally, with the help of hashing, our method and DSH can save memory load and retrieval time by almost four orders of magnitude. (iv) In comparison to the only deep hashing competitor DSH, our method has a better performance under several metrics. Especially, our method can achieve a good behavior even with quite compact hash code. For instance, our 12-bit hashing result surpasses 64-bit DSH in mAP. It demonstrates that our unified prototype hash code set is suitable for fast large-scale retrieval task. Moreover, our method utilizes a point-to-point training rather than pairwise loss in DSH, avoid tedious sample building step and weak training guidance.

5 Conclusion

In this paper, we introduce a novel deep hashing method for sketch-based image retrieval. Our method adopts a prototype hash code set for constraining feature representation. A deep hashing architecture is specially designed for two different domains, sketches and natural photos respectively. By means of mapping different domains into a common hamming space, our method achieves good performances with very compact binary codes. Extensive experiments across large-scale retrieval benchmarks demonstrate that our method outperforms all non-deep and deep methods under several metrics. In general, our method exhibits promising result in fast and efficient retrieval via compact binary codes learning.

References

1. Canny, J.: A computational approach to edge detection. In: Readings in Computer Vision, pp. 184–203. Elsevier (1987)
2. Dalal, N., Triggs, B.: Histograms of oriented gradients for human detection. In: IEEE Computer Society Conference on Computer Vision and Pattern Recognition 2005. CVPR 2005, vol. 1, pp. 886–893. IEEE (2005)
3. Datar, M., Immorlica, N., Indyk, P., Mirrokni, V.S.: Locality-sensitive hashing scheme based on p-stable distributions. In: Proceedings of the Twentieth Annual Symposium on Computational Geometry, pp. 253–262. ACM (2004)
4. Eitz, M., Hays, J., Alexa, M.: How do humans sketch objects? ACM Trans. Graph. **31**(4), 44-1 (2012)
5. Gong, Y., Lazebnik, S.: Iterative quantization: a procrustean approach to learning binary codes. In: IEEE Conference on Computer Vision and Pattern Recognition (CVPR) 2011, pp. 817–824. IEEE (2011)
6. Hu, R., Collomosse, J.: A performance evaluation of gradient field hog descriptor for sketch based image retrieval. Comput. Vis. Image Underst. **117**(7), 790–806 (2013)
7. Kingma, D., Ba, J.: Adam: A method for stochastic optimization. arXiv preprint arXiv:1412.6980 (2014)
8. Kulis, B., Darrell, T.: Learning to hash with binary reconstructive embeddings. In: Advances in Neural Information Processing Systems, pp. 1042–1050 (2009)

9. Lai, H., Pan, Y., Liu, Y., Yan, S.: Simultaneous feature learning and hash coding with deep neural networks. In: Proceedings of the IEEE Conference on Computer Vision and Pattern Recognition, pp. 3270–3278 (2015)
10. LeCun, Y., Bengio, Y., et al.: Convolutional networks for images, speech, and time series. Handb. Brain Theory Neural Netw. **3361**(10), 1995 (1995)
11. Li, G., Liu, J., Jiang, C., Zhang, L., Lin, M., Tang, K.: Relief R-CNN: utilizing convolutional features for fast object detection. In: Cong, F., Leung, A., Wei, Q. (eds.) ISNN 2017. LNCS, vol. 10261, pp. 386–394. Springer, Cham (2017). https://doi.org/10.1007/978-3-319-59072-1_46
12. Li, W.J., Wang, S., Kang, W.C.: Feature learning based deep supervised hashing with pairwise labels. arXiv preprint arXiv:1511.03855 (2015)
13. Liu, L., Shen, F., Shen, Y., Liu, X., Shao, L.: Deep sketch hashing: Fast free-hand sketch-based image retrieval. In: Proceedings of CVPR, pp. 2862–2871 (2017)
14. Liu, W., Wang, J., Ji, R., Jiang, Y.G., Chang, S.F.: Supervised hashing with kernels. In: IEEE Conference on Computer Vision and Pattern Recognition (CVPR) 2012, pp. 2074–2081. IEEE (2012)
15. Lowe, D.G.: Distinctive image features from scale-invariant keypoints. Int. J. Comput. Vis. **60**(2), 91–110 (2004)
16. Norouzi, M., Blei, D.M.: Minimal loss hashing for compact binary codes. In: Proceedings of the 28th International Conference on Machine Learning (ICML-11), pp. 353–360 (2011)
17. Qi, Y., Song, Y.Z., Zhang, H., Liu, J.: Sketch-based image retrieval via siamese convolutional neural network. In: IEEE International Conference on Image Processing (ICIP) 2016, pp. 2460–2464. IEEE (2016)
18. Saavedra, J.M.: Sketch based image retrieval using a soft computation of the histogram of edge local orientations (s-helo). In: IEEE International Conference on Image Processing (ICIP) 2014, pp. 2998–3002. IEEE (2014)
19. Saavedra, J.M., Barrios, J.M., Orand, S.: Sketch based image retrieval using learned keyshapes (LKS). In: BMVC, vol. 1, p. 7 (2015)
20. Sangkloy, P., Burnell, N., Ham, C., Hays, J.: The sketchy database: learning to retrieve badly drawn bunnies. ACM Trans. Graph. (TOG) **35**(4), 119 (2016)
21. Shechtman, E., Irani, M.: Matching local self-similarities across images and videos. In: IEEE Conference on Computer Vision and Pattern Recognition 2007. CVPR 2007, pp. 1–8. IEEE (2007)
22. Szegedy, C., Vanhoucke, V., Ioffe, S., Shlens, J., Wojna, Z.: Rethinking the inception architecture for computer vision. In: Proceedings of the IEEE Conference on Computer Vision and Pattern Recognition, pp. 2818–2826 (2016)
23. Wang, F., Kang, L., Li, Y.: Sketch-based 3D shape retrieval using convolutional neural networks. In: IEEE Conference on Computer Vision and Pattern Recognition (CVPR) 2015, pp. 1875–1883. IEEE (2015)
24. Weiss, Y., Torralba, A., Fergus, R.: Spectral hashing. In: Advances in Neural Information Processing Systems, pp. 1753–1760 (2009)
25. Wu, X., Kamata, S.i., Ma, L.: Supervised two-step hash learning for efficient image retrieval. In: 2017 4th Asian Conference on Pattern Recognition. IEEE (2017)
26. Xia, R., Pan, Y., Lai, H., Liu, C., Yan, S.: Supervised hashing for image retrieval via image representation learning. In: AAAI, vol. 1, p. 2 (2014)
27. Yu, Q., Liu, F., Song, Y.Z., Xiang, T., Hospedales, T.M., Loy, C.C.: Sketch me that shoe. In: IEEE Conference on Computer Vision and Pattern Recognition (CVPR) 2016, pp. 799–807. IEEE (2016)
28. Yu, Q., Yang, Y., Song, Y.Z., Xiang, T., Hospedales, T.: Sketch-a-net that beats humans. arXiv preprint arXiv:1501.07873 (2015)

29. Zeiler, M.D., Fergus, R.: Visualizing and understanding convolutional networks. In: Fleet, D., Pajdla, T., Schiele, B., Tuytelaars, T. (eds.) ECCV 2014. LNCS, vol. 8689, pp. 818–833. Springer, Cham (2014). https://doi.org/10.1007/978-3-319-10590-1_53
30. Zhang, H., Liu, S., Zhang, C., Ren, W., Wang, R., Cao, X.: Sketchnet: Sketch classification with web images. In: Proceedings of the IEEE Conference on Computer Vision and Pattern Recognition, pp. 1105–1113 (2016)

Improved Nuclear Segmentation on Histopathology Images Using a Combination of Deep Learning and Active Contour Model

Lei Zhao[1], Tao Wan[1(✉)], Hongxiang Feng[2], and Zengchang Qin[3(✉)]

[1] School of Biomedical Science and Medical Engineering, Beijing Advanced Innovation Centre for Biomedical Engineering, Beihang University, Beijing, China
taowan@buaa.edu.cn
[2] Department of General Thoracic Surgery, China Japan Friendship Hospital, Beijing, China
[3] Intelligent Computing and Machine Learning Lab, School of ASEE, Beihang University, Beijing, China
zcqin@buaa.edu.cn

Abstract. Automated nuclear segmentation on histopathological images is a prerequisite for a computer-aided diagnosis system. It becomes a challenging problem due to the nucleus occlusion, shape variation, and image background complexity. We present a computerized method for automatically segmenting nuclei in breast histopathology using an integration of a deep learning framework and an improved hybrid active contour (AC) model. A class of edge patches (nuclear boundary), in addition to the two usual classes - background patches and nuclei patches, are used to train a deep convolutional neural network (CNN) to provide accurate initial nuclear locations for the hybrid AC model. We devise a local-to-global scheme through incorporating the local image attributes in conjunction with region and boundary information to achieve robust nuclear segmentation. The experimental results demonstrated that the combination of CNN and AC model was able to gain improved performance in separating both isolated and overlapping nuclei.

Keywords: Convolutional neural network · Active contour model
Nuclear segmentation · Histopathology

1 Introduction

With the recent advent of whole slide digital scanners and advances in computational power, it is now possible to use digitized histopathological images and

T. Wan and Z. Qin—This work was supported in part by the National Natural Science Foundation of China under award No. 61401012.

© Springer Nature Switzerland AG 2018
L. Cheng et al. (Eds.): ICONIP 2018, LNCS 11306, pp. 307–317, 2018.
https://doi.org/10.1007/978-3-030-04224-0_26

computer-aided image analysis to facilitate breast diagnosis and prognosis [12]. An automatic, high-throughput image analysis system usually requires accurate and robust nuclei segmentation as the first and critical step. Nuclear segmentation becomes a difficult problem particularly for routinely stained hematoxylin and eosin (H&E) slides due to the nucleus occlusion or overlapping, shape variation, inter- and intra-nucleus inhomogeneity, background complexity, and image artifacts [13].

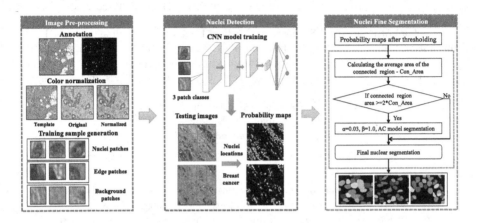

Fig. 1. Workflow of the presented nuclear segmentation method

A large variety of segmentation techniques to tackle these problems have been applied to histopathological images [6,9]. Active contour (AC) models remain the most popular methods in segmenting medical images [5]. For instance, the classic Chan-Vese model [1] computed the gray level of homogeneity derived from image foreground and background. Hybrid AC models incorporating boundary, region, and shape information, have been approved to be effective methods to segment nuclear structures. These methods depended on the choice of initialization which limit their ability in segmenting multiple touching or overlapping cells. Recently, many nuclei detection methods using deep learning strategies have been developed on histopathological images. Xu *et al.* [14] utilized a staked sparse autoencoder to learn high-level features to distinguish nuclei and non-nuclei. In [13], a deep convolutional neural network (CNN) was used to generate a probability map served as shape initializations for a deformable model. These learning-based methods could provide accurate initial seed points for subsequent application of segmentation models. This motivates our work on building a fine nuclear segmentation method by fusing CNN technique and AC model.

In this work, we present a computerized image-based method for automatically segmenting nuclei in digitalized breast histopathology. The workflow of the presented method is depicted in Fig. 1. We integrate a deep learning framework and an improved hybrid active contour model to perform a robust nuclear segmentation. A light fully convolutional neural network model is employed to

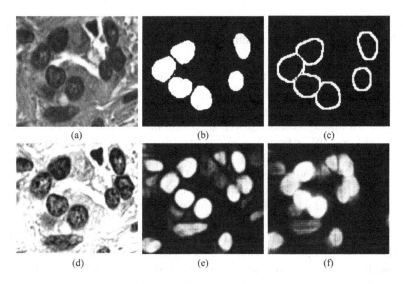

Fig. 2. (a) Original histopathology image; (b) Binary image with manually labeled nuclei boundaries; (c) Nuclei contours using the morphological erosion operation; (d) Enhanced images using de-convolution operation and histogram equalization; (e) Nuclei detection with edge patches; (f) Nuclei detection without edge patches.

reduce the computational burden of training deep CNN model and applicable to large-scale datasets. In order to accurately detect touching or overlapping nuclei, we introduce a third class of edge patches (nuclear boundary), in addition to the two usual classes - background patches (outside all nuclei) and nuclei patches as input samples, to train the CNN model. The probability map obtained from the trained CNN model is to form an initial curves close to the real nuclear boundary, enabling to accelerate the computation of the following AC segmentation, thus solving the problem that general AC models are sensitive to the initial contour curves. Moreover, we integrate the local image attributes, reflecting the intensity homogeneity of the nuclear regions, with region and boundary information into the devised hybrid AC model to partition touching nuclei as well as isolated nuclei from the image background. The combination of CNN and AC model could be useful in building an image analysis tool for computational histopathology.

2 Methodology

2.1 Image Pre-processing

Color Normalization: There are undesirable color variations between digital tissue images often caused by differences in stain vendors, staining protocols, scanning parameters, and illumination. Color normalization has been proved to be able to improve tissue segmentation by maintaining color constancy in digitalized pathology meanwhile preserving biological structure information present in the images. We utilized a color-map based quantile normalization method [6] at a pixel basis to reduce color differences across the histopathological images.

Training Sample Generation: We created three classes of training samples, referred as nuclei, edge, and background through a semi-automated method. The nuclei patches were generated based on manually labeled nuclear boundaries. We then applied a morphological erosion operation on the binary images (see Fig. 2(b)) resulting from the manually delineated images to form a three-pixel width nuclei contours shown in Fig. 2(c). The edge patches were selected around these nuclei boundaries. Subsequently, we used a de-convolutional operation to find the erosion stain components in order to obtain non-nuclei pixels. A histogram equalization method was employed to further improve the contrast between the nuclei and non-nuclei pixels (see Fig. 2(d)). The background patches were chosen based on the detected non-nuclei pixels. All the three types of patches were the same size with 64×64 pixels. Different sizes of patches were also considered. The patch size was determined by the trade-off between the coverage of large nuclei as well as the useful context information and the computation complexity. The training dataset consisted of positive samples (nuclei patches) and negative samples (edge and background patches). Unlike the general training process, we added the edge patches as the negative samples in order to improve nuclei detection performance. Figures 2(e) and (f) show two examples of nuclei detection results using and without using edge patches as training samples. we noted that the edge patches allowed a better learning process for the nuclei boundaries within the CNN model, particularly useful for these overlapping nuclei with connected margins.

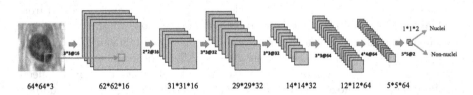

Fig. 3. Illustration of the architecture for the presented fully convolutional neural network

2.2   Nuclei Detection

We adopted a CNN-based architecture to train a patch-based classifier to distinguish nuclei or non-nuclei patches. The nuclei detection task was converted into a patch-based classification problem. A probability map on a pixel basis for the entire histopathological image was computed based on the classification result of each patch centered pixel. As mentioned before, we introduced the edge patches describing nuclei boundaries into the training negative samples in addition to the usual binary of foreground (inside any nucleus) and background (outside every nucleus), thus allowing a better identification of overlapping nuclei.

Algorithm 1. The H-minima based nuclei marker identification

Input: Gradient map $\mathbf{G}_{map}$, area threshold t_{area}, disk size d_{size}

Output: Nuclei markers $\mathbf{M}$

1. $\mathbf{M} = \emptyset$, $h = 1$
2. **repeat**
3.   Given $\mathbf{G}_{map}$ using H-minima to suppress noise.
4. $\mathbf{M}_{curr} \leftarrow$ Region minima (h_{map})
5. $\mathbf{M}_{curr} \leftarrow$ Remove small regions $(\mathbf{M}_{curr}, t_{area})$
6. $\mathbf{M}_{curr} \leftarrow$ Remove overlapping $(\mathbf{M}, \mathbf{M}_{curr}, d_{size})$
7. $\mathbf{M} = \mathbf{M} \bigcup \mathbf{M}_{curr}$
8. $h = h + 1$
9. **until** $\mathbf{M}_{curr} = \emptyset$

We presented a fully convolutional neural network model based on the AlexNet [8], which contained only the convolutional layers (see Fig. 3). We replaced the pooling layers and fully connected layers by the convolutional layers in order to reduce the size of the input at each layer. This led to a large reduction in the number of network parameters and a avoidance of over-fitting problem. As the network became deeper and the input size shrank, we increased the number of kernels so that essentially each layer was computed in the same amount of time, which was similar as the scheme for selecting the number of kernels used in the VGG model [10]. This scheme offered a good ability to predict the computational time of the training and testing processes. Further, we used the batch normalization layers before the activation layers to avoid over-fitting [3]. After the patch-based classification, we created a pixel-level probability map for the whole histopathological image using a bilinear up-sampling operation. A gradient map $\mathbf{G}_{map}$ was adopted to iteratively identify the nuclei markers. Algorithm 1 shows the iteration process using the H-minima transform to mitigate the noise interference due to the image artifacts [7].

2.3   Nuclei Fine Segmentation

In the fine segmentation phase, we devised an improved hybrid AC model via fusing both boundary- and region-based information within a local-to-global strategy to achieve a robust nuclei segmentation in breast histopathology. For the initialization of AC model, we used the previously detected nuclei markers, rather than using the manual delineation or automated approaches, such as uniform griding, watershed algorithm [5]. Although these methods are easy to implement, the arising over-segmentation problem leads to many inaccurate non-nuclei segments. The CNN detection method provided better performance in real nuclei detection on the histopathological images. The probability maps from the CNN results after thresholding were used as initial shapes for AC model. The hybrid AC model was then driven via a level set method, in which the energy function to be minimized could be expressed as:

$$E(\psi) = -a \int_{\Omega} (A - \eta)\mathbf{H}(\psi)d\Omega + b \int_{\Omega} \xi|\nabla\mathbf{H}(\psi)|d\Omega, \tag{1}$$

where Ω represents image domain, ψ is the zero set of embedding function representing the active contour $\mathcal{C} = \{y|\psi(y) = 0\}$, where points inside and outside of $\mathcal{C}$ have positive and negative ψ values. A denotes the image to be segmented, $\xi = \xi(|\nabla|A)$ is a boundary feature map obtained from image gradient, and $\mathbf{H}(\psi)$ is the regularized approximation of Heaviside function. The constants a and b are predefined weights to balance the two terms. The first term defines the region term and η is a parameter indicating the lowest value of gray level of the target object. We assumed here that the target object of nucleus had relatively high gray level values. If this was not the case, a simple gray-level remapping technique could be applied to achieve it. The second term of geodesic active contour function guides the contours to attach to the regions with high image gradients.

Moreover, we incorporated the local image information into the hybrid AC model in order to overcome the disadvantage of inhomogeneous intensity distribution appearing in the initial contours. The local term E_l derived from the local image statistical information can be defined as: $E_l = \int_{in(\mathcal{C})} (\zeta(A)) - A) - d_1)^2 d\mathcal{C} + \int_{out(\mathcal{C})} (\zeta(A)) - A) - d_2)^2 d\mathcal{C}$, where $\zeta(A)$ is an averaging filter, d_1 and d_2 are the intensity averages of the difference image $(\zeta(A) - A)$ inside contour $in(\mathcal{C})$ and outside contour $out(\mathcal{C})$, respectively. This local term had been used in our previous work and obtained a superior segmentation performance in refining the boundaries of multiple overlapping nuclei [5]. This global-to-local scheme improved the hybrid AC model for better segmenting both individual and overlapping nuclei in the breast histopathological images.

3   Experimental Results and Discussion

3.1   Experimental Design

Data Description: We collected a dataset containing 137 histopathological images scanned at $40\times$ magnification from 89 patients who diagnosed with breast cancer. All the images were de-identified and H&E stained. We extracted 141 regions of interest (ROIs) from the data cohort. Among these ROIs, approximately 16,000 breast cancer nuclei were manually annotated by an expert histopathologist. The annotated data were separated into training dataset (80%) and testing dataset (20%). We also ensured that the training and testing samples were not from the same patient study simultaneously.

Data Augmentation: Because of unbalance between the number of nuclei and background patches, the classification could be biased towards the background (non-nuclei). Therefore, a data augmentation method was applied to the nuclei patches by rotating the patches with $90°, 180°$ and $270°$, thus producing three times of positive samples. In the experiment, we used $48,000$ positive and $120,000$ negative samples for breast cancer histopathology.

Parameters Setting: Our CNN model was implemented under a parallel computing platform CUDA using a Caffe framework [4]. The learning rate of model was 0.001, and epoch was 10. The learning process was optimized via a stochastic gradient descent method [4]. In the nuclei detection task, the parameter of initial depth for the H-minima transform was set as $h_0 = 1$, $t_{area} = 3$, and $d_{size} = 3$. The AC model was empirically tuned to achieve the best segmentation performance. In all the experiments, we used $a = 0.03$, $b = 1.0$, $\eta = \{0.47 - 0.50\}$ in the nuclei fine segmentation process.

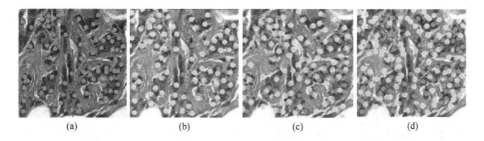

  (a)   (b)   (c)   (d)

Fig. 4. Nuclei detection results. (a) Ground truth; (b) The presented CNN-based method; (c) The deconvolution method [11]; (d) The MPAV method [2]. Green dots and yellow dots represent the manual annotated nuclei centers and the detected nuclei centers, respectively (Color figure online).

Table 1. Quantitative comparison of nuclei detection methods.

Method	CNN-based method	Deconvolution [11]	MPAV [2]
Precision	**0.85**	0.82	0.81
Recall	**0.78**	0.74	0.72
F-measure	**0.82**	0.78	0.76

3.2 Evaluation Metrics

The performance of the presented CNN-based nuclei detection method was quantified in terms of three popular metrics of precision, recall, and F-measure, which were defined as: $Precision = \frac{TP}{TP+FP}$, $Recall = \frac{TP}{TP+FN}$, $F - measure = 2 \times \frac{Precision \times Recall}{Precision+Recall}$, where TP, FP, FN are the true positive, false positive, and false negative. F-measure is the harmonic mean of precision and recall. In this work, TP was defined as the correctly identified nuclei, in which the center of the nuclei was within the distance of 15 pixels to the center of manually annotated nuclei.

  For the nuclei fine segmentation, three measures, including Dice similarity coefficient (DSC), F-measure, and average Housdroff distance (AveHd), were used to quantitatively evaluate the segmentation performance. The DSC metric can be expressed as: $DSC = \frac{2|\Omega_{sr} \cap \Omega_{gt}|}{|\Omega_{sr}|+|\Omega_{gt}|}$, where Ω_{sr} and Ω_{gt} are the areas enclosed by the automated segmentation ($\mathcal{C}_{sr}$) and manual segmentation ($\mathcal{C}_{gt}$), respectively. The closer the DSC value is to 1, the more similar the automated nuclear segmentation is to the manual reference segmentation. The Housdroff distance, which penalizes distance of furthest pixels on contours of two shapes, is calculated as: $AveHd = \frac{1}{|\mathcal{C}_{sr}|} \sum_{\mathcal{C}_{sr}} [\min_{\mathcal{C}_{gt}} \|z_{sc} - z_{gt}\|]$, where $|\mathcal{C}_{sr}|$ is the total number of points on $\mathcal{C}_{sr}$, z_{sc} and z_{gt} are the points on the automated segmentation contour and the corresponding closest points on the manual segmentation, respectively. AveHd reflects the average error between the automated and manual segmentations.

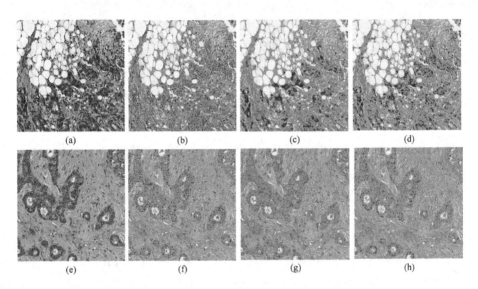

Fig. 5. Segmentation results. (a)(e) Original images; (b)(f) The presented method; (c)(g) The Chan-Vese model [1]; (d)(h) The Otsu's method [15]. The green contours are superimposed on the original images

3.3   Nuclei Detection Performance

The presented CNN-based nuclei detection method was compared with the other two recent reference approaches. The deconvolution method was a region-based segmentation method, in which the nuclei were detected through the color deconvolution and a generalized fast radial symmetry transform [11]. The multi-pass adaptive voting approach (MPAV) utilized the symmetric property of nuclear boundary and selected gradient from edge fragments to perform voting for a potential nucleus location [2]. For a fair comparison, we used their original implementation code to generate the detection results.

Figure 4 demonstrates the visual results using three methods. By observing the figures, we noted that the presented CNN-based method achieved the best detection results, where the position of the detected nuclei were closely matched with the manual detection. Both reference methods produced more non-nuclei detection and miss the real nuclei locations. In addition, we conducted a quantitative evaluation using the three metrics of precision, recall, and F-measure shown in Table 1. The measurement values were obtained using the histopathological images in the testing set. The evaluation provided a consistency with the visual results. The devised CNN-based method achieved the best quantitative results in all three metrics. The main reason was that in the training procedure, we added the edge patches to allow the CNN model to learn the nuclei boundaries and improve the detection performance in the overlapping nuclei, which further enhanced the following nuclear segmentation.

Table 2. Quantitative comparison of nuclei segmentation methods.

	CNN+AC	C-V [1]	Otsu [15]	Xing [13]	Kumar [9]
DSC	**0.78**	0.72	0.58	0.68	0.75
F-measure	**0.85**	0.81	0.78	0.84	0.72
AveHd	6.22	8.32	8.54	7.14	**6.04**

3.4   Nuclear Segmentation Performance

For the nuclei fine segmentation, we qualitatively compared our results with two classic region-based segmentation methods. The Chan-Vese (C-V) model employed the global image statistics inside and outside the evolving curve [1]. The Otsu's binarization method is one of the most popular thresholding methods used in image segmentation [15]. The visual evaluation of our segmentation method was performed across all the testing images. The segmentation results are illustrated in Fig. 5. Our presented method outperformed the C-V model [1] and the Otsu's method [15] in terms of good nuclei separation and smooth contour enclosure. We noted that the C-V model (see Figs. 5(c) and (g)) failed to handle the nuclear overlapping problem due to the inaccurate curve convergence on the outer boundaries enclosing multiple clumped nuclei. The Otsu's method yielded worse segmentation results compared with the other approaches, especially for the challenging cases with crowded nuclei (see Fig. 5(a)).

Furthermore, we quantitatively compared our segmentation method with two alternative deep learning based approaches. Xing *et al.* [13] introduced a deep learning-based framework for nucleus segmentation with shape preservation. Kumar *et al.* [9] developed a deep learning based technique, in which a CNN model was used to produce a ternary map, to perform a nuclear segmentation. Since the implementations of these two methods are not available, we used the original measure results reported in their papers which were also validated using the breast histopathological images. Table 2 shows the comparison results evaluated using DSC, F-measure, and AveHd. Our segmentation method achieved the highest DSC, F-measure, and second lowest AveHd. This indicated that combination of deep learning technique and the improved active contour model could lead to superior nuclear segmentation, particularly useful in accurately separating multiple overlapping nuclei with a high density on digital histopathological images.

4 Conclusions and Future Work

We presented a computerized method for automated nuclear segmentation using an integration of a deep learning framework and an improved hybrid active contour model on histopathological images. Three classes of patches, including nuclei, edge, and background, were utilized to train a modified fully convolutional neural network model. The edge patches, containing nuclear pixels, were able to find inter-nuclear boundaries irrespective of the configuration of the crowded nuclei. Thus, the CNN-based nuclei detection method could help in solving the problem of general AC models which were sensitive to the initial contour shapes. Both qualitative and quantitative evaluation demonstrated that the presented method outperformed other alternative deep learning based methods and classic AC models. This suggested that the combination of CNN and AC model could provide improved nuclear segmentation in separating clumped nuclei as well as chromatin sparse nuclei.

References

1. Chan, T., Vese, L.: Active contours without edges. IEEE Trans. Image Process. **10**(2), 266–277 (2001)
2. Cheng, L., Xu, H., Xu, J., Gilmore, H., Mandal, M., Madabhushi, A.: Multi-pass adaptive voting for nuclei detection in histopathological images. Sci. Rep. **6**, 33985 (2016)
3. Ioffe, S., Szegedy, C.: Batch normalization: Accelerating deep network training by reducing internal covariate shift. In: International Conference on Machine Learning, pp. 448–456 (2015)
4. Jia, Y., et al.: Caffe: convolutional architecture for fast feature embedding. In: ACM International Conference on Multimedia, pp. 675–678 (2014)
5. Jing, J., Wan, T., Cao, J., Qin, Z.: An improved hybrid active contour model for nuclear segmentation on breast cancer histopathology. In: IEEE International Symposium on Biomedical Imaging, pp. 1155–1158 (2016)
6. Kothari, S., Phan, J., Stokes, T., Wang, M.: Pathology imaging informatics for quantitative analysis of whole-slide images. J. Am. Med. Inf. Assoc. **20**(6), 1099–1108 (2013)
7. Koyuncu, C., Akhan, E., Ersahin, T., Cetin-Atalay, R., Gunduz-Demir, G.: Iterative H-minima-based marker-controlled wathershed for cell nucleus segmentation. Cytometry **89A**, 338–349 (2016)
8. Krizhevsky, A., Sutskever, I., Hinton, G.E.: Imagenet classification with deep convolutional neural networks. In: International Conference on Neural Information Processing Systems, pp. 1097–1105 (2012)
9. Kumar, N., Verma, R., Sharma, S., Bhargava, S., Vahadane, A., Sethi, A.: A dataset and a technique for generalized nuclear segmentation for computational pathology. IEEE Trans. Med. Imaging **36**(7), 1550–1560 (2017)

10. Simonyan, K., Zisserman, A.: Very deep convolutional networks for large-scale image recognition. Comput. Vis. Pattern Recogn. **1**, 1–14 (2015)
11. Taheri, S., Fevens, T., Bui, T.D.: Robust nuclei segmentation in cyto-histopathological images using statistical level set approach with topology preserving constraint. In: SPIE Medical Imaging, pp. 1–10 (2017)
12. Wan, T., Cao, J., Chen, J., Qin, Z.: Automated grading of breast cancer histopathology using cascaded ensemble with combination of multi-level image features. Neurocomputing **229**, 34–44 (2017)
13. Xing, F., Xie, Y., Yang, L.: An automatic learning-based framework for robust nucleus segmentation. IEEE Trans. Med. Imaging **35**(2), 550–566 (2016)
14. Xu, J., et al.: Stacked sparse autoencoder (SSAE) for nuclei detection on breast cancer histopathology images. IEEE Trans. Med. Imaging **35**(1), 119–130 (2016)
15. Xue, J., Titterington, D.: t-tests, F-tests and Otsu's methods for image thresholding. IEEE Trans. Image Process. **20**(8), 2392–2396 (2011)

Hierarchical Online Multi-person Pose Tracking with Multiple Cues

Chuanzhi Xu and Yue Zhou$^{(\boxtimes)}$

School of Electronic Information and Electrical Engineering,
Shanghai Jiao Tong University, Shanghai 200240, China
{ChuanzhiXu,zhouyue}@sjtu.edu.cn

Abstract. Multi-person articulated pose tracking is a newly proposed computer vision task which aims at associating corresponding person articulated joints to establish pose trajectories. In this paper, we propose a region-based deep appearance model combined with an LSTM pose model to measure the similarity between different identities. A novel hierarchical association method is proposed to reduce the time consumption for deep feature extraction. We divide the association procedure into two stages and extract deep feature only when the pairs of identities are difficult to distinguish. Extensive experiments are conducted on the newly released multi-person pose tracking benchmark: PoseTrack. The results show that the tracking accuracy gains an obvious improvement when adopting multiple association cues, and the hierarchical association method could improve the tracking speed obviously.

Keywords: Multi-person pose tracking · Hierarchical association
Region-based deep network · LSTM pose model

1 Introduction

Multi-person articulated pose tracking (MPT) is a newly proposed computer vision task and aims at associating corresponding person articulated joints in consecutive video frames to get the pose trajectories. MPT is the basic work for human action recognition and human behavior prediction, and has a wide application in robotics, surveillance and human-computer interaction.

Multi-person pose tracking is a similar problem with multi-object tracking (MOT), but there are still many differences between them. MOT focuses on the surveillance scenarios, and objects in such situation move smoothly and regularly. The goal of MOT is to get consistent trajectory for each object. However, MPT focuses on more complicated situations such as unconstrained dance and sports videos, so there are more challenges such as severe occlusion, deformation, similar appearance and non-linear motion, and the objective of MPT is to associate corresponding pose joints and get the pose flow. Considering MOT is a deeply explored computer vision task, some inspirations could be token from it for MPT.

© Springer Nature Switzerland AG 2018
L. Cheng et al. (Eds.): ICONIP 2018, LNCS 11306, pp. 318–328, 2018.
https://doi.org/10.1007/978-3-030-04224-0_27

Following traditional online MOT methods, which could be formulated as tracking by detection framework [12], MPT could be solved by tracking by estimation technique, i.e., human articulated points are previously estimated in each frame and then the pose flows are built by associating corresponding joints. This task could be formulated as an assignment problem and the key issue under this framework is to establish reliable affinity matrix. Variety of measurements have been explored in MOT field. [19] adopts the deep neutral network to build appearance model and measure the appearance similarity between objects. [2] employs the Kalman Filter to measure position similarity. [16] proposes an interaction model to explore the group behavior. Inspired by previous works, similar attempts have also been introduced into MPT field. [21] proposes the pose distance to measure pose similarity in consecutive frames. [20] measures the motion similarity by optical flow. In recent years, Long Short-Term Memory(LSTM) [8] network has shown its strong ability on sequential problems and their special structures make them gain the ability to learn long term dependencies. [5] introduces an LSTM model to fuse different affinities and calculate the final matching likelihood. In this paper, we propose a novel LSTM pose model to learn the limb motion information and predict the limb angles, and then the pose affinity is measured by the limb angles similarity.

Appearance model is a key factor for affinity computation. In complex and crowded scenarios, many objects are presented with similar appearance and may be partly occluded. The result is that the tracker can not associate the objects consistently. With recent development in convolutional neutral network (CNN), deep feature has shown its strong representative ability. Some previous works [2,5,18,19] have introduced CNN to extract deep feature for appearance similarity computation, but it is time consuming to extract deep feature for each person at each frame. In this paper, we propose a region-based deep CNN for part-based appearance feature extraction. This model could generate reliable similarity score even if the persons are occluded with each other and improve the tracking accuracy significantly. To reduce the time consumption for feature extraction, we propose a novel hierarchical association method. The association procedure is divided into two stages, at first stage, the pose tracklets and detected poses are divided into easy and hard associations according to the basic affinity matrix which is established based on simple metrics. At second stage, the region-based deep appearance network is employed to extract deep feature for appearance similarity computation and get more distinguishable scores for hard associations. In this way, the part-based deep feature is adopted only when pairs of identities are difficult to distinguish.

In this paper, we focus on multi-person pose tracking task, and our main contributions are summarized as below:

1. We propose a region-based deep network to extract part-based appearance feature, and in this way, our tracker could associate correctly even if the objects are partly occluded;
2. We propose a novel LSTM pose model to learn the limb motion information and predict limb angles which is then adopted for pose similarity computation;

3. To reduce the time consumption for deep feature extraction, we propose a hierarchical association method and extract deep feature only when pairs of identities are difficult to distinguish. We conduct extensive experiments on PoseTrack [1] benchmark, and the results show that our tracker could improve the tracking accuracy and reduce the time consumption significantly.

2   Our Proposed Approach

2.1   Problem Formulation

Under tracking by detection framework, the online multi-person pose tracking task could be formulated as an assignment problem [4] between pose tracklets and new pose joints. Suppose that there are m existing pose tracklets and n detected poses at frame t. A matrix $C_t \in R^{m*n}$ indicates the costs to assign tracklets with corresponding objects. The objective is to find the optimal solution with minimal cost. Suppose $m < n$, we first expand C_t to a square matrix with dummy rows, then formulate this assignment problem as below:

$$min \sum_{i=1}^{n} \sum_{j=1}^{n} c_{ij} x_{ij}$$

$$s.t. \quad \sum_{i=1}^{n} x_{ij} = 1 \quad j = 1, 2, ..., n \tag{1}$$

$$\sum_{j=1}^{n} x_{ij} = 1 \quad i = 1, 2, ..., n$$

where c_{ij} is an element of cost matrix. x_{ij} is a binary variable indicating whether to associate i-th tracklet with j-th pose. We adopt the famous $O(n^3)$ Hungarian algorithm [10] to address above problem and accomplish fast online association, and the main issue under this framework is to get reliable cost matrix. During tracking, we calculate the similarities between pairs of identities, and the opposite numbers of the similarity scores are regarded as the cost scores.

2.2   Multi-person Pose Estimation

Our main focus in this paper is multi-person pose tracking based on predetected human joints, and we adopt top-down method to accomplish multi-person pose estimation. The Faster RCNN [15] detector is firstly employed to detect the positions and scales of persons in each frame, then the Pyramid Network [22] is adopted to estimate pose keypoints for each person.

2.3   Hierarchical Multi-person Pose Tracking

In recent years, CNN has shown its strong representative ability in variety of computer vision tasks. To get reliable affinity for association, a simple idea may be introducing CNN for deep appearance feature extraction, but such method is time consuming owning to the procedure for frequent forward propagations. We propose a novel hierarchical association method to address above problem. In most cases, the movement of same identities in consecutive frames is minor, and adopting spatial constraints such as pose box IOU (intersection area over union area) could get a satisfactory association result. There is no need to extract deep feature for these objects. However, when occlusion, camera motion or severe deformation happens, spatial constrains are no longer reliable, and deep appearance feature should be employed in such situation to get robust affinity score. Based on above criteria, the association procedure is divided into two stages as demonstrated in Fig. 1. At first stage, the basic affinity matrix is established based on basic association metric and divided into easy associations and hard associations according to the basic score. At second stage, deep association metric is applied to the hard associations to obtain more distinguishable affinity score. Finally, Hungarian algorithm is adopted based on easy association matrix and modified hard association matrix to get the final association result.

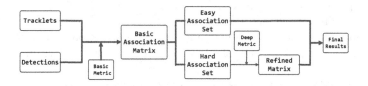

Fig. 1. The framework of hierarchical association.

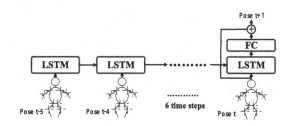

Fig. 2. The network structure of LSTM pose model.

Basic Similarity Metrics. We adopt pose IOU and LSTM pose model as basic similarity metrics to accomplish fast association. Pose IOU acts as the basic spatial constrain and is the IOU between corresponding pose boxes which are square boxes centered at the joint points. We select 7 joints lying on head and body region to compute pose IOU as they are more stable compared with limb joints. The width of pose box is set to 60 pixels during experiment. In this place, we did not design velocity model as most motions in MPT are always non-linear, and adopting history velocity information could cause identity switch easily. When the objects are difficult to distinguish by spatial constrains, our tracker would rely more on pose and region-based appearance similarity.

Beyond that, we propose a novel LSTM pose model to measure the pose similarity of corresponding identities. As demonstrated in Fig. 2, the pose model is composed of a 6-time-step LSTM unit followed by one fully-connected layer. A residual connection is added to reduce the difficulty for network training. During tracking, the network takes the tangent values of 8 limb angles in 6 consecutive frames as inputs and predicts the limb angles in next frame. The time step is set to 6 as most occlusions on PoseTrack dataset last less than six frames. The network is trained based on PoseTrack [1] training dataset. Pose similarity is calculated by the euclidean distance of the limb angles and then restricted to [0,1] by negative exponential function. The final basic affinity is the average value of pose IOU and pose similarity.

Deep Appearance Similarity Metric. To get reliable affinity even if when there exist occlusion, camera motion or severe deformation, we propose a region-based deep network for deep appearance feature extraction. The pipelines are demonstrated in Fig. 3, at first, the Fast RCNN detector and Pyramid Network are employed to accomplish multi-person detection and pose estimation. Then the region-based deep network is adopted to extract three types of deep features based on input image patches and body part positions.

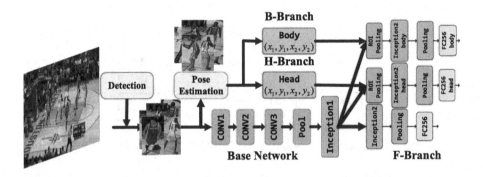

Fig. 3. The framework of region-based deep network

The region-based deep network could be divided into two components, i.e. the base network and the part feature extraction network. The base network is composed of three convolutional layers followed by one inception unit. The input image patches are first resized into 96 * 96 and then processed by base network to generate a shared feature map with spatial size of 24 * 24. The following part feature extraction network is composed of three branches denoted as F-Branch, H-Branch and B-Branch, which extract full object (FO), head region

(HR) and body region (BR) appearance feature respectively. The main structure of these three branches is an inception unit [17] followed by pooling layer and 256-dimension fully-connected layer. However, different from the F-Branch, which takes the entire shared feature map as input, B-Branch and H-Branch adopt the ROI pooling layer to extract appearance feature from corresponding region in the shared feature map. The ROI pooling layer first maps the scale and position of the body or head region from original image patch to the shared feature map and obtains the corresponding ROI windows, then divides the $h * w$ ROI window into an $H * W$ grid of sub-window with approximate size $h/H * w/W$ and maxpools the values in each sub-window into corresponding output grid cell [6]. The adoption of ROI pooling method significantly reduces the time consumption of the additional part-based model as they extract features from the shared feature maps.

For each joint keypoint, there exist a confidence score generated during detection and pose estimation. We calculate a confidence score for each region which is the average value of the scores of all the joints belong to it. During experiment, the region with highest confidence score is employed for similarity computation. We select the body and head region for part feature extraction as they are more stable compared with the limb regions. A three-layer BP network is trained to measure the appearance affinity which takes the deep features of two corresponding identities as inputs and outputs the affinity in range of [0,1].

To make our network gain the ability to distinguish different persons, we collect a large person re-identification datasets for network training including Market-1501 [23], PRID [7], CUHK02 [11] and Shinpuhkan2014 [9]. The network is trained hierarchically, firstly the base network and F-Branch are trained together with softmax classification loss for 60k iterations. Then we freeze the weights of base network, replace F-Bracnch with H-Branch and train the weights for H-Branch. Same procedure is applied to B-Branch for weight training. Finally three branches are integrated together and the whole network is fine-tuned with a low learning rate to get the final model.

Hierarchical Association. At first, the basic association matrix (BAM) is established based on basic similarity metric. Then a simple double thresholds method is applied to divide BAM, let $A_B \in R^{m*n}$ indicate the BAM between m pose tracklets and n detected poses in current frame. A_i means the i-th row of A_B, if the maximal value of A_i is greater than τ_1 and the difference value between the largest two values is larger than τ_2, the i-th pose tracklet and corresponding pose with maximal score would be regarded as easy association. After establishing the easy association set, the remaining tracklets and poses are regarded as hard associations.

Under above rules, there might exist two or more trajectories fighting for same one object in the easy association set, which means there exist objects that are not detected because of severe occlusion, and the missed objects would be reconstructed by the history information in the tracklets. For hard associations, the region based deep network would be employed to calculate the appearance similarity between pairs of identities. A linear SVM is trained for affinity fusion which takes two types of similarities as inputs and outputs the final similarity in range of [0,1]. The Hungarian algorithm is adopted based on the easy and modified hard association matrix to get the final association relationship.

3 Experiment

3.1 Dataset and Evaluation Metrics

The PoseTrack benchmark [1] is a newly released dataset for video-based multi-person pose tracking, and contains 550 video sequences covering many different scenarios such as sports, dancing and driving. Three different tasks are supplied and we focus on the multi-person pose tracking task only. This benchmark provides an online evaluation server to quantify the performance of different trackers on the test dataset. The multi-object tracking accuracy (MOTA) [3] is adopted as the evaluation metric which penalizes the false positives, false negatives and identity switches during tracking.

As the multi-person pose estimation results would affect the tracking accuracy under tracking by detection framework, the mean average precision (mAP) [13] metric is adopted to indicate the precision of the pose estimation results.

3.2 Experiment Results

Double Thresholds Selection. To get robust thresholds to divide the basic association matrix, we test our tracker with grid search method on the PoseTrack val dataset. The results are demonstrated in Fig. 4. With the increase of τ_1 and τ_2, the MOTA score gradually become stable. According to the results, the final threshold τ_1 is set to 0.7 and τ_2 is set to 0.3 to obtain a competitive tracking speed and accuracy.

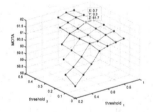

Fig. 4. Double thresholds selection on PoseTrack val dataset

Time Performance. To explore the effect of the hierarchical association metric and different cues, five types of tracker are tested on the PoseTrack val dataset, and the tracking speed (dose not contain the time consumption for detection and pose estimation) is shown in Table 1. All the results are tested on the Intel E5V3 CPU and Quadro M4000 GPU with the python interface of Caffe. Metrics mean the association methods. G denotes traditional global association and H denotes the hierarchical association method. The association cue I means pose box IOU, L means LSTM pose model, F means deep appearance model without B-Branch and H-Branch and R means region-based deep appearance model. The results in Table 1 show the hierarchical association metric significantly reduces the time consumption for deep feature extraction and the tracking speed increases from 2.85 fps to 9.41fps. Beyond that, there is just minor speed loss for the added part-based model as we adopt the ROI pooling method to extract region-based features from the shared feature maps.

Table 1. Tracking speed analysis

Metrics	G	G	H	H	G
Cues	I	I+L	I+L+F	I+L+R	I+L+R
Speed(fps)	47.8	25.01	11.08	9.41	2.85

Ablation Studies. To explore the effect of the LSTM pose model and region-based deep appearance feature, we do ablation studies on PoseTrack val dataset. The results are demonstrated in Table 2. The tracker's name in Table 2 has same meaning with that in Table 1. After selecting correct double thresholds, hierarchical association metric could get similar performance with global association metric. The results also show that the adoption of LSTM pose model and region-based deep appearance feature could improve the tracking accuracy obviously.

Table 2. Multi-person pose tracking results on posetrack val dataset

Cues	Metric	MOTA Head	MOTA Shou	MOTA Elb	MOTA Wri	MOTA Hip	MOTA Knee	MOTA Ankle	MOTA Total	MOTP Total
I	G	58.8	66.3	57.8	51.9	54.5	56.0	50.5	**56.7**	51.1
I+L	G	62.4	69.3	56.9	52.2	59.3	57.0	50.5	**58.5**	51.2
I+L+F	H	63.0	70.7	62.0	55.4	58.3	59.7	53.5	**60.6**	51.3
I+L+R	H	64.1	71.8	63.1	56.5	59.4	60.8	54.6	**61.7**	51.3
I+L+R	G	64.2	72.0	63.3	56.6	59.5	61.0	54.8	**61.8**	51.3

Comparison. To compare with other trackers, we test our tracker on PoseTrack test dataset and upload the results to the online evaluation server. As shown in Table 3, the accuracy of our tracker is competitive compared with other trackers and yields higher MOTA score. The joints MOTA score of some trackers could not be found as the evaluation server only reports the overall performance and the author does not publish their methods. Our tracker dose not perform better than FlowTrack [20], and it is mostly due to their high mAP score i.e., they employ more accurate detection and pose estimation methods and obtain more accurate pose joints. If performing on same detection results, our tracker could get more competitive accuracy score. More details can be found on the online leaderboard with challenge 3: https://posetrack.net/leaderboard.php.

Table 3. Multi-person pose tracking results on posetrack test dataset

Methods	MOTA Head	MOTA Shou	MOTA Elb	MOTA Wri	MOTA Hip	MOTA Knee	MOTA Ankle	**MOTA Total**	MOTP Total	mAP Total
FlowTrack [20]	-	-	-	-	-	-	-	57.81	-	74.57
ProTracker [5]	-	-	-	-	-	-	-	51.82	-	59.56
PoseFlow [21]	52.0	57.4	52.8	46.6	51.0	51.2	45.3	50.98	16.9	62.95
ML-Lab [14]	-	-	-	-	-	-	-	40.77	-	70.33
HMPT(ours)	51.9	59.7	50.8	48.5	52.0	52.7	47.6	51.89	40.4	63.73

Some sampled tracking results are demonstrated in Fig. 5, and the numbers following '#' denote the frame numbers in corresponding video sequence. Different identities are distinguished with different colors. The results demonstrate that our tracker could handle the occlusion problem successfully.

(a) #37	(b) #45	(c) #64	(d) #80
(e) #22	(f) #36	(g) #51	(h) #74

Fig. 5. Sampled tracking result in PoseTrack test dataset.

4   Conclusion

In this paper, we propose a region-based deep CNN for part-based appearance feature extraction, and this makes our tracker gain the ability to associate correctly even if the objects are partly occluded. Besides, we propose a novel pose LSTM model to measure the pose similarity. The hierarchical association method is proposed to reduce the time consumption for deep feature extraction. The results in PoseTrack benchmark show the hierarchical association method could reduce the time consumption significantly and multiple association cues could improve the tracking accuracy obviously.

Acknowledgments. This work is supported by National High-Tech R&D Program (863 Program) under Grant 2015AA016402.

References

1. Andriluka, M., et al.: PoseTrack: a benchmark for human pose estimation and tracking. arXiv preprint arXiv:1710.10000 (2017)
2. Bae, S.H., Yoon, K.J.: Robust online multi-object tracking based on tracklet confidence and online discriminative appearance learning. In: Proceedings of the IEEE Conference on Computer Vision and Pattern Recognition, pp. 1218–1225. IEEE Press, New York (2014)
3. Bernardin, K., Stiefelhagen, R.: Evaluating multiple object tracking performance: the CLEAR MOT metrics. Eurasip J. Image Video Process. **2008**(1), 246–309 (2008)
4. Emami, P., Pardalos, P.M., Elefteriadou, L., Ranka, S.: Machine learning methods for solving assignment problems in multi-target tracking. arXiv preprint arXiv:1802.06897 (2018)
5. Girdhar, R., Gkioxari, G., Torresani, L., Paluri, M., Tran, D.: Detect-and-track: efficient pose estimation in videos. arXiv preprint arXiv:1712.09184 (2017)
6. Girshick, R.: Fast R-CNN. In: 2015 IEEE International Conference on Computer Vision (ICCV), pp. 1440–1448. IEEE Press, New York (Dec 2015)
7. Hirzer, M., Beleznai, C., Roth, P.M., Bischof, H.: Person re-identification by descriptive and discriminative classification. In: Heyden, A., Kahl, F. (eds.) SCIA 2011. LNCS, vol. 6688, pp. 91–102. Springer, Heidelberg (2011). https://doi.org/10.1007/978-3-642-21227-7_9
8. Hochreiter, S., Schmidhuber, J.: Long short-term memory. Neural Comput. **9**(8), 1735 (1997)
9. Kawanishi, Y., Wu, Y., Mukunoki, M., Minoh, M.: Shinpuhkan 2014: a multi-camera pedestrian dataset for tracking people across multiple cameras. In: 20th Korea-Japan Joint Workshop on Frontiers of Computer Vision, vol. 5. Citeseer (2014)
10. Kuhn, H.W.: The hungarian method for the assignment problem. Naval Res. Logistics **52**(1), 7–21 (2005)
11. Li, W., Wang, X.: Locally aligned feature transforms across views. In: Computer Vision and Pattern Recognition, pp. 3594–3601. IEEE Press, New York (2013)
12. Luo, W., et al.: Multiple object tracking: a literature review. arXiv preprint arXiv:1409.7618 (2014)
13. Pishchulin, L., et al.: DeepCut: joint subset partition and labeling for multi person pose estimation. In: Proceedings of the IEEE Conference on Computer Vision and Pattern Recognition, pp. 4929–4937. IEEE Press, New York (2016)
14. PoseTrack: Posetrack leader board. https://posetrack.net/leaderboard.php

15. Ren, S., Girshick, R., Girshick, R., Sun, J.: Faster R-CNN: towards real-time object detection with region proposal networks. IEEE Trans. Pattern Anal. Mach. Intell. **39**(6), 1137–1149 (2017)
16. Sadeghian, A., Alahi, A., Savarese, S.: Tracking the untrackable: learning to track multiple cues with long-term dependencies. arXiv preprint arXiv:1701.01909, **4**(5) 6 (2017)
17. Szegedy, C., et al.: Going deeper with convolutions. In: Proceedings of the IEEE Conference on Computer Vision and Pattern Recognition, pp. 1–9. IEEE Press, New York (2015)
18. Wang, B., et al.: Joint learning of convolutional neural networks and temporally constrained metrics for tracklet association. In: Proceedings of the IEEE Conference on Computer Vision and Pattern Recognition Workshops, pp. 1–8. IEEE Press, New York (2016)
19. Wojke, N., Bewley, A., Paulus, D.: Simple online and realtime tracking with a deep association metric. In: 2017 IEEE International Conference on Image Processing (ICIP), pp. 3645–3649. IEEE Press, New York (2017)
20. Xiao, B., Wu, H., Wei, Y.: Simple baselines for human pose estimation and tracking. arXiv preprint arXiv:1804.06208 (2018)
21. Xiu, Y., Li, J., Wang, H., Fang, Y., Lu, C.: Pose Flow: Efficient online pose tracking. arXiv preprint arXiv:1802.00977 (2018)
22. Yang, W., Li, S., Ouyang, W., Li, H., Wang, X.: Learning feature pyramids for human pose estimation. In: 2017 IEEE International Conference on Computer Vision (ICCV), vol. 2, pp. 1290–1299. IEEE Press, New York, October 2017
23. Zheng, L., Shen, L., Tian, L., Wang, S., Wang, J., Tian, Q.: Scalable person re-identification: A benchmark. In: IEEE International Conference on Computer Vision, pp. 1116–1124. IEEE Press, New York (2015)

Fast Portrait Matting Using Spatial Detail-Preserving Network

Shaofan Cai[1], Biao Leng[1($\boxtimes$)], Guanglu Song[1], and Zheng Ge[2]

[1] School of Computer Science and Engineering, Beihang University,
Beijing 100191, China
lengbiao@buaa.edu.cn
[2] Graduate School of Information, Production and Systems, Waseda University,
Kitakyushu 169-8050, Japan

Abstract. Image matting plays an important role in both computer vision and graphics applications. Natural image matting has recently made significant progress with the assistance of powerful Convolutional Neural Networks (CNN). However, it is often time-consuming for pixel-wise label inference. To get higher quality matting in an efficient way, we propose a well-designed SDPNet, which consists of two parallel branches—Semantic Segmentation Branch for half image resolution and Detail-Preserving Branch for full resolution, capturing both the semantic information and image details, respectively. Higher quality alpha matte can be generated while largely reducing the portion of computation. In addition, Spatial Attention Module and Boundary Refinement Module are proposed to extract distinguishable boundary features. Extensive Experiments show that SDPNet provides higher quality results on Portrait Matting benchmark, while obtaining 5x to 20x faster than previous methods.

Keywords: Portrait · Fast matting · Detail-preserving · Deep learning

1 Introduction

Matting refers to the problem of accurate foreground estimation in images and videos. It is one of the key techniques in many image editing and film production applications. Mathematically, the input image can be modeled as a convex combination of a foreground and background colors as follows [7]:

$$I_i = \alpha_i F_i + (1 - \alpha_i) B_i \tag{1}$$

where I_i, F_i, B_i and α_i denote the natural RGB image, foreground, background color and alpha matte at pixel i respectively. Thus, for a three-channel color image, at each pixel, there are 7 unknown values but only 3 known values.

Given an input image I, finding F, B, and α simultaneously without any user interaction makes natural matting problem highly ill-posed. Image matting

© Springer Nature Switzerland AG 2018
L. Cheng et al. (Eds.): ICONIP 2018, LNCS 11306, pp. 329–339, 2018.
https://doi.org/10.1007/978-3-030-04224-0_28

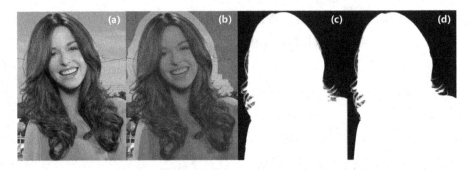

Fig. 1. (a) Image from our Synthetic dataset(1280 × 960 pixels). (b) Trimap. Red color stands for definite foreground and blue color stands for definite background. The rest of the region stands for unknown. (c) Result of Our SDPNet. The running time is 40ms on GPU. (d) Our labeled groundtruth.

techniques [5,11] require a trimap (or strokes) indicating definite foreground, definite background and unknown region. Traditional matting algorithms can be divided into two classes, color sampling based methods and matting affinity based methods. The limitation of these methods is that the distinguishing feature largely rely on color. When the color distributions overlay between the foreground and background, it is really tough for such approaches to generate clear alpha matte without low-frequency "smearing" or high-frequency "chunky" artifacts. To overcome this problem, recently deep learning based methods are proposed for image matting. Instead of relying primarily on color information, CNN also extracts structure and semantic information, which helps to produce high quality alpha matte (Fig. 1(c)).

Although CNN [10] provides powerful assistance for image matting, amount of the huge parameter and calculation make it expensive for multi-megapixel images produced by digital cameras. Shen et al. [14] proposed an automatic matting with the help of semantic segmentation [12]. But their approach has a high computational complexity. Zhu et al. [20] designed a fast and effective method for portrait matting. It can realize real-time matting on the mobile phone for a low-resolution image. However, their approaches fail to distinguish tiny details in the hair areas because they downsample the input size of image to 128 × 128. When the resolution of input image get higher, the speed of inference will be largely limited and it is not detail-preserving.

In this paper, we focus on fast portrait matting techniques with decent prediction accuracy. To achieve our goal, we propose a network, named Spatial Detail-Preserving Network(SDPNet). Different from previous single branch matting network [14,20], our SDPNet uses two branch to utilize processing efficiency of low-resolution images and high inference quality of high-resolution ones. The idea is that low-resolution images can go through the full semantic segmentation network first for a coarse score map. The second branch is used to capture details structure to refine the coarse semantic map. Then the output of two branches will be aggregated to generate a high quality alpha matte. We also consider the impact of different pixels in full-resolution feature map to improve matting performance. Our contributions in this work are as follows:

- A Spatial Detail-Preserving network (SDPNet) is proposed, which utilizes semantic and structure information in lower-resolution branch along with details from higher-resolution branch efficiently.
- Further more, we present Spatial Attention Module to improve the quality of feature map via spatial embedding. Boundary Refinement Module is adopted to refine the boundary of feature map produced by Semantic Segmentation Branch.
- Experiments show that our proposed method achieves 5x+ speed of inference. SDPNet can run at resolution 800 × 600 in speed of 40 fps while accomplishing high-quality portrait alpha matte.

2   Related Work

2.1   Natural Image Matting

Natural image matting is crucial for image and video editing, but it remains challenging because it is a severely underconstrained problem. Interactive image matting aims to predict alpha matte in unknown regions. [7] tried to apply Gaussian mixture models on both background and foreground. To infer the alpha matte in the unknown regions, closed-form matting [11] uses a matting Laplaian matrix, under a color line assumption. Large-Kernel Laplacian [9] helps accelerating matting Laplacian computation. Shared matting [8] was the first real-time matting algorithm running on modern GPUs by shared sampling. Inter-Pixel Information Flow Matting [1] proposed a purely affinity-based natural image matting method.

Recently, deep-learning based methods have shown great potential on solving computer vision tasks. DCNN [6] is the first attempt to apply deep learning on image matting problem. They used a relatively shallow neural network to deal with patches of images, with the result of closed-form and KNN matting as extra input. Xu [18] released a large matting dataset with high-quality foreground and alpha matte. Then they trained an encoder-decoder structure network on this dataset.

2.2   Semantic Segmentation

Traditional semantic segmentation methods adopt hand-craft feature to learn the representation. Recently, CNN based methods largely improve the performance. FCN [12] is the pioneer work to use fully convolution layers in semantic segmentation task. Encoder-decoder structures [2] can restore the feature map from higher layers with spatial information from lower layers. ICNet [19] incorporates multi-resolution branches under label guidance to achieve realtime inference without significantly reducing performance.

3   Proposed Algorithm

As shown in Fig. 2, the proposed SDPNet consists of two branches, Semantic Segmentation Branch, and Detail-Preserving Branch, which respectively captures the structure and details components of the input image. Specially, the input size of Detail-Preserving branch is full resolution ($h \times w$), and the input size of Semantic Segmentation Branch is lower resolution (e.g. $\frac{h}{2} \times \frac{w}{2}$), with input image height h and width w. Given a high-resolution image and trimap, each branch has different functionalities. The Semantic Segmentation Branch provides the roughly boundary and semantic information of the image from lower resolution. The Detail-Preserving Branch captures the detail information, such as points, lines or edges, from full resolution. Finally, the feature maps from two branches are fused together, resulting in a high quality alpha matte.

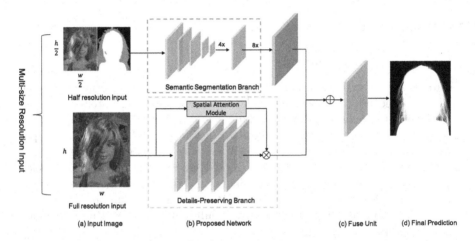

Fig. 2. Overall architecture of SDPNet. It contains two branches, Semantic Segmentation Branch and Detail-Preserving Branch and a feature fusing module. The Semantic Segmentation Branch (Sect. 3.1) generates a rough boundary mask from half resolution and the Detail-Preserving Branch (Sect. 3.2) captures details and structures from full resolution. Detail-Preserving Branch contains a Spatial Attention Module. Finally SPDNet fuses results from two branches by Feature Fusing Unit. The whole SDPNet is end-to-end trainable.

3.1   Semantic Segmentation Branch

Image resolution is the most critical factor that affect speed, since above analysis shows a half-resolution image only uses nearly quarter time compared to the full-resolution one. A naive approach is to directly use small-resolution image as input. We downsample images with ratios 1/2 and feed the resulting images into our Semantic Segmentation Branch. The detail structure of Semantic Segmentation Branch is shown in Table. This sub-network consists of an Encoder and a Decoder. Similar to Unet [3], we employ skip connections in encoder-decoder network. The encoder network consists of one convolutional layer and 11 resnext [17] blocks. The decoder network uses deconvolution as upsampling module. At each stage after upsampling, the feature maps are fed to Boundary Refinement Modules, which will be illustrated later.

3.2 Detail-Preserving Branch

The tiny structures and details components of image will be destroyed during downsampling operations, such as max pooling or convolution with stride 2. Hence, we design a Detail-Preserving Branch to capture low-level features that are missing in the half-resolution branch. We can limit the number of convolutional layers since half-resolution branch already catches most semantically information. Here we use only three convolutional layers with kernel stride size 3×3 and stride 1 to extract low-level features. The Details structure of this branch is shown in Fig. 3.

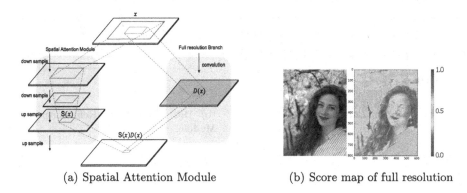

(a) Spatial Attention Module   (b) Score map of full resolution

Fig. 3. (a) Details of structures in spatial attention module. (b) The score map generated by spatial attention module.

3.3 Spatial Attention Module

Spatial Attention Module aims at improving Detail-Preserving Branch features. Following previous attention mechanism in [15], we apply Spatial Attention Module in Detail-Preserving Branch. The module's target is to output scores for each pixel of feature maps. Given the input image I and trimap T with height h and width w, max pooling are performed several times to increase the receptive field rapidly after a small number of convolution layers. After reaching the lowest resolution, the global information is then expanded by a symmetrical upsample operations. We use linear interpolation up sample the output after one 1×1 convolution layer with stride 1. The number of upsampling module is the same as max pooling to keep the output score map size the same as the input feature map. Then we use a sigmoid layer to normalize the output score maps range to $[0, 1]$. The full module is illustrated in Fig. 3(a). It also shows that the consecutive up-sample and down-sample operations can expand receptive filed. Experiment in is conducted to verify this.

3.4 Boundary Refinement Module

We propose a Boundary Refinement Block, schematically depicted in Fig. 4(b). The feature maps after upsampling go through the Boundary Refinement Block, which is designed to model the boundary alignment as a residual structure. More specially, we use $\widetilde{S}$ denote refined score map: $\widetilde{S} = S + \mathcal{R}(S)$, where S is the coarse score map, and $\mathcal{R}(\cdot)$ is residual branch. After refinement, the boundary information is embedded in its output feature map, as show in Fig. 4(a).

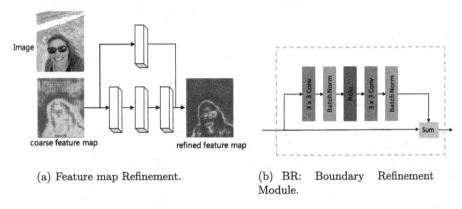

(a) Feature map Refinement.   (b) BR: Boundary Refinement Module.

Fig. 4. (a) Refine feature map from coarse to fine. (b) Components of the boundary refinement blocks (BR)

4 Experiments

In this section, we evaluate the performance of SDPNet on publicly available 2K Portrait Matting Dataset [14] and our Synthetic Portrait Matting Dataset.

4.1 Datasets

2K Portrait Matting Dataset: We choose the primary dataset from [14], which is collected from Flickr. We evaluate the proposed method on the benchmark dataset. This dataset collects 2000 portrait image with labeled alpha matte as ground truth. These images are split into the training and testing sets with 1700, 300 images respectively.

Synthetic Portrait Matting Dataset: We further evaluate our method using real-world examples. We download some pictures, whose background color is pure, from Internet and manually label the trimap. With the selected portrait images, we labele alpha values with intensive user interaction tools provided by

[16] to make sure they are with high quality. The alpha matte is calculated while labeling. After this labeling process, we collect 200 image with high-quality alpha mattes. These images are randomly split into the training and testing sets with 150 and 50 images respectively. We also download some background pictures in real scenes. We randomly sample N background images in them and composite the portrait foreground onto those background images. Finally we got 20,000 (N = 100) training portrait images and 50 (N = 1) test images.

4.2 Implementation Details

Inspired by the work [4], we use the "poly" learning rate policy in which current learning rate is defined as the base one multiplying $(1 - \frac{iter}{max_{iter}})^{power}$. We set base learning rate to 0.001 and power to 0.9. Momentum and weight decay are set to 0.9 and 0.0001 respectively. The proposed network is trained on the training set above. To avoid overfitting, we randomly crop a 480 × 480 patch and this patch can cover the unknown region in the trimap. In order to generate trimaps for training, we randomly dilate the alpha matte by random size to make our network more robust to different quality of trimap. For data augmentation, we adopt random flip and random resize between 0.75 and 1.5 for all images, and additionally add random rotation between −45 and 45°. We also apply random Gamma transforms to increase color variation.

4.3 Accuracy Measure

We select the gradient error and mean squared error to measure matting quality, which can be expressed as:

$$G(\alpha^p, \alpha^{gt}) = \frac{1}{T} \sum_i \| \nabla \alpha_i^p - \nabla \alpha_i^{gt} \| \tag{2}$$

$$MSE(\alpha^p, \alpha^{gt}) = \frac{1}{T} \sum_i (\alpha_i^p - \alpha_i^{gt})^2 \tag{3}$$

where α^p is the predicted alpha matte and α^{gt} is corresponding ground truth. T is the number of pixels in unknown region of given trimap. ∇ is the operator to compute gradients. Specially, alphamatting [13] points out that the correlation of SAD and MSE with the percption of average human observer is rather low, Gradient Error, which is more reliable, outperforms both of other two metric with a higher correlation.

4.4 Ablation Study on Synthetic Portrait Matting Datasets

In this subsection, we will step-wise decompose our approach to revel the effect of each component. In the following experiments, we evaluate all comparisons on Synthetic Portrait Matting dataset.

Ablation for Boundary Refinement Module: To refine the coarse feature scores after upsampling, we use our Boundary Refinement Module to refine score map. As show in Table 1, this module further improves the performance on two metrics – – MSE and gradient error. It reduces gradient error from 24.70 to 22.92 and MSE from 0.0133 to 0.0113.

Ablation for Detail-Preserving Branch: By contrast, the proposed SDP-Net is motivated by the decomposition of a image signal into structure and details. For fair comparison, we keep the same amount of calculations of the single Semantic Segmentation Branch's and Two Branch Network, show as Table 1. Especially, gradient error decreases dramatically from 24.7 to 20.45, which is an obvious improvement.

Ablation for Spatial Attention Module: We evaluate the effectiveness of spatial attention learning mechanism. As show in Table 1, the network trained using spatial attention module consistently outperform the networks without it, which proves the effectiveness of our method.

Table 1. The quantitative comparisons of proposed SDPNet on the Synthetic Portrait testing dataset. **SS:** Semantic Segmentation. **BR:** Boundary Refinement Module. **SA:** Spatial Attention Module. **DP:** Detail-Preserving Branch.

Method	Grad Error	MSE
SS Branch	24.70	0.0133
SS Branch + BR	22.92	0.0113
SS Branch + DP Branch	20.45	0.0126
SS Branch + DP Branch + BR	20.12	0.0113
SS Branch + DP Branch + BR +SA	**19.63**	**0.0107**

4.5 Comparison with State-of-the-Art Methods on 2k Portrait Matting Dataset

To further confirm the performance of our method, we also compare our methods with others. We visually and quantitatively evaluate our methods in 2k-Portrait Matting Dataset [14].

Quantitative Analysis. In experiments, we quantitatively evaluate the SDP-Net on 2k Portrait Matting Dataset [14] and compared it with DAPM [14] and LDN+FB [20]. We also use FCN [12] to generate trimap, then using closed-form [11] to calculate alpha matte. As show in Table 2, our method achieves lower gradient error than other two deep learning based methods.

Running Time. We evaluate our method and state-of-the-art methods on the same PC with an Intel(R) Core i7 CPU and a Nvidia Titan X GPU. Table 3 shows speed comparison between our method and other methods. Running time

Table 2. Results on 2k-Portrait Matting of [14]. DAPM means the approach of Deep Automatic Portrait Matting in [14]. LDN +FB means the approach in [20].

Method	Trimap-FCN [12] + Closed-form [11]	DAPM [14]	LDN+FB [20]	Ours
Grad ($\times 10^{-3}$)	4.14	3.03	7.40	**2.48**

Table 3. Speed comparison with other methods. Running time for a 800 × 600 image. All the method run by their publicly available scripts except for DIM [18], which we implement as its paper. **G:**GPU. **C:**CPU

Method	Closed-form [11]	Shared [8]	Info [1]	DIM [18](G)	Ours(C)	Ours(G)
Time (sec)	9.88	63.65	9.15	0.23	**1.76**	**0.024**

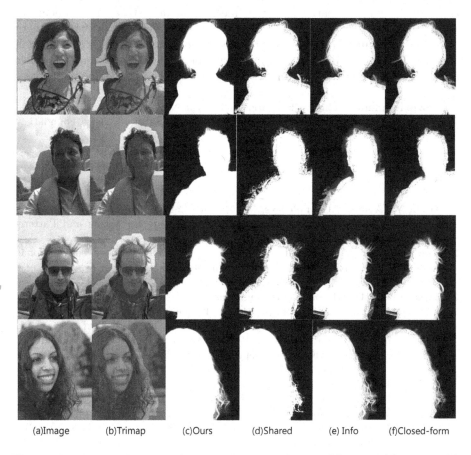

(a)Image	(b)Trimap	(c)Ours	(d)Shared	(e) Info	(f)Closed-form

Fig. 5. Visual comparisons on 2k portrait matting dataset. (a) Image (b) Trimap (c) Ours (d) Shared-Matting [8] (e) Information-flow [1] (e) Closed-form [11]

for a 800 × 600 image, our SDPNet is nearly 5.6 times faster than closed-form [11], 36.11 times faster than shared-matting [8] and 5.19 times faster than Information-flow [1] matting on CPU. DIM [18] achieves state of the art performance in public available test set, but it is very time-consuming for a large resolution input. SDPNet is almost 10 times faster than DIM, while still generate alpha matte with fine details. Visually Comparison is showed in Fig. 5.

5   Conclusion

This paper proposes the Spatial Detail-Preserving Network (SDPNet) for fast portrait matting. SDPNet can simultaneously capture semantic structure and low-level details by its network design, which contains two branches: Semantic Segmentation Branch for lower resolution and Detail-Preserving Branch for full resolution. With the spatial attention mechanism and stage-wise refinement, our approach can capture the discriminative features for portrait matting. Our experimental results show that the proposed approach indeed takes less time for inference. Besides, SDPNet can also improve the quality of alpha matte, which shows our approach is comparable with the state-of-the-art matting methods.

Acknowledgements. This work is supported by the National Natural Science Foundation of China (No. 61472023) and Beijing Municipal Natural Science Foundation (No. 4182034).

References

1. Aksoy, Y., Aydın, T.O., Pollefeys, M., Zürich, E.: Designing effective inter-pixel information flow for natural image matting. In: Computer Vision and Pattern Recognition (CVPR) (2017)
2. Badrinarayanan, V., Kendall, A., Cipolla, R.: SegNet: a deep convolutional encoder-decoder architecture for image segmentation. IEEE Trans. Pattern Anal. Mach. Intell. **39**(12), 2481–2495 (2017)
3. Barkau, R.L.: UNET: one-dimensional unsteady flow through a full network of open channels. Technical report, Hydrologic Engineering Center Davis CA (1996)
4. Chen, L.C., Papandreou, G., Kokkinos, I., Murphy, K., Yuille, A.L.: DeepLab: semantic image segmentation with deep convolutional nets, atrous convolution, and fully connected CRFs. arXiv preprint arXiv:1606.00915 (2016)
5. Chen, Q., Li, D., Tang, C.K.: KNN matting. IEEE Trans. Pattern Anal. Mach. Intell. **35**(9), 2175–2188 (2013)
6. Cho, D., Tai, Y.-W., Kweon, I.: Natural image matting using deep convolutional neural networks. In: Leibe, B., Matas, J., Sebe, N., Welling, M. (eds.) ECCV 2016. LNCS, vol. 9906, pp. 626–643. Springer, Cham (2016). https://doi.org/10.1007/978-3-319-46475-6_39
7. Chuang, Y.Y., Curless, B., Salesin, D.H., Szeliski, R.: A bayesian approach to digital matting. In: Proceedings of the 2001 IEEE Computer Society Conference on Computer Vision and Pattern Recognition, CVPR 2001, vol. 2, p. 2. IEEE (2001)

8. Gastal, E.S., Oliveira, M.M.: Shared sampling for real-time alpha matting. In: Computer Graphics Forum, vol. 29, pp. 575–584. Wiley Online Library (2010)
9. He, K., Sun, J., Tang, X.: Fast matting using large kernel matting laplacian matrices. In: 2010 IEEE Conference on Computer Vision and Pattern Recognition (CVPR), pp. 2165–2172. IEEE (2010)
10. Krizhevsky, A., Sutskever, I., Hinton, G.E.: Imagenet classification with deep convolutional neural networks. In: Advances in Neural Information Processing Systems, pp. 1097–1105 (2012)
11. Levin, A., Lischinski, D., Weiss, Y.: A closed-form solution to natural image matting. IEEE Trans. Pattern Anal. Mach. Intell. **30**(2), 228–242 (2008)
12. Long, J., Shelhamer, E., Darrell, T.: Fully convolutional networks for semantic segmentation. In: Proceedings of the IEEE Conference on Computer Vision and Pattern Recognition, pp. 3431–3440 (2015)
13. Rhemann, C., Rother, C., Wang, J., Gelautz, M., Kohli, P., Rott, P.: A perceptually motivated online benchmark for image matting. In: IEEE Conference on Computer Vision and Pattern Recognition, CVPR 2009, pp. 1826–1833. IEEE (2009)
14. Shen, X., Tao, X., Gao, H., Zhou, C., Jia, J.: Deep automatic portrait matting. In: Leibe, B., Matas, J., Sebe, N., Welling, M. (eds.) ECCV 2016. LNCS, vol. 9905, pp. 92–107. Springer, Cham (2016). https://doi.org/10.1007/978-3-319-46448-0_6
15. Wang, F., et al.: Residual attention network for image classification. arXiv preprint arXiv:1704.06904 (2017)
16. Wang, J., Agrawala, M., Cohen, M.F.: Soft scissors: an interactive tool for realtime high quality matting. ACM Trans. Graph. (TOG) **26**(3), 9 (2007)
17. Xie, S., Girshick, R., Dollár, P., Tu, Z., He, K.: Aggregated residual transformations for deep neural networks. In: 2017 IEEE Conference on Computer Vision and Pattern Recognition (CVPR), pp. 5987–5995. IEEE (2017)
18. Xu, N., Price, B., Cohen, S., Huang, T.: Deep image matting. In: Computer Vision and Pattern Recognition (CVPR) (2017)
19. Zhao, H., Qi, X., Shen, X., Shi, J., Jia, J.: ICNet for real-time semantic segmentation on high-resolution images. arXiv preprint arXiv:1704.08545 (2017)
20. Zhu, B., Chen, Y., Wang, J., Liu, S., Zhang, B., Tang, M.: Fast deep matting for portrait animation on mobile phone. In: Proceedings of the 2017 ACM on Multimedia Conference, pp. 297–305. ACM (2017)

EEG Signal Analysis in 3D Modelling to Identify Correlations Between Task Completion in Design User's Cognitive Activities

Muhammad Zeeshan Baig$^{(\boxtimes)}$ and Manolya Kavakli

VISOR (Virtual and Interactive Simulations of Reality) Research Group,
Department of Computing, Macquarie University, Sydney, Australia
baig.mzeeshan@gmail.com, manolya.kavakli@mq.edu.au

Abstract. Modelling software applications vary from construction to gaming, but learning a modelling software and becoming a skilled user takes a long time and effort. Reducing the time to learn a modelling software is an important topic in human-computer interaction (HCI). To develop futuristic computer-aided design (CAD) systems that require little or no training, it is important to study the user-dependent factors that affect the system performance directly and indirectly by analysing the cognitive activity of the users. In this research, we have presented a new method to segment the EEG data: we segmented designer's actions and then used it to align with the EEG data, while they draw a 3D object in AutoCAD. We video recorded the design activities and Electroencephalography (EEG) signals while users were performing the task. The mean EEG power of the alpha, beta, theta and gamma bands has been used to analyse the designer behaviour. We found that the users who completed the experiment in a short time-frame were performing more physical actions than perceptual and conceptual actions. Participants with low Completion Time (CT) participants perform 30% more actions per minute than high-CT participants. EEG analysis demonstrated that the task completion time (CT) was negatively correlated with physical actions. Alpha-and beta-band analysis showed that low-CT participants were more comfortable in performing physical action and high-CT participants are relaxed in performing conceptual actions.

Keywords: Cognitive activity · CAD · EEG · HCI · 3D modelling

1 Introduction

Skills and expertise are developed after learning basic techniques and practising those techniques over time, but to define those activities during a design task is quite difficult. Even skilled users cannot articulate what kind of techniques are involved in performing a certain task and how they are using these techniques.

© Springer Nature Switzerland AG 2018
L. Cheng et al. (Eds.): ICONIP 2018, LNCS 11306, pp. 340–352, 2018.
https://doi.org/10.1007/978-3-030-04224-0_29

Our aim is to understand the behaviour of a skilled designer to guide a novice and also to use it to develop next-generation design and modelling systems.

3D modelling or CAD/CAM tools has a great impact on design efficiency [6], but it requires a specific set of skills, training and experience to master these tools. Skilled users are able to capitalize on their design skills at the early stages compared to novice users, but little research has been done on analysing the stages in conceptual design process. One way to tackle this problem is to study the cognitive processes behind the designer's actions. Protocol analysis is one major method to examine the cognitive activity of a designer, but these methods mainly focus on designer's actions and do not incorporate the mental or emotional state of the designer.

In the last few years, psycho-physiological methods have been used to analyse and understand the science behind a designer's actions. This increase in psycho-physiological research is due to the widespread use of non-invasive, inexpensive and easy to use psycho-physiological equipment. Researchers have used electrocardiograms (ECG) [20], galvanic skin response (GSR) [22], eye-tracking [12], gesture analysis [28], and electroencephalography (EEG) [21] to study the behaviour of design protocols and process. The most popular technique reported in the literature to analyse design process is verbal protocol analysis, but it has some limitations [2]. These limitations are exploited when analysing non-reportable processes such as creativity, judgement or task insights [17], so other techniques for analysing a designer's actions must be investigated.

Due to the limitations of verbal protocol analysis, alternative techniques to study a designer's actions have been introduced including sketching [14], gesture analysis [13] and eye-tracking [12]. Some researchers use modelling tools and techniques by analysing the activity through psycho-physiological signals [18]. In this research, we present a new approach to use EEG signals' segmentation in the analysis of the designer's cognitive activity. The aim of our research is to prove the idea that the content-oriented approach for analysing a designer's activities can benefit from overlying EEG signals to understand the cognitive behaviour of the designer.

The research questions addressed in this paper are:

1. What is the relationship between task completion and mental effort?
2. What are the factors affecting the task completion of a designer with 3D modelling?
3. What are the relationships between alpha, beta, theta, and gamma bands activities and task completion?

These questions are addressed by the following research tasks:

1. To monitor the cognitive states of designers as they perform a certain 3D modelling task.
2. To investigate the individual modelling behaviour involved in 3D object modelling and develop a methodology to understand the behaviour.
3. To validate the modelling behaviour through EEG signal analysis.

4. To determine if there is a correlation between designer's performance and psychological signals.

The rest of the paper is organized as follows. Section 2 provides a brief survey of the literature related to a designer's actions and psychological signal analysis. Section 3 explains the methodology. The coding scheme and its implementation are described in Sect. 4 followed by EEG signal analysis in Sect. 5. Section 6 concludes the paper by summarizing our findings.

2 Psychological Signals and Mental States

Identifying cognitive states of a user is a very complex assignment and has been studied extensively in the field of psychology and the cognitive science. With the advancement in the field of clinical psychology, devices such as an EEG headsets, ECG and GSR have become more accessible and easy to use. With these devices, we can analyse the user behaviour using quantitative techniques. The relationship between psycho-physiological signals and mental states has been studied over the last few years [19]. Many researchers have used heart-rate variability (HRV) to predict the user's mental states. [25] have found an association between mental stress level and low-and high-frequency bands. Hjortskov et al. [11] also tried to measure mental stress when users are in a demanding and unfriendly working environment. Some researchers also used HRV to compare resting and working conditions and found distinct differences [16].

EEG signals are also used in recent literature to measure cognitive load. Researchers have reported an increase in theta-band activity and a decrease in alpha-band activity with increasing task complexity [7]. In video games, an increase in front mid-line theta and a decrease in posterior alpha is observed as the game proceeds [23]. The same kind of relationship has been observed in visual scanning tasks [10]. The decrease in alpha activity has been found in high cognitive load tasks but no significant variation was found in low-and medium-load tasks in flight-simulation experiments [26]. Nguyen et al. also used EEG along with the heart rate to measure a designer's mental states and found that mental effort is lowest at the highest mental stress, and no significant differences were observed at low-and medium-mental stress levels [21].

In our research, we have used EEG signals to investigate the user's mental activity while the users perform a 3D object modelling task. The theta band over the parietal lobe has been observed to associate with low-and high-cognitive load tasks better than the beta and alpha bands [5]. The beta band relates to cortical activation, an increase in beta-band activity over the occipital lobe has also been observed in high-visual-attention tasks [8]. After reviewing the literature, we identified a hypothesis that alpha-band activity is negatively, and theta-and beta-band activity is positively, correlated with mental effort.

3   Methodology

3.1   Experimental Setup

The experimental task was to design a 3D table with three parts: a base, a pillar and a top. In the literature, 5–10 participants were used to analyse the cognitive activity [18, 21, 22], so a total of eight participants were selected for the experiment. All of them are computer-science students at Macquarie University. The ages of the participants range from 21 to 30 years. The participants have no prior experience of using AutoCAD or any other 3D modelling tool. As all the participants are novices, so we use task completion time to divide the participants. Task completion is one of the fundamental usability metrics that is used to quantify effectiveness of an interface [24]. The participants who completed the task quickly are called low completion time (low-CT) participants, and who take a long time to finish the task are called high completion time (high-CT) participants. The experiment was approved (Approval no. 5201700784) by the Faculty of Science and Engineering Human Research Ethics Sub-Committee, Macquarie University. Each subject was given a tutorial of 10 min before the experiment. A video log has been created for each subject and EEG signals were also recorded. For analysing EEG signals, we used an off-the-shelf research edition of the Emotiv EEG headset, which has 14 channels and records signals at a sampling rate of 128 Hz. The electrode placement is based on the international 10–20 system. The subjects were given an open-ended task to depict the real-world setting, which results in different cognitive processing and analysis strategies. A picture of the experimental setup is shown in Fig. 1.

Fig. 1. The experimental setting

3.2   Experimental Procedures

The process can be divided into 5 experimental procedures:

1. Information about the experiment was given to each subject along with the consent form. After reading and signing the consent form, the experimenter gives a walk-through of the experiment and some instructions to minimize the body and head movements.
2. The EEG headset was placed on the head of the subject and the experimenter made sure that all channels were in good contact with the skull. The subjects were asked to rest for two minutes with their eyes open with hands on their laps, and after that the subjects were asked to start drawing.

3. After the completion of modelling task, the subjects were asked to fill in a questionnaire about the experiment.
4. EEG data were filtered and segmented with the help of the video log.
5. The mean power for the different EEG bands was calculated using Welch power spectral density.

4   Coding Scheme

To examine the differences between the subjects mental states and their task completion time (CT), we used a coding scheme that allows us to assign codes to the cognitive actions of the designers using the video recordings. This coding scheme is an extension of the coding scheme used by Kavakli et al. [14].

4.1   Codes for Modelling Actions

To analyse a certain set of cognitive actions, there are two approaches: the process-oriented approach and the content-oriented approach. We have used the retrospective protocol analysis method that lies under the content-oriented approach, similar to the one used by Suwa and Tversky [27]. We have categorized the actions of the subjects into three groups: Physical, Perceptual and Conceptual. There is also a functional category, but for this experiment the subjects were already given a function (ie. "Table") therefore, this category was not used in the analysis. The perceptual and physical actions present the visual information and the conceptual and functional actions presents non-visual information. The modelling actions of each participants were coded for each cognitive segment.

Physical Actions. Physical actions are all the actions that involved in drawing new objects, tracing over the sheet and copying previously drawn elements, paying attention to previously drawn elements etc. In this paper, we have defined three group of physical actions: D-actions (Drawing, coping), L-actions (paying attention to previous design) and M-actions (movements on design depictions). The details of the Physical actions are given in Table 1.

Table 1. Sub-codes for D-actions, L-actions and M-actions

D-Actions	L-actions	M-Actions
Pd: Drawing depictions	Pl: Viewing (Camera Manipulation)	Pm: Moving depictions
	Pl: Manipulating depictions	

Perceptual Actions. The actions that are related to visual features of the objects and spatial relations among them are known as perceptual actions (P-Actions). Perceptual actions have a further eight categories, but in this research we have defined only three perceptual actions as shown in Table 2.

Table 2. Sub-codes for Perceptual (P-actions) and Conceptual (C-actions)

P-Actions related to implicit space	P-actions related to features	C-Actions
Ps: Selecting depictions	Pc: Colouring depictions	Ct: Thinking
	Psd: Deleting depictions	Cr: Reading
		Ci: Idle state

Conceptual Actions. Conceptual actions are those actions that are used to retrieve knowledge, previous similar cases or setup goals. In this research, we only examined the retrieval of knowledge and represent these as C-actions. We defined three conceptual actions as shown in Table 2.

4.2   Coding of Modelling Actions

We examined the actions of the participants who completed the task in different time windows. We found that the participants with a low task completion time (Users A, B, C) have an average action rate of 20 action/minute, which is 30% higher than for high-CT participants (Users D, E, F). High-CT subjects have an average action rate of 14 actions/minute.

We analysed the video records and assigned codes to each action performed by participants. The tasks were grouped together to find the ratios between CT levels and cognitive actions. Table 3 shows a summary of the modelling actions performed based on task completion time.

Table 3. Summarized action performance comparison based on task completion time

Perceptual	Low-CT	High CT	Physical	Low-CT	High CT	Conceptual	Low-CT	High CT
Ps	22%	17%	Pd	18%	13%	Ct	14%	18%
Pc	9%	9%	Pm	22%	15%	Ci	2%	2%
Psd	0%	1%	Pl	10%	7%	Cr	3%	19%
Total	**31%**	**27%**	**Total**	**51%**	**34%**	**Total**	**19%**	**39%**

The statistical results (chi-squared test, $X^2 < c$, $p < 0.05$) show that there are significant differences in the modelling actions of low-CT and high-CT users. The maximum difference is observed in conceptual tasks. Low-CT users performed 1.5 times as many physical actions as high-CT users. High-CT users spend a large amount of time on conceptual actions. They were relying more on the experimental instructions rather than their short-term memory. The rate of conceptual actions for high-CT users was twice as high as for low-CT users. The biggest difference was in reading the experiment instructions from the handouts.

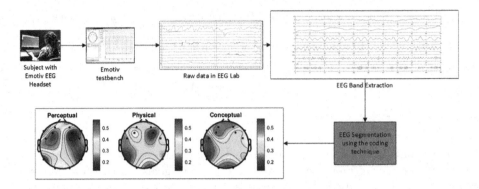

Fig. 2. EEG data processing steps

5   EEG Preprocessing and Signal Analysis

EEG data is very prone to noise and artefacts because the signals are highly sensitive and can be easily contaminated by various artefacts such as eye blinking, speaking, and other muscle movements. There are also some external sources of noise, for instance electrical line noise which can be removed by applying a notch filter at that particular frequency (i.e. 50 or 60 Hz). Noise and artefact removal from EEG data is one of the main parts of EEG signal processing. The EEG data of six participants were used for EEG analysis, as the data of the two participants was corrupted with noise. The preprocessing was done in MATLAB 2017 using the EEGLAB toolbox [4]. The baseline was removed from the EEG signal and low-pass filtering at a cut-off frequency of 45 Hz was performed using a linear-phase FIR filter. EEG signals were then high-pass filtered at a cut-off frequency of 0.1 Hz and notch filtered at 50 and 60 Hz using a linear phase FIR filter. The order of the filter in all cases was 300. After filtering the data, Independent Component Analysis (ICA) decomposition was performed to detect and remove artefacts such as eye blinking and muscle movements. The artefacts were removed manually. Once the data was clean enough, we calculated the Welch power spectral density (PSD) with a window size of 128 without overlapping. Data from all electrodes were incorporated in the analysis. A block diagram of all the EEG data processing steps is shown in Fig. 2. After calculating the PSD, the mean power for each band has been extracted as shown in Figs. 3–5. We have divided the dataset based on the completion time (CT) of the users for comparison. The threshold was set at 190 seconds by calculating the mean and standard deviation of the completion time. Figure 3 shows the alpha-band energy in perceptual, physical and conceptual actions. The subjects on the left side of the figure have a low task CT, whereas the subjects on the right side of the figure have a high task CT. As alpha-band activity is associated with cognitive functions such as task performance preparation [3], language comprehension and memory [9], so a change in alpha activity indicates a change in the cognitive

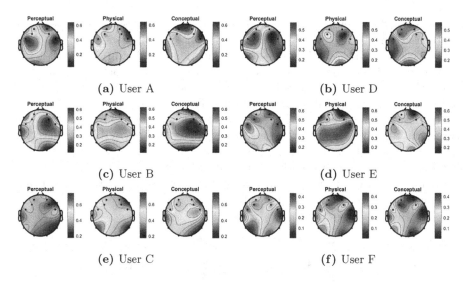

Fig. 3. Mean alpha activity at different segments

activity. In the literature, researchers have established that task complexity is inversely related to alpha-band activity [7].

In Fig. 3, the alpha power in the conceptual segment is higher than in the physical segment for low-CT subjects, which means that the subjects are more relaxed in performing the conceptual tasks or they have spent less time in conceptual tasks than in other segments. Subjects A and B have performed more physical tasks and the corresponding alpha-band activity is less in that segment, meaning that their attention highly focus on physical actions. For the users with a high-CT, the alpha activity is less in the conceptual actions (excluding User D) which is a sign that they were more relaxed or comfortable when they were thinking or reading handouts compared to others while performing modelling actions. The variation in alpha activity is higher in the frontal cortex than in other regions for low-CT users. The frontal cortex of the brain is responsible for higher mental functions such as concentration, planning, and problem solving [1], so it is also an indication that low-CT users were performing more cognitive activity than high-CT users. For subjects with high-CT, alpha activity variations were more in the motor, temporal and parietal cortex (electrodes T7, FC6, and P4) than in the frontal cortex. These locations are more associated with voluntary motor functions [1]. The continuous activation on the left frontal cortex (electrodes FC5 and F3) in the perceptual segment of high-CT subjects is possibly due to the continuous eye movements because of the nature of the experiment. We have also observed that the change in alpha activity for users with low-CT is quicker than for subjects with high-CT. The reason behind this response may be that the low-CT subjects change the design stages very rapidly compared to the other subjects. Figure 4 shows the beta activity of the three segments. The beta band usually relates to alertness and has a very low amplitude [15].

By looking at the beta activity response in Fig. 4, we can say that on average the beta activity is high in physical segments for low-CT users and in the perceptual segments in high-CT users. As we mentioned above in Sect. 2 that the beta activity directly relates to concentration, the high beta activity is an indication that the concentration is high in the physical segments compared to the perceptual and conceptual segments of low-CT users. The higher concentration can be due to the higher number of physical actions performed by low-CT users as seen in Table 3. For high-CT users, on average, the beta activity is lower in physical segments than in other segments, and this finding can also relate to the fact that fewer actions are performed in physical sections. Users with low-CT are more attentive in physical segments whereas users with high-CT concentrate more on perceptual and conceptual activities.

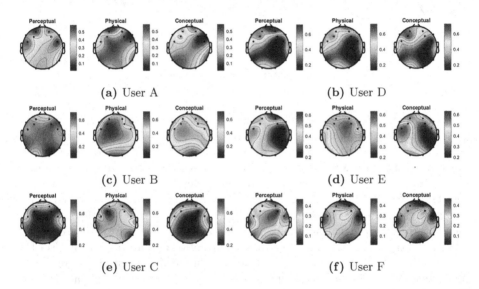

(a) User A   (b) User D

(c) User B   (d) User E

(e) User C   (f) User F

Fig. 4. Mean beta activity at different segments

The theta-band response is observed in adult individuals who are in a state of focus and is also associated with memory performance and functional processes [9]. We observe that the theta-band activity varies in each segment for low-CT users, especially for users A and B, whereas for other users the relative change is very small as shown in Fig. 5. This is also an indication that the focus or attention level of users with low-CT varied based on what actions they were performing. We have also observed that, by looking at the gamma-band activity and comparing it with the actions performed in each segment, the average gamma-band activity is low in segments where actions are more and high in segments with fewer actions.

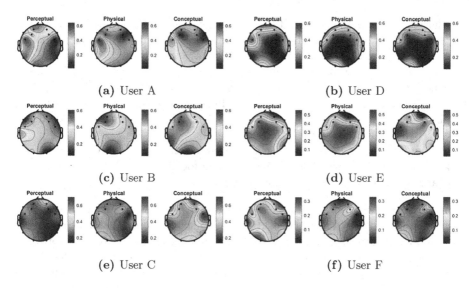

Fig. 5. Mean theta activity at different segments

6 Conclusion

In this paper, we have proposed a new method to segment EEG signals for understanding cognitive actions and their relation to brain activities. For this purpose, we have conducted an experiment in which each user had to draw a 3D table and we have used video recording and EEG signals to analyse the user's cognitive activities. All the participants had no prior experience of designing 3D objects in AutoCAD, so we have used task completion time as a measure to differentiate between designers. We have analysed the reasons for why some users completed the task earlier than the other users.

We have used a coding scheme designed by Suwa et al. [27] to analyse the quantified designer's actions. The coding analysis for EEG segmentation provides two advantages: the first is that we can see the EEG power variation in different segments, and the second is that we can use the results of coding analysis and compare with EEG power to easily track the cause of particular behaviour. From video recording, all the actions were decoded and divided into three segments: Perceptional, Physical and Conceptual actions. These segments were used to segment out the EEG data. We have analysed the alpha, beta, theta and gamma activity of the users. The findings from our data analysis are listed below:

1. Low-CT users performed 1.5 times more physical actions, which gave them the advantage of drawing quickly.
2. The rate of conceptual actions for high-CT users was twice as high as for low-CT users. This slows the overall design process.
3. The action rate per minute for low-CT users is 30% higher than for high-CT users. This is an indication that they are utilizing their short-term memory more efficiently.

4. The alpha band shows that low-CT users were comfortable in performing physical tasks whereas high-CT users were not relaxed in physical segments as their mean alpha-band power was high.
5. The maximum variation in the frontal cortex was found in low-CT users, which indicates that they were using their short-term memory more.
6. From beta activity, we have found that low-CT users were more attentive to physical segments, whereas the attention of high-CT users was focused on perceptual and conceptual actions.
7. We have found more variation in theta-band activity for low-CT users than for high-CT users, which indicates that the focus level of low-CT users was changing in relation to the action performed.

From the above analysis, we have concluded that if a user would utilize short-term memory more, reducing their attention to the conceptual actions and performing more physical actions instead, then their performance would improve.

In future work, we will use the above mentioned finding to develop a feedback system that will motivate the user to perform more physical actions and less conceptual actions to help them complete the tasks. We will also expand our research to analysis of novices and experts to see whether coding-based cognitive analysis and EEG segmentation can be used to analyse and compare the design behaviour of novices and experts and improve novices' performance. We also plan to use a high-end EEG system to increase the quality of the EEG data.

References

1. Al-Qahtani, A., Nasir, A., Shakir, M.Z., Qaraqe, K.A.: Cognitive impairments in human brain due to wireless signals and systems: an experimental study using EEG signal analysis. In: 2013 IEEE 15th International Conference on e-Health Networking, Applications & Services (Healthcom), pp. 1–3. IEEE (2013)
2. Chiu, I., Shu, L.: Potential limitations of verbal protocols in design experiments. In: ASME 2010 International Design Engineering Technical Conferences and Computers and Information in Engineering Conference, pp. 287–296. American Society of Mechanical Engineers (2010)
3. De Graaf, T.A., Gross, J., Paterson, G., Rusch, T., Sack, A.T., Thut, G.: Alpha-band rhythms in visual task performance: phase-locking by rhythmic sensory stimulation. PloS one 8(3), e60035 (2013)
4. Delorme, A., Makeig, S.: EEGLAB: an open source toolbox for analysis of single-trial eeg dynamics including independent component analysis. J. Neurosci. Methods 134(1), 9–21 (2004)
5. Fairclough, S.H., Venables, L., Tattersall, A.: The influence of task demand and learning on the psychophysiological response. Int. J. Psychophysiol. 56(2), 171–184 (2005)
6. Gal, U., Lyytinen, K., Yoo, Y.: The dynamics of it boundary objects, information infrastructures, and organisational identities: the introduction of 3D modelling technologies into the architecture, engineering, and construction industry. Eur. J. Inf. Syst. 17(3), 290–304 (2008)
7. Gevins, A., et al.: Monitoring working memory load during computer-based tasks with EEG pattern recognition methods. Hum. Fact. 40(1), 79–91 (1998)

8. Gola, M., Magnuski, M., Szumska, I., Wróbel, A.: EEG beta band activity is related to attention and attentional deficits in the visual performance of elderly subjects. Int. J. Psychophysiol. **89**(3), 334–341 (2013)
9. Gruzelier, J.H.: EEG-neurofeedback for optimising performance. i: a review of cognitive and affective outcome in healthy participants. Neurosci. Biobehav. Rev. **44**, 124–141 (2014)
10. Gundel, A., Wilson, G.F.: Topographical changes in the ongoing eeg related to the difficulty of mental tasks. Brain Topogr. **5**(1), 17–25 (1992)
11. Hjortskov, N., Rissén, D., Blangsted, A.K., Fallentin, N., Lundberg, U., Søgaard, K.: The effect of mental stress on heart rate variability and blood pressure during computer work. Eur. J. Appl. Physiol. **92**(1–2), 84–89 (2004)
12. Kahneman, D., Egan, P.: Thinking, Fast and Slow, vol. 1. Farrar, Straus and Giroux, New York (2011)
13. Kavakli, M., Boyali, A.: Design: robust gesture recognition in conceptual design, sensor analysis and synthesis. In: Gulrez, T., Hassanien, A.E. (eds.) Advances in Robotics and Virtual Reality, pp. 201–225. Springer, Heidelberg (2012). https://doi.org/10.1007/978-3-642-23363-0_9
14. Kavakli, M., Suwa, M., Gero, J., Purcell, T.: Sketching interpretation in novice and expert designers. In: Visual and Spatial Reasoning in Design, vol. 99, pp. 209–219. Citeseer (1999)
15. Klimesch, W., Sauseng, P., Hanslmayr, S.: EEG alpha oscillations: the inhibition-timing hypothesis. Brain Res. Rev. **53**(1), 63–88 (2007)
16. Kristiansen, J., et al.: Stress reactions to cognitively demanding tasks and open-plan office noise. Int. Arch. Occup. Environ. Health **82**(5), 631–641 (2009)
17. Kuusela, H., Pallab, P.: A comparison of concurrent and retrospective verbal protocol analysis. Am. J. Psychol. **113**(3), 387 (2000)
18. Liu, Y., Ritchie, J.M., Lim, T., Kosmadoudi, Z., Sivanathan, A., Sung, R.C.: A fuzzy psycho-physiological approach to enable the understanding of an engineer affect status during cad activities. Computer-Aided Design **54**, 19–38 (2014)
19. Mandryk, R.L., Atkins, M.S.: A fuzzy physiological approach for continuously modeling emotion during interaction with play technologies. Int. J. Hum.-Comput. Stud. **65**(4), 329–347 (2007)
20. Nguyen, T.A., Zeng, Y.: Analysis of design activities using EEG signals. In: ASME 2010 International Design Engineering Technical Conferences and Computers and Information in Engineering Conference, pp. 277–286. American Society of Mechanical Engineers (2010)
21. Nguyen, T.A., Zeng, Y.: A physiological study of relationship between designer mental effort and mental stress during conceptual design. Comput.-Aided Des. **54**, 3–18 (2014)
22. Nguyen, T.A., Zeng, Y.: Effects of stress and effort on self-rated reports in experimental study of design activities. J. Intell. Manuf. **28**(7), 1609–1622 (2017)
23. Pellouchoud, E., Smith, M.E., McEvoy, L., Gevins, A.: Mental effort-related eeg modulation during video-game play: Comparison between juvenile subjects with epilepsy and normal control subjects. Epilepsia **40**(s4), 38–43 (1999)
24. Sauro, J., Kindlund, E.: A method to standardize usability metrics into a single score. In: Proceedings of the SIGCHI Conference on Human Factors in Computing Systems, pp. 401–409. ACM (2005)
25. Schächinger, H., Weinbacher, M., Kiss, A., Ritz, R., Langewitz, W.: Cardiovascular indices of peripheral and central sympathetic activation. Psychosom. Med. **63**(5), 788–796 (2001)

26. Sterman, M., Mann, C.: Concepts and applications of EEG analysis in aviation performance evaluation. Biol. Psychol. **40**(1–2), 115–130 (1995)
27. Suwa, M., Purcell, T., Gero, J.: Macroscopic analysis of design processes based on a scheme for coding designers' cognitive actions. Des. Stud. **19**(4), 455–483 (1998)
28. Tang, Y., Zeng, Y.: Quantifying designer mental stress in the conceptual design process using kinesics study. In: DS 58–9: Proceedings of ICED 09, The 17th International Conference on Engineering Design, vol. 9, Human Behavior in Design, Palo Alto, CA, USA, 24–27 August 2009

Intelligent Educational Data Analysis with Gaussian Processes

Jiachun Wang, Jing Zhao$^{(\boxtimes)}$, Shiliang Sun, and Dongyu Shi

Department of Computer Science and Technology, East China Normal University,
3663 Zhongshan Road, Shanghai 200241, People's Republic of China
{jzhao,slsun}@cs.ecnu.edu.cn

Abstract. As machine learning evolves, it is significant to apply machine learning techniques to the intelligent analysis on educational data and the establishment of more intelligent academic early warning system. In this paper, we use Gaussian process (GP)-based models to discover valuable inherent information in the educational data and make intelligent predictions. Specifically, the mixtures of GP regression model is adopted to select personalized key courses and the GP regression model is applied to predict the course scores. We conduct experiments on real-world data which are collected from two grades in a certain university. The experimental results show that our approaches can make reasonable analysis on educational data and provide prediction information about the unknown scores, thus helping to make more precise academic early warning.

Keywords: Academic early warning · Key course selection
Course score prediction · Gaussian process regression
Mixtures of Gaussian processes

1 Introduction

Academic early warning (AEW), which can give prompt warnings to the students who have poor grades, is very popular in many colleges and universities. It is recently discussed in [1] as one of the computational education problems. Effective warning strategies will help students to study more purposely and effectively. However, most existing AEW systems just work with simple statistical methods which can only discover the explicit information in data. For example, the students who have failed more than ten credits in a semester will be warned. Further, the existing AEW systems are only able to send a single warning message to students without any advice. With the development of machine learning (ML), it is very meaningful to upgrade the AEW systems through ML models to provide students with smarter and more personalized learning advice. For instance, some intelligent analysis of educational data, such as key course selection and course score prediction, will provide much valuable information for students and educators. Therefore, discovering useful information implicit in

© Springer Nature Switzerland AG 2018
L. Cheng et al. (Eds.): ICONIP 2018, LNCS 11306, pp. 353–362, 2018.
https://doi.org/10.1007/978-3-030-04224-0_30

educational data and then making predictions through some machine learning methods is significant and practical [2–4].

Recently, a method of key course selection for AEW based on Gaussian progress regression (GPR) model [5] is proposed. The selected key course can warn the students to pay more attention to some specific courses. GPR model is a typical probabilistic model for nonlinear regression [6,7]. However, when used for key course selection, GPR assumes that the data are generated from a single mode and provides the universal key courses for all students in a class, which is unreasonable because different students may have different weaknesses in their studies. For dealing with specific problems, several GPR-based improvements were also proposed, such as sparse GPs [8,9] and mixtures of GP regression (MGPR) [10–12]. Among them, MGPR model is able to capture the multimodal characteristics of data, which provides methods for discriminately data analysis. For the case of key course selection, the mixture components in the MGPR model can represent different groups of students. In this paper, we employ MGPR model to make personalized key course selection for different kind of students. Moreover, we use the GPR model to predict the courses scores by utilizing the historical scores of key courses.

Key course selection can be regarded as feature selection in the machine learning area [5]. There are deterministic methods and Bayesian probabilistic models for feature selection. Bayesian probabilistic models can provide uncertainty estimate for features. Thus features with much higher weights are more likely to be the key courses. Further, the MGPR model is very suitable for this selection task because of the following two facts. Firstly, like GPR, MGPR has an elegant form of modeling nonlinear mappings as well as the flexibility of choosing kernel functions, which would capture the relationship between courses and warning results. Secondly, MGPR is capable of dealing with multimodal data which correspond to different groups of students. Specifically, we use MGPR model with several automatic relevance determination (ARD) kernels to select key courses. The length-scales as hyperparameters in the ARD kernels can be learned through the model selection procedure, which represents the importance of different courses. Compared with GPR, MGPR can provide personalized and diversified key courses, since it can obtain different ARD parameters for different components.

Course score prediction can be regarded as regression in the machine learning area. As described above, GPR is an effective method for regression, and has been widely used and developed [13–16]. The relationship between the historical course scores and the unknown course scores in the next semester can be learned through the GPR model. Besides models, another key to prediction is feature design. Considering that every student may choose multiple courses and the courses vary from person to person even in the same group, we use the selected key courses as the input features of the regression model. In addition, in order to jointly learn a prediction model on different courses over a semester, the predictive courses are encoded as input features as well. Specifically, we calculate the correlation coefficient between the predictive courses and the historical key courses as the features for the predictive courses. We will introduce the details of feature construction in the later sections.

We conduct experiments on real-world data which include academic performance of two grades throughout their university period. We test our approach on two practical tasks, i.e., key course selection for every semester and course score prediction for every next semester. The experimental results show that our approach can discover implicit information like key courses in the educational data and provide reasonable prediction of unknown scores, which is of significance for both students and educators.

The remainder of this paper is organized as follows. Section 2 introduces the MGPR model which is used to select key courses for different groups of students. In Sect. 3, we present the method of predicting the course scores with GPR model. Experimental results and analysis are provided in Sect. 4. Finally, we conclude this paper and discuss the future work in Sect. 5.

2   Key Course Selection with MGPRs

MGPR models multimodal data by using varying covariances for different mixture components, which can automatically adjust the number of mixture components by integrating a Dirichlet process prior [17]. Suppose there are N training points $\{\mathbf{x}_n, y_n\}_{n=1}^N (\mathbf{x}_n \in \mathbf{R}^D)$, and each GP component has a support set of M training points $(M < N)$. The kernel $k_k(\mathbf{x}_i, \mathbf{x}_j)$ is a covariance function used to capture the relationship within the kth GP support set $\mathbf{I}_k$, which is usually specified as a squared exponential kernel with hyperparameters $\boldsymbol{\theta}_k$. In this paper, we use the ARD kernel with different hyperparameters for different components, which is defined as:

$$\kappa(\mathbf{x}, \mathbf{x}') = \sigma_f^2 \exp\left\{ -\frac{1}{2} \sum_{d=1}^D \frac{1}{\ell_d^2} (x_d - x_d')^2 \right\}, \tag{1}$$

where ℓ_d is the length-scale of the covariance and σ_f^2 is the signal variance. Such a covariance function implements automatic relevance determination, since the inverse of the length-scale determines how relevant an input feature is: if the length-scale is very large, the covariance of the input will become almost independent, effectively removing it from the inference. Thus, we select the key courses according to the values of the corresponding length-scales.

Given the input $\mathbf{x}$ and the component indicator variable $z = k$, the distribution of an observed output is expressed as

$$p(y|\mathbf{x}, z = k, \mathbf{w}_k, r_k) = \mathcal{N}(y|\mathbf{w}_k^\top \boldsymbol{\phi}_k(\mathbf{x}), r_k^{-1}), \tag{2}$$

where $\boldsymbol{\phi}_k(\mathbf{x})$ represents the connections between $\mathbf{x}$ and the support set $\mathbf{I}_k$, which is formulated as $\boldsymbol{\phi}_k(\mathbf{x}) = [k_k(\mathbf{x}, \mathbf{x}'_1), k_k(\mathbf{x}, \mathbf{x}'_2), ..., k_k(\mathbf{x}, \mathbf{x}'_M)]^\top (\mathbf{x}'_i \in \mathbf{I}_k)$. The weight vector $\mathbf{w}_k$ with a Gaussian prior $\mathcal{N}(\mathbf{w}_k|0, \mathbf{U}_k^{-1})$ shows the responsibility of the kth Gaussian. The precision matrix $\mathbf{U}_k$ is set to $K_k + \sigma_{kb}^2\mathbf{I}$, where σ_{kb}^2 is the Gaussian noise variance and

$$K_k = [\boldsymbol{\phi}_k(\mathbf{x}_1), \boldsymbol{\phi}_k(\mathbf{x}_2), ..., \boldsymbol{\phi}_k(\mathbf{x}_N)]. \tag{3}$$

Thus the hyperparameters of ARD kernel for the kth component are denoted as $\boldsymbol{\theta}_k = \{\{\ell_{kd}\}, \sigma_{kf}^2, \sigma_{kb}^2\}$. The inverse variance r_k has a Gamma prior distribution with parameters a_0 and b_0.

The MGPR distinguishes from GPR by introducing the indicator variable z and multiple GP components. MGPR can be regarded as a more general model than GPR, which assembles multiple results of GPR in a principled way. The ARD kernels for the k components provide the basis for key course selection for different groups of students, where the support sets $\mathbf{I}_k$ divide the students into groups.

The distribution of an input point for a mixture component is given by a Gaussian distribution with a full covariance [18],

$$p(\mathbf{x}|z = k, \boldsymbol{\mu}_k, \mathbf{R}_k) = \mathcal{N}(\mathbf{x}|\boldsymbol{\mu}_k, \mathbf{R}_k^{-1}), \tag{4}$$

where parameters $\boldsymbol{\mu}_k$ and $\mathbf{R}_k$ are embodied by the Gaussian and Wishart distributions $\boldsymbol{\mu}_k \sim \mathcal{N}(\boldsymbol{\mu}_0, \mathbf{R}_0^{-1})$ and $\mathbf{R}_k \sim \mathcal{W}(\mathbf{W}_0, \nu_0)$, respectively.

The distribution of the indicator z given the input $\mathbf{x}$ is calculated according to Bayesian rules,

$$p(z|\mathbf{x}) = \frac{p(z|\overline{\boldsymbol{\nu}})p(\mathbf{x}|z)}{\sum_z p(z|\overline{\boldsymbol{\nu}})p(\mathbf{x}|z)}, \tag{5}$$

where the prior distribution of z is a discrete distribution with parameters denoted as $\overline{\boldsymbol{\nu}} = \{\nu_1, ..., \nu_\infty\}$ from a Dirichlet process with the hyperparameter α_0. Denote all the hidden variables as $\Omega = \{\{\nu_k\}, \{\boldsymbol{\mu}_k\}, \{\mathbf{w}_k\}, \{\mathbf{R}_k\}, \{z_i\}, \{r_k\}\}$. The joint distribution of all the random variables is given by

$$p(X, Y, \Omega) = \prod_{k=1}^{\infty} p(\nu_k)p(\boldsymbol{\mu}_k)p(\mathbf{w}_k)p(\mathbf{R}_k)p(r_k)$$
$$\prod_{i=1}^{N} p(z_i|\overline{\boldsymbol{\nu}})p(\mathbf{x}_i|z_i, \boldsymbol{\mu}_{z_i}, \mathbf{R}_{z_i})p(y_i|\mathbf{x}_i, z_i, \mathbf{w}_{z_i}, r_{z_i}).$$

With the above model assumptions, the MGPR model can be trained by the variational expectation maximization (EM) method [10]. It is implemented through alternatively inferring variational posterior distributions of latent variables and optimizing hyperparameters and support sets.

The advantage of MGPR is to capture the multimodal characteristics of data, which provides methods for discriminately data analysis. For the case of key course selection, the mixture components in the MGPR model represent the different groups of students, and the learned groups of ARD kernel hyperparameters offer personalized key course information for different groups of students.

3   Course Score Prediction

Course score prediction is a regression problem. We use the GPR model to predict the course scores, where the input features are constructed by historical key course scores and the correlation coefficients between the predictive course and historical key courses.

3.1 Feature Construction

For making course score prediction, the historical course scores can be directly used as input. However, in this case, the dimension of input features will be very high which may include redundant information [19]. What's more, a separate GPR model should be trained for each predictive course, which is not appropriate. In order to reduce the dimension of input features, the key course scores instead of all the course scores are used as input. In addition, in order to jointly train a GPR model for all the predictive courses over a semester, we encode the predictive course as the additional input features. As all we can obtain from the data are the historical scores of all the courses, we calculate the well-known correlation coefficient between the predictive course and historical key course scores as the codes for the predictive course. The correlation coefficient between two random variables is calculated as

$$r(X,Y) = \frac{Cov(X,Y)}{\sqrt{Var[X]Var[Y]}}. \tag{6}$$

3.2 Course Score Prediction with GPR

The GPR model assumes that data are generated from a GP. As a regression model, it can be used to make prediction. With the GPR model being trained, the joint distribution of the observed outputs $\mathbf{y}$ and the test outputs $\mathbf{f}^*$ given the observed inputs X and test inputs X^* is

$$\begin{bmatrix} \mathbf{y} \\ \mathbf{f}^* \end{bmatrix} \sim \mathcal{N}\left(0, \begin{bmatrix} K(X,X) + \sigma_b^2 I & K(X,X^*) \\ K(X^*,X) & K(X^*,X^*) \end{bmatrix}\right). \tag{7}$$

Similar to Eq. (3), K is the covariance matrix calculated by the kernel function $\kappa(\mathbf{x}, \mathbf{x}')$. After deriving the conditional distribution, we arrive at the key predictive distribution for the GPR model as

$$p(\mathbf{f}^*|X, \mathbf{y}, X^*) = \mathcal{N}(\bar{\mathbf{f}}^*, \text{cov}(\mathbf{f}^*)), \tag{8}$$

where

$$\bar{\mathbf{f}}^* = K(X^*,X)[K(X,X) + \sigma_b^2 I]^{-1}\mathbf{y}, \tag{9}$$
$$\text{cov}(\mathbf{f}^*) = K(X^*,X^*) - K(X^*,X)[K(X,X) + \sigma_b^2 I]^{-1}K(X,X^*). \tag{10}$$

Thus, the mean $\bar{\mathbf{f}}^*$ is used as the prediction result of course score for the next semester.

4 Experiments

In this section, we will analyze the key courses selected by MGPR models on real-world data and make prediction for the course scores in the next semester by using the historical key course scores.

4.1 Dataset and Setups

The educational dataset we use is the scores and warning results (failed credits) of seven semesters for the Grade 2010 and Grade 2011 students in the department of computer science and technology of a university. There are two classes for each grade, namely pedagogical class and regular class. Therefore, we have totally 28 semesters of scores from 4 classes. The student number of each class is different. For Grade 2010, there are 47 students in the pedagogical class and 51 students in the regular class, while Grade 2011 has 23 and 52 students, respectively. Since the original data are incomplete and the number of courses chosen by different students varies in each semester, we employ the nearest neighbor (NN) data-filling method to reconstruct the data.

The component number of MGPR is set to three, which means that we treat the students as three groups, and thus we will obtain three groups of key courses. Both the maximum EM iteration number and the maximum variational inference iteration number are set to 25.

For course score prediction with GPRs, we use the first six key courses per semester selected by the MGPR and their correlation coefficients with the predictive course as the input features. Therefore, for the $i + 1$th semester, the dimension of input features is $12 * i$. The maximum iteration number for optimizing the GPR model is set to 1000.

For all the experiments, we preprocess the data to make them suitable for the zero-mean GP based models. In order to construct zero-mean output in our experiment, we change the zero-value outputs to be negative values. Moreover, we scale the input into the range of $[0, 1]$.

4.2 Experimental Results and Analysis

Key Course Selection with MGPRs. As mentioned, MGPRs can help to make personalized key course selection. It is implemented by assuming multiple mixture components. We use the course scores and warning results from different classes and different grades to train MGPR models. In this experiments, we obtain three groups of key courses through the MGPR model with three components. We report three groups of the most critical courses for the pedagogical and regular class in Grade 2010 in Table 1. The full name of the courses corresponding to the abbreviations in the table is provided in the appendix.

From the table, we can find the MGPR model gives diverse key courses which correspond to different students. The corresponding students are expressed as support sets in the MGPR. Note that there may exist the same key courses for different groups, e.g., the same 'C' in Table 1. Seen from Table 1, for instance, in the pedagogical class, the courses 'C Programming (C)' and 'Education (Edu.)' are key courses for all students. This is convincing because 'C' is the basis of programming which is very important for computer science students, and 'Edu.' is a professional course for pedagogical students in order to cultivate their teaching ability. Further from Table 1, we find that the third group of students is poor in 'English (Eng.)', and the first two groups of students are advised to pay more

Table 1. The most critical courses for the pedagogical and regular class in Grade 2010 selected by MGPRs.

	Sem. 1	Sem. 2	Sem. 3	Sem. 4	Sem. 5	Sem. 6	Sem. 7
2010-pedagogical-class	ICSP	Java	C	AM	DLP	AI	Edu.
	C	CPP	C	PMS	Edu.	AM	Edu.
	C	Eng.	Psy.	COA	Edu.	Eng.	DLP
2010-regular-class	C	AM	OCMH	COAP	Eng.	CPP	LA
	PE	CP	CP	OS	DM	XML	MC
	C	CP	CP	AA	DM	XML	Eng.

attention to 'Advanced Mathematics (AM)'. In addition, 'Introduction to Computer Science and Practice (ICSP)' is a key course for the first group. A similar analysis can also be performed in the regular class in Grade 2010 according to Table 1. Thus, MGPR model has offered personalized study advice, which can help students to make study plan more purposely and effectively.

Course Score Prediction. In this part, we show the course score prediction results for the Grade 2011 by using the key course information learned from Grade 2010 and the correlation information between courses. In order to demonstrate the importance of key course information, we compare the experimental results using different historical course scores, i.e., using all courses and using key courses.

The average prediction errors of the course scores for different semesters are listed in Table 2, where the root mean squared error (RMSE) is calculated. Seen from the table, all the results using key courses except for the fourth semester performs better than that using all courses. In addition, we show the average prediction errors of the failed course numbers per student in different semesters in Table 3. The results in terms of RMSE implies a similar conclusion to Table 2. In order to display the prediction performance of our approach better, we also list the last five predicted rankings and the corresponding true rankings. The rankings on 'C' in the fifth semester are shown in Table 4, which tells that the obtained prediction results can provide valuable information for AEW.

Table 2. Average prediction errors of the course scores in different semesters in terms of RMSE.

	Sem. 2	Sem. 3	Sem. 4	Sem. 5	Sem. 6	Sem. 7
Use all courses	23.44	14.11	**15.26**	15.06	16.33	11.40
Use key courses	**15.50**	**12.69**	16.49	**13.24**	**15.26**	**9.24**

Table 3. Average prediction errors of the failed course numbers per student in different semesters in terms of RMSE.

	Sem. 2	Sem. 3	Sem. 4	Sem. 5	Sem. 6	Sem. 7
Use all courses	3.13	3.88	**3.03**	5.14	3.12	1.88
Use key courses	**2.27**	**3.34**	3.43	**4.92**	**2.00**	**1.18**

Table 4. The last five true rankings and predicted rankings of the students on 'C' course in the fifth semester.

True rankings	19	20	21	22	23
Predicted rankings	20	23	19	21	22

5 Conclusion

We have proposed two GP based educational data analysis methods. Firstly, we select personalized key courses of different groups of students by MGPR model which is capable to capture the multimodal characteristics of data. The weights of courses are determined by the parameters of the ARD kernels. Secondly, we use the GPR model to intelligently predict the courses scores by utilizing the historical scores of key courses and the correlation coefficient between the predictive course and historical key course scores. From the experimental results on the real-world data, we conclude that our approaches can make reasonable analysis of key courses and provide useful prediction information about the unknown scores. The information obtained by the above two methods can warn students to pay more attention to some specific courses and help educators to make proper teaching policies, which makes significant progress in AEW systems and educational data analysis. In the future work, we will consider using more features about courses or students to make more accurate prediction.

Acknowledgments. The first two authors Jiachun Wang and Jing Zhao are joint first authors. The corresponding author is Jing Zhao. This work is sponsored by Shanghai Sailing Program, NSFC Project 61673179 and Shanghai Knowledge Service Platform Project (No. ZF1213).

Appendix

The following is the comparison table for abbreviations and full names of different courses.

Abbreviation	Full Name
AA	Abstract Algebra
AI	Artificial Intelligent
AM	Advanced Mathematics
C	C Programming
COA	Computer Organization and Architecture
COAP	Computer Organization and Architecture Practice
CP	College Physics
CPP	Computer Programming Practice
DLP	Digital Logic and Practice
DM	Discrete Mathematics
Edu.	Education
Eng.	English
ICSP	Introduction to Computer Science and Practice
LA	Linear Algebra
MC	Modular Class
OCMH	Outline of Chinese Modern History
OS	Operating System
PE	Physical Education
PMS	Probability and Mathematical Statistics
Psy.	Psychology
XML	XML Programming

References

1. Sun, S.: Computational education science and ten research directions. Commun. Chin. Assoc. Artif. Intell. **9**(5), 15–16 (2015)
2. Hsu, T.: Research methods and data analysis procedures used by educational researchers. Int. J. Res. Method Educ. **28**(2), 109–133 (2005)
3. Weng, C.: Mining fuzzy specific rare itemsets for education data. Knowl. Based Syst. **24**(5), 697–708 (2011)
4. Limprasert, W., Kosolsombat, S.: A case study of data analysis for educational management. In: International Joint Conference on Computer Science and Software Engineering, pp. 1–5. IEEE, New York (2016)

5. Yin, M., Zhao, J., Sun, S.: Key course selection for academic early warning based on Gaussian processes. In: Yin, H., et al. (eds.) IDEAL 2016. LNCS, vol. 9937, pp. 240–247. Springer, Cham (2016). https://doi.org/10.1007/978-3-319-46257-8_26
6. Rasmussen, C.E., Williams, C.K.I.: Gaussian Process for Machine Learning. MIT Press, Cambridge (2006)
7. Ghahramani, Z.: Probabilistic machine learning and artificial intelligence. Nature **521**(7533), 452–459 (2015)
8. Titsias, M.K.: Variational learning of inducing variables in sparse Gaussian processes. In: Proceedings of the 12th International Conference on Artificial Intelligence and Statistics, pp. 567–574. JMLR, Cambridge (2009)
9. Zhu, J., Sun, S.: Single-task and multitask sparse Gaussian processes. In: Proceedings of the International Conference on Machine Learning and Cybernetics, pp. 1033–1038. IEEE, New York (2013)
10. Sun, S., Xu, X.: Variational inference for infinite mixtures of Gaussian processes with applications to traffic flow prediction. IEEE Trans. Intell. Transp. Syst. **12**(2), 466–475 (2011)
11. Sun, S.: Infinite mixtures of multivariate Gaussian processes. In: Proceedings of the International Conference on Machine Learning and Cybernetics, pp. 1011–1016. IEEE, New York (2013)
12. Luo, C., Sun, S.: Variational mixtures of Gaussian processes for classification. In: Proceedings of 26th International Joint Conference on Artificial Intelligence, pp. 4603–4609. Morgan Kaufmann, San Francisco (2017)
13. Álvarez, M.A., Lawrence, N.D.: Computationally efficient convolved multiple output Gaussian processes. J. Mach. Learn. Res. **12**(May), 1459–1500 (2011)
14. Wilson, A.G., Knowles, D.A., Ghahramani, Z.: Gaussian process regression networks. In: Proceedings of the 29th International Conference on Machine Learning, pp. 599–606. ACM, New York (2012)
15. Damianou, A., Lawrence, N.: Deep Gaussian processes. In: International Conference on Artificial Intelligence and Statistics, pp. 207–215. JMLR, Cambridge (2013)
16. Wang, Y., Chaib-draa, B.: KNN-based kalman filter: an efficient and non-stationary method for Gaussian process regression. Knowl. Based Syst. **114**(10), 148–155 (2016)
17. Rasmussen, C.E., Ghahramani, Z.: Infinite mixtures of Gaussian process experts. Adv. Neural Inf. Process. Syst. **14**(4), 881–888 (2002)
18. Meeds, E., Osindero, S.: An alternative infinite mixture of Gaussian process experts. Adv. Neural Inf. Process. Syst. **18**(1), 883–890 (2006)
19. Bolón-Canedo, V., Sánchez-Maroño, N., Alonso-Betanzos, A.: Recent advances and emerging challenges of feature selection in the context of big data. Knowl. Based Syst. **86**(5), 33–45 (2015)

Dual-Convolutional Enhanced Residual Network for Single Super-Resolution of Remote Sensing Images

Xuewei Li[1], Hongqian Shen[1], Chenhan Wang[2], Han Jiang[2], Ruiguo Yu[1(✉)],
Jianrong Wang[1], and Mankun Zhao[1]

[1] School of Computer Science and Technology, Tianjin University, Tianjin, China
{lixuewei,hongqianshen,rgyu,wjr,zmk}@tju.edu.cn
[2] Beijing AXIS Technology Company Limited, Beijing, China
{gabriel,hahn}@signcl.com

Abstract. The image super-resolution aims to recover a high-resolution image using a single or sequential low-resolution images. The super resolution methods based on deep learning, especially the deep convolutional neural network, have achieved good results. In this paper,we propose Dual-Convolutional Enhanced Residual Network (DCER) for remote sensing images based on residual learning, which concatenates the feature maps of different convolutional kernel sizes (3×3, 5×5). On the one hand, it can learn more high-frequency detail information by combining the local details of different scales; on the other hand, it reduces network parameters and greatly shorten the training time. The experimental results show that DCER achieves favorable performance of accuracy and visual performance against the state-of-the-art methods with the scale factor 2x, 4x and 8x.

Keywords: Dual-Convolutional Enhanced Residual Network (DCER)
Single super-resolution · Remote sensing images

1 Introduction

High-resolution (HR) images contain more information and better accuracy, which is helpful to the further mining, understanding and processing of image contents. The spatial resolution of the satellite remote sensing image depends on the accuracy of the sensor, and the improvement of the performance of the imaging system is accompanied with an expensive manufacturing cost. Super-resolution (SR) [10] is to use a single or sequential low-resolution (LR) images with complementary information to obtain high-resolution images, which is of great significance to improve image quality.

At present, there are many researches on super-resolution of images. The super-resolution technology is divided into three categories: interpolation, multi-frame reconstruction and learning-based methods. The early interpolation

© Springer Nature Switzerland AG 2018
L. Cheng et al. (Eds.): ICONIP 2018, LNCS 11306, pp. 363–372, 2018.
https://doi.org/10.1007/978-3-030-04224-0_31

method [5] generally uses an interpolation kernel to estimate the value of an unknown pixel in a high-resolution grid. Because the interpolation kernel cannot adapt to the image and does not introduce effective high-frequency information. Therefore, the upsampled image is easy to blur, and many sharp edge details cannot be retained. The super-resolution technology based on multi frame reconstruction [1,6,9,15,16] uses the prior information of the image to establish a mathematical model or gradually improve the HR by using an iterative method. Due to the need to fuse information from multiple frames of LR, the applicability and accuracy of the motion model will greatly affect the reconstruction effect. The learning-based super-resolution algorithm [17,20] searches for the mapping relationship between LR and the corresponding HR image by the training dataset, and finds the optimal solution for LR.

Recently, with the development of deep neural networks, Dong et al. [2] proposed a Super-Resolution Convolutional Neural Network (SRCNN) with three convolutional layers for SR. Kim et al. [7] proposed a Very Deep Convolutional Networks (VDSR), which increased the network depth by cascading small filters and learning residuals. But since both SRCNN [2] and VDSR [7] need to use the bicubic interpolation as the upsampling operator to enlarge images to the size of the target, thus it increased the unnecessary computing cost. Dong et al. [3] proposed Accelerating the super-resolution convolutional neural network (FSRCNN) to construct the SR images at the last step by deconvolution. Kim et al. [8] used a deep recursive neural network which got better performance at a small number of parameters. Kolte et al. [11] used the Laplacian Pyramid Super-Resolution Network (LapSRN) to progressively extract features and carry out upsampling, which then computed the output of the previous layer into the loss function, and supervised the results of each stage. Ledig et al. [12] used 16 blocks deep ResNet (SRResNet) to construct loss function, Lim et al. [14] proposed Enhanced deep residual networks (EDSR) which also applied residual network for SR. EDSR [14] removed batch normalization from SPResNet [12], simplifying the network while improving its performance, which won the championship in the NTIRE2017 Super Resolution Challenge.

Multi-frame reconstruction has been widely used in remote sensing image SR [13,21], and multi-frame reconstruction needs to obtain multiple images of the same scene. However, in many specific satellite applications, multiple images of the same scene are difficult to obtain because of the different angle and resolution of the sensor, or the long revisit time of the same sensor. Therefore, single frame SR is more suitable for remote sensing. The traditional based on learning method [19] is mostly based on low-level feature design, and reconstruction results are poor in complex space remote sensing images. Therefore, Dual-Convolutional Enhanced Residual Network (DCER) is proposed to better reconstruct super-resolution images by combining local image details at different scales.

The main contributions of this paper can be summarized in the following two points:

(1) DCER is applied to the single super-resolution of remote sensing images to enhance the details of remote sensing images.

(2) DCER with different convolutional kernels (3×3, 5×5) can combine the local details of different scales, and uses 1×1 convolutional layer to reduce the dimensions, and speed up network training.

2 Related Works

In recent years, the deep neural network, after a rapid development, has shown its superior capability in the field of computer vision. Dong et al. [2] first proposed SRCNN for SR, and established the end-to-end mapping between LR and HR. SRCNN consists of three operations:patch extraction and representation, non-linear mapping and reconstruction. As the network deepens, the advanced features learned on the one hand will become much more. On the other hand, the correlation between gradients in backward propagation will become increasingly worse, and the degradation of the network will lead to greater errors. He et al. [4] presented a residual learning framework to train deeper networks. The Residual block was implemented by shortcut connection to stack the input and output. Adding element-wise overlays does not add extra parameters and computational complexity to the network. When the number of network layers increases, the residual learning framework can solve the gradient disappearance and over fitting problem of deep convolutional network.

SRResnet [12] and EDSR [14] applied a residual learning to the super resolution. Each residual block structure in SRResnet [12] has two convolutional layers. The batch normalization and Relu activation functions are used between the convolutional layers. Batch normalization is to process the mean and variance of each training batch data with the addition of translation and scaling parameters, but it destroys the spatial information of the image to a certain extent and offsets the gain brought about by the network depth. Given that, EDSR [14] removed batch normalized modules in the residual structure. Since the batch normalization layer consumes memory, after removing the batch normalization, EDSR [14] stacks more convolutional layers and extracts more features at each layer, resulting in better performance. The final version of EDSR [14] has 32 residual blocks, each convolutional layer with 256 channels. The huge amount of network parameters leads to a slow network training. For upsampling factor 3x and 4x, initializing model parameters with pre-trained 2x network parameters. Even though this can reduce the training time of the high upsampling factor model, the network training time per epoch still takes a lot of time.

Increasing the depth or width of the network model will inevitably increase the computational complexity. In order to prevent over-fitting and increased computational complexity, Szegedy et al. [18] proposed an efficient deep neural network architecture called GoogLeNet. The core of the GoogLeNet architecture is the Inception module, with a depth of 22 layers. In terms of model depth, in order to avoid the problem of gradient disappearance, GoogLeNet increases two loss at different depths; and when it comes to the width of the model, the Inception architecture consists of filter sizes 1×1, 3×3, 5×5. With different sizes of convolutional kernels, the sizes of the receptive fields are also different, and the final splicing merges features of different scales. Concating the feature maps of different convolutional kernel sizes fuses local detail information. In order to reduce the huge computational complexity of the 5×5 convolutional kernel, the 1×1 convolutional kernel is used for dimensionality reduction, and then multi-scale convolution is performed separately. Therefore, the Inception not only improves the ability of convolution to extract features, but also eases the bottleneck of computing resources.

This paper is based on the use of multi-scale convolutional filters to improve EDSR [14], shorten the network training time, and achieve excellent performance.

3   Proposed Methods

In this section, we will specifically describe the proposed DCER network and the training details of the network.

3.1   Dual-Convolutional Enhanced Residual Networks

The EDSR [14] has been applied to super resolution which has achieved excellent results. It has restored image details better than other methods. The application of residual learning in EDSR [14] increases the number of network layers, and at the same time, more convolutional kernels are used on each layer, resulting in a huge amount of parameters. To solve this problem, we proposes Dual-Convolutional Enhanced Residual Networks. In Fig. 1, (a) is the residual block of EDSR [14], and (b) is the residual block presented in this paper. Firstly, we change the single 3×3 convolutional kernel to the 3×3 and 5×5 convolutional kernels to jointly extract features. Secondly, after the Relu activation, 1×1 convolutional kernel is used to reduce the dimension of the previous 256 features to 64. Then, two convolutional operations are respectively performed using 3×3, 5×5 dual-channel networks, and finally feature fusion is performed. Comparing the parameter quantities of a block (only calculating the convolutional weight parameters and ignoring the deviation parameters), EDSR [14] has 1179648 parameters and DCER has 851968 parameters, so in a residual block, DCER reduces the parameter amount of about 21.78%.

As for the network depth, the deeper the network is, the greater the receptive field is, that is, the larger the mapping range of the output result is, the higher the globality and the abstractness are. Deep networks are mostly used for image classification and recognition. For super-resolution, no specific features and high abstract features are needed to determine the final output. On the other hand, as network layers deepen, the network model is likely to degenerate and reduce the effective dimension of the model. So this paper sets up 16 residual blocks for network. The overall network structure is shown in Fig. 2.

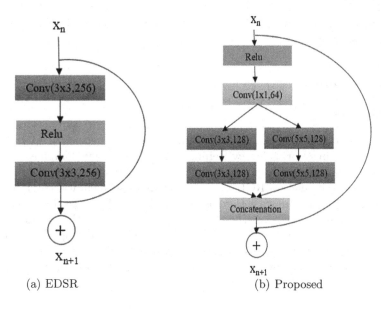

(a) EDSR   (b) Proposed

Fig. 1. Comparison of residual blocks in EDSR and ours

3.2 Training Details

During the training phase, we downsample high-resolution images to corresponding low-resolution images, which form training-label sample pairs. The sample size and total number of different sampling times can be referred to in Sect. 4.1. The activation function is Relu. Batchsize is 2. The initial learning rate is set to 0.001, with inverse time decay, setting the globalstep to 2500. The loss function uses L1 Loss. Although the mean squared error is the most important measure used in image quality assessment, the image output by minimizing the mean squared error is too smooth, and the PSNR could not effectively capture the complex features of the human visual system [22]. All experiments are carried

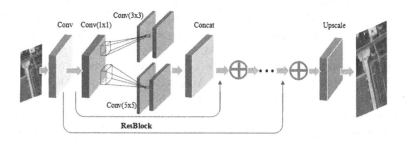

Fig. 2. The architecture of the proposed DCER network structure. A low-resolution image goes through the first convolutional layer, and then enters 16 residual learning blocks. The high-resolution image is obtained by transposed convolutions

on an Inter E5-2690 CPU 2.6 GHz with 188 GB RAM and Tesla K80 using the tensorflow package.

4   Experiments

4.1   Datasets

The experimental data in this paper is remote sensing image with spatial resolution of 1.07 m/pixel provided by Beijing AXIS Technology Company Limited. According to the spectral response of the objects, three channels are selected to synthesize three channels of RGB images. The image segmentation step is based on the image size, ensuring that there is no cross between images. The segmented image randomly selects 20% as the testing set and the rest is the training set. In the experiment, considering the size of the image after x8 upsampling, we reduce the size of original image, so the image input sizes of different upsampling factors are different, as shown in Table 1. All images are down-sampled to get the corresponding low-resolution image.

Table 1. Dataset of different up sampling factor

Scale	LR size	Training samples	Testing samples
x2	64×64	13600	3400
x4	64×64	10000	2500
x8	32×32	10000	2500

4.2   Comparisons with the State-of-the-Arts

To confirm the ability of the proposed network, we compare our network with four state-of-the-art SR algorithms:Bicubic, SRCNN [2], VDSR [7], EDSR [14]. The quantitative results with the scale factor 2x, 4x and 8x are shown in the Table 2. Image reconstruction quality uses PSNR and SSIM as two evaluation indicators, and PSNR calculation method is as shown in formula (1) and formula (2). MAX is 255, and C, H, and W are the number of channels, height, and width of the image, respectively. $X(i, j, k)$ and $Y(i, j, k)$ represent network output results and high-resolution image pixel values. Among these methods, DCER has the best performance with the highest PSNR and SSIM over $2, 4, 8$ scale. We also present the qualitative results in Fig. 3. By visual comparison, DCER performs well in SR applications of remote sensing images. Compared with other methods, the edges of the HR images obtained by DCER are clear, and the details of the image texture are obviously enhanced, and the edge structure features of the images are well preserved.

$$PSNR = 10 \times lg(\frac{MAX^2}{MSE}) \tag{1}$$

$$MSE = \frac{1}{C \times H \times W} \times \sum_{i=1}^{C}\sum_{j=1}^{H}\sum_{k=1}^{W} ||X(i,j,k) - Y(i,j,k)||^2 \tag{2}$$

Table 2. Mean PSNR (dB) and SSIM over different scale

Scale	Bicubic PSNR/SSIM	SRCNN PSNR/SSIM	VDSR PSNR/SSIM	EDSR PSNR/SSIM	DCER PSNR/SSIM
x2	31.366/0.87	32.037/0.88	32.977/0.90	33.975/0.91	**34.155/0.92**
x4	26.503/0.70	27.443/0.71	28.077/0.74	28.880/0.77	**29.000/0.78**
x8	21.989/0.48	24.182/0.54	24.501/0.56	26.053/0.63	**26.098/0.64**

4.3 Network Parameters and Training Time

DCER is improved based on the network structure of EDSR [14]. In this section, we analyze the network parameter settings of EDSR [14] and DCER, and compare the parameter capacity. In addition, referring to the EDSR [14] paper and code, we implement and apply the data set of this paper to train and test the model. By reason of diverse the input sizes of different sampling multiple networks, we list the number of model parameters of different sampling factors, time of each epoch training and their corresponding training results, as shown in Table 3.

Table 3. Comparison of parameters and training time

Scale	Method	Parameters	Training time/epoch (min)	Mean PSNR (dB)
x2	EDSR	1179648	85.11	33.975
x2	DCER	854968	60.05	34.155
x4	EDSR	1179648	65.05	28.880
x4	DCER	854968	40.04	29.000
x8	EDSR	1179648	21.32	26.053
x8	DCER	854968	11.20	26.098

In the training process, we test all the test data after each training epoch to observe the training result. As shown in Fig. 4, comparing the training/test results of EDSR [14] and DCER at different sampling factor per epoch, it can be clearly seen that DCER can reach a steady trend quickly, and is higher than EDSR [14] in each round of PSNR evaluation indicators. This illustrates the effectiveness of DCER network extraction features.

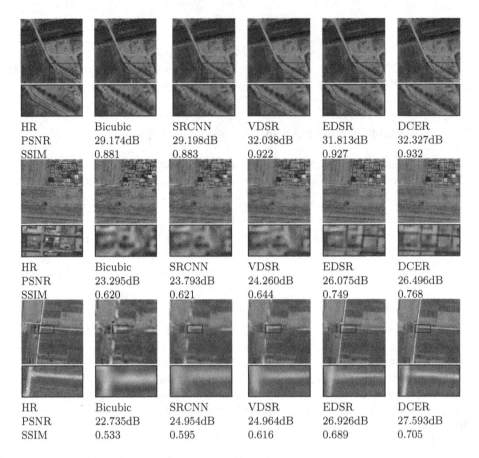

Fig. 3. Qualitative comparison of our models with other works on scale different super-resolution, 2x, 4x, 8x upsampling from top to bottom

5 Conclusion

In this paper, we propose a Dual-Convolutional Enhanced Residual network for single super-resolution of Remote Sensing Images. DCER combines different scale features in the basic blocks, and uses local residual learning between each module to improve the model degradation caused by the depth of the network. During training and testing models on remote sensing image datasets, compared with other methods, DCER has achieved very good results. In particular, when compared to the EDSR [14] and DCER training process in each epoch, DCER has a stronger ability to extract features and can achieve overall fast convergence.

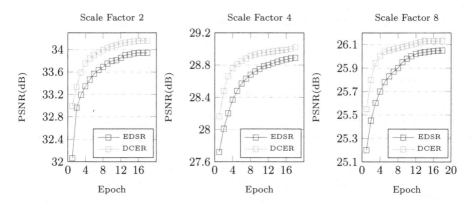

Fig. 4. Qualitative comparison of the results of EDSR and DCER (our models) on each training epoch

References

1. Alam, M.S., Bognar, J.G., Hardie, R.C., Yasuda, B.J.: Infrared image registration and high-resolution reconstruction using multiple translationally shifted aliased video frames. IEEE Trans. Instrum. Meas. **49**(5), 915–923 (2000)
2. Dong, C., Loy, C.C., He, K., Tang, X.: Learning a deep convolutional network for image super-resolution. In: Fleet, D., Pajdla, T., Schiele, B., Tuytelaars, T. (eds.) ECCV 2014. LNCS, vol. 8692, pp. 184–199. Springer, Cham (2014). https://doi.org/10.1007/978-3-319-10593-2_13
3. Dong, C., Chen, C.L., Tang, X.: Accelerating the super-resolution convolutional neural network. In: IEEE Conference on Computer Vision and Pattern Recognition, pp. 1646–1654 (2016)
4. He, K., Zhang, X., Ren, S., Sun, J.: Deep residual learning for image recognition. In: 2016 IEEE Conference on Computer Vision and Pattern Recognition (CVPR), pp. 770–778, June 2016
5. Hou, H., Andrews, H.: Cubic splines for image interpolation and digital filtering. IEEE Trans. Acoust. Speech Signal Process. **26**(6), 508–517 (1978)
6. Irani, M., Peleg, S.: Improving resolution by image registration. CVGIP Graph. Model. Image Process. **53**(3), 231–239 (1991). http://www.sciencedirect.com/science/article/pii/104996529190045L
7. Kim, J., Lee, J.K., Lee, K.M.: Accurate image super-resolution using very deep convolutional networks. In: 2016 IEEE Conference on Computer Vision and Pattern Recognition (CVPR), pp. 1646–1654, June 2016
8. Kim, J., Lee, J.K., Lee, K.M.: Deeply-recursive convolutional network for image super-resolution. In: 2016 IEEE Conference on Computer Vision and Pattern Recognition (CVPR), pp. 1637–1645 (2016)
9. Kim, S.P., Bose, N.K., Valenzuela, H.M.: Recursive reconstruction of high resolution image from noisy undersampled multiframes. IEEE Trans. Acoust. Speech Signal Process. **38**(6), 1013–1027 (1990)
10. Kolte, R., Arora, A.: Image super-resolution. Lap Lambert Acad. Publ. **3**(10), 7195–7199 (2013)

11. Lai, W.S., Huang, J.B., Ahuja, N., Yang, M.H.: Deep Laplacian pyramid networks for fast and accurate super-resolution. In: IEEE Conference on Computer Vision and Pattern Recognition, pp. 5835–5843 (2017)
12. Ledig, C., et al.: Photo-realistic single image super-resolution using a generative adversarial network. In: 2017 IEEE Conference on Computer Vision and Pattern Recognition (CVPR), pp. 105–114 (2017)
13. Li, L., Chen, Y., Xu, T., Liu, R., Shi, K., Huang, C.: Super-resolution mapping of wetland inundation from remote sensing imagery based on integration of back-propagation neural network and genetic algorithm. Remote Sens. Environ. **164**, 142–154 (2015)
14. Lim, B., Son, S., Kim, H., Nah, S., Lee, K.M.: Enhanced deep residual networks for single image super-resolution. In: IEEE Conference on Computer Vision and Pattern Recognition Workshops, pp. 1132–1140 (2017)
15. Patti, A.J., Sezan, M.I., Murat, T.A.: Superresolution video reconstruction with arbitrary sampling lattices and nonzero aperture time. IEEE Trans. Image Process. **6**(8), 1064–76 (1997)
16. Schultz, R.R., Stevenson, R.L.: Extraction of high-resolution frames from video sequences. IEEE Trans. Image Process. **5**(6), 996–1011 (1996)
17. Shu Zhang, S.Y.Z.: Example-based super-resolution. IEEE Computer Graphics and Applications (2013)
18. Szegedy, C., et al.: Going deeper with convolutions. In: IEEE Conference on Computer Vision and Pattern Recognition, pp. 1–9 (2015)
19. Yang, J., Wright, J., Huang, T.S., Ma, Y.: Image super-resolution via sparse representation. IEEE Trans. Image Process. **19**(11), 2861–2873 (2010)
20. Yang, J., Wright, J., Huang, T., Ma, Y.: Image super-resolution as sparse representation of raw image patches. In: IEEE Conference on Computer Vision and Pattern Recognition, CVPR 2008, pp. 1–8 (2008)
21. Zhang, H., Yang, Z., Zhang, L., Shen, H.: Super-resolution reconstruction for multi-angle remote sensing images considering resolution differences. Remote Sens. **6**(1), 637–657 (2014)
22. Zhang, L., Zhang, L., Mou, X., Zhang, D.: A comprehensive evaluation of full reference image quality assessment algorithms. In: 2012 19th IEEE International Conference on Image Processing, pp. 1477–1480 (2012)

Thyroid Nodule Segmentation in Ultrasound Images Based on Cascaded Convolutional Neural Network

Xiang Ying[1], Zhihui Yu[1], Ruiguo Yu[2(✉)], Xuewei Li[2], Mei Yu[2], Mankun Zhao[2], and Kai Liu[2]

[1] School of Software, Tianjin University, Tianjin, China
{xiang.ying,yzh2012}@tju.edu.cn
[2] School of Computer Science and Technology, Tianjin University, Tianjin, China
{rgyu,lixuewei,yumei,zmk,kedixa}@tju.edu.cn

Abstract. Based on U-shaped Fully Convolutional Neural Network (UNET), Convolutional Neural Network (CNN) classifier and Deep Fully Convolutional Neural Network (FCN), this paper proposes a thyroid nodule segmentation model in form of cascaded convolutional neural network. In this paper, we study the segmentation of thyroid nodules from two aspects, segmentation process and model structure. On the one hand, the research of the segmentation process includes the gradual reduction of the segmentation region and the selection of different model structures. On the other hand, the research of model structures includes the design of network structure, the adjustment of model parameters and so on. And the experiment shows that our thyroid nodule segmentation in ultrasound images has a good performance, which is superior to the current algorithms and can be used as a reference for the diagnosis of the doctor.

Keywords: Thyroid ultrasound image
Image semantic segmentation · Fully convolutional neural network

1 Introduction

Thyroid nodules refer to a common lesion in the endocrine system [9], and among them, thyroid malignant tumor is treated with the greatest attention. Thyroid cancer has no obvious clinical symptoms, so doctors often need to analyze the patient's case comprehensively if there being a goiter in clinic. Thyroid ultrasound examination is the best choice among imaging methods for judging benign and malignant thyroid nodules, which is easy to operate and repeat with no wounds, no ionizing radiations, high speed and low price.

For the reason that medical images are endowed with abundant information, radiologists make a diagnosis for the thyroid nodule by observing a series of features of thyroid nodules such as aspect ratio and margin shape. However, due to the diversity of cognitive capability, subjective experience and fatigue

© Springer Nature Switzerland AG 2018
L. Cheng et al. (Eds.): ICONIP 2018, LNCS 11306, pp. 373–384, 2018.
https://doi.org/10.1007/978-3-030-04224-0_32

degree, different doctors may come to different classifications for only one thyroid ultrasound image. Besides, low contrast and speckle noises in ultrasound images also affect the doctor's diagnosis. Therefore, computer-aided identification is critical to medical diagnosis, whose main mode is image segmentation. And Semantic segmentation, which has always been an important research field in computer vision and image processing, has made remarkable progress with the development of deep learning.

Before deep learning was widely used for processing images, Texton Forest [5] and Random Forest based classifiers [2] were common methods for semantic segmentation. Since then, Convolutional Neural Network (CNN) has been widely used in image semantic segmentation. The traditional CNN method is patch classification, which will produce a large number of repeated calculations, thus leading to a low computational efficiency. Moreover, each pixel can only be classified according to the local features contained by the patch, which results in the lack of reliability and accuracy for classification. However, after the concept of Fully Convolutional Networks (FCN) [7] was put forward, a series of semantic segmentation algorithms have emerged one after another which facilitate image semantic segmentation. Under the guidance of FCN, this paper, based on the understanding and analysis of thyroid ultrasound image, puts forward a segmentation model of thyroid nodules, namely, layer-by-layer segmentation. This model can respectively segment the ROI (Region of interest) of the ultrasound image, the targeted nodal area and the edge of the nodule, thus obtaining the ultrasound image with segmented thyroid nodule margins. The input of the algorithm is thyroid ultrasound image containing the nodule when the final output is the accurate segmentation result of the thyroid nodule in ultrasound images.

In this paper we aim to work out the algorithm of layer-by-layer segmentation, the selection and adjustment of neural network in ROI extraction algorithm, as well as the selection and adjustment of nodule segmentation algorithm. The working process is shown in Fig. 1. This algorithm can automatically segment nodules in thyroid ultrasound images.

2   Related Work

Medical image segmentation is a hot issue in image processing, and its accuracy directly influences the validity of following processing. Traditional medical image methods include Level Set-based method, which was applied by Chunming Li et al. to cardiac MRI image segmentation [6] and glandular staining images [12]. However, due to the low contrast, speckle echoes, blurred margins of thyroid nodules and shadows of calcification points in the thyroid ultrasound images, Level Set-based method is insufficient in accuracy for segmenting thyroid nodules. But in recent years, owing to the rapid development of deep learning, CNN in the field of image segmentation has achieved a far higher accuracy and efficiency than those of traditional methods [3]. And Jinlian Ma et al. were the first to use CNN for the nodule segmentation of thyroid ultrasound images [8], with a satisfactory accuracy.

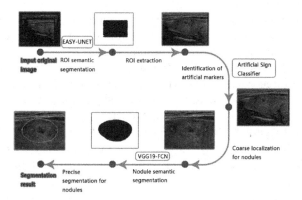

Fig. 1. System architecture: First, we input the original ultrasonic image, then use the EASY-UNET to extract ROI, and then locate the nodules based on the artificial markers, and finally use VGG19-FCN to accurately locate the nodules and output the results

After the convolutional layer, the traditional CNN uses the feature vector of fixed length obtained from the fully connected layer for classification. However, the Fully Convolutional Neural Network (FCN) [7] by Jonathan Long replaced the fully connected layer with the convolutional layer. And then deconvolutional layer was used to up-sample the feature map of the last convolutional layer to make it restored to the size of the original image, thus generating a prediction for each pixel and preserving the spatial information in the original image. Finally, FCN made the classification from pixel to pixel on the feature map that had been up-sampled. Huang Lin et al. applied FCN to Ct image segmentation of osteosarcoma [4]. And the stacked fully convolutional networks put forward by Leibi et al. utilized medical image segmentation [1] and achieved good performance in chest X-ray, Echocardiographic images and histological images. What's more, Ronneberger's U-net model [10] based on FCN won the champion in ISBI cell tracking challenge 2015. It is obvious that FCN has a significant potential in the field of medical image segmentation.

In recent years, the medical image semantic segmentation technology is to directly segment the original image, which is expected that the neural network will learn the differences between nodules and other parts. However, there are various kinds of other information in ultrasound images, which easily cause interference to the segmentation of nodules. For that reason, this paper tries to segment the ultrasound image layer by layer in three steps to achieve a more accurate segmentation.

3   Research Methods

3.1   ROI Extraction from Ultrasound Images

The ultrasound image is composed of the region of interest (ROI) and background regions. The ROI contains important diagnostic information, while the background region contains a large number of highlighted letters and symbols that may interfere with the detection of human markers. In this case, this paper firstly extract the ROI from input ultrasound images.

ROI is a rectangular region located near the center of the whole image, accounting for a large area. Despite that the ROI of ultrasonic images acquired from different kinds of diasonographs have some slight differences in aspect ratio, square measure and hues, etc., the discrepancies between background regions and themselves are similar, thus facilitating the study of neural network. In this paper, we propose a simple FCN model based on U-net for ROI semantic segmentation of thyroid ultrasound images. And the network structure is illustrated in Table 1.

Table 1. Simple full convolution neural network

Layer name	Output shape	Activation
conv1_1	$224 \times 224 \times 64$	ReLU
conv1_2	$224 \times 224 \times 64$	ReLU
pool1	$112 \times 112 \times 128$	
conv2_1	$112 \times 112 \times 128$	ReLU
conv2_2	$112 \times 112 \times 128$	ReLU
pool2	$56 \times 56 \times 256$	
conv3_1	$56 \times 56 \times 256$	ReLU
conv3_2	$56 \times 56 \times 256$	ReLU
pool3	$28 \times 28 \times 256$	
conv4_1	$28 \times 28 \times 512$	ReLU
conv4_2	$28 \times 28 \times 512$	ReLU
pool4	$14 \times 14 \times 512$	
conv5_1	$14 \times 14 \times 1024$	ReLU
conv5_2	$14 \times 14 \times 1024$	ReLU
pool5	$7 \times 7 \times 1024$	
deconv1	$14 \times 14 \times 512$	
deconv2	$28 \times 28 \times 256$	
deconv3	$56 \times 56 \times 128$	
deconv4	$112 \times 112 \times 128$	
deconv5	$224 \times 224 \times 2$	

In this paper, five times convolution and pooling are used to extract the deep features from the input image. The essence of convolution is the dot product of the filter (also called convolution kernel) and the local region of the input data, as shown in Formula 1.

$$y[m,n] = x[m,n] \times h[m,n]$$
$$= \sum_j \sum_i x[i,j] \times h[m-i, n-j] \qquad (1)$$

The feature map of the original image can be obtained by sliding the filter on the original image. The depth of the feature map corresponds to the number of convolutional kernels, representing different features of the original image. Besides, there is a positive correlation between the size of the convolutional kernel and the number of the convolution's receptive fields and parameters. Large convolutional kernels could extract more features, but it also contains more parameters to reduce the training speed and computational efficiency of the model. Such being the case, two continuous 3×3 convolutional kernels are used in our experiment to replace 5×5 convolutional kernels, which can reduce the complexity of convolutional kernels and improve the computational efficiency without changing the size of receptive field.

The essence of pooling is sampling. The input feature image is compressed under an algorithm. And using maximum value is one of the common pooling algorithms, as shown in Formula 2.

$$y = \max(a, b, c, d) \qquad (2)$$

Pooling can reduce the number of parameters while increasing the receptive field of the subsequent convolutional layers, thus reducing the complexity of the model. Furthermore, the output of the pooling layer can stay still when the pixels of the input image have some slight displacements in the neighborhood. Taking this into consideration, pooling has the advantage of anti-disturbance effect, which can improve the robustness of the network.

After five successive down-sampling, the resolution of the feature image is lower than that of the original image, so the pixel category can not be restored in the original image. To solve this problem, the resolution of the feature image is gradually restored to the same value with that of the original image after up-sampling five times. And during restoration, each down-sampling feature image is interpolated with the corresponding up-sampling feature image in proper order so as to supplement the lost details in the process of pooling.

The mask images produced by the model often have irregular margins. After analysis and observation, the difference between the irregular boundary and the label mask could be found as 2–3 pixels. Since some large nodules in the ROI almost occupy the whole region, the artificial mark would appear on the edge of the ROI. And if a small number of pixels with irregular boundaries are discarded, cross marks may be missed, resulting in a failure of localization. Such being the case, that segmented ROI contains all nodal areas is ensured by finding the largest margin of the mask picture in this paper.

3.2   The Algorithm of Segmentation for Thyroid Nodules

When ultrasound doctors make ultrasonic examination for patients, they will make artificial marks in the upper, lower, left and right parts of the ultrasonic image, according to the ROI obtained in the previous section. Firstly, we adopt the method of combining image processing with convolutional neural network, and then find the artificial marks in ROI and determine the specific location of thyroid nodules in ultrasound images according to the location of the marks.

The manual sign recognition and boundary adjustment method [13] is used by us to locate the thyroid nodules in ultrasound images, the region of the thyroid nodule cut out in this method is shown in Fig. 2. Most of the areas in the image are part of the thyroid nodule, but some nodules have calcification points which would cause acoustic shadows in the ultrasound image thus leading to the fusion of the margin and the shadow outside the nodule, as is shown in Fig. 2(a). Besides, some nodules are difficult to identify due to the blur of the margin itself, as shown in Fig. 2(b). Therefore, the margin segmentation of thyroid nodules is a challenging job.

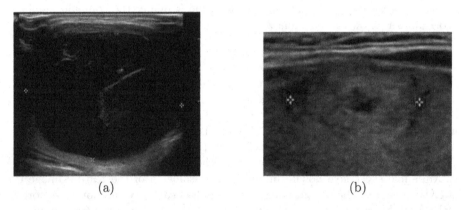

(a)   (b)

Fig. 2. (a) The fusion of acoustic shadows and the shadow around the nodule, (b) The blurred boundary of nodule

A deep VGG19 [11] is selected in this paper, as the down-sampling layer to extract the deep features from the nodule's image. Besides, the three pooling layers in the up-sampling process and the down-sampling process are respectively interpolated to obtain the details of the shallow features, and then a fully convolutional neural network for thyroid ultrasound images is constructed. The network structure is illustrated in Table 2.

The down-sampling process based on VGG19 model is divided into five convolutional steps, in which the first two convolutions replace a 5×5 convolution kernels with two 3×3 convolution kernels, and in the last four steps a 11×11 convolution kernel is replaced by four 3×3 convolution kernels. With the increase of the number of convolution layers, the required parameters are less than that

Table 2. Full convolution neural network based on VGG19

Layer name	Output shape	Activation
conv1_1	$224 \times 224 \times 64$	ReLU
conv1_2	$224 \times 224 \times 64$	ReLU
pool1	$112 \times 112 \times 64$	
conv2_1	$112 \times 112 \times 128$	ReLU
conv2_2	$112 \times 112 \times 128$	ReLU
pool2	$56 \times 56 \times 128$	
conv3_1	$56 \times 56 \times 256$	ReLU
conv3_2	$56 \times 56 \times 256$	ReLU
conv3_3	$56 \times 56 \times 256$	ReLU
conv3_4	$56 \times 56 \times 256$	ReLU
pool3	$28 \times 28 \times 256$	
conv4_1	$28 \times 28 \times 512$	ReLU
conv4_2	$28 \times 28 \times 512$	ReLU
conv4_3	$28 \times 28 \times 512$	ReLU
conv4_4	$28 \times 28 \times 512$	ReLU
pool4	$14 \times 14 \times 512$	
conv5_1	$14 \times 14 \times 1024$	ReLU
conv5_2	$14 \times 14 \times 1024$	ReLU
conv5_3	$14 \times 14 \times 1024$	ReLU
conv5_4	$14 \times 14 \times 1024$	ReLU
pool5	$7 \times 7 \times 1024$	
conv6	$7 \times 7 \times 1024$	ReLU
conv7	$7 \times 7 \times 1024$	ReLU
conv8	$7 \times 7 \times 2$	ReLU
deconv1	$14 \times 14 \times 512$	
deconv2	$28 \times 28 \times 256$	
deconv3	$224 \times 224 \times 2$	

of the single 11×11 convolution kernel and the complexity of the model is reduced, under the circumstance of ensuring the down-sampling receptive field. And the multi-layer convolution implies more activation functions, which make the decision function more discernible and have a better capability to distinguish different categories.

In this network, the last three fully-connected layers of VGG19 are replaced by convolutional layers, the sizes of whose convolution kernels are $7 \times 7 \times 4096$, $1 \times 1 \times 4096$ and $1 \times 1 \times 2$. The last layer has a convolution depth of 2, resulting from that our thyroid nodule segmentation is essentially a problem of binary classification for pixel. The resolution of the input image is reduced by 2, 4, 8, 16 and 32 times respectively after down-sampling. And the feature maps in the fifth, fourth and third layer are up-sampled by three times of deconvolution correspondingly, in which the first two up-samples are interpolated with pool4 and pool3 one by one. So that the detailed information lost during the down-sampling could be complemented. Finally the final mask image is obtained and the margin of the node is drawn in the image of the nodal area by the mask image.

4   The Experimental Results

All thyroid ultrasound images in our experiment were obtained from Tianjin Medical University Cancer Institute and Hospital and the number of the images is 1000. To train and test the model, all these 1000 ultrasound images were labeled with the ROI and nodule margins, under the guidance of ultrasound doctors and radiologists in Tianjin Medical University Cancer Institute and Hospital. The data set of our experiment is finally established with the training set containing 800 images and the testing set containing 200 images.

Intersection-over-Union (IoU) is a commonly used method of computing accuracy in image processing. It is the overlap rate between the generated image and the original image, that is, the ratio of the intersection and union of two images. Ideally, there would be a complete overlap with a ratio of 1. And IoU is defined by the Formula 3.

$$IoU = \frac{area(A) \cap area(B)}{area(A) \cup area(B)} \tag{3}$$

where A is the area of manual segmentation, and B represents the region of automatic segmentation by model. This method is used in calculating the accuracy of both ROI extraction and nodule segmentation.

4.1   ROI Extraction from Ultrasonic Image

The performance of the ROI semantic segmentation model in the training set is illustrated in Fig. 3. The accuracy of the model changes as the increase of training epoches, which is demonstrated in Fig. 4(a). We can see from the figure that in the early stage of training, the accuracy of the model increases sharply, then it tends to be stable at about 100 epoches, and finally stabilizes at about 96%.

In order for a straightforward comparison, the image processing method is used for ROI segmentation of the thyroid ultrasound images. As there is a circle of black margin between the background region and the ROI, we scan from the center of the image and calculate the sum of the pixel values of each line. When the value is less than a certain threshold value, the ROI margin could be confirmed to be here. However, in the nodules with calcification points, for the reason that the acoustic shadow is also a large area of pixels in the color of pure

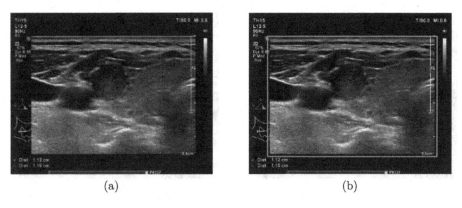

Fig. 3. (a) is the original thyroid ultrasound image, (b) is the ROI extracted from (a)

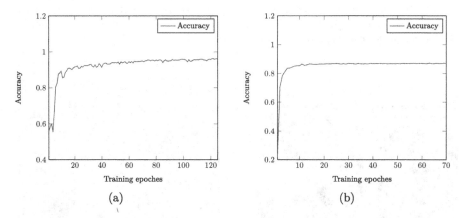

Fig. 4. (a) is the changing trend of the accuracy of the ROI semantic segmentation model, (b) is the trend of the accuracy of thyroid nodule segmentation in nodular area

black, it is easy to be misjudged as the boundary thus resulting in some of the nodules segmented outside the ROI, as shown in Fig. 5(a). But in this paper, The segmentation model performs well on this kind of images, as is shown in Fig. 5(b).

4.2   The Algorithm of Segmentation for Thyroid Nodule

In the thyroid nodule segmentation experiment, we use the manual sign recognition and boundary adjustment method to locate the thyroid nodule. As for Fig. 6(a) where there are four marks, the algorithm draws a rectangular box according to the estimated boundary position given by the four marks, the area in the rectangular box being the position of the nodule. And as for Fig. 6(b) where there are only two marks, the algorithm estimates the upper and lower boundaries of the nodule according to the relative position of the two marks, and then gives out a square box, in which the nodule is coarsely located.

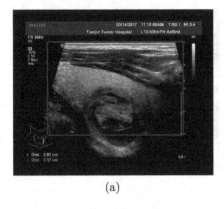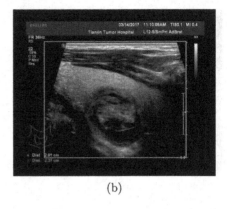

(a)   (b)

Fig. 5. Comparison of image processing method and semantic segmentation method for segmentation of ROI: (a) is a poor result of image processing method for segmentation of ultrasound image ROI, (b) is the result of segmenting the same ultrasound image ROI by semantic segmentation method.

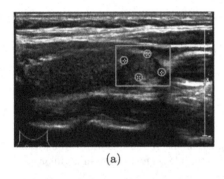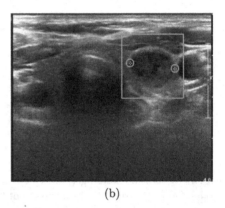

(a)   (b)

Fig. 6. Nodule coarsely located based on identified markers: (a) is the nodular region coarsely located based on four marks, (b) is the nodular region coarsely located based on two marks

After that, we use the parameters pre-trained by VGG19 in the ImageNet data set to speed up and optimize the learning efficiency of the model. The accuracy of the testing set varies with the training epochs as shown in Fig. 4(b). We can see from the figure that in the early stage of training, for the existence of pre-trained parameters, the accuracy of the testing set increases rapidly but fluctuates greatly. After the analysis of our team, that results form that the ImageNet data set used in parameter pretraining is inconsistent with the ultrasound image data set of thyroid nodules in this experiment. But then, after 350 times of training, the accuracy rate gradually levels off and finally becomes stabilized at 87%.

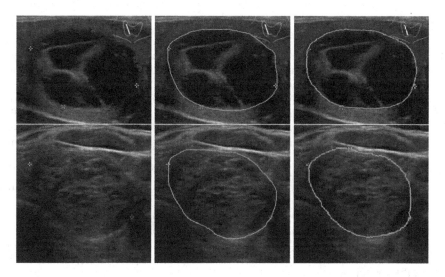

Fig. 7. Nodular edge segmentation: The first column is the original nodular region, the second column is the edges of the nodules drawn under the guidance of a doctor, the third column is the nodular edges obtained by the method in this paper.

In accordance with the method for segmenting thyroid nodules proposed in this paper, the margin of the thyroid nodule located with artificial marks is segmented, with its outcomes shown in Fig. 7. Compared with the margins drawn under the guidance of the doctor which are showed in the second column, the margin images of the thyroid nodule by means of the segmentation algorithm presented in the third column, are not accurate enough in details. However, our method can accurately describe the shape characteristics and the aspect ratio of nodules, both of which plays a crucial role in the identification of benign and malignant nodules and the diagnosis of doctors.

5   Conclusion

The aspect ratio and margin shape of the nodules in thyroid ultrasound images have a significant impact on the diagnosis of the doctor. The method in this paper can not only help doctors to find the specific location of nodules, but also provide the diagnosis references such as the margin, shape and aspect ratio of the nodule. In the experimental results, the segmentation results of nodules with complicated margin details are not accurate enough, which makes us realize that FCN still has some demerits for an accurate segmentation and it is difficult to restore all details by up-sampling and interpolation, so we will focus on the study of edge segmentation fineness in the future. This paper is aimed to provide a promising thread for scholars in this field to explore more deeply and find a better method for the margin segmentation of thyroid nodules.

References

1. Bi, L., Kim, J., Kumar, A., Fulham, M., Feng, D.: Stacked fully convolutional networks with multi-channel learning: application to medical image segmentation. Vis. Comput. **33**(6–8), 1061–1071 (2017)
2. Bosch, A., Zisserman, A., Munoz, X.: Image classification using random forests and ferns. In: IEEE International Conference on Computer Vision, pp. 1–8 (2007)
3. Garcia-Garcia, A., Orts-Escolano, S., Oprea, S., Villena-Martinez, V., Garcia-Rodriguez, J.: A review on deep learning techniques applied to semantic segmentation. arXiv preprint arXiv:1704.06857 (2017)
4. Huang, L., Xia, W., Zhang, B., Qiu, B., Gao, X.: MSFCN-multiple supervised fully convolutional networks for the osteosarcoma segmentation of CT images. Comput. Methods Programs Biomed. **143**, 67–74 (2017)
5. Johnson, M., Shotton, J., Cipolla, R.: Semantic texton forests for image categorization and segmentation. In: Criminisi, A., Shotton, J. (eds.) Decision Forests for Computer Vision and Medical Image Analysis. Advances in Computer Vision and Pattern Recognition. Springer, London (2013). https://doi.org/10.1007/978-1-4471-4929-3_15
6. Liu, Y., Li, C., Guo, S., Song, Y., Zhao, Y.: A novel level set method for segmentation of left and right ventricles from cardiac mr images. In: 2014 36th Annual International Conference of the Engineering in Medicine and Biology Society (EMBC), pp. 4719–4722 (2014)
7. Long, J., Shelhamer, E., Darrell, T.: Fully convolutional networks for semantic segmentation. In: IEEE Conference on Computer Vision and Pattern Recognition, pp. 3431–3440 (2015)
8. Ma, J., Wu, F., Zhao, Q., Kong, D., et al.: Ultrasound image-based thyroid nodule automatic segmentation using convolutional neural networks. Int. J. Comput. Assist. Radiol. Surg. **12**(11), 1895–1910 (2017)
9. Paschou, S., Vryonidou, A., Goulis, D.G.: Thyroid nodules: guide to assessment, treatment and follow-up. Maturitas **92**, 79–85 (2016)
10. Ronneberger, O., Fischer, P., Brox, T.: U-Net: convolutional networks for biomedical image segmentation. In: Navab, N., Hornegger, J., Wells, W.M., Frangi, A.F. (eds.) MICCAI 2015. LNCS, vol. 9351, pp. 234–241. Springer, Cham (2015). https://doi.org/10.1007/978-3-319-24574-4_28
11. Simonyan, K., Zisserman, A.: Very deep convolutional networks for large-scale image recognition. arXiv preprint arXiv:1409.1556 (2014)
12. Wang, C., Bu, H., Bao, J., Li, C.: A level set method for gland segmentation. In: Computer Vision and Pattern Recognition Workshops, pp. 865–873 (2017)
13. Yu, R., et al.: Localization of thyroid nodules in ultrasonic images. In: Chellappan, S., Cheng, W., Li, W. (eds.) WASA 2018. LNCS, vol. 10874, pp. 635–646. Springer, Cham (2018). https://doi.org/10.1007/978-3-319-94268-1_52

An Adjustable Dynamic Self-Adapting OSEM Approach to Low-Dose X-Ray CT Image Reconstruction

Jinghua Sheng[1(✉)], Bin Chen[1], Bocheng Wang[1,2], Yangjie Ma[1], Qingqiang Liu[1], and Weixiang Liu[1]

[1] College of Computer Science, Hangzhou Dianzi University, Hangzhou, Zhejiang, China
j.sheng@ieee.org, 1275695021@qq.com,
wangboc@zjicm.edu.cn, 2468320459@qq.com,
172050037@hdu.edu.cn, 619986206@qq.com
[2] Communication University of Zhejiang, Hangzhou, Zhejiang, China

Abstract. Low-dose CT imaging has been applied in modern medical practice, because it can greatly reduce the radiation for patients. However, with the decrease of radiation dose, noise level is getting much higher. The widely used traditional filtered back-projection (FBP) method is not competent for dealing with the low-dose CT projection data because it lacks of consideration on noise characteristic. Therefore, the statistical iteration algorithm which can consider the noise characteristics is gradually taken into account. But the slow convergence speed and heavy time-consuming limits its application in clinic. Many researchers are also working on the state-of-the-art iterative algorithms, so that they can adapt to low dose CT reconstruction and greatly reduce time. In this paper, we first analyzed the noise characteristics of low-dose CT projection. Then, considering the statistical property of noisy sinogram and the superiority and inferiority of the iterative algorithm, we proposed an adjustable dynamic self-adapting OSEM method (ADSA-OSEM). This method combines variable subset strategy with the least squares merit algorithm applied to maximum likelihood function on OSEM algorithm instead of fixed subsets of traditional OSEM method. A simulation study is performed to test the effectiveness and advantage of the proposed method by comparing with FBP and traditional OSEM method. Through flexible adjustment of the adaptive parameters, results show the new method has greater performance in reconstructed image quality with fewer iterations, the granularity noise and streak-like artifacts could be well suppressed.

Keywords: Low-dose CT · Noise characteristic · Image reconstruction
Iteration algorithm · ADSA-OSEM

1 Introduction

The x-ray computed tomography (CT) has been used in a wide range of fields since its advent, especially in the diagnosis of disease. However, more and more people are worried about its high levels of radiation now [1]. Therefore, to ensure the image

© Springer Nature Switzerland AG 2018
L. Cheng et al. (Eds.): ICONIP 2018, LNCS 11306, pp. 385–396, 2018.
https://doi.org/10.1007/978-3-030-04224-0_33

quality meets the clinical requirements also reduce the radiation dose as far as possibly, with the principle of as low as reasonably achievable (ALARA) has become a leading research direction. Low dose CT technology development of the next generation is very urgent [2]. In clinical practice, the most effective and practical methods of obtaining low dose CT are scanning at low current intensity [3]. However, the low intensity of current causes the "photon starvation" phenomenon in detectors, which brings a lot of noise to projection data and resulting in a significant decline in the quality of the reconstructed image [4]. As we know that the projection data received by the detector bins is in Poisson distribution. Thus, the photon counting caused by photon starvation will introduce Poisson noise [5]. In addition, electronic noise should not be ignored as well. Zeng et al. studied the low dose simulation CT technology based on high dose scanning. They suggested that the low-dose CT noise include Poisson noise and electronic noise [6]. By analyzing the statistical characteristics of low dose CT from repeated scanning data, Lu et al. proposed that the low-dose projection data, after systematic calibration and logarithmic transformation, is subject to a spatial nonstationary Gauss distribution, and there is a nonlinear relationship between the mean and the variance [7].

Generally, CT image reconstruction technologies include filtered back projection and iterative reconstruction. However, in low dose CT image reconstruction, FBP method is highly affected by noise and the reconstruction effect is poor. The iterative method can reconstruct better images because of considering the noise characteristics. For example, Maximum likelihood expectation maximization (MLEM) is a good algorithm, which can address the noise well. Unfortunately, the slow convergence speed and heavy time-consuming limit its application in clinic. Hudson proposed the OSEM algorithm, which is based on MLEM to accelerate the convergence speed by introducing an order-subset technology. The division of subsets can be divided into different projection angles [8]. Our previous work had studied the ordered subset with different subset division. It is considered that different subset partition can greatly affect the quality and time cost of image reconstruction. In previous study, the number of subsets can be varied from more to less in each complete iteration, so that high-frequency parts of the image can be restored more quickly in the early stages. In the later iteration process, fewer subsets can reduce the introduction of noise and make the image smoother [9]. In this paper, for low dose condition, we proposed an adjustable dynamic self-adapting OSEM method, which can flexibly deal to image iteration reconstruction. By taking full account of low dose noise characteristics, the experimental results showed that the new algorithm can reconstruct the images with high quality and reduce the number of iterations.

2 Methods

2.1 Low-Dose Noise Model and Simulation Procedure

As we know that the projection data received by the detector bins is subject to Poisson distribution. Thus, the Poisson distribution is used for noise modeling. At the same time, there will be electronic noise in the process of signal transmission. Therefore, the two aspects should be considered in the noise model of low dose CT. In view of that, Zeng et al. simulate low-dose projection by add noise to the raw sinogram. A statistical model of CT transmission data by energy integrating detection can be described as a statistically independent Poisson distribution and a statistically independent Gaussian distribution [6]. The formula can be expressed as follows:

$$\hat{I} = Poisson(\lambda) + Gaussian\left(m_e, \delta_e^2\right) \tag{1}$$

where $\hat{I}$ is the noisy transmission data and λ is the number of photons passing through the body of patient. The value of λ is determined by the mAs value. m_e and δ_e^2 are the mean and variance of the electronic noise, respectively. Generally, m_e in each scan are usually calibrated to be zero. On the other hand, δ_e^2 can be estimated from the sample variance of a series of dark current measurements [10].

Then, we based on Zeng's low-dose CT simulation procedure [6], depending on the Shepp-Logan phantom of Matlab. The following low dose simulation steps were listed:

1. Obtain the raw sinogram data P_{raw}, shown on the left of Fig. 1, by normal CT scan from the Shepp-Logan phantom.
2. Transform the raw projection data P_{raw} into the transmission one:

$$T_{raw} = \exp(-P_{raw}).$$

3. Calculate the low-dose transmission data $\hat{I}_{ld,sim} = I_0 * T_{raw}$, where I_0 is the incident fluxes of photon.
4. Inject Poisson and Gaussian noise into $\bar{I}_{ld,sim}$, $I_{ld,sim} = Poisson(\bar{I}_{ld,sim}) + Gaussian (m_e, \delta_e^2)$.
5. Achieve the desirable low-dose projection by applying logarithm to transform I_0 and $I_{ld,sim}$.
6. $P_{ld,sim} = \log(I_0/I_{ld,sim})$, which is shown on the right of Fig. 1.

Fig. 1. Original raw projection (left) and low-dose projection (right)

2.2 Adjustable Dynamic Self-Adapting OSEM Method

In general, the maximum likelihood expectation maximization (MLEM) algorithm have been found to produce very good results in applications [11]. Equation (2) describes the classical MLEM iterative algorithm.

$$\hat{x}_j^{n+1} = \frac{\hat{x}_j^n}{\sum_{i=1}^M c_{ij}} \sum_{i=1}^M c_{ij} \frac{y_i}{\sum_{l=1}^N c_{il} \hat{x}_l^n} \qquad (2)$$

where $\hat{x}_j$ is the jth element of the reconstructed image, n is iteration number, i is projection number, and j is the pixel number. However, it requires a large number of iterations, because of the slowly convergence speed. The OSEM algorithm is an MLEM algorithm acting on a subset S of the projections at the time. Compared with MLEM, the OSEM algorithm divides the projection data into a finite set of ordered subsets, the MLEM algorithm is used in each subset to update the image. All the subsets are used in a complete iteration. Actually, the OSEM algorithm updates n times in one iteration process, thus the image convergence speed could be increased to a great extent. It can be written as followed:

$$\hat{x}_j^{n+1} = \frac{\hat{x}_j^n}{\sum_{i \in S_t} c_{ij}} \sum_{i \in S_t} c_{ij} \frac{y_i}{\sum_{l=1}^N c_{il} \hat{x}_l^n} \qquad (3)$$

The subsets S_t in (3) may correspond properly to groups of projections. The problem of subset partition and subset projection arrangement of OSEM has great influence on it. A subset of data can be composed of 1, 2 or more directions of projection data. When a subset of OSEM contains all projections, it is equivalent to MLEM method. The high frequency part of the reconstruction process will be evenly distributed and it's hard to be recovered. When the number of subsets equals to the total number of projection angles, it is equivalent to the algebraic reconstruction method (ART). In the case of low dose CT projection, the reconstructed image is easily covered by noise. Nowadays, under the trend of low dose CT scanning, normal OSEM is no longer able to meet the precise and speed in clinical requirements. In this paper, we proposed an adjustable dynamic self-adapting OSEM (ADSA-OSEM) based on the normal OSEM. This method adjusts the number of projections and iteration step size in each sub-iteration. Meanwhile, the EM algorithm combined maximum likelihood and least squares merit function. Therefore, in the process of reconstruction, the part of high frequency would be restored in the early iteration and the introduction of noise could be minimized in the later stage. The following formulas reasoning can be divided into two parts. Equation (4–6) are the OSEM method for analyzing the subset. The latter part is the analysis of the combination of EM and least squares. Based on (2), another equivalent form of it could be expressed in (4).

$$\hat{x}_j^{n+1} = \hat{x}_j^n + \frac{\hat{x}_j^n}{\sum_{i=1}^M c_{ij}} \sum_{i=1}^M c_{ij} \left[\frac{y_i}{\sum_{l=1}^N c_{il} \hat{x}_l^n} - 1 \right] \qquad (4)$$

Next, through an analogy between the block iterative EM algorithm [13] and OSEM algorithm. We have developed an extended algorithm, the algorithm updates the current image estimate using only a portion of the projection data, the algorithm is as follows:

$$\hat{x}_j^{n+1} = \hat{x}_j^n + \mu_k \frac{\hat{x}_j^n}{\sum_{i=1}^M c_{ij}} \sum_{i \in S_t} c_{ij} \left[\frac{y_i}{\sum_{l=1}^N c_{il}\hat{x}_l^n} - 1 \right] \tag{5}$$

In (5), a relaxation factor is variable for different subset to control convergence behavior, it can be computed as $\mu_k = f(t_w, k)$, k is the index of the whole cycle, t_w denotes the scaling factor. It can be defined as:

$$t_w = min_j \frac{\sum_{i \in S_t} c_{ij}}{\sum_{i=1}^M c_{ij}}$$

The following part introduces the least square method to the iterative algorithm of OSEM. We know that the iterative reconstruction algorithm is based on the Poisson distribution model. From [11], it indicates that estimation of image vectors by using maximum likelihood function, which $L(x)$ is subject to non-negative constraints on X. The formula is as follows:

$$L(X) = \sum_{i=1}^M \left[y_i log \sum_{j=1}^N c_{ij}x_j - \sum_{j=1}^N c_{ij}x_j \right] \tag{6}$$

where X is the N × 1 activity image vector. Next, with a penalized function, Eq. (7) could be represented as follows

$$argmaxG(X) = argmax[L(X) - \gamma F(X)] \tag{7}$$

where $F(X)$ is a penalty function, γ is a positive weight coefficient. $F(X)$ is the quadratic sum of the difference between the estimated value and the actual projection value. So, $F(X)$ is formed as follows:

$$F(X) = \sum_{i=1}^M \left(\sum_{j=1}^N c_{ij}x_j - y_i \right)^2 \tag{8}$$

Equation (8) is characterized by fast convergence, but the accuracy will be reduced to a certain extent. From (7), the former is maximum likelihood, it is mainly devoted to the good fitting of X. The latter contributes to penalize the roughness of reconstructed image by penalty function penalty function $F(X)$. In this case, γ is crucial to the speed of convergence and image smoothness.

The next step is aimed to maximize $G(X)$, in general, the formula can be obtained as follows

$$\frac{\partial G(X)}{\partial x_j} = \sum_{i=1}^M \left[y_i \frac{c_{ij}}{\sum_{l=1}^N c_{il}x_l} - c_{ij} - 2c_{ij}\gamma \left(\sum_{l=1}^N c_{il}x_l - y_i \right) \right] \tag{9}$$

With Kuhn-Tucker conditions solving (9) and combined (5), the new formula named ADSA-OSEM can be written as

$$\hat{x}_j^{n+1} = \hat{x}_j^n + \mu_k \frac{\hat{x}_j^n \sum_{i \in S_t} c_{ij} y_i \left(\frac{1}{\sum_{l=1}^{N} c_{il} x_l} + 2\gamma \right)}{\sum_{i=1}^{M} c_{ij} \left(1 + 2\gamma \sum_{l=1}^{N} c_{il} x_l \right)} \tag{10}$$

On the one hand, accelerating the reconstruction speed through variable subsets. Meanwhile, this algorithm can modify iteration steps according to different subsets instead of fixed subsets of OSEM method. The principle to adjust the number of subsets is based on that the number of subsets in a complete iteration is a non-increasing sequence from more to less. Secondly, μ_k in the algorithm controls the convergence degree according to the ratio of subset information to total information in each iteration. On the other hand, γ control least squares penalty is applied to the reconstruction process. It achieves better image smoothness by controlling the weight of the penalty function, which can suppress noise of low dose projection very well. As described above, the algorithm can respond well to the effect of high noise in low dose cases by adjusting the dynamic parameters and adjusting the constraint coefficient of the penalty function. In the following simulation experiments, we will explore and verify this algorithm from multiple perspectives.

3 Simulation Study

The computer simulations were performed to evaluate the proposed method by using the 2D Sheep-Logan head phantom. In the simulation, the phantom was set as 512×512 pixels. A total of 729 parallel beam projections were simulated at each projection view. The angles of projection are $0°$–$179°$ with step length of $2°$. Thus, all the projections were collected as a sinogram of 729×90 array size. In order to verify our algorithm, we compared it with normal OSEM (NOSEM), FBP methods.

The angles of projections were grouped into 5 subsets, with 18 angles of projection data in each subset. With regard to FBP, we applied the classical Ram-Lak kernel as the filter of FBP. In the ADSA-OSEM method, in order to illustrate the low-dose CT image reconstruction quality was affected by diverse subsets data subdivision, we deliberately set up many tests in the sequence of different number of subsets. During the experiments, we manually set different number of iterations correspondingly. In these simulation experiments, we selected several representative experiments to conduct a comprehensive analysis. For example, we set the sequence of subsets in the way of A, $A \in \{18 \rightarrow 10\}$. In this case, the first number in the braces means total number of subsets in the first iteration and the second number is the total number of subsets in the second iteration.

Considering that we need quantitative indicators to evaluate the quality of reconstructed images and the convergence rate. Here we applied three criterions to measure: signal noise rate (SNR), normalized average absolute distance (NAAD) and normalized mean square distance (NMSD). SNR is used to evaluate the SNR of reconstructed images from low-dose CT projections. Both NAAD and NMSD are used to evaluate the convergence rate. The smaller the value, the better the convergence, these equations are followed:

$$\text{SNR} = 10 \log \left\{ \frac{\sum_{i=1}^{M} \left(\hat{f}_i - m_{\hat{f}} \right)^2}{\sum_{i=1}^{M} \left(\hat{f}_i - f_i \right)^2} \right\} \tag{11}$$

$$\text{NAAD} = \frac{\sum_{i=1}^{M} \left| \hat{f}_i - f_i \right|}{\sum_{i=1}^{M} |f_i|} \tag{12}$$

$$\text{NMSD} = \left[\frac{\sum_{i=1}^{M} \left(\hat{f}_i - f_i \right)^2}{\sum_{i=1}^{M} \left(f_i - m_f \right)^2} \right]^{\frac{1}{2}} \tag{13}$$

where $\hat{f}_i$ and f_i are the value of pixel i of the reconstructed image and origin image. $m_{\hat{f}}$ and m_f are separately the value of mean in reconstructed image and origin image. M is the total sum of pixels.

In the process of simulation experiment, first of all, according to low-dose noise model, the low-dose sinogram was produced by the simulation of parallel beam scan, which is shown in Fig. 1. The parameters setting of the model were illustrated as follows: $m_e = 0$, $\delta_e^2 = 0.05$ and incident fluxes $I_0 = 1e5$, within the range of low-dose scan [12]. In this case, in order to illustrate the influences of variable subsets, the parameter γ_k was set to 0. The reconstructed image in FBP, NOSEM, ADSA-OSEM (sequence A), with 2 numbers of iterations, were presented as followed in Fig. 2:

From the Fig. 2, one can see that reconstructed image by FBP is the worst one, its strip artifact and granularity noise of background are very strong. In addition, we found that with the same number of iterations, the resolution of reconstructed image with ADSA-OSEM is higher than NOSEM. At the same time, the quantitative parameters are following shown in Table 1.

Furthermore, in order to make a clearer comparison of the reconstructed image details, the zoomed images of ROI are shown in Fig. 3. Presented by the Fig. 3, the FBP result is the worst, with strong noise stripe and granularity noise. Comparison between pictures in the center and the right one in Fig. 3, image outline of the center is clearer than that of right one, which showing better resolution with more high frequency components are restored.

At the same time, the comparison of vertical profiles through the different reconstructed images are shown in Fig. 4. Obviously, FBP method has poor effect on low dose CT projection reconstruction. The image convergence effect of NOSEM is not good to ADSA-OSEM image. Furthermore, on the condition of ADSA-OSEM (sequence A), NOSEM with more iterations was applied. After several experiments, we found that when the number of iterations is 5, the two methods achieved the same effect, the result is shown in Fig. 5.

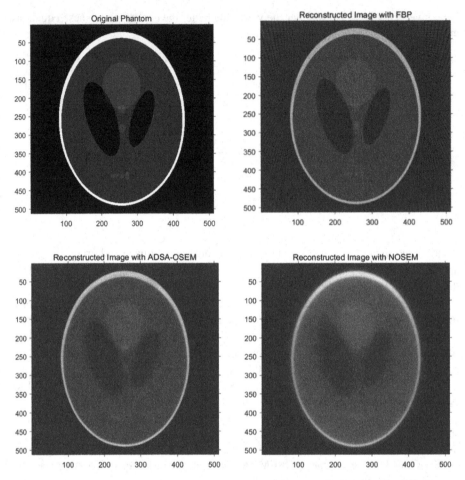

Fig. 2. Original Shepp-Logan phantom and reconstructed images with FBP, NOSEM, ADSA-OSEM

Table 1. Quantitative parameters of Fig. 2

	NMSD	NAAD	SNR
ADSA-OSEM	0.3953	0.3264	9.3071
NOSEM	0.4451	0.3797	8.2774
FBP	0.5232	0.5812	7.281

On the condition of low-dose CT scan, the ADSA-OSEM algorithm is different from NOSEM, which applies different subset numbers to accelerate convergence at each iteration in ADSA-OSEM. From the above test, the number of iterations reduced 3 times the speed increased by nearly 60%. On the other hand, we applied a sequence of B in ADSA-OSEM, B $\in \{18 \rightarrow 10 \rightarrow 10 \rightarrow 9\}$. Correspondingly, the number of

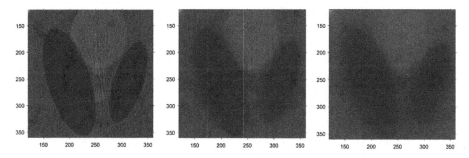

Fig. 3. Zoomed images of ROI from the reconstructed images with FBP (left), ADSA-OSEM (center) and NOSEM (right) in Fig. 2.

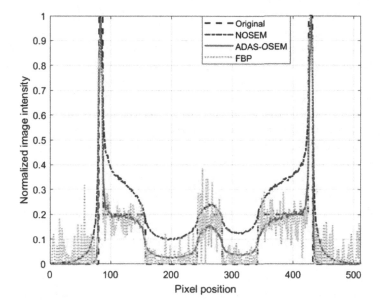

Fig. 4. Vertical profiles through the original image and reconstructed image with FBP, NOSEM (2 iterations) and ADSA-OSEM (sequence A, 2 iterations).

iterations of NOSEM is 4 times. But the result is similar to the sequence of A, when the iterations reached 7 times, the reconstructed image can converge same level to the ADSA-OSEM with sequence B and it can save 3 times of iterations. For more illustrations, a sequence C was chosen, $C \in \{18 \rightarrow 10 \rightarrow 9\}$. What's the difference to the method mentioned above is that the number of subsets of NOSEM algorithm are a fixed number, such as: 10. Through our simulation experiments, we found that ADSA-OSEM only used 3 iterations, and the same quality of NOSEM needed 4 iterations. What's more, we had also carried out several experiments with similar sequences, the results showed that when the number of NOSEM subset is 10, ADSA-OSEM need less 1–2 iterations than the NOSEM method.

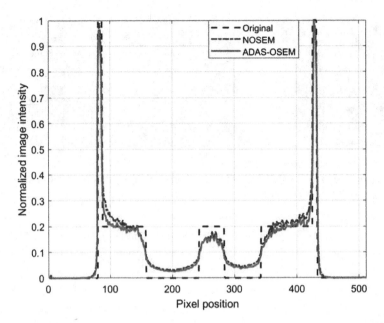

Fig. 5. Vertical profiles through the original image and reconstructed image with NOSEM (5 iterations) and ADSA-OSEM (sequence A, 2 iterations).

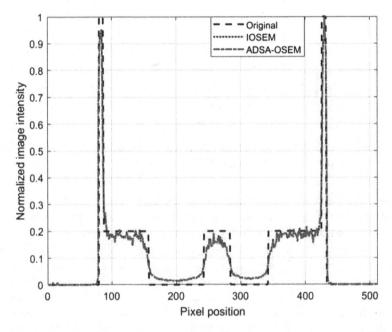

Fig. 6. Vertical profiles through the original image and reconstructed image with IOSEM (5 iterations) and ADSA-OSEM (4 iterations).

Next, we compared our algorithm with an ordinary variable subset OSEM method (IOSEM). In the following example, the sequence of ADSA-OSEM was set as { 18 → 10 → 10 → 9} and the sequence of IOSEM was set as { 18 → 10 → 10 → 9 → 5}, μ_0 = 14, γ_k was a group of dynamic variable constants with k, where k is the index of iterations. The figure of vertical profiles is showed in Fig. 6. One can intuitively see that the effect of image convergence is basically consistent when the number of iterations is respectively 4 and 5.

Similarly, related quantitative indicators is showed in Table 2.

From Table 2, it illustrates that the image quality is almost the same or even slightly better under the condition with one time difference in the number of iterations. On the condition of low-dose CT scanning, reducing the number of iterations means saving more time and improving the efficiency of clinical examination, when the good reconstructed image could be acquired. Thus, the ADSA-OSEM algorithm shows better performance and it has a wider foreground of applications.

Table 2. Quantitative parameters of the test in Fig. 6.

	NMSD	NAAD	SNR
ADSA-OSEM	0.3780	0.3132	9.6962
NOSEM	0.3812	0.3163	9.6227

4 Discussions and Conclusion

The CT image reconstruction algorithm has been developed for a long time. However, with the increasing demand for low-dose CT, some algorithms have gradually lost their dominant position. For example, the FBP algorithm is too sensitive to noise. The traditional iterative reconstruction algorithm is good at dealing with noise, such as the MLEM algorithm, but it costs a lot of time. The OSEM algorithm divides the projections into several subsets in one iteration and each subset contains the same numbers of projection, which accelerates the convergence rate and speed of reconstructed images to a great extent.

In this paper, we have studied the characteristics of low dose CT and proposed a new algorithm based on the normal OSEM algorithm. The ADSA-OSEM algorithm takes a division of subsets into account in each iteration. For different subsets, adjusting the parameters and adjusting the penalty function coefficients so that can adapt to low dose CT imaging with fast speed. Of course, with the introduction of some extreme low dose CT or its application to practical CT machine, it is necessary to make great adjustments and improvements to this algorithm.

With the enhancement of computing power, many neural network-based deep learning methods have also been applied to low-dose CT imaging and denoising problems [14, 15]. In the future, we will explore the combination of our method and deep learning theory to synthesize the advantages of artificial intelligence so that developing a new low-dose CT imaging technology.

References

1. Brenner, D.J., Hall, E.J., Brenner, D.J., et al.: Computed tomography—an increasing source of radiation exposure. N. Engl. J. Med. **357**(22), 2277–2284 (2007)
2. Prasad, K.N., Cole, W.C., Haase, G.M.: Radiation protection in humans: extending the concept of as low as reasonably achievable (ALARA) from dose to biological damage. Br. J. Radiol. **77**(914), 97–99 (2004)
3. Kalender, W.A., Wolf, H., Suess, C., et al.: Dose reduction in CT by anatomically adapted tube current modulation: principles and first results. In: Krestin, G.P., Glazer, G.M. (eds.) Advances in CT IV. Springer, Heidelberg (1998). https://doi.org/10.1007/978-3-642-72195-3_4
4. Yazdi, M., Beaulieu, L.: Artifacts in spiral X-ray CT scanners: problems and solutions. Int. J. Biol. Med. Sci. **4**(3), 135–139 (2008)
5. Deledalle, C.A., Tupin, F., Denis, L.: Poisson NL means: unsupervised non local means for Poisson noise. In: IEEE International Conference on Image Processing, pp. 801–804. IEEE (2010)
6. Dong, Z., Huang, J., Bian, Z., et al.: A simple low-dose X-ray CT simulation from high-dose scan. IEEE Trans. Nucl. Sci. **62**(5), 2226–2233 (2015)
7. Lu, H., Hsiao, I.T., Li, X., et al.: Noise properties of low-dose CT projections and noise treatment by scale transformations. In: Nuclear Science Symposium Conference Record, vol. 3, pp. 1662–1666. IEEE (2001)
8. Hudson, H.M., Larkin, R.S.: Accelerated image reconstruction using ordered subsets of projection data. IEEE Trans. Med. Imaging **13**(4), 601–609 (2002)
9. Sheng J, Liu D.: An improved maximum likelihood approach to image reconstruction using ordered subsets and data subdivisions. In: Design & Test Symposium. IEEE Xplore, pp. 43–46 (2004)
10. Hsieh, J.: Computed tomography: principles, design, artifacts, and recent advances. SPIE, Bellingham (2009)
11. Shepp, L.A., Vardi, Y.: Maximum likelihood reconstruction for emission tomography. IEEE Trans. Med. Imaging **1**(2), 113–122 (1982)
12. Hu, Z., Zhang, Y., Liu, J., et al.: A feature refinement approach for statistical interior CT reconstruction. Phys. Med. Biol. **61**(14), 5311–5334 (2016)
13. Byrne, C.L.: Block-iterative methods for image reconstruction from projections. IEEE Trans. Image Process. **5**(5), 792–794 (1996)
14. Chen, H., Zhang, Y., Zhang, W., et al.: Low-dose CT via convolutional neural network. Biomed. Opt. Express **8**(2), 679–694 (2017)
15. Wolterink, J.M., Leiner, T., Viergever, M.A., et al.: Generative adversarial networks for noise reduction in low-dose CT. IEEE Trans. Med. Imaging **36**(12), 2536–2545 (2017)

DeepITQA: Deep Based Image Text Quality Assessment

Hongyu Li$^{(\boxtimes)}$, Fan Zhu, and Junhua Qiu

AI Lab, ZhongAn Information Technology Service Co., Ltd., Shanghai, China
{lihongyu,zhufan,qiujunhua}@zhongan.io

Abstract. To predict the OCR accuracy of document images, text related image quality assessment is necessary and of great value, especially in online business processes. Such quality assessment is more interested in text and aims to compute the quality score of an image through predicting the degree of degradation at textual regions. In this paper, we propose a deep based framework to achieve image text quality assessment, which is composed of three stages: text detection, text quality prediction, and weighted pooling. Text detection is used to find potential text lines and the quality is solely estimated on detected text lines. To predict text line quality, we train a deep neural network model with our synthetic samples. The overall text quality of an image can be computed through pooling the quality of all detected text lines by way of weighted averaging. The proposed method has been tested on two benchmarks and our collected pictures. Experimental results show that the proposed method is feasible and promising in image text quality assessment.

Keywords: Image quality assessment · Text detection · Text quality
Deep neural network

1 Introduction

With the pervasive use of smart devices in our daily life, mobile captured document images are often required to be submitted in business processes of Internet companies. For the purpose of intelligent analysis of such document images, image text quality assessment (ITQA, for short) are generally needed before text detection and text recognition since the performance of document recognition and analysis is highly dependent on text quality of acquired images.

The text recognition accuracy of mobile captured document images is often decreased with the low text quality due to artifacts introduced during image acquisition [15], which probably hinders the following business process severely. Different from traditional image quality assessment [3,7], image text quality assessement is closely related to text, where the major concern is word/text. Inspired by this observation, we expect to only compute the quality score on interesting textual areas, which is more beneficial and practical than computing on the whole area in ITQA.

© Springer Nature Switzerland AG 2018
L. Cheng et al. (Eds.): ICONIP 2018, LNCS 11306, pp. 397–407, 2018.
https://doi.org/10.1007/978-3-030-04224-0_34

Current studies mainly focus on document images, which is also called DIQA (short for document image quality assessment). Many no-reference algorithms have been developed to estimate document image quality. According to the difference of feature extraction, these methods can be categorized as two groups: metric-based assessment and learning-based assessment.

The metric-based methods are usually based on hand-crafted features that have shown to correlate with the OCR accuracy. Around 30 degradation-specific quality metrics have been proposed to measure noise and character shape preservation [10]. Although much progress has been made in metric-based assessment, there still exists a clear problem. Features used in existing methods are generally extracted from square image patches, many of which do not have visual meaning involving character/text. Therefore, the resultant features, probably containing much noise, are suboptimal for image text quality assessment.

The learning-based methods take advantage of learning techniques, such as [8,12], to extract discriminant features for different types of document degradations. In [8], the authors proposed a deep learning approach for document image quality assessment, which crops an image into patches and then uses the CNN to estimate quality scores for selected patches. However, the strategy of selecting text patches is based upon the simple technique, Otsu's binarization, which often can not work well for images with complicated background.

To ensure that the text legibility of an image is sufficient for character or text analysis, we propose a no-reference image text quality assessment framework, based on deep learning techniques. The proposed method manages to take advantage of valid text lines as significant character patches, and modifies a deep neural network to describe the text quality model involving text lines. The image quality with respect to text is obtained through pooling quality scores of valid text lines by way of weighted averaging.

To train the text quality model, it is necessary to collect for training enough text line samples containing text quality labels. However, the publicly available datasets have only ground truth quality for documents, not for text lines. As a result, we design a way of generating training data with quality labels in this work. In the generated dataset, blur is considered as the main factor of affecting image text quality, consistent with the real-world captured image samples where blur seems the most common issue [1,2]. To do this, we model the Gaussian blur to smooth textual regions and produce images with blurry text.

The main advantages of this proposed method is that it can not only deal with pure document images but also scene images or document images with complicated background. The proposed method has been tested on two benchmarks and our private dataset. Experimental results demonstrate that the proposed method is feasible and promising in image text quality assessment.

2   Methodology

In this paper, we propose a deep based framework, *DeepITQA*, for image text quality assessment. This framework can be divided into three stages: text detection, text quality prediction, and weighted pooling. Different stages of the DeepITQA framework are described in detail in the consecutive subsections.

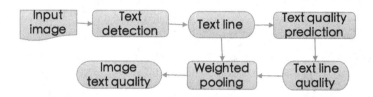

Fig. 1. The proposed framework for image text quality assessment.

2.1   Framework

A high-level overview of our framework is illustrated in Fig. 1. In this framework, an image is first fed into a text detector to find significant and valid text lines and then text line quality is assessed with a text quality prediction network. The overall image text quality is eventually computed with a weighted pooling strategy on the basis of text line areas.

During the text quality prediction stage, we modify a residual network (ResNet) [6] to predict text quality of text lines. The network model will directly output a quality prediction score for input text lines. To train this network model, we need to synthesize some training samples with ground truth labels involving text quality. According to the point of view in [15], blur seems the most common issue in mobile captured images, which suggests that detecting the blur degradation is more attractive and useful in practical applications. As a result, our synthetic data is mainly produced under different levels of blur degradations.

2.2   Text Detection

Before applying a text quality measure, it is necessary to extract meaningful features to describe the attributes of the image. Motivated by the fact that the most significant characteristic for text quality assessment is character/text in images, we are supposed to replace an image with text lines that contain characters during text quality assessment. Text detection aims to find meaningful text lines with a text detector.

Comprehensive reviews about text detection can be found in survey papers [16,19]. Previous text detection approaches [4,14] have already obtained promising performances on various benchmarks and deep neural network based algorithms [14,17,18] are becoming the mainstream in this field. In [14], the connectionist text proposal network (CTPN) is proposed to accurately localize text lines in natural image. Exploring rich context information of image, CTPN is powerful in detecting extremely ambiguous text and works reliably on multi-scale and multilanguage text without further post-processing. Due to its high efficiency and good performance in text detection, CTPN is selected as the text detector in this work, as presented in Fig. 2a. In this work, CTPN is directly utilized for detection without finetuning.

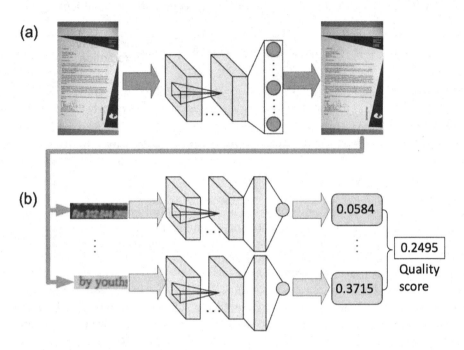

Fig. 2. Illustration of our method.

2.3 Text Quality Prediction

To predict text quality, it is straightforward to cast text quality assessment as a regression problem since the estimation is essentially to predict a scalar. It has been proved in [8] that deep features are effective in document image quality assessment, therefore we employ a deep neural network to extract significant features of text lines for quality prediction, as shown in Fig. 2b.

In the prediction network, the early layers are based on any standard architecture truncated before the classification layer. An auxiliary regression layer, whose output is a neuron, is added behind the early layers for estimation. In our method, the ResNet [6] is adopted as a base, where the extracted deep features are of 512 dimensions, due to its good performance in feature representation, but other networks should also produce good results.

Loss Function. To predict text line quality, the estimation loss adopts Euclidean loss for quality regression, which is stacked with the customized ResNet. Specifically, the estimation loss $\mathcal{L}_q$ is defined as

$$\mathcal{L}_q = \|\mathbf{Q} - \overline{\mathbf{Q}}\|_2^2, \tag{1}$$

where $\mathbf{Q}$ and $\overline{\mathbf{Q}}$ are respectively the predicted and ground truth quality.

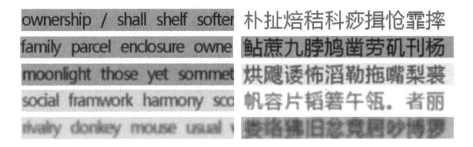

Fig. 3. Examples of synthetic text lines with different quality.

Training Strategy. As shown in Fig. 2, the detection and prediction networks are sequentially concatenated and are thus two separate models in our framework. This allows us to independently train each network with its own set of training parameters.

The prediction network is optimized with synthetic text lines labeled with ground truth quality. The convolutional layers of the prediction network are initialized using the ResNet weights pre-trained on ImageNet. The weights for the regression layer are randomly initialized under a uniform distribution in the range $(-0.1, 0.1)$.

In order to optimize the prediction network, the Adam optimizer is utilized with a learning rate 1e-5. A weight decay of 2e-4 is applied to all layers and the dropout with probability 0.5 is used after global pooling in the prediction network.

Data Generation for Training. To train the prediction network model, it is required that the training samples must contain labels involving quality scores of text lines. However, the publicly available datasets have only ground truth quality for documents, not for text lines. For example, the DIQA dataset [9] provides OCR accuracies of documents as ground truth quality scores. As a result, we need to collect text lines with text quality labels to train the prediction model.

In this work, we design a new way of generating training data with quality labels. Since blur seems the most common issue in mobile captured images, we model the Gaussian blur to smooth textual regions and produce images with blurry text. The whole process is briefly introduced as follows:

Step1: create characters to form clear text lines with the standard size of 40×400, and label text lines with quality score 1;

Step2: blur each text line using a Gaussian function with a random kernel size $s \in [3, 30]$;

Step3: label each blurred text line with ground truth quality $q_t = 1 - s/32$.

In each produced text line, there are about 10 Chinese characters or 5 English words, and the backgrounds are of seven different intensities. In total, we synthesize 100,000 text lines for training, and some examples are shown in Fig. 3. Our ground truth q_t has been normalized and is inverse proportional to the kernel size s in the synthetic data.

Table 1. Comparison on the document-wise protocol.

	Median LCC	Median SROCC
Moments [5]	0.8197	0.8207
Proposed	**0.9175**	**0.9429**

2.4   Weighted Pooling

For an image, the overall quality $\hat{q}$ with regard to text is defined as the weighted pooling of the quality of all text lines in this image. It can be computed in the following form,

$$\hat{q} = \sum_j w_j q_t(j), \tag{2}$$

where w_j is a weight on the j-th text line of the image. The weight is linearly proportional to the text line area, $w_j = \frac{R_t(j)}{\sum_k R_t(k)}$, where $R_t(j)$ represents the area of the j-th text line in the image. In this pooling strategy, we expect to put more emphasis on large textual regions that attract more attention after all.

3   Experiments

To evaluate the performance of the proposed method, we conducted experiments of image text quality assessment on two public benchmarks and some complicated images we collected.

3.1 Datasets and Protocols

The **DIQA** dataset [9] contains a total of 175 color images with resolution 1840 × 3264. These images are captured from 25 documents with different levels of blur degradations. In **SmartDoc-QA** [11], there are 30 different documents used to capture 4260 images, where 142 different images are captured per document.

To compute the correlation between the predicted quality scores and ground truth OCR accuracies, we use the Linear Correlation Coefficient (LCC) and the Spearman Rank Order Correlation Coefficient (SROCC) as evaluation metrics.

3.2 Comparative Analysis

In our experiments, the prediction model obtained with our synthetic data is directly applied to all detected textlines from each document image. The overall image text quality is computed in the fashion of the weighted pooling.

We first conducted experiments on the DIQA dataset. The median LCC and SROCC computed independently document-wise are respectively 0.9175 and 0.9429. We compared the proposed method with the Moments based method [5] that computes the correlation coefficients as well in view of the Tesseract OCR accuracy. As shown in Table 1, the proposed method achieves the higher median LCC and SROCC than the Moments based approach that takes advantage of hand-crafted features.

To avoid the bias towards the good results in terms of the document-wise evaluation protocol, we directly computed one LCC (0.8082) and one SROCC (0.8560) on all document images in the DIQA dataset. Table 2 shows that our method performs better in SROCC than the other two metric-based methods: MetricNR [10] and Focus [13]. In addition, the SmartDoc-QA dataset was also evaluated on all images over the average accuracy. The LCC and SROCC are respectively 0.7506 and 0.8045. The higher SROCC demonstrates that the trained network model has a good ability in describing the monoticity of predicted text quality scores. The possible reason causing the low LCC is that the ground truth for training in our synthetic samples is linearly related to the kernel size, but these two benchmarks use the OCR accuracy as the ground truth for text quality, which is overwhelmingly dependent to the OCR engines.

Table 2. Comparison on all document images

	LCC	SROCC
MetricNR [10]	**0.8867**	0.8207
Focus [13]	0.6467	N/A
Proposed	0.8082	**0.8560**

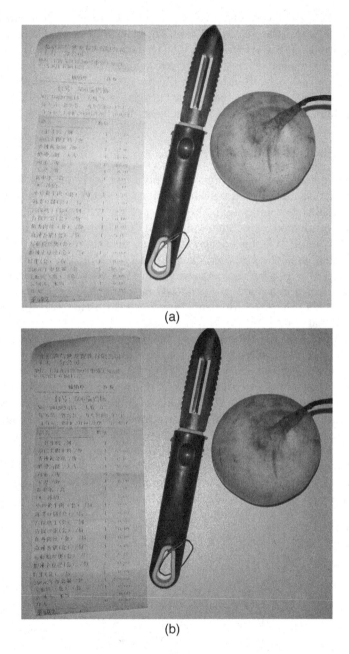

Fig. 4. Two collected images with a document and two objects.

3.3   Other Results

The main advantage of our method is that it can effectively assess the quality of complicated images with regard to text. Figure 4 is a typical case, where the top image presents a blurry document with low text quality and two objects look clear with high dynamic range; on the contrary, the document has good text legibility at the bottom image but those objects have bad contrast and seem not too good in quality. If text areas are the regions of interest, the right image will be considered of higher quality than the left one, which is consistent with our subjective judgement of image quality with respect to text. These two images can be well evaluated with text quality of 0.0893 and 0.3636 respectively, where previous state-of-the-art methods for document image quality assessment can not work in a good way.

Figure 5 presents another three natural images with text we captured, and their predicted text quality scores are 0.9546, 0.7631, and 0.4061, respectively, from left to right. All these results are consistent with our subjective judgement of image text quality, *good*, *normal*, and *poor*, respectively, from left to right.

It is also worth noting that text detector plays an important role in computing final image text quality since the weighting strategy is based on the area of detected text lines. For instance, for the same image, if only a clear and big text line is found with a detector, but both clear and blurry text lines are extracted with the other detector, two overall image text quality scores, computed with the weighted pooling strategy, will be totally different. That is, the more blurry text lines the detector finds, the lower the overall image text quality is. But, in essence, as long as all potential text lines, clear and blurry, are correctly detected, the overall score strategy can work in a regular way.

Fig. 5. Three collected scene pictures with different levels of text quality.

4   Conclusion and Future Work

This paper proposes a deep based framework to estimate image text quality, mainly including three stages: text detection, text quality prediction, and weighted pooling. Our motivation is to compute the image quality only on text lines that can be found with a text detector. To predict text line quality, we train a deep network model with synthetic data. The final image text quality is obtained with a weighted pooling strategy on the basis of detected text lines. Refining the prediction network model to improve the LCC value on image text quality assessment is our future work.

References

1. Bong, D.B.L., Khoo, B.E.: Blind image blur assessment by using valid reblur range and histogram shape difference. Signal Process. Image Commun. **29**(6), 699–710 (2014)
2. Bong, D.B.L., Khoo, B.E.: Objective blur assessment based on contraction errors of local contrast maps. Multimedia Tools Appl. **74**(17), 7355–7378 (2015). https://doi.org/10.1007/s11042-014-1983-5
3. Bosse, S., Maniry, D., Wiegand, T., Samek, W.: A deep neural network for image quality assessment. In: 2016 IEEE International Conference on Image Processing (ICIP), September 2016, pp. 3773–3777 (2016)
4. Buta, M., Neumann, L., Matas, J.: Fastext: Efficient unconstrained scene text detector. In: IEEE International Conference on Computer Vision, pp. 1206–1214 (2015)
5. De, K., Masilamani, V.: Discrete orthogonal moments based framework for assessing blurriness of camera captured document images. In: Vijayakumar, V., Neelanarayanan, V. (eds.) Proceedings of the 3rd International Symposium on Big Data and Cloud Computing Challenges (ISBCC – 16'). SIST, vol. 49, pp. 227–236. Springer, Cham (2016). https://doi.org/10.1007/978-3-319-30348-2_18
6. He, K., Zhang, X., Ren, S., Sun, J.: Deep residual learning for image recognition. In: 2016 IEEE Conference on Computer Vision and Pattern Recognition (CVPR), pp. 770–778, June 2016
7. Hou, W., Gao, X., Tao, D., Li, X.: Blind image quality assessment via deep learning. IEEE Trans. Neural Netw. Learn. Syst. **26**(6), 1275–1286 (2015)
8. Kang, L., Ye, P., Li, Y., Doermann, D.: A deep learning approach to document image quality assessment. In: IEEE International Conference on Image Processing, pp. 2570–2574 (2014)
9. Kumar, J., Chen, F., Doermann, D.: Sharpness estimation for document and scene images. In: International Conference on Pattern Recognition, pp. 3292–3295 (2013)
10. Nayef, N.: Metric-based no-reference quality assessment of heterogeneous document images. In: SPIE Electronic Imaging, pp. 94020L–94020L-12 (2015)
11. Nayef, N., Luqman, M.M., Prum, S., Eskenazi, S., Chazalon, J., Ogier, J.M.: SmartDoc-QA: a dataset for quality assessment of smartphone captured document images - single and multiple distortions. In: International Conference on Document Analysis and Recognition, pp. 1231–1235 (2015)
12. Peng, X., Cao, H., Natarajan, P.: Document image quality assessment using discriminative sparse representation. In: Document Analysis Systems, pp. 227–232 (2016)

13. Rusinol, M., Chazalon, J., Ogier, J.M.: Combining focus measure operators to predict OCR accuracy in mobile-captured document images. In: IAPR International Workshop on Document Analysis Systems, pp. 181–185 (2014)
14. Tian, Z., Huang, W., He, T., He, P., Qiao, Y.: Detecting text in natural image with connectionist text proposal network. In: European Conference on Computer Vision, pp. 56–72 (2016)
15. Ye, P., Doermann, D.: Document image quality assessment: a brief survey. In: International Conference on Document Analysis and Recognition, pp. 723–727 (2013)
16. Ye, Q., Doermann, D.: Text detection and recognition in imagery: a survey. IEEE Trans. Pattern Anal. Mach. Intell. **37**(7), 1480–1500 (2015)
17. Zhang, Z., Zhang, C., Shen, W., Yao, C., Liu, W., Bai, X.: Multi-oriented text detection with fully convolutional networks, pp. 4159–4167 (2016)
18. Zhou, X., et al.: EAST: an efficient and accurate scene text detector. In: 2017 IEEE Conference on Computer Vision and Pattern Recognition, CVPR 2017, Honolulu, HI, USA, 21–26 July 2017, pp. 2642–2651 (2017)
19. Zhu, Y., Yao, C., Bai, X.: Scene text detection and recognition: recent advances and future trends. Front. Comput. Sci. **10**(1), 19–36 (2016)

High Efficient Reconstruction
of Single-Shot Magnetic Resonance T_2
Mapping Through Overlapping Echo
Detachment and DenseNet

Chao Wang[1], Yawen Wu[1], Xinghao Ding[1], Yue Huang[1], and Congbo Cai[1,2]([✉])

[1] Fujian Key Laboratory of Sensing and Computing for Smart City,
School of Information Science and Engineering,
Xiamen University, Xiamen 361005, Fujian, China
cbcai@xmu.edu.cn
[2] Fujian Provincial Key Laboratory of Plasma and Magnetic Resonance,
Department of Electronics Science, Xiamen University, Xiamen 361005, Fujian, China

Abstract. Rapid and quantitative magnetic resonance T_2 imaging plays an important role in medical imaging field. However, the existing quantitative T_2 mapping method are usually time-consuming and sensitive to motion artifacts. Recently, a novel single-shot quantitative parameter mapping method based on overlapped-echo detachment technique has been proposed by us, but an efficient reconstruction algorithm is necessary. In this paper, a multi-stage DenseNet was utilized to reconstruct single-shot T_2 mapping efficiently. The contributions of the paper mainly include the following aspects. First, an end-to-end neural network is proposed, which can directly obtain the reconstructed images without any secondary processing. Second, DenseNet was introduced into the reconstruction network to better reuse the features. Third, a weighted Euclidean loss function is proposed, which can be better used for image reconstruction.

Keywords: Magnetic resonance imaging (MRI)
Single-shot T_2 mapping · Reconstruction
Deep learning · DenseNet

1 Introduction

Quantitative information of the relaxation properties of different tissue and organization can be represented by quantitative parameter MR mapping (T_1 mapping, T_2 mapping, T_2* mapping etc.) which has been wildly used in clinical MRI. The irrelevant effects can be eliminated via quantitative parameter MR mapping [1], and make it possible to compare between different research result. Especially, quantitative T_2 mapping has a decisive application value in clinical practice, and has draw more and more attention in the diagnosis of many diseases, such as neuro-degenerative diseases, multiple sclerosis and epilepsy [2], etc.

© Springer Nature Switzerland AG 2018
L. Cheng et al. (Eds.): ICONIP 2018, LNCS 11306, pp. 408–418, 2018.
https://doi.org/10.1007/978-3-030-04224-0_35

However, quantitative T_2 mapping often needs a long acquisition time, and the artifacts caused by consciously or unconsciously movements can affect the quality of imaging. Thus real-time imaging cannot be achieved.

Recently, a novel ultrafast quantitative parameters mapping method based on overlapping echo detachment (OLED) technique has been proposed, and applied in obtaining single-shot T_2 mapping [3] and diffusion mapping [4]. OLED method aims to reconstruct a clean image from the phase-aliased image via efficient reconstruction algorithm, which can reduce the sampling time of T_2 mapping from minutes to milliseconds and has strong immunity to motion. To detach the overlapped echoes as well as reconstruct the clear T_2 mapping, a separation algorithm based on some priors such as sparsity and structural similarity was proposed (called as echo-detachment-based method). However, this algorithm has very limited reconstruction efficiency, which still impedes its application in clinical practice.

Deep learning, especially convolutional neural networks (CNN), has received increasing attention in image processing and many other fields because of its powerful nonlinear mapping capability and ultrafast forward propagation process. Especially in the area of medical image processing, deep CNN have approached or even surpassed human experts in many aspects [5–7]. Therefore, a reconstruction method [8] based on residual neural network (ResNet) [9] has been proposed and can achieve the better reconstruction results compared to traditional optimization-based method. In this method, none of the priori constraint is required, the network is used to study the complicated nonlinear mapping from phase-aliased image to clean image, thus reduces the reconstruction time to milliseconds. However, the loss function in this method is inappropriate, and it leads the network to pay too much attention to the area with larger T_2 values during the training process. Besides, we need a guided filter [10] in the postprocessing stage to remove the residual noise, and the parameters such as the radius of the guide filter may affect the quality of reconstructed results under the different noise levels.

An end-to-end neural network based on DenseNet [11] was utilized in this paper to further improve the reconstruction of OLED without the help of guided filter.

2 Proposed Method

2.1 Pulse Sequence

In this paper, we use the OLED pluse sequence shown in the Fig. 1. Two small flip angle α RF pulses produce two echo signals which have different echo time. G_1 is the first echo-shifting gradients (ESGs), G_2 second, and G_{cr} is defined as crusher gradients along three directions. β is the flip angle of the refocusing pulse.

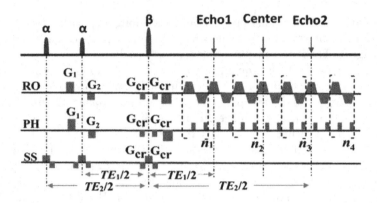

Fig. 1. Single-shot OLED sequence. TE_1 and TE_2 represent the first and the second echo time respectively.

After the refocusing pulse β, we can get the following signal formula.

$$S_{OLED} = S_1(TE_1) + S_2(TE_2) + S_3\left(\frac{TE_2}{2}\right)$$

$$S_1(TE_1) = \frac{1}{2} \int \rho(\boldsymbol{r}) |\sin\alpha \cdot \cos\alpha| \cdot (1 - \cos\beta) e^{-TE_1/T_2(\boldsymbol{r})} d\boldsymbol{r}$$

$$S_2(TE_2) = \frac{1}{4} \int \rho(\boldsymbol{r}) |\sin\alpha| \cdot (1 + \cos\alpha) \cdot (1 - \cos\beta) e^{-TE_2/T_2(\boldsymbol{r})} d\boldsymbol{r} \qquad (1)$$

$$S_3(TE_2/2) = \frac{1}{4} \int \rho(\boldsymbol{r}) |\sin\alpha| \cdot (1 - \cos\alpha) \cdot (1 - \cos\beta) e^{-TE_1/T_2(\boldsymbol{r})} d\boldsymbol{r}$$

where $\rho(\boldsymbol{r})$ means spin density at position $\boldsymbol{r}$. As can be seen from the above formula, there are actually three echo signals, $S_1(TE_1)$, $S_2(TE_2)$ and $S_3(TE_2/2)$, with different modulation phases. We observe that the amplitude of the double-spin echo signal $S_3(TE_2/2)$ is relatively small at the small RF pulse angle, so we use a Gaussian filter to filter out the double-spin echo during the data pre-processing. Thus, we get only two spin echoes in k-space.

2.2 Dataset

Because it is difficult to obtain real training dataset for potential reconstruction of ultrafast MR images, the simulated data pairs were utilized to train the neural network, as described in the previous works [8]. After the network is trained, real data pairs was fed to the network and we can get the forward propagation result. Each simulation sample consist of 300 different shapes (ellipse, lines, squares, etc.) with random size, T_2 values and spin density distribution. The simulated OLED training data was obtained using SPROM (Simulation with PRoduct Operator Matrix) developed by our group [12], and produced through using the same pulse sequence and the same parameters as the real data. Figure 2 shows the simulated data pairs and real brain data pairs. They all have a size of 256×256 pixel after Fourier transform with zero padding.

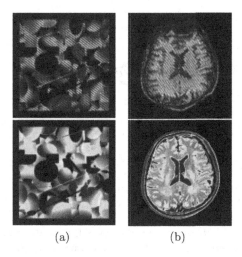

(a)   (b)

Fig. 2. (a) The simulated data pairs. (b) The real brain data pairs.

As far as we know, the reason why the simulated dataset suitable for real brain dataset is as follows. First, we make the statistics between two dataset be relatively similar as much as possible. Second, simulated dataset and real dataset have the same modulation pattern because they use the same parameters and sequence, therefore, they have the same nonlinear mapping relationship between phase-aliased images and clean images theoretically. Third, in a relatively small reception field, simulation data and real data share a similar textures and structural features [8].

2.3 Network Architecture

As is shown in Fig. 3, the network architecture is a two-stage DenseNet. The first stage of the network is built for reconstruction, and the second stage for fine-tuning. The entire process is an end-to-end mapping without any additional operations. The reconstruction network consists of two dense blocks. Each dense block has three bottleneck layers. The growth rate of channel is 64. At the end of last bottleneck layer of each dense block, we reserve the transition layer, which is used in original paper to reduce the size and number of feature maps. The fine-tuning network has similar structure as reconstruction network except that fine-tuning network uses only one dense block.

The input of the network is a two-channel data, the first channel is the real part of the input plural image and the second channel the imaginary part. In the training phase, we randomly crop the full FOV image into 64×64 patches (each patch can be overlapped) and send it to the network. This process can not only save computing resources, but also make a data augmentation. In testing phase, we directly feed the full FOV images to network and get the reconstructed T_2 mapping. Since the proposed network does not have a fully connected layer, it is

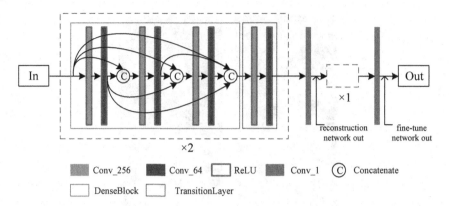

Fig. 3. Network Architecture. Conv_256 means the convolution has an output channel of 256 feature maps, with 1×1 convolution kernel; Conv_64 and Conv_1 means the convolution has an output channel of 64 and 1 feature maps respectively, with 3×3 convolution kernel. The feature maps in all layers have the same size as the input image.

feasible to use the different input size for training phase and testing phase respectively [13]. We chosen the ReLU [15] activation function and did not use the BN [14] operation. RMSProp [16] algorithm were chosen. The size of batch is 16, and the weight and bias is initialized with Xavier [17] and constant respectively.

2.4   Loss Function

The loss function used by the network is a weighted Euclidean distance, which can be expressed as a formula as following:

$$L = \frac{1}{N} \sum_{i=1}^{N} \|(f(X^i, W, \theta) - Y^i) \cdot Y_{change}\|_F^2 \tag{2}$$

where $\cdot$ means Hadamard product, N denotes the number of data in one mini-batch, $f(\cdot)$ denotes the nonlinear mapping of deep convolutional neural network, X denotes OLED images, Y denotes the ground truth label images, W and θ denote the weight and bias. Y_{change} can be expressed as a formula as following:

$$Y_{change} = \begin{cases} \dfrac{1}{a} & if \quad y_{ij} \leq a \\ \dfrac{1}{y_{ij}} & if \quad y_{ij} > a \end{cases} \tag{3}$$

where α is a hyperparameter, in experiment we set it 0.05, y_{ij} is the pixel T_2 value of Y at i^{th} row j^{th} column.

3   Experimental Results and Analysis

We performed experiments on a whole-body 3T MRI instrument. Before the experiment, we obtain the informed consent from volunteers, and the authorized MRI protocols from local research ethics committees. Simulated data were generated using SPROM on a windows computer. Both the simulated data and the real brain data were preprocessed using MATLAB R2016b software. The architecture is implemented using TensorFlow framework [18]. The reconstruction was performed on a Linux computer with one NVIDIA GTX1080 GPU.

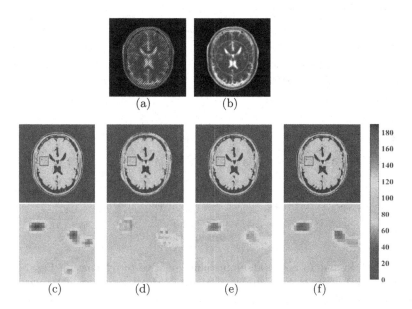

Fig. 4. Simulated MR brain images $(22 \times 22 \, \text{cm}^2)$ gained by using single-shot overlapping echo detachment sequence. We set the acquisition matrix and sw the same value as we used before. The echo chain of the OLED and EPI sequences has a duration time 90 ms. The perpendicular and horizontal line are phase-encoded and frequency-encoded respectively. Prior to the fast Fourier transform, the image matrix will be expand shape with 256×256 by zero padding. (a) Original OLED amplitude image. (b) SE-EPI image. (c) Reference T_2 mappings. (d–f) T_2 mappings reconstructed by echo-detachment-based method (d), ResNet & guided filter (e) and proposed method (f).

It spends about 6 h to train our network. The training process was alternated. In first 50k iterations, the reconstruction network is optimized, while in the next 50k iterations, the fine-tuning network is optimized, and in last 50k iterations, both network is optimized. In order to evaluate the proposed method, we make an compare on the result between the proposed and those previous methods, i.e. echo-detachment-based and ResNet & guided filter methods.

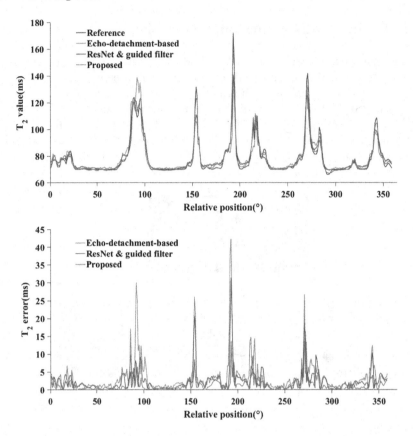

Fig. 5. T_2 values and T_2 errors(absolute value) relative to reference of Fig. 4 (c)–(f) trace on the red circle in Fig. 4(b). It starts at the east direction of the circle and rotates counterclockwise.

3.1   Reconstruction on Numerical Human Brain Data

The results of numerical human brain simulation data were shown in Figs. 4 and 5. The reconstructed T_2 mappings from almost all methods are very consistent with the reference T_2 mapping (Fig. 4c–f). However, the method we proposed gives more details and less noise-like artifacts at many regions. Besides, echo-detachment-based method, ResNet & guided filter method and the proposed method gives the mean error of T_2 values 3.3141 ms, 2.3346 ms, 1.4503 ms respectively. As T_2 errors shown in Fig. 5, proposed method outperforms others obviously which has smaller T_2 errors.

3.2   Reconstruction on in Vivo Human Brain Data

We obtain the overlapped echo detachment images and SE images from four healthy volunteers with age ranging from 26 to 44 years. 17–20 brain slices per

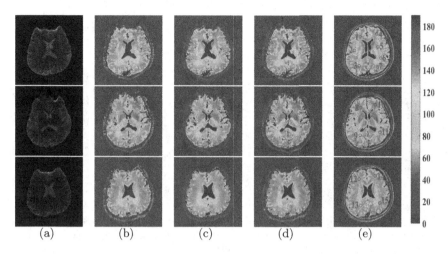

Fig. 6. The reconstruction of MR images ($22 \times 22\,\mathrm{cm}^2$) of a human brain. (a) OLED original images. (b–d) Reconstructed T_2 mappings by echo-detachment-based method (b), ResNet & guided filter (c) and proposed method (d). (e) Reference T_2 mappings.

volunteer were acquired. Scanning for motion-corrupted was done at the same time. Both sequences have the same parameters and conditions such as acquisition matrix and FOV and thickness of slice. Especially, we set $sw = 751\,\mathrm{Hz/pixel}$, $\alpha = 50°$, and $\Delta TE \approx 45\,\mathrm{ms}$ for OLED sequence. While in conventional SE sequence, a different sw is used with its value $201\,\mathrm{Hz/pixel}$, and we have four different echo time with their value is $35\,\mathrm{ms}, 50\,\mathrm{ms}, 70\,\mathrm{ms}$, and $90ms$ respectively, thus we got a total scan time $17\,\mathrm{min}$.

T_2 mappings reconstructed for real human brain data are shown in Fig. 6. Quantitative analysis is shown in Fig. 7, we can see that almost all methods are consistent with SE mapping in many areas. However, the method we proposed outperforms others in most regions of interest (ROIs). The mean error of the mean T_2 values between SE and each method is $1.4638\,\mathrm{ms}$, $1.4342\,\mathrm{ms}$ and $0.9659\,\mathrm{ms}$ respectively. The expanded images are shown in Fig. 8.

3.3   Discussion

For numerical human brain data, the model is relative simple compared with real human brain, and ResNet & guided filter method also show quite well reconstruction result. However, the proposed method still shows obvious improvement on reconstruction accuracy and image fidelity. In Fig. 4(f) at expanded image, DenseNet method shows better performance at tiny cerebrospinal fluid areas, this areas generally have larger T_2 values, as can also be seen in Fig. 5. This phenomenon reflects that DenseNet does play a role to some extent at feature reuse and leads the network to find some useful features at final reconstruction. On the other hand, the weighted Euclidean loss function also gives the higher reconstruction accuracy at the regions with the relatively smaller T_2 values (Fig. 5 at $160°$–$180°$, $200°$–$210°$, $325°$–$335°$).

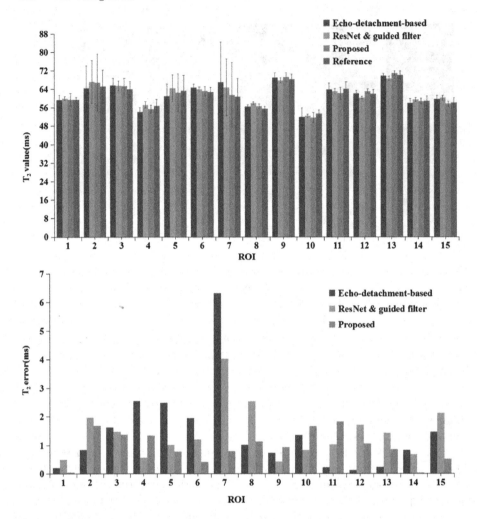

Fig. 7. Mean T_2 values with standard deviations and mean T_2 errors (absolute value) relative to reference for 15 ROIs circled out by red circles and numbered in Fig. 6(e) for above methods.

For in vivo human brain data, we can see from Fig. 6(b–e) that the deep-learning-based methods have a relatively high quality and low mean error than echo-detachment-based method. Moreover, the quantitative analysis in Fig. 7 also show that DenseNet method can give the higher reconstruction accuracy in the most of ROIs. In Fig. 8(f), we can see especially in red rectangles marked regions that the reconstructed images share more similar structural features and texture details with reference both in large T_2 value regions and in small T_2 value regions, which also reflect that the proposed method has strong generalization and robustness.

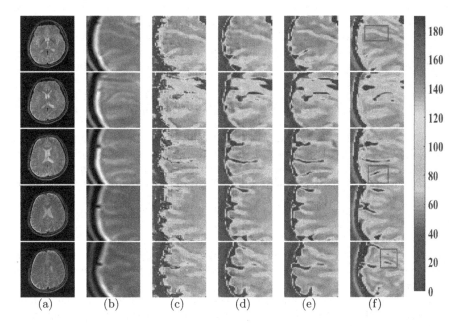

Fig. 8. Magnifying images for five different slices selected from 21 MR images of a human brain T_2 mapping. (a) Full FOV SE images. (b) Magnifying SE images. (c–f) Magnifying T_2 mappings from echo-detachment-based method (c), ResNet & guided filter (d), proposed method (e) and reference T_2 mappings (f). Magnifying ROIs were circled out by red rectangles show in (a). (Color figure online)

4   Conclusion

In this paper, a new reconstruction method based on multi-stage DenseNet was proposed for single-shot T_2 mapping through OLED sequence. The reconstructed results show that the proposed method outperforms the traditional echo-detachment-based reconstruction method and ResNet & guided filter reconstruction method, not only on fidelity of T_2 values, but also on texture fidelity of T_2 images. Without the utilizing of guided filter, the DenseNet method would be more robust to the different noise level.

Acknowledgement. This work was supported by National Natural Science Foundation of China; Grant numbers: 81671674 and 61571382.

References

1. Ma, D., et al.: Magnetic resonance fingerprinting. Nature **495**(7440), 187–192 (2013)
2. Townsend, T.N., Bernasconi, N., Pike, G.B., Bernasconi, A.: Quantitative analysis of temporal lobe white matter t2 relaxation time in temporal lobe epilepsy. Neuroimage **23**(1), 318–324 (2004)

3. Cai, C., et al.: Single-shot t2 mapping through overlapping-echo detachment (OLED) planar imaging. IEEE Trans. Biomed. Eng. **64**(10), 2450–2461 (2017)
4. Ma, L., et al.: Motion-tolerant diffusion mapping based on single-shot overlapping-echo detachment (OLED) planar imaging. Magn. Reson. Med. **80**(1), 200–210 (2018)
5. Pereira, S., Pinto, A., Alves, V., Silva, C.A.: Brain tumor segmentation using convolutional neural networks in MRI images. IEEE Trans. Med. Imaging **35**(5), 1240–1251 (2016)
6. Dou, Q., Chen, H., Jin, Y., Lin, H., Qin, J., Heng, P.-A.: Automated pulmonary nodule detection via 3D convnets with online sample filtering and hybrid-loss residual learning. In: Descoteaux, M., Maier-Hein, L., Franz, A., Jannin, P., Collins, D.L., Duchesne, S. (eds.) MICCAI 2017. LNCS, vol. 10435, pp. 630–638. Springer, Cham (2017). https://doi.org/10.1007/978-3-319-66179-7_72
7. Han, S.S., Kim, M.S., Lim, W., Park, G.H., Park, I., Chang, S.E.: Classification of the clinical images for benign and malignant cutaneous tumors using a deep learning algorithm. Journal of Investigative Dermatology (2018)
8. Cai, C., et al.: Single-shot t2 mapping using overlapping-echo detachment planar imaging and a deep convolutional neural network. Magnetic resonance in medicine (2018)
9. He, K., Zhang, X., Ren, S., Sun, J.: Deep residual learning for image recognition. In: Proceedings of the IEEE Conference on Computer Vision and Pattern Recognition, pp. 770–778 (2016)
10. He, K., Sun, J., Tang, X.: Guided image filtering. IEEE Trans. Pattern Anal. Mach. Intell. **35**(6), 1397–1409 (2013)
11. Huang, G., Liu, Z., Weinberger, K.Q., van der Maaten, L.: Densely connected convolutional networks. In: Proceedings of the IEEE Conference on Computer Vision and Pattern Recognition, vol. 1, p. 3 (2017)
12. Cai, C., Lin, M., Chen, Z., Chen, X., Cai, S., Zhong, J.: Sprom-an efficient program for NMR/MRI simulations of inter-and intra-molecular multiple quantum coherences. Comptes Rendus Physique **9**(1), 119–126 (2008)
13. He, K., Zhang, X., Ren, S., Sun, J.: Spatial pyramid pooling in deep convolutional networks for visual recognition. IEEE Trans. Pattern Anal. Mach. Intell. **37**(9), 1904–1916 (2015)
14. Ioffe, S., Szegedy, C.: Batch normalization: Accelerating deep network training by reducing internal covariate shift. arXiv preprint arXiv:1502.03167 (2015)
15. Glorot, X., Bordes, A., Bengio, Y.: Deep sparse rectifier neural networks. In: Proceedings of the Fourteenth International Conference on Artificial Intelligence and Statistics, pp. 315–323 (2011)
16. Tieleman, T., Hinton, G.: Rmsprop gradient optimization (2014). http://www.cs.toronto.edu/tijmen/csc321/slides/lecture_slides_lec6.pdf
17. Glorot, X., Bengio, Y.: Understanding the difficulty of training deep feedforward neural networks. In: Proceedings of the Thirteenth International Conference on Artificial Intelligence and Statistics, pp. 249–256 (2010)
18. Abadi, M., et al.: TensorFlow: large-scale machine learning on heterogeneous distributed systems. arXiv preprint arXiv:1603.04467 (2016)

Cascaded Deep Hashing for Large-Scale Image Retrieval

Jun Lu[1] and Li Zhang[1,2(✉)]

[1] School of Computer Science and Technology and Joint International,
Research Laboratory of Machine Learning and Neuromorphic Computing,
Soochow University, Suzhou 215006, Jiangsu, China
zhangliml@suda.edu.cn
[2] Provincial Key Laboratory for Computer Information Processing Technology,
Soochow University, Suzhou 215006, Jiangsu, China

Abstract. It is very crucial for large-scale image retrieval tasks to extract effective hash feature representations. Encouraged by the recent advances in convolutional neural networks (CNNs), this paper presents a novel cascaded deep hashing (CDH) method to generate compact hash codes for highly efficient image retrieval tasks on given large-scale datasets. Specifically, we ingeniously utilize three CNN models to learn robust image feature representations on a given dataset, which solves the issue that categories with poor feature representation have a fairly low retrieval precision. Experimental results indicate that CDH outperforms some state-of-the-art hashing algorithms on both CIFAR-10 and MNIST datasets.

Keywords: Image retrieval · Convolutional neural networks · Hash code
Image representation

1 Introduction

In recent years, multimedia data including images have being produced on the Internet every day, making it extremely hard to retrieve similar data from a large-scale database. Content-based image retrieval (CBIR) is a popular image retrieval method, which searches for similar images according to compare the content of images [1–3]. The main steps in CBIR include image representation and similarity measurement. Along this research track, the most challenging issue is to improve the "semantic gap" between the pixel-level information captured by machines and semantics from human perceptions [3, 4].

Recent studies [5–8] revealed that the deep features obtained by convolutional neural networks (CNNs) are more suitable for computer vision tasks, which is a significant breakthrough compared with traditional methods using hand-crafted features [1, 2, 9]. Better effect of deep features gives the credit to advantages of deep CNNs which can learn high-level abstractions in images. But deep features are high-dimensional, which makes it unwise to directly compute the similarity between two high-dimensional vectors. For a large scale image database, it is an undesirable method would consume a lot of time and computing resources.

© Springer Nature Switzerland AG 2018
L. Cheng et al. (Eds.): ICONIP 2018, LNCS 11306, pp. 419–429, 2018.
https://doi.org/10.1007/978-3-030-04224-0_36

Hashing approaches have been turned out to be more appropriate when images need to be retrieved from a large-scale image database, because of its fast speed for searching process and low memory costs [10–17]. Projecting the high-dimensional data into a low-dimensional space, hashing methods can generate compact binary codes that approximately preserve the data structure in the original space. Binary codes are easy to store and compare, which dramatically reduces the computational and memory cost. Hashing algorithms consist of two groups: data-independent and data-dependent methods [10–17].

Most of early researchers pay more attention to data-independent methods which employ random hash functions to map data points to similar hash codes. The most representative one is the locality-sensitive hashing (LSH) [10] and its variants [11], which use random projections to produce binary codes. However, data-independent methods are unpractical because they would produce long codes.

Fortunately, data-dependent hashing methods through machine learning have shown their effectiveness in overcoming the issue mentioned above [12–17]. The data-dependent methods can better access compact and short hash codes from the large-scale data. In general, these techniques are made up of two parts: (1) Generating visual descriptor feature vectors from images; and (2) Encoding vectors into binary hash codes by implementing projection and quantization steps. Existing data-dependent hash methods can be further split into supervised (semi-supervised) and unsupervised methods. The unsupervised methods only utilize the training data without labels to acquire hash functions, which encode neighborhood relation of samples from a certain metric space into the Hamming space [12, 13]. For instance, Spectral Hashing (SH) [12] tries to preserve the similarity structures defined in the original space.

Supervised methods boost hash codes by taking advantage of label information to learn more complex semantic similarity [14–17]. In the inspiration of deep learning, some researchers utilized deep architectures for hash learning under the supervised framework. Xia et al. [15] proposed a hashing method based on the supervised data to acquire binary hashing codes through deep learning. Although this approach is proven effective, it consumes too much computational time and considerable storage space for the input of a pair-wised similarity matrix of data. Very recently, Lin et al. [16] put forward an effective method that based on a deep CNN model to learn simultaneously binary codes and image representation when the image data are labeled.

There is such a phenomenon in this method that the retrieval performance is closely related with the classification accuracy of deep CNN models. The categories which can be recognized well by a CNN model also have a high retrieval performance, but the categories with low classification accuracies have a fairly low retrieval precision. Thus, the images recognized bad could reduce the efficiency of image retrieval.

In order to address the issue mentioned above, a novel and effective cascaded deep hashing (CDH) algorithm based on multiple CNNs is developed for the task of large-scale image retrieval. Different from other supervised methods (such as [16]), we use three CNNs, a global CNN and two local CNNs, to generate binary codes. The global CNN is used to recognize the label of images and generate candidate binary codes. The two local CNNs can improve the representation ability of deep features, especially the categories with poor classification ability.

The rest of this paper is as follows. Section 2 elaborates on details of CDH. Section 3 compares CDH with several state-of-the-art methods and reports experimental results. Finally, we conclude this paper in Sect. 4.

2   Our Method

Recent studies have proved that deep hashing methods using CNN can achieve better results in content-based image retrieval [15, 16]. But the precision of image retrieval depends on the classification accuracy of CNN models, and the categories with low classification accuracies have a fairly low retrieval precision. That means hash-like binary codes learned from deep features of poor representation are inefficient for image retrieval tasks in this case. In order to improve this situation, we present a cascaded deep hashing (CDH) method for hash code learning.

We expect that CDH could raise the classification accuracy of the categories with poor classification ability, which makes the hash-like binary codes of all categories have a good representation ability to be used for retrieving.

2.1   Cascaded Models

Consider a large-scale image database consisting of c categories as $X = \{X_i\}_{i=1}^c$, where X_i represents the set of ith category. Let the label set of images be $Y = \{1, 2, \cdots, c\}$. Further, we partition X into two subsets, the validation set $X_{validation}$ and the training set X_{train}.

Figure 1 shows the training framework of CDH. From Fig. 1, we can see that the CDH model includes three CNNs, one global model CNN_1, and two local models CNN_2 and CNN_3. The training data for CNN_1 is the whole training set X_{train}. The training data for both CNN_2 and CNN_3 are subsets of X_{train}, which are dependent on the classification results of CNN_1.

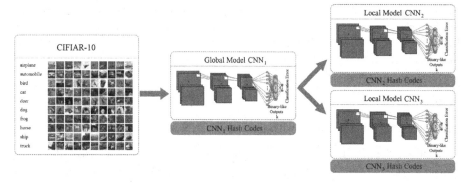

Fig. 1. The training framework of CDH

First, we determine the structure of CNNs a priori, so that the network would have a good classification ability on X_{train}. A typical CNN architecture is given in Fig. 2, which is usually composed of convolution layers, pooling layers and fully connected layers. Then the global model CNN_1 can be obtained by training this network on the training set X_{train}. Second, we apply CNN_1 to the subset $X_{validation}$ and calculate the classification accuracy of each class in $X_{validation}$. Let $CP = \{p_1, p_2, \cdots, p_c\}$ be the set of classification accuracy for all class, where p_i is the classification accuracy for class i. Sort elements in the set CP in descending order and define the sorted CP as $CP_S = \{p_{s_1}, p_{s_2}, \cdots, p_{s_c}\}$.

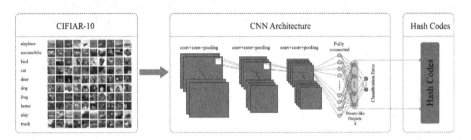

Fig. 2. A typical CNN architecture

Next, we consider how to divide the set X into two subsets X_{good} and X_{bad}, where X_{good} consists of samples belonging to categories have higher classification accuracies, and $X_{bad} = X - X_{good}$. Thus, we need to determine a threshold to separate categories at first. If

$$i^* = \underset{i=2,\cdots,c-1}{\operatorname{argmax}} (p_{s_i} - p_{s_{i-1}}) \qquad (1)$$

we can partition CP_S to two subsets $\{p_{s_1}, p_{s_2}, \cdots, p_{s_{i*}}\}$ and $\{p_{s_{i*+1}}, \cdots, p_{s_c}\}$. Therefore, the corresponding subset $Y_{good} = \{s_1, \cdots, s_{i*}\} \subseteq Y$ represents the set of categories with good classification ability. The remaining categories construct the subset $Y_{bad} = Y - Y_{good}$. Correspondingly, the samples in X_{good} belong to the classes in Y_{good}, and those in X_{bad} to the classes in Y_{bad}.

Then, we train the same CNN on X_{good} and X_{bad} to obtain the local model CNN_2 and CNN_3, respectively.

2.2   Learning Binary Codes with Cascaded CNN Models

Lin analyzed the deep CNN and showed that the final outputs of the classification layer rely on a set of k hidden attributes with each attribute on or off [16]. It means images having the same label would induce similar binary activations. According to the above point of view, it is an effective way to learn hash-like binary codes by binarizing the k activations by a threshold $\theta \in \mathbb{R}$. As shown in Fig. 2, we set the latent layer with k nodes in front of the output layer in the network.

For an image $\mathbf{x}_i$, we denote the output vector of the latent layer by $\mathbf{o}_i = [o_1^i, \ldots, o_k^i]^T \in \mathbb{R}^k$. Then, the binary codes of image $\mathbf{x}_i$ can be represented as $\mathbf{h}_i = [h_1^i, h_2^i, \cdots, h_k^i] \in \{0, 1\}^k$, where

$$h_j^i = \begin{cases} 1, & \text{if } o_j^i \geq \theta \\ 0, & \text{otherwise} \end{cases} \tag{2}$$

We must decide an appropriate value for the threshold θ for different database, which makes the binary codes more effective for image retrieval.

Using the above mentioned method, we use CNN_1, CNN_2 and CNN_3 to generate hash codes for X, X_{good} and X_{bad}, respectively. Let H_{global}, H_{good} and H_{bad} represent the hash code sets that are obtained from X, X_{good} and X_{bad}, respectively.

2.3 Image Retrieval

For a query image, our goal is to search similar images from the given dataset. The method in [16] directly retrieves the query image in one hash code database, or H_{global}. However, we should strength the representation ability of the images, especially those recognized bad by CNN_1. To fulfill this idea, we use a cascaded search method to retrieval the similar images. Figure 3 shows the retrieval process.

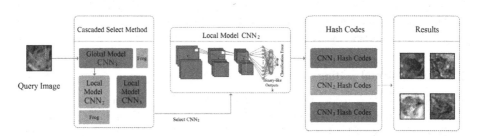

Fig. 3. The retrieval process of CDH

Given a query image $\mathbf{x}$, we input it into CNN_1 and receive the output as the prediction result denoted as y_{global}. If y_{global} is a component of Y_{good}, we use CNN_2 to get the prediction label of $\mathbf{x}$ and define it as y_{good}. If y_{global} is in Y_{bad}, we input $\mathbf{x}$ to CNN_3 to obtain the prediction y_{bad}.

Since we have generated three hash code sets: H_{global}, H_{good} and H_{bad}, from which we need to select an appropriate hash code set for image retrieval. The final hash code set H_{goal} is defined as

$$H_{goal} = \begin{cases} H_{good}, & \textit{if } y_{global} = y_{good} \\ H_{bad}, & \textit{if } y_{global} = y_{bad} \\ H_{global}, & \textit{otherwise} \end{cases} \tag{3}$$

Once H_{goal} is determined, we can generate the hash code **h** for **x** using the corresponding CNN model. For example, if $H_{goal} = H_{good}$, then we use the local model CNN_2 to generate **h**. Moreover, retrieval is carried out in H_{goal}.

Suppose we need to search out t images that are most similar to **x**. The Hamming distance between the hash code of query image and that of any training sample is taken as their similarity. The smaller the Hamming distance is, the higher level the similarity of the two images is. The candidates are ranked in ascending. We select the top t images as the results of retrieval.

3   Experiments

To verify the effectiveness of our proposed method, we perform experiments on two image datasets, MNIST [18] and CIFAR-10 [19]. In the following, we first describe datasets and experimental settings, and then analyze experimental results.

3.1   Datasets

MNIST Dataset [18] contains 70 K 28×28 gray scale images belonging to 10 categories of handwritten Arabic numerals from 0 to 9. There are 60,000 training images, and 10,000 test images.

CIFAR-10 Dataset [19] contains 60 K 32×32 color tiny images which are categorized into 10 classes (6 K tiny images per class). Each image belongs to one of the 10 classes in a single-label dataset.

3.2   Experimental Setting

The basic network is made up of three convolution-pool layers and three fully connected layers sequencely. The size of filters in convolution layers is 3×3 and the stride is 1. There are $64, 64,$ and 128 filters in the three convolution layers, respectively. Each convolution layer follows a pool layer with a stride of 2. Besides, the first fully connected layer contains 500 nodes, the second (latent layer) has k (the hash code length) nodes and the third (output layer) has c nodes (the label number).

To illustrate the effectiveness of our retrieval method, we compare CDH with six typical hashing methods: DLBHC [16], CNNH+ [15], KSH [17], BRE [14], LSH [11], and SH [12]. We evaluate the retrieval procedure by a Hamming ranking-based criterion. Given a query image, we find the t images with the smallest Hamming distance between it and training samples. The average precision (AP) for this query image is as

$$Precison@t = \frac{\sum_{i=1}^{t} Rel(i)}{t} \tag{4}$$

where $Rel(i)$ is the ground truth relevance between a query **x** and the ith ranked image [16]. Here, we consider only the category label in measuring the relevance so $Rel(i) \in \{0, 1\}$, where $Rel(i) = 1$ if the query and the ith image have the same label; otherwise $Rel(i) = 0$. The mean retrieval precision (MRP) is used to measure the retrieval ability of these methods, which is the mean of AP on all query images.

3.3   Experimental Results on CIFAR-10 Dataset

A. Performance of Image Classification

When training the CNN model on CIFAR-10, the output layer is set as 10-way softmax to predict 10 object categories. In the latent layers, we fit the nodes of neurons k range from 16 to 64 to measure the performance of the latent layer embedded in the deep CNN model. The stochastic gradient descent (SGD) method is adopted to train CNN with 150 iterations and a learning rate of 0.01 on the CIFAR-10 dataset.

Table 1 gives MRP of three models. As shown in Table 1, the local models can effectively improve the classification performance of X_{good} and X_{bad} compared with the global model.

Table 1. MPR (%) of three models on data partition of CIFAR-10 dataset.

Data subset	Model		
	CNN_1	CNN_2	CNN_3
X_{train}	88.96	–	–
X_{good}	91.53	93.35	–
X_{bad}	82.21	–	88.96

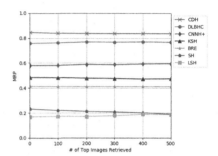

Fig. 4. MRP vs. top t images with 48 bits on CIFAR10

B. Performance of Images Retrieval

In this experiment, we map images to the hash codes from 16 to 64 for image retrieval measured with the hamming distance. To compare with traditional hashing approaches in hand-craft representation, 512-dimensional generalized search tree (GIST) features are extracted from each image [20].

Table 2 shows the MRP of the top 500 returned images with different lengths of hash codes, where the best results are in bold. Figure 4 shows MRP regarding to various numbers of the top images received from compared methods. From experimental results, we can see that CDH obviously has the best experimental results among the compared methods, including unsupervised and supervised ones.

We also investigate more details for the relationship between classification accuracy and retrieval accuracy. We take the hash codes of 48 as an example, as shown in Fig. 5.

In Fig. 5(a), we can see that categories with slightly lower classification accuracy obtained by the global model CNN_1 have the bigger difference between the mean classification accuracy (MCP) and the mean retrieval accuracy. Figure 5(b) shows that we can improve the classification accuracy to enhance the deep feature representation ability with the help of local models CNN_2 and CNN_3. Inspection on Fig. 5(b) indicates that those categories with poor classification accuracy also receive pretty good retrieval accuracy by CNN_3. Figure 5(c) reports that CDH can generally obtain higher retrieval accuracy in the cascade way.

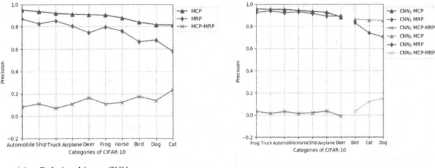

(a)　Relationship on CNN_1　　　　　　(b)　Relationship on CNN_2 and CNN_3

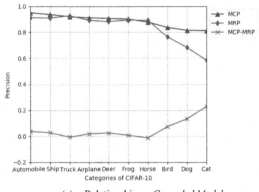

(c)　Relationship on *Cascaded Model*

Fig. 5. Relationship between mean classification prediction (MCP) and MRP of each category on different models.

Table 2. MAP (%) with various number of bits on the CIFAR-10 dataset

Method	16 bits	32 bits	48 bits	64 bits
CDH	**82.08**	**82.96**	**83.40**	**83.78**
DLBHC	73.39	74.18	75.16	75.99
CNNH+	58.63	58.94	59.31	59.98
KSH	41.03	41.78	42.07	42.22
BRE	19.22	19.88	20.43	20.51
SH	19.63	19.87	19.99	20.06
LSH	15.66	16.21	16.44	16.51

Figure 6 shows the top images retrieved by our method CDH and the state-of-the-art method DLBHC. CDH can successfully retrieve images with relevant categories and similar appearance. It can be easily found that the images retrieved by CDH are more appearance-relevant according to our empirical eyeball checking, which makes CDH have better performance.

3.4   Experimental Result on MNIST Dataset

A. Performance of Image Classification

To transfer the deep CNN to the dataset of MNIST, we modify the latent layer to 10-way softmax to predict 10 digit classes and k is also set from 16 to 64. We then train our cascaded model on the MNIST dataset. In Table 3, we list the classification accuracy of three models on different parts of MNIST.

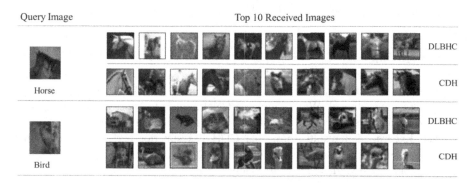

Fig. 6. The retrieval process of CDH

Table 3. MPR (%) of three models on data partition of MNIST dataset.

Data subset	Model		
	CNN_1	CNN_2	CNN_3
X_{train}	99.41	–	–
X_{good}	99.49	99.72	–
X_{bad}	99.23	–	99.51

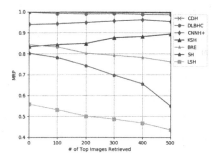

Fig. 7. MRP vs. top t images with 48 bits on MNIST.

B. Performance of Images Retrieval

In order to make a comparison fairly with other hashing methods, we unify the evaluation method that retrieves the relevant images by hash codes from 16 to 64 and using the Hamming distance. We still use the 512-d GIFT features for traditional hashing learning approach. We can see the precision (MRP) of top 500 returned image with different lengths of hash codes in Table 4, where the best results are in bold. It can be seen that our method has excellent results no matter how many images are retrieved. Figure 7 gives the relationship curves of the precision (MRP) vs. the number of the top retrieved samples. CDH can also stand out when compared with other fine methods.

Table 4. MAP (%) with various number of bits on the MNIST dataset

Method	16 bits	32 bits	48 bits	64 bits
CDH	**99.53**	**99.54**	**99.56**	**99.57**
DLBHC	98.12	98.34	98.63	98.83
CNNH+	90.34	90.89	91.23	91.89
KSH	84.93	86.23	88.57	90.04
BRE	72.33	74.58	76.62	78.92
SH	48.91	49.98	51.12	53.43
LSH	42.14	44.33	45.58	46.63

4   Conclusions

We present a cascaded framework to generate compact and short hashing codes for large-scale image retrieval. We use multiple CNN models to boost feature expression ability of images, so that our hash-like binary codes are more suitable for image retrieval. Experimental results show that CDH has superior performance over the previous best retrieval results, which has an elevation of about 8% and 1% on the CIFAR-10 and MNIST datasets, respectively.

Acknowledgments. This work was supported in part by the National Natural Science Foundation of China under Grants No. 61373093, No. 61402310, No. 61672364 and No. 61672365, by the Soochow Scholar Project of Soochow University, by the Six Talent Peak Project of Jiangsu Province of China, by the Postgraduate Research & Practice Innovation Program of Jiangsu Province (No. SJCX18_0846), and by the Graduate Innovation and Practice Program of colleges and universities in Jiangsu Province.

References

1. Lowe, D.G.: Distinctive image features from scale-invariant keypoints. Int. J. Comput. Vis. **60**(2), 91–110 (2004)
2. Qiu, G.: Indexing chromatic and achromatic patterns for content-based colour image retrieval. Pattern Recogn. **35**(8), 1675–1686 (2002)
3. Smeulders, A.W.M., Worring, M., Santini, S., Gupta, A., Jain, R.: Content-based image retrieval at the end of the early years. IEEE Trans. Pattern Anal. Mach. Intell. **22**(12), 1349–1380 (2000)
4. Wan, J., et al.: Deep learning for content-based image retrieval: a comprehensive study. In: Proceedings of the 22nd ACM International Conference on Multimedia, pp. 157–166. ACM (2014)
5. Krizhevsky, A., Sutskever, I., Hinton, G.E.: ImageNet classification with deep convolutional neural networks. In: Advances in Neural Information Processing Systems, pp. 1097–1105 (2012)
6. Girshick, R., Donahue, J., Darrell, T., Malik, J.: Rich feature hierarchies for accurate object detection and semantic segmentation. In: Proceedings of the IEEE Conference on Computer Vision and Pattern Recognition, pp. 580–587 (2014)

7. Oquab, M., Bottou, L., Laptev, I., Sivic, J.: Learning and transferring mid-level image representations using convolutional neural networks. In: Proceedings of the IEEE Conference on Computer Vision and Pattern Recognition, pp. 1717–1724 (2014)
8. Razavian, A.S., Azizpour, H., Sullivan, J., Carlsson, S.: CNN features off-the-shelf: an astounding baseline for recognition. In: Proceedings of the IEEE Conference on Computer Vision and Pattern Recognition, pp. 806–813 (2014)
9. Bay, H., Tuytelaars, T., Van Gool, L.: SURF: speeded up robust features. In: Leonardis, A., Bischof, H., Pinz, A. (eds.) ECCV 2006. LNCS, vol. 3951, pp. 404–417. Springer, Heidelberg (2006). https://doi.org/10.1007/11744023_32
10. Gionis, A., Indyk, P., Motwani, R.: Similarity search in high dimensions via hashing. In: International Conference on Very Large Data Bases, pp. 518–529. Morgan Kaufmann Publishers Inc. (1999)
11. Raginsky, M., Lazebnik, S.: Locality-sensitive binary codes from shift-invariant kernels. In: Advances in Neural Information Processing Systems, pp. 1509–1517 (2009)
12. Weiss, Y., Torralba, A., Fergus, R.: Spectral hashing. In: Advances in Neural Information Processing Systems, pp. 1753–1760 (2008)
13. Gong, Y., Lazebnik, S., Gordo, A., Perronnin, F.: Iterative quantization: a procrustean approach to learning binary codes for large-scale image retrieval. IEEE Trans. Pattern Anal. Mach. Intell. **35**(12), 2916–2929 (2013)
14. Kulis, B., Darrell, T.: Learning to hash with binary reconstructive embeddings. In: Advances in Neural Information Processing Systems, pp. 1042–1050 (2009)
15. Xia, R., Pan, Y., Lai, H., Liu, C., Yan, S.: Supervised hashing for image retrieval via image representation learning. In: AAAI (2014)
16. Lin, K., Yang, H.F., Hsiao, J.H., Chen, C.S.: Deep learning of binary hash codes for fast image retrieval. In: Proceedings of the IEEE Conference on Computer Vision and Pattern Recognition, pp. 27–35 (2015)
17. Liu, W., Wang, J., Ji, R., Jiang, Y.G., Chang, S.F.: Supervised hashing with kernels. In: Proceedings of the IEEE Conference on Computer Vision and Pattern Recognition, pp. 2074–2081 (2012)
18. Lecun, Y., Cortes, C.: The MNIST database of handwritten digits. http://yann.lecun.com/exdb/mnist/ (2010)
19. Krizhevsky, A., Hinton, G.: Learning multiple layers of features from tiny images. Technical report 1 (4), p. 7. University of Toronto (2009)
20. Hellerstein, J.M.: Generalized search tree. In: Liu, L., Özsu, M.T. (eds.) Encyclopedia of Database Systems, pp. 1222–1224. Springer, Boston (2009). https://doi.org/10.1007/978-1-4899-7993-3

Text-Independent Speaker Verification from Mixed Speech of Multiple Speakers via Using Pole Distribution of Speech Signals

Toshiki Tagomori, Kazuya Matsuo, and Shuichi Kurogi[✉]

Kyushu Institute of Technology, Tobata, Kitakyushu, Fukuoka 804-8550, Japan
tagomori.toshiki821@mail.kyutech.jp,
{matsuo,kuro}@cntl.kyutech.ac.jp
http://kurolab.cntl.kyutech.ac.jp/

Abstract. This paper presents a method of text-independent speaker verification from mixed speech of multiple speakers via using pole distribution of speech signals. The poles of speech signal derived from all-pole speech production model are obtained via a neural net called bagging CAN2 (competitive associative net 2) for learning efficient piecewise linear approximation of nonlinear function. We show an analysis that poles of mixed speech are expected to be composed of the poles farther from zeros of ARMA (autoregressive moving average) models of constituent speeches. By means of experiments using unmixed and mixed speeches, we show the distribution of the poles of speeches has two typical regions: one involves poles which change suddenly with the change of the speech from unmixed to mixed, and the other involves poles which change continuously with the change of the mixing weight, which is considered to support the analysis. We execute experiments of speaker verification, and obtain the following properties of recall and precision as measures of verification performance: the recall decreases suddenly with the change of the speech from unmixed to mixed, while the precision does not decreases so much with the decrease of SNR (signal to noise ratio) until below 0 dB. Finally, we show the usefulness of the present method.

Keywords: Text-independent speaker verification
Mixed speech of multiple speakers · Pole distribution of speech signals

1 Introduction

This paper presents a method of text-independent speaker verification from mixed speech of multiple speakers via using pole distribution of speech signals. Here, the poles of speech signal is derived from all-pole speech production model [1,2], and we obtain them by competitive associative nets (CAN2s). Here, a single CAN2 is an artificial neural net for learning efficient piecewise linear approximation of nonlinear function [3]. We have shown that feature vectors of pole

© Springer Nature Switzerland AG 2018
L. Cheng et al. (Eds.): ICONIP 2018, LNCS 11306, pp. 430–440, 2018.
https://doi.org/10.1007/978-3-030-04224-0_37

distribution extracted by bagging (bootstrap aggregating) CAN2 extract nonlinear and time-varying features of the speaker stably than a single CAN2 [4]. Here, although the most common way to characterize the speech signal in the literature is short-time spectral analysis, such as Linear Prediction Coding (LPC) and Mel-Frequency Cepstrum Coefficients (MFCC), they extract spectral features of the speech from each of consecutive interval frames spanning 10–30 ms [1], while the CAN2 obtains PLPCs (piecewise linear predictive coefficients) corresponding to poles of a speech spanning about 1 ms (with the prediction order $k = 8$ and sampling rate 8 kHz as shown below). Thus, a single feature vector of LPC and MFCC corresponds to a kind of average of multiple PLPCs obtained by the CAN2. Namely, the CAN2 learns more precise information on the speech signal than conventional methods. For example, the CAN2 is able to reproduce vowel signals with very high precision, while the LPC cannot [5].

So far, we have shown the effectiveness of the feature vector of the distribution of poles obtained by bagging CAN2 in several speaker recognition tasks, such as single step speaker recognition [4], flexible multistep speaker verification [6] and speaker detection [7]. Although these researches have shown the improvement of the performance and the flexibility on speaker recognition, they have been evaluated with unmixed speech sounds of multiple speakers. So, in this paper, we deal with mixed speech of multiple speakers and examine the property and the performance in a simple and basic task of text-independent and single step speaker verification. This task has a close relationship with cocktail party problem [8] studied researches on psychoacoustics, auditory scene analysis, and attention. Although the researches focus mainly on binaural speech perception and localization, the present method uses speech sounds from a single channel. There are also speech processing researches for speaker recognition from mixed speech of multiple speakers, such as a method using MFCC and statistical decision theory [9]. Here, MFCC is considered to have a relationship with the poles of speech signal, which we would like to clarify much more in our future research. Furthermore, some additional functions, such as the above decision theory, multistep processing developed in [6], may be necessary for practical use, which is also for our future research studies.

This paper is for clarifying basic ability of speaker verification from mixed speech of multiple speakers by means of using feature vectors of pole distribution. We describes the system of speaker verification using bagging CAN2 and pole distribution of speech signals in Sect. 2. Next, we analyze the pole distribution of mixed speech of multiple speakers in speaker verification task in Sect. 3. We show experimental results and analysis in Sect. 4, followed by the conclusion in Sect. 5.

2   Speech Processing System for Speaker Verification Using Bagging CAN2 and Pole Distribution of Speech Signals

Figure 1 shows the present speaker verification system. In the same way as general speaker recognition systems [1], it consists of four steps: speech data acquisition, feature extraction, pattern matching, and making a decision. Different from other research studies, we use feature vectors of pole distribution, q, obtained from speech signal by means of bagging CAN2, which we denote $CAN2^{[PLPC]}$ for obtaining PLPCs. Furthermore, we utilize bagging CAN2s for learning q to execute speaker verification, which we denote $CAN2^{[s_i]}$ for the verification of a speaker s_i. The system is almost the same as the one presented in [4], but we execute speaker verification from mixed speech of multiple speakers and examine the ability and the performance.

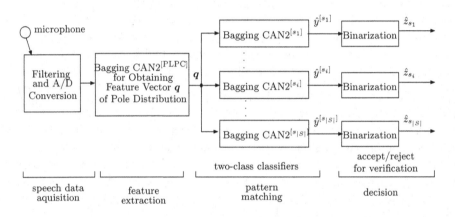

Fig. 1. Diagram of speaker verification system using CAN2s

2.1   Model of Speech Signal Production and Pole Distribution

The most standard model of the speech production is all-pole model, or AR (autoregressive) model described as follows (see [1,2] for details); a speech signal y_t at a discrete time t is modeled by a linear combination of its past values $\boldsymbol{x}_t = (y_{t-1}, y_{t-2}, \cdots, y_{t-k})^\top$ as

$$y_t = \boldsymbol{\alpha}^\top \boldsymbol{x}_t + \beta u_t. \tag{1}$$

Here, $\boldsymbol{\alpha} = (\alpha_1, \alpha_2, \cdots, \alpha_k)^\top$ and β represent linear coefficients, the input u_t is unknown, and k is called prediction order. It is supposed that $\boldsymbol{\alpha}$ and β do not change for a short period of time, while they change during a speech.

By means of z-transform of (1) and partial fraction expansion, we have

$$Y(z) = \sum_{m=1}^{k} \frac{c_m}{1 - p_m z^{-1}} = \sum_{m=1}^{k} \frac{c_m}{1 - r_m \exp(j\theta_m)z^{-1}}, \tag{2}$$

where j represents the imaginary unit, i.e. $j^2 = -1$, and the pole $p_m = r_m \exp(j\theta_m)$ has the magnitude $r_m \geq 0$ and the argument $\theta_m \in [0, 2\pi]$. Now, we evenly divide the regions of the magnitude, $[0, r_{max}]$, and the argument, $[0, \pi]$, into n_r and n_θ regions, respectively, and obtain the regions $R_i^{[r]} \times R_l^{[\theta]}$ given by

$$R_i^{[r]} = [i\Delta r, (i+1)\Delta r) \tag{3}$$
$$R_l^{[\theta]} = [l\Delta\theta, (l+1)\Delta\theta) \tag{4}$$

for $\Delta r = r_{max}/n_r$, $\Delta\theta = \pi/n_\theta$, $i = 0, 1, 2, \cdots, n_r - 1$ and $l = 0, 1, 2, \cdots, n_\theta - 1$. Then, by counting the number of poles in each region up during a certain period of time, we obtain $k_q = n_r \times n_\theta$ dimensional feature vector of pole distribution, $q = (q_{00}, q_{10}, \cdots, q_{n_r-1, n_\theta-1})^\top$, where the n $(= i + n_r l)$th element, q_{il}, of q corresponds to the region $R_i^{[r]} \times R_l^{[\theta]}$ for $i = \text{mod}(n, n_r)$ and $l = \text{floor}(n/n_r)$. Here, $\text{mod}(\cdot)$ and $\text{floor}(\cdot)$ represent the mod and the floor function, respectively. Note that we neglect the poles with negative imaginary part because the pole distribution is symmetric with respect to the real axis on the z-plane. Furthermore, we also neglect the poles on the real axis because we could not have obtain the effectiveness in the experiments shown below.

2.2 Pole Distribution and Speaker Verification via Using Bagging CAN2

We utilize bagging CAN2 to obtain a number of piecewise linear coefficients of speech signal (see [4] for details of single and bagging CAN2). Let $D^{[\text{train}]} = \{(\boldsymbol{x}_t, y_t) | t \in I^{\text{train}}\}$ be a training data set consisting of $\boldsymbol{x}_t = (y_{t-1}, y_{t-2}, \cdots, y_{t-k})$ and y_t of speech signal given by (1) whose linear coefficient $\boldsymbol{\alpha}$ changes from period to period in time. The bagging CAN2 for obtaining PLPCs, or CAN2$^{[\text{PLPC}]}$, has a number, b, of single CAN2s, or CAN2$^{[\text{PLPC},i]}$ $(i = 1, 2, \cdots, b)$, and CAN2$^{[\text{PLPC},i]}$ learns the ith bag $D^{[\text{train},i]}$ generated via resampling with replacement from $D^{[\text{train}]}$. After the learning, CAN2$^{[\text{PLPC},i]}$ with N units has associative matrices $\boldsymbol{M}_c^{[i]} = (M_{c0}^{[i]}, M_{c1}^{[i]}, \cdots, M_{ck}^{[i]})$ $(c = 1, 2, \cdots, N)$ to execute linear approximation of y_t by

$$y_t^{[i]} = \boldsymbol{M}_c^{[i]} \tilde{\boldsymbol{x}}_t = \boldsymbol{M}_c^{[i]} (1, y_{t-1}, y_{t-2}, \cdots, y_{t-k})^\top. \tag{5}$$

From (1) and (5), we can expect that CAN2$^{[\text{PLPC},i]}$ obtains $(M_{c1}^{[i]}, \cdots, M_{ck}^{[i]}) \approx \boldsymbol{\alpha}^\top$ and $M_{c0}^{[i]} \approx \beta u_t$. Then, the approximation of the poles $p_m = r_m \exp(j\theta_m)$ in (2) can be obtained from $\boldsymbol{M}_c^{[i]}$. By means of counting the obtained poles up in the regions given by (3) and (4), we derive feature vector q reflecting the pole distribution.

For learning $\boldsymbol{q}$ to verify a speaker, bagging CAN2 is also employed. Here, bagging CAN2 basically is for solving regression problems but we use it as a two-class classifier by means of binarizing the output of the CAN2. Let $S = \{s_1, s_2, \cdots, s_{|S|}\}$ be a set of speakers, and $Q^{[s]}$ be the set of feature vectors $\boldsymbol{q}$ obtained from a speaker s. Then, the verification function that a learning machine should learn to predict is given by

$$y^{[s]} = f^{[s]}(\boldsymbol{q}) = \begin{cases} 1, \text{ if } \boldsymbol{q} \in Q^{[s]} \\ -1, \text{ if } \boldsymbol{q} \notin Q^{[s]} \end{cases} \tag{6}$$

We train bagging CAN2 for speaker s, or $\text{CAN2}^{[s]} = \cup_{i \in \{1,2,\cdots,b^{[s]}\}} \text{CAN2}^{[s,i]}$, with training bags $Q^{[s',i]}$ generated by the resampling with replacement from $Q^{[s']}$ for $s' \in S$. Then, we execute the verification with $\text{CAN2}^{[s,i]}$ after the learning by binarizing the mean output of $\text{CAN2}^{[s,i]}$ as

$$\hat{z}^{[s]} = \begin{cases} 1, \text{ if } \hat{y}^{[s]} = \dfrac{1}{b^{[s]}} \sum_{i=1}^{b^{[s]}} \boldsymbol{M}_c^{[s,i]} \tilde{\boldsymbol{q}} \geq 0, \\ -1, \text{ otherwise}, \end{cases} \tag{7}$$

where $\hat{z}^{[s]} = 1$ indicates accept, and -1 reject for the input test vector $\boldsymbol{q}$. Here, $\tilde{\boldsymbol{q}} = \left(1, \boldsymbol{q}^\top\right)^\top = (1, q_1, q_2, \cdots, q_{k_q})^\top$. Thus, $\text{CAN2}^{[s_i]}$ is expected to verify the speaker s correctly after learning $\left(\beta u_t, \boldsymbol{\alpha}^\top\right)$ by bagging ensemble of $\frac{1}{b^{[s]}} \sum_{i=1}^{b^{[s]}} \boldsymbol{M}_c^{[s,i]}$.

3   Speaker Verification from Mixed Speech of Multiple Speakers

Here, we analyze speaker verification from mixed speech of multiple speakers. By means of inverse z-transform of (2) for sampling period T_S, we have

$$y_t = \sum_{m=1}^{k} c_m (r_m)^{T_S t} \exp(j\, \theta_m T_S t) \tag{8}$$

for $t = 0, 1, 2, \cdots$. From the discretization of the pole space with small Δr and $\Delta \theta$ corresponding to (3) and (4), we can approximate the above y_t by

$$y_t \simeq \sum_{i=0}^{n_r-1} \sum_{l=0}^{n_\theta-1} \sum_{m=1}^{k} q_{ilm} c_m (i\Delta r)^{T_S t} \exp\left(j(l\Delta \theta)T_S t\right), \tag{9}$$

where q_{ilm} is 1 when the pole (r_m, θ_m) is in the region $R_i^{[r]} \times R_l^{[\theta]}$, and 0 otherwise. Thus, the $n\ (= i + n_r l)$th element of the feature vector $\boldsymbol{q}$ of pole distribution is given by $q_{il} = \sum_{m=1}^{k} q_{ilm}$.

Next, let us examine a mixed speech of two speakers s_1 and s_2 weighted by γ and $1 - \gamma$, respectively, for $0 \le \gamma \le 1$, i.e.

$$y_t^{[s_1,s_2,\gamma]} = \gamma y_t^{[s_1]} + (1 - \gamma)y_t^{[s_2]}, \tag{10}$$

where $y_t^{[s_1]} = \sum_{m=1}^{k} \alpha_m^{[s_1]} y_{t-m}^{[s_1]} + \beta^{[s_1]} u_t^{[s_1]}$ and $y_t^{[s_2]} = \sum_{m=1}^{k} \alpha_m^{[s_2]} y_{t-m}^{[s_2]} + \beta^{[s_2]} u_t^{[s_2]}$. Then, from (9) and the left hand side of (10), we have

$$y_t^{[s_1,s_2,\gamma]} \approx \sum_{i=0}^{n_r-1} \sum_{l=0}^{n_\theta-1} \sum_{m=1}^{k} q_{ilm}^{[s_1,s_2,\gamma]} c_m^{[s_1,s_2,\gamma]} (i\Delta r)^{T_S t} \exp\left(j(l\Delta\theta)T_S t\right), \tag{11}$$

and from (9) and the right hand side of (10), we have

$$y_t^{[s_1,s_2,\gamma]} \approx \sum_{i=0}^{n_r-1} \sum_{l=0}^{n_\theta-1} \sum_{m=1}^{k} \left(\gamma q_{ilm}^{[s_1]} c_m^{[s_1]} + (1-\gamma) q_{ilm}^{[s_2]} c_m^{[s_2]}\right) (i\Delta r)^{T_S t} \exp\left(j(l\Delta\theta)T_S t\right). \tag{12}$$

Here, (11) indicates an approximation of mixed speech signal, while (12) the sum of the approximation of two speech signals. Since the number of non-zero $q_{ilm}^{[s_1,s_2,\gamma]}$, $q_{ilm}^{[s_1]}$ and $q_{ilm}^{[s_2]}$, respectively, is k for all i and l, the number k of non-zero $q_{ilm}^{[s_1,s_2,\gamma]}$ are considered be $q_{ilm}^{[s_1\gamma]}$ and $q_{ilm}^{[s_2\gamma]}$ with k largest values among $\gamma c_m^{[s_1]}$ and $(1 - \gamma)c_m^{[s_2]}$ because they minimize the difference of (11) and (12). Thus, the poles corresponding to large $\gamma c_m^{[s_1]}$ of a speaker s_1 are expected to appear in the pole distribution $q_{ilm}^{[s_1,s_2,\gamma]}$ of mixed speech. Furthermore, it may be effective to execute a learning of pole distribution obtained from mixed speech in order to neglect the pole distribution corresponding to large $(1 - \gamma)c_m^{[s_2]}$ of other speaker s_2. Incidentally, it is expected that small $c_m^{[\cdot]}$ corresponds to the zeros of ARMA (autoregressive moving average) model, which is the most general speech production model more precise than all-pole model given by (1) (see [10] for details of speech production models). Thus, the poles with larger $c_m^{[\cdot]}$ are expected to correspond to the poles farther from zeros.

4 Experiments

4.1 Experimental Setting

We have recorded speech data sampled with $8\,\mathrm{kHz}$ of sampling rate and 16 bits of voltage resolution in a silent room of our laboratory. They are from seven speakers (2 female and 5 mail speakers): $S = \{\mathrm{fHS}, \mathrm{fMS}, \mathrm{mKK}, \mathrm{mKO}, \mathrm{mMT}, \mathrm{mNH}, \mathrm{mYM}\}$. We use ten digits (words) of Japanese pronunciations: $D = \{$ /zero/, /ichi/, /ni/, /san/, /shi/, /go/, /roku/, /nana/, /hachi/, /kyu/ $\}$. For each speaker and each digit, ten samples are recorded on different time and date among two months, where we denote the index set by $L = \{1, 2, \cdots, 10\}$. Let $y_t^{[s,d,l]}$ denote the spoken digit signal for $s \in S$, $d \in D$ and $l \in L$, and $X^{[s]} = \{y_t^{[s,d,l]} | d \in D, l \in L\}$ be the dataset for $s \in S$.

For each speech signal $y_t^{[s,d,l]} \in X^{[s]}$, we have made mixed speech $y_t^{[s,d,l,s',d',l',\gamma]} = \gamma y_t^{[s,d,l]} + (1 - \gamma)y_t^{[s',d',l']}$ for $\gamma = 0.8, 0.6, 0.4, 0.2$, and s', d' and l' are selected randomly from $S\backslash\{s\}$, D and L, respectively. Here, before making the mixture, we have normalized the power of a speech signal $y_t^{[\cdot]}$ ($t = 0, 1, \cdots, T - 1$) as $y_t^{[\cdot]} := y_t^{[\cdot]}/\sqrt{\sum_{i=0}^{T-1} \left(y_i^{[\cdot]}\right)^2/T}$. Therefore, SNR (signal to noise ratio), or $20\log_{10}\gamma/(1-\gamma)$, for $\gamma = 0.8$, 0.6, 0.4 and 0.2 is about 12, 3.5, -3.5 and -12 [dB], respectively. Furthermore, when two signals have different lengths, we have added $y_t^{[\cdot]} = 0$ for shorter signal so that two speech signals will have the same length. Let $X^{[s,\gamma]}$ denote the set of $y_t^{[s,d,l,s',d',l',\gamma]}$ for $\gamma = 0.8$, 0.6, 0.4, 0.2, and $X^{[s,1]}$ be the same as $X^{[s]}$.

In order to obtain pole distribution of $y_t \in X^{[s,\gamma]}$ for $t = 0, 1, 2, \cdots$, we have trained bagging CAN2 with input-output pairs $(\boldsymbol{x}_t, y_t)$ for $\boldsymbol{x}_t = (y_{t-1}, y_{t-2}, \cdots, y_{t-k})$, $t = k, k+1, \cdots$ and $k = 8$. We have employed bagging CAN2 with 20 bags resampled with bagsize ratio of $\alpha = 0.7$ and $N = 24$ units for each constituent single CAN2. (see [11] for bagsize ratio) For the feature vector $\boldsymbol{q}$ of pole distribution, we use $r_{\max} = 2$, $n_r = 2$ and $n_\theta = 18$ in (3) and (4), and then the dimensionality of $\boldsymbol{q}$ is $k_q = 36$. Let $Q^{[s,\gamma]}$ denote the set of $\boldsymbol{q}$ obtained from $X^{[s,\gamma]}$. In order to estimate the performance in text-independent speaker verification, we have employed bagging CAN2 with 40 bags, bagsize ratio $\alpha = 1.6$, and $N = 100$ units for each constituent single CAN2, to learn the relationship from $\boldsymbol{q} \in Q^{[s,\gamma]}$ to s. In order to use test data different from training data, we have employed OOB (out-of-bag) estimate, with which we can evaluate the performance of bagging CAN2 in speaker verification tasks (see [11] for details of OOB estimate).

4.2   Example of Pole Distribution and Feature Vector

We show an example of the distribution of poles p_m and feature vector $\boldsymbol{q}$ in Fig. 2. We can see that the distribution of poles p_m of the speaker fHS changes to the one for speaker mKK with the change of γ from 1 to 0. Here, the number of poles in the region around $r_m \simeq 1$ and $\theta_m \simeq 2\pi/3$ increases rapidly at the change of γ from 1 to 0.8, where the distribution for $\gamma = 0.8$ seems more similar to that for $\gamma = 0$ than for $\gamma = 1$. This sudden change of the poles can be expected when there are poles with large $c_m^{[s_2]}$ in the region as mentioned in Sect. 3.

This sudden change of the pole distribution corresponds to the sudden change of green bars at $n = 22$ and 24, which corresponds to $\theta_m = (n/n_r)(\pi/n_\theta) = 11\pi/18$ and $2\pi/3$, respectively, for the feature vector $\boldsymbol{q}$. Here, the green bar represents the poles $p_m = r_m \exp(j\theta_m)$ with the magnitude $r_m < 1$, while red bar with $r_m \geq 1$. Here, by means of looking closely at the red bars for $n \leq 15$, we can see that they change gradually for the change of γ from 1 to 0, which is expected to contribute to continuous performance in verification task shown in the next section.

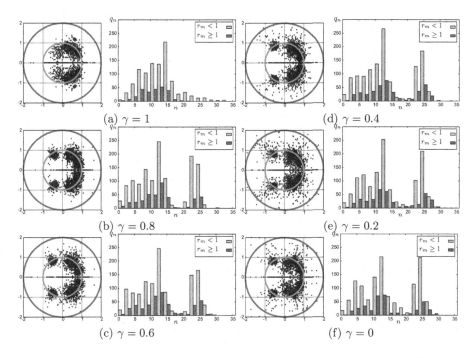

Fig. 2. Example of the distribution of poles p_m (left) and feature vector q (right) of mixed speech of two speakers fHS and mKK having pronounced /san/ mixed with $\gamma = 1$, 0.8, 0.6, 0.4, 0.2, 0.

4.3   Experimental Result of Speaker Verification of Mixed Speech

We have examined the performance of speaker verification from mixed speech of multiple speakers via OOB estimate from the learning and prediction of $q \in Q^{[s,\gamma]}$ to s.

For performance measure of verification, we have obtained precision $r_{\text{prec}} = \text{TP}/(\text{TP} + \text{FP})$ and recall $r_{\text{recall}} = \text{TP}/(\text{TP} + \text{FN})$, where we regard acceptance and rejection as positive and negative classes, respectively, in two class classification. The experimental result is shown in Fig. 3. Here, we can see that the largest mean value of r_{prec} and r_{recall} has been achieved for unmixed speech data with $\gamma = 1$ followed by with $\gamma = 0.8$, 0.6, 0.4 and 0.2. This continuous order of the performance is considered to be owing that the pole distribution changes continuously with the change of γ. Next, a big decrease of the recall r_{recall} has occurred for the change of γ from 1 to 0.8. Here, r_{recall} indicating the ratio of the correct acceptance to all (positive) speech uttered by the reference speaker decreases with the increase of γ from 1 to 0.8. This is supposed to be owing that in the situation where we are detecting speech periods of a reference speaker from mixed speech with other speakers', we may not be able to detect all periods of reference speaker's speech because the power of the speech changes from time to time although the average power is normalized. Namely, a decrease

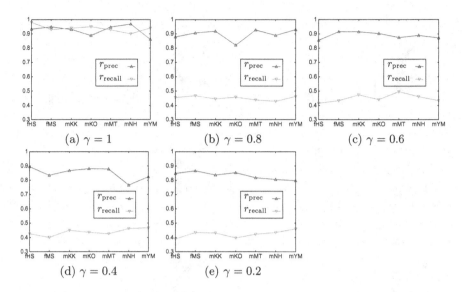

Fig. 3. Experimental result of r_{prec} and r_{recall} in speaker verification from mixed speech. The mean value for $\gamma = 1$, 0.8, 0.6, 0.4, 0.2 is $r_{\mathrm{prec}} = 0.926$, 0.896, 0.889, 0.850, 0.832, and $r_{\mathrm{recall}} = 0.938$, 0.451, 0.451, 0.439, 0.424, respectively.

of r_{recall} is supposed to be inevitable in speaker verification from mixed speech except for $\gamma \simeq 1$. On the other hand, the decrease of the precision r_{prec} seems small for the decrease of γ from 0.8 to 0.2. This property is considered to be obtained by the present method using pole distribution feature vectors. Namely, from the analysis in Sect. 3, this property is considered to be achieved by the learning machine which learns the feature vector q corresponding to the poles with large $\gamma c_m^{[s_1]}$ of the reference speaker s_1 and neglect other speaker's feature of $(1 - \gamma) c_m^{[s_2]}$.

The above result of the precision r_{prec} from 0.896 to 0.832 for γ from 0.8 (SNR of 12 dB) to 0.2 (SNR of -12 dB), seems competitive with other related methods. Namely, the method using MFCCs and statistical decision theory [9] achieved error rate $r_{\mathrm{err}} = 22.6\%$ in three-speaker environment, and a method for robust speaker recognition using SVD has achieved the correct recognition accuracy $r_{\mathrm{acc}} = 32\%$ for SNR of 0 dB [12]. Although the present speaker verification task from mixed speech of two speakers is different from the tasks of the above two methods, it may be remarkable that our method works in much noisy environments with SNR below 0 dB, e.g. by means of obtaining the error rate by $r_{\mathrm{err}} = (\mathrm{FP} + \mathrm{FN})/(\mathrm{TP} + \mathrm{TN} + \mathrm{FP} + \mathrm{FN})$ and the accuracy by $r_{\mathrm{acc}} = 1 - r_{\mathrm{err}}$, the present method has achieved the error rate $r_{\mathrm{err}} = 18.9\%$ or the accuracy $r_{\mathrm{acc}} = 81.1\%$ for SNR of -12 dB (or $\gamma = 0.2$). Incidentally, there are a number of research tasks on cocktail party problem executed in environments with SNR below 0 dB [8], while the tasks are complicated and we could not have clarified the correspondence and the difference with the present experiments.

Finally, the result of $r_{prec} = 0.832$ and $r_{recall} = 0.424$ for $\gamma = 0.2$ can be interpreted as that $r_{prec} = 83.2\%$ of accepted speeches are correct although only $r_{recall} = 42.4\%$ of the reference speaker's speeches are accepted correctly. These performance in text-independent speaker verification from mixed speech will be useful for speaker detection, or the act of detecting a specific speaker in an audio stream [2], and our probabilistic multistep prediction methods developed for unmixed speech [7] are considered to be applicable.

5 Conclusion

We have presented a method of text-independent speaker verification from mixed speech of multiple speakers via using pole distribution of speech sounds. An analysis of the formula of mixed speech shows that poles of mixed speech are expected to be composed of the poles farther from zeros of ARMA models of constituent speeches. The distribution of the poles of speeches obtained in the experiment has two typical regions: one involves poles which change suddenly with the change from unmixed ($\gamma = 1$) to mixed ($\gamma = 0.8$) speeches, and the other involves poles which change continuously with the change of the mixing weight γ from 1 to 0.2, which is considered to support the analysis. By means of experiments of speaker verification, we have shown that the recall decreases suddenly with the change from unmixed to mixed speeches, or r_{recall} changes from 0.938 to 0.451 for the change of γ from 1 to 0.8 (12 dB). On the other hand, the precision does not decreases so much with the decrease of SNR until -12 dB, or r_{prec} changes from 0.896 to 0.832 for the change of γ from 0.8 (12 dB) to 0.2 (-12 dB).

One of the application of the present method will be speaker detection to find out a specific speaker in an audio stream, which will be in our future research.

References

1. Campbell, J.P.: Speaker recognition: a tutorial. Proc. IEEE **85**(9), 1437–1462 (1997)
2. Beigi, H.: Fundamentals of Speaker Recognition. Springer, New York (2011). https://doi.org/10.1007/978-0-387-77592-0
3. Kurogi S., Ueno T., Sawa M.: A batch learning method for competitive associative net and its application to function approximation. In: Proceedings of SCI 2004, vol. V, pp. 24–28 (2004)
4. Kurogi, S., Mineishi, S., Sato, S.: An analysis of speaker recognition using bagging CAN2 and pole distribution of speech signals. In: Wong, K.W., Mendis, B.S.U., Bouzerdoum, A. (eds.) ICONIP 2010, part I. LNCS, vol. 6443, pp. 363–370. Springer, Heidelberg (2010). https://doi.org/10.1007/978-3-642-17537-4_45
5. Kurogi S., Nedachi N.: Reproduction and recognition of vowels using piecewise linear predictive coefficients obtained by competitive associative nets. In: Proceedings of SICE- ICCAS2006, CD-ROM (2006)

6. Sakashita, S., Takeguchi, S., Matsuo, K., Kurogi, S.: Probabilistic prediction for text-prompted speaker verification capable of accepting spoken words with the same meaning but different pronunciations. In: Hirose, A., Ozawa, S., Doya, K., Ikeda, K., Lee, M., Liu, D. (eds.) ICONIP 2016, part IV. LNCS, vol. 9950, pp. 312–320. Springer, Cham (2016). https://doi.org/10.1007/978-3-319-46681-1_38

7. Sakata, K., Sakashita, S., Matsuo, K., Kurogi, S.: Speaker detection in audio stream via probabilistic prediction using generalized GEBI. In: Hirose, A., Ozawa, S., Doya, K., Ikeda, K., Lee, M., Liu, D. (eds.) ICONIP 2016, part IV. LNCS, vol. 9950, pp. 302–311. Springer, Cham (2016). https://doi.org/10.1007/978-3-319-46681-1_37

8. Bronkhorst A.W.: The cocktail-party problem revisited: early processing and selection of multi-talker speech. Atten. Percept. Psychophys. (2015). https://doi.org/10.3758/s13414-015-0882-9

9. Wang, Y., Sun, W.: Multi-speaker recognition in cocktail party problem. In: Proceedings of International Conference on Communications, Signal Processing, and Systems arXiv:1712.01742 (2017)

10. Bimbot, N., et al.: A tutorial on text-independent speaker verification. J. Appl. Signal Process. **2004**, 430–451 (2004)

11. Kurogi, S.: Improving generalization performance via out-of-bag estimate using variable size of bags. J. Jpn. Neural Netw. Soc. **16**(2), 81–92 (2009)

12. Aldhaheri, W.R., Al-Saadi, F.E.: Robust text-independent speaker recognition with short utterance in noisy environment using SVD as a matching measure. J. King Saud Univ. Comput. Inf. Sci. Arch. **17**, 25–44 (2004)

13. Kurogi, S., Sato, S., Ichimaru, K.: Speaker recognition using pole distribution of speech signals obtained by bagging CAN2. In: Leung, C.S., Lee, M., Chan, J.H. (eds.) ICONIP 2009, part I. LNCS, vol. 5863, pp. 622–629. Springer, Heidelberg (2009). https://doi.org/10.1007/978-3-642-10677-4_71

Deep Adaptive Update of Discriminant KCF for Visual Tracking

Xin Ning[1,2], Weijun Li[1(✉)], Weijuan Tian[2], Xuchi[2], Dongxiaoli[1,2], and Zhangliping[1,2]

[1] Institute of Semiconductors,
Chinese Academy of Sciences, 100083 Beijing, China
{ningxin,Wjli,Dongxiaoli,zliping}@semi.ac.cn
[2] Cognitive Computing Technology Wave Joint Lab, 100083 Beijing, China
{tianweijuan,xuchi}@wavewisdom-bj.com

Abstract. In order to solve the challenges of In-plane/Out-of-plane Rotation (IPR/OPR), fast motion (FM) and occlusion (OCC), a new robust visual tracking framework combining an adaptive template update strategy and tracking validity evaluation, named (AU_DKCF) is presented in this paper. Specifically, the proposed appearance discriminant models are firstly used to determine the tracking validity, and then a new adaptive template update strategy is introduced, which provides an efficient update mechanism to distinguish IPR/OPR from FM and OCC states, and furthermore, a new visual tracking framework AU_DKCF is presented, which combines object detection to distinct FM and OCC states. We implement two versions of the proposed tracker with the representations from both conventional hand-crafted and deep convolution neural networks (CNNs) based features to validate the strong compatibility of the algorithm. Experiment results demonstrate the state-of-the-art performance in tracking accuracy and speed for processing the cases of IPR/OPR, FM and OCC.

Keywords: Visual tracking · Kernelized correlation filters
Discriminant model · Convolution neural network · Object detection

1 Introduction

Recently, computer vision is developing very fast, and meanwhile, visual tracking has significant academic value and wide application prospect, which make it rapidly and successfully applied in many fields [1, 2], e.g., smart surveillance, robotic services, medical imaging, etc. Many scholars and institutions have conducted lots of researches with demonstrated success, however, IPR/OPR, FM and OCC are still the bottlenecks that restrict the wide application of visual tracking.

Kernelized correlation filters (KCF) [3] attracts many scholars' attention for the efficient computation ability and superior performance in object tracking. KCF creatively utilizes the fast Fourier transform to conduct template training and object detection on the densely-sampled samples, and accordingly, achieves the real-time tracking. However, no extra processing is adopted for cases of FM and OCC. And meanwhile, each frame is used for template update to reduce the effects of few training

© Springer Nature Switzerland AG 2018
L. Cheng et al. (Eds.): ICONIP 2018, LNCS 11306, pp. 441–451, 2018.
https://doi.org/10.1007/978-3-030-04224-0_38

samples, while plentiful background information is added into the template, leading to the rectangle drift and performance decline.

Aiming at the problems above, the latest tracking methods mainly make promotions from two aspects: one is the optimization and design of object features, and the other is the design of robust classifiers.

The optimization and design of object features is an important research clue, which focus on the design of the object representation. In [4–8], new feature representations are achieved based on traditional features. Considering the superiority of feature representation in deep learning, deep convolutional neural networks (CNNs) are adopted to mine high-level convolutional features, and the tracking performance is further improved [9–13]. To be specific, Ma et al. [9] exploit features from pre-trained deep CNNs and learn adaptive CFs on several CNN layers to improve tracking accuracy and robustness. Wang et al. [10] present a sequential training method for CNN that is regarded as an ensemble with each channel of the output feature map as an individual base learner. These methods validate the strong capacity of CNNs for the target representation at the cost of time consumption and high requirements of computational resources.

In summary, the feature design methods make certain achievements. However, the tracking algorithms aim to distinguish background and object region, and the learned features only represent the object vision information, and the constraint for background is not referenced, which causes the quite limited discrimination of the template. Accordingly, how to obtain discriminative features is still a difficulty.

The design of robust classifiers used to construct optimization mechanism and restrict the samples for template update, and thus the detection effect is assured. In order to fully utilize the advantage that Structured output SVM can deal with complex outputs like trees, sequences, or sets rather than class labels, Hare et al. [14] employ this algorithm in the visual tracking for the first time and improve tracking accuracy considerably. In order to further reduce computation complexity, Ning et al. [15] propose a dual linear structured SVM (DLSSVM) algorithm, which approximates nonlinear kernels with explicit feature maps. DLSSVM improves tracking performance significantly, while its tracking speed is not fast enough for real-time applications. Consequently, Wang et al. [16] propose a high-confidence update strategy, which uses samples with high confidence, and thus avoids the introduction of background information. Nevertheless, the varying object shape results in smaller confidence, and thus the shape variation can't be captured.

The two kinds of methods above aim to design more complex features and classifiers, which improves the tracking accuracy, and yet undoubtedly increase the complexity, and thus restrict the real-time performance of visual tracking. Meanwhile, few approaches are used for fast motion and occlusion.

2   Proposed Works

In this part, we describe the proposed new framework AU_DKCF in details, illustrated in Fig. 1. This framework is divided into four parts: (1) Initial information extraction, tracking template and discriminant model training. (2) Appearance discriminant model for tracking quality evaluation. (3) Adaptive template update strategy definition. (4) Determination of Object states as IPR/OPR, FM and OCC for long-term and high-speed tracking. Specifically, the training and update of appearance discriminant model is marked in light blue, and the adaptive template update strategy is marked in light orange in Fig. 1.

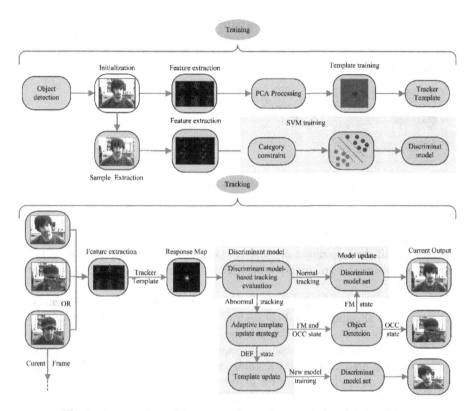

Fig. 1. An overview of the proposed new framework for visual tracking.

2.1   Tracking Evaluation with Appearance Discriminant Models

In order to test if the tracked object is a ghost, a new tracking validity index based on appearance discriminant models is proposed, which provides the confidence of tracking results in a certain appearance state, and thus contributes to efficient object tracking. The specifics of proposed tracking validity index are as follows. Furthermore, the object state corresponding to the proposed appearance discriminant models and the update process are detailedly shown in Fig. 2.

Step 1. Object detection is conducted, and then the initial object information is obtained, including the initial position and size of the object.

Step 2. Certain step size is firstly utilized to densely sample multiple sample images. And then, assuming the overlap threshold being T_{ovp} between the obtained

Discriminant appearance state and model update

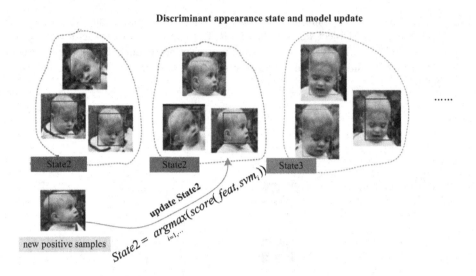

Fig. 2. Discriminant appearance state and model update strategy

samples and the object, the overlap is computed accordingly. Specifically, if it is more than T_{ovp}, the sample label is set as 1. Otherwise, 0.

Step 3. With the inspiration of PCA-net [17], the dimension reduction of HOG features for all samples are further conducted to obtain the low-dimensional representation *feat*.

Step 4. For all training samples, step 3 is repeated to obtain low-dimensional representation of all samples. And the discriminant model svm_1 is learned and added into the model set SVM_{obj}, each of which corresponds to one appearance state.

Step 5. Aiming at the easily confused samples, category constraint is introduced into the objective function of SVM learning. Specifically, if the following misclassifications occur, that is: (1) $feat_{obj}$ of the object is misclassified as background, namely that the score on the object is smaller than that on background; (2) $feat_{bgd}$ of the background is misclassified as object, namely that the score belonging to background is smaller than that to object. In order to correct the above misclassifications, the category constraint is introduced as Eq. (1).

$$score(feat_{obj}, svm_1) - score(feat_{bgd}, svm_1) > 0 \qquad (1)$$

Equation (1) assures that the positive samples shouldn't be classified into the background, and meanwhile, background samples not into object class, and thus the misclassifications occurring in the training set can be effectively corrected, and the discriminative ability of classifier is improved.

Step 6. KCF is used to obtain tracking result, and then the classification result $score(feat, SVM_{obj})$ that descriptor *feat* corresponding to the tracking area on SVM_{obj} are obtained to judge if the current tracking state trk_sta is normal, namely the current rectangle contain the true object, which is determined as Eq. (2).

$$trk\_sta = \begin{cases} normal, & if\ score(feat, SVM_{obj}) > T_{sco} \\ abnormal, & otherwise \end{cases} \quad (2)$$

where T_{sco} is the threshold of tracking results on the discriminant model, and $score(feat, SVM_{obj})$ is defined as follows.

$$score(feat, SVM_{obj}) = max(score(feat, svm_1), \ldots, score(feat, svm_n)) \quad (3)$$

where n is the current number of appearance discriminant models. Equation (3) shows what appearance state the object is in now.

Specifically, if trk_sta is normal, it means that the tracking result is reliable; Otherwise, it indicates that the tracking is abnormal and the object is in one of the three states, IPR/OPR, FM and OCC.

Step 7. If the object is in abnormal tracking state, and further judged in IPR/OPR state by the method in Sect. 2.2, namely that the object appearance changes largely, then steps 1–5 is repeated to learn new discriminant model svm_2, which is added into SVM_{obj}; Otherwise, the tracking results is retained for subsequent model update.

2.2 Minimum-Based Template Update Strategy

During object tracking, it can be observed that the tracking confidence may present an upward parabola in several continuous frames, which can be interpreted to two cases: (1) The object shape first deviates from the original template, and then recovers similarly to template gradually, such as in-plane rotation. (2) The object shape deviates far away from the original template, and meanwhile the background exists an area similar to the template, which causes the tracking drifting to the background, such as the out-of-plane rotation.

Further analysis indicates that the object shape changes in the two cases above, and the template needs to be updated immediately. Consequently, the adaptive update strategy is defined as follows: the tracking confidence of continuous several frames present as an upward parabola. Figure 3 shows the curves of response value vs frame num in the two above cases of object rotation. The specific steps of proposed update strategy are as follows.

Step 1. KCF is applied to the inputted frame to obtain tracking results and corresponding confidence.

Step 2. The above step is repeated to extract the confidence set for continuous num frames $Conf = \{conf_1, \ldots, conf_{num}\}$, and then the variation trend is analyzed. The update flag F indicates that if the tracking result corresponds to the minimal confidence. F is determined by Eq. (4).

$$F = \begin{cases} 1, & if(\arg\min(Conf) == num - 1) \\ 0, & otherwise \end{cases} \quad (4)$$

Equation (4) shows that if the variation trend $Conf$ presents as an upward parabola, namely that the first $num - 1$ frames decreases, while $conf_{num}$ increases, namely that the minimum occurs at $conf_{num-1}$, and then F is set as 1. Otherwise, F is 0.

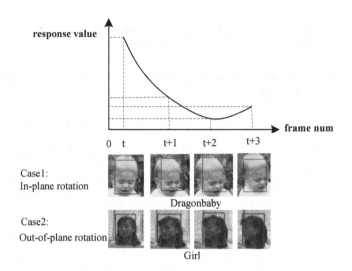

Fig. 3. Response value vs frame num in two cases of object rotation

Step 3. If F is 1, the tracking result corresponding to the minimum is used for template update. Furthermore, considering that the template possesses memory property with a certain weight coefficient, the following $num - 1$ frames are all used for template update.

Step 4. If the template is updated, the confidence for all of the num frames is removed. Otherwise, the last $num - 1$ frames are retained when the minimum occurs at Fra_{num}, while only $conf_{num}$ is retained in other cases. And then, steps 1–4 are repeated. New confidence set $Conf'_{num}$ is updated as follows.

$$Conf'_n = \begin{cases} \{\}, & if\ (\arg\min((Conf) == num - 1) \\ \{conf_2, \ldots, conf_{num}\}, & if(\arg\min((Conf) == num) \\ \{conf_{num}\}, & otherwise \end{cases} \quad (5)$$

2.3   Object Tracking New Framework

In this paper, an adaptive updated discriminant KCF tracking new framework is presented, which is able to supervise tracking condition, and meanwhile capture and distinguish the abnormal tracking state from IPR/OPR, FM and OCC, and furthermore, the loss recovery mechanism is designed to assure a long-term and high-speed tracking. The specifics are as follows.

Step 1. Input image and utilize the approach proposed in Sect. 2.1 until the tracking state is abnormal, namely the current object is in one of the three states, IPR/OPR, FM and OCC.

Step 2. Utilize the proposed update strategy in Sect. 2.2 to obtain F. $F = 1$ means that the object in IPR/OPR state, and it is required to update the template. $F = 0$ means that little effect is made on the object shape variation, and thus the object is in state of FM or OCC.

Step 3. Conduct object detection. If the object is correctly detected, it means that the object is in FM state, and the detected object, rather than tracking object is used for subsequent tracking. Otherwise, it means that the object is in OCC state. In this case, assuming that occlusion has the property of small displacement in object motion and short duration span, the tracking result is kept as that of the last frame in following several frames until the tracking state is normal or the object is correctly detected.

3 Experiments and Analyses

In order to demonstrate the validity of the proposed framework, we first analyze AU_DKCF on 50 videos in OTB-100 dataset [18]. Then, several sequences with IPR/OPR, FM and OCC attributes are respectively used to evaluate the robustness of our framework. One-pass evaluation (OPE) is adopted for evaluation, where the mean precision indicates the percentage of frames in which the estimated locations are within 20 pixels compared to the ground-truth positions. The success scores are defined as the average of the success rates corresponding to the sampled overlap threshold 0.6. Meanwhile, FPS indicates frame number per second.

3.1 Datasets and Experimental Setups

We implement experiment on the OTB-100 benchmark datasets. All sequences are annotated with 11 attributes which cover various challenges, including scale variation (SV), OCC, illumination variation (IV), motion blur (MB), IPR/OPR, FM, and etc. These attributes are useful for characterizing the behavior of trackers. Our tracker is implemented for AU_DKCF and DAU_DKCF with a 3.10 GHz CPU (Fig. 4).

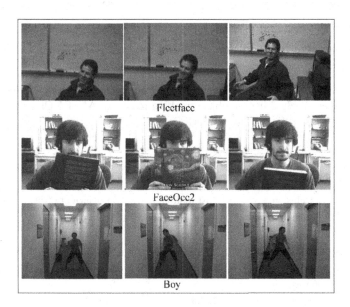

Fig. 4. Some example frames of OTB-100 video sequences

The experiment parameter settings are as follows. (1) Frame number n used to judge if the minimum exists is set as 4. (2) The overlap threshold T_{ovp} is set to 0.7. (3) T_{sco} is also set to 0.7.

3.2 Experiments with New Framework

Table 1 list the results of our AU_DKCF and DAU_DKCF as well as contrastive methods. It is obvious that relatively high performance is obtained, which is superior to those in contrastive methods. The reasons lie in that the proposed AU_DKCF and DAU_DKCF distinguishes the object state once tracking failure, and meanwhile utilizes different mechanisms to recover the lost object, and thus realizes the long-term and efficient tracking. AU_DKCF runs faster than 143 fps, while DAU_DKCF runs faster than 71 fps.

Table 1. Object tracking results of proposed method and contrastive methods on OTB-50

Method	OPE		FPS
	Mean precision (20 px) (%)	Success rate (%)	
KCF [3]	73.20	52.00	172.00
DCF [3]	72.80	50.40	292.00
LMCF [16]	83.90	62.40	85.23
Struck [19]	65.60	38.38	20.00
TLD [20]	60.80	33.40	28.00
ORIA [21]	45.70	23.10	9.00
AU_DKCF	**85.40**	**64.30**	**143.00**
DAU_DKCF	**88.00**	**66.10**	**71.47**

3.3 Experiments with Sequence Attributes

In this section, the experiment of object tracking on OTB-100 dataset with specific attributes, including IPR/OPR, FM and OCC are conducted to demonstrate robustness to the challenges. Table 2 list the object tracking results of AU_DKCF and contrastive methods on videos with specific attributes.

Table 2. Object tracking results of proposed methods and contrastive methods on OTB-100 with sequence attributes

Method	Occlusion	Fast motion	In-plane/out-of-plane rotation
	Mean precision (%)	Mean precision (%)	Mean precision (%)
KCF [3]	74.90	65.00	74.00
Struck [19]	58.40	53.90	52.10
TLD [20]	58.30	57.60	51.20
LMCF [16]	62.70	–	58.85
AU_DKCF	**76.60**	**73.80**	**81.9**

From Table 2, it is noticed that a large improvement is achieved with AU_DKCF regarding IPR/OPR, FM and OCC. Under the abnormal states, AU_DKCF is able to distinguish IPR/OPR from the three attributes by using the proposed adaptive template update strategy. Afterwards, the discrimination and failure recovery mechanism is made out for the object in FM state, which focus on object re-detection. In the case of OCC, the properties of small displacement and short duration are applied to recover the object. Consequently, the tracking accuracy are undoubtedly improved, and the tracking continuity is also assured. However, the methods like KCF, TLD, CT and LMCF fail to judge which attribute causes the tracking failure in real-time, and meanwhile too much time is spent on the object recovery.

3.4 Qualitative Results

To visualize the impact our framework makes on tracking performance, we show examples with attributes IPR/OPR, FM and OCC of each baseline method compared to our framework on sample videos from OTB-100 in Fig. 5. It can be easily observed that the proposed new tracking framework achieves better performance, especially in processing videos with attributes IPR/OPR, FM and OCC.

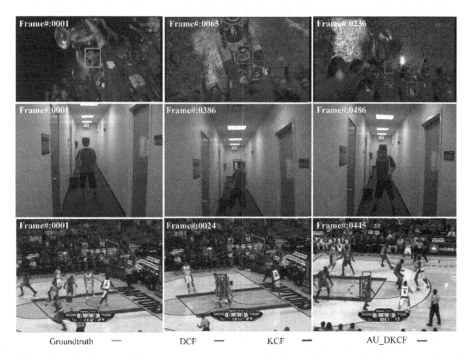

Fig. 5. Tracking results of two baseline CF trackers compared to proposed AU_DKCF. Videos are (from top to bottom): Soccer, Boy and Basketball

4 Conclusion

On the foundation of adequate analysis, study and experiment, following conclusions are drawn from this paper. (1) A tracking evaluation approach based on discriminant model is proposed, which judges if the current tracking state is normal, in order to recovery the object immediately. (2) A new minimum-based model update strategy is presented, which selects samples with various shapes to improve the generalization ability. (3) A new object tracking framework, with the fusion of tracking evaluation and template update, is proposed, which discriminates the tracked object state in real-time, and meanwhile, finds the lost object by making out different recovery mechanisms, and furthermore, achieves the efficient tracking with high precision and speed. In addition, our further work will devote ourselves to combining the existing methods with deep learning to research more complicated visual tracking.

Acknowledgements. This work was supported by the National Nature Science Foundation of China (No. 61572458).

References

1. Yilmaz, A., Javed, O., Shah, M.: Object tracking: a survey. ACM Comput. Surv. **38**(4), 1–45 (2006)
2. Cannons, K.: A review of visual tracking. Technical report CSE-2008-07, York University, Canada (2008)
3. Henriques, J.F., Caseiro, R., Martins, P., et al.: High-speed tracking with kernelized correlation filters. IEEE Trans. Pattern Anal. Mach. Intell. **37**(3), 583–596 (2015)
4. Danelljan, F.M., Shahbaz, K., Felsberg, M., Van de Weijer, J.: Adaptive color attributes for real-time visual tracking. In: CVPR, pp. 1090–1097. IEEE (2014)
5. Jia, X., Lu, H., Yang, M.-H.: Visual tracking via adaptive structural local sparse appearance model. In: CVPR, pp. 1822–1829. IEEE (2012)
6. Kalal, Z., Mikolajczyk, K., Matas, J.: Tracking-learning detection. IEEE Trans. Pattern Anal. Mach. Intell. **34**(7), 1409–1422 (2012)
7. Kwon, J., Lee, K.M.: Visual tracking decomposition. In: CVPR, pp. 1269–1276. IEEE (2010)
8. Ross, D.A., Lim, J., Lin, R.-S., Yang, M.-H.: Incremental learning for robust visual tracking. Int. J. Comput. Vision **77**(1–3), 125–141 (2008)
9. Hong, S., You, T., Kwak, S., Han, B.: Online tracking by learning discriminative saliency map with convolutional neural network. In: Blei, D., Bach, F. (eds.) ICML, pp. 597–606 (2015)
10. Ma, C., Huang, J.-B., Yang, X., Yang, M.-H.: Hierarchical convolutional features for visual tracking. In: ICCV, pp. 3074–3082 (2015)
11. Qi, Y., et al.: Hedged deep tracking. In: CVPR, pp. 4303–4311. IEEE (2016)
12. Wang, L., Ouyang, W., Wang, X., Lu, H.: Visual tracking with fully convolutional networks. In: ICCV, pp. 3119–3127. IEEE (2015)
13. Wang, L., Ouyang, W., Wang, X., Lu, H.: STCT: sequentially training convolutional networks for visual tracking. In: CVPR (2016)
14. Hare, S., Saffari, A., Torr, P.H.: Struck: structured output tracking with kernels. In: ICCV, pp. 263–270. IEEE (2011)

15. Ning, J., Yang, J., Jiang, S., Zhang, L., Yang, M.-H.: Object tracking via dual linear structured SVM and explicit feature map. In: CVPR, pp. 4266–4274. IEEE (2016)
16. Wang, M., Liu, Y., Huang, Z.: Large margin object tracking with circulant feature maps. In: CVPR, Honolulu, HI, USA (2017)
17. Chan, T.H., Jia, K., Gao, S., et al.: PCANet: a simple deep learning baseline for image classification? IEEE Trans. Image Process. **24**(12), 5017–5032 (2015)
18. Wu, Y., Lim, J., Yang, M.H.: Online object tracking: a benchmark. In: CVPR, pp. 2411–2418 (2013)
19. Hare, S., Saffari, A., Torr, P.: Struck Structured output tracking with kernels. In: ICCV (2011)
20. Kalal, Z., Mikolajczyk, K., Matas, J.: Tracking-learning detection. In: TPAMI (2012)
21. Wu, Y., Shen, B., Ling, H.: Online robust image alignment via iterative convex optimization. In: CVPR (2012)

Evolution of Images with Diversity and Constraints Using a Generative Adversarial Network

Aneta Neumann[✉], Christo Pyromallis, and Bradley Alexander

Optimisation and Logistics, School of Computer Science, The University of Adelaide,
Adelaide, Australia
aneta.neumann@adelaide.edu.au

Abstract. Generative Adversarial Networks (GANs) are a machine learning approach that have the ability to generate novel images. Recent developments in deep learning have enabled a generation of compelling images using generative networks that encode images with lower-dimensional latent spaces. Nature-inspired optimisation methods has been used to generate new images. In this paper, we train GAN with aim of generating images that are created based on optimisation of feature scores in one or two dimensions. We use search in the latent space to generate images scoring high or low values feature measures and compare different feature measures. Our approach successfully generate image variations with two datasets, faces and butterflies. The work gives insights on how feature measures promote diversity of images and how the different measures interact.

1 Introduction

In recent years, Generative Adversarial Networks (GANs) [13] have been used to map low dimensional real-valued latent vector into images. There has been work in using GANs to generate and mix novel images [24] and perform style transfer [10], along with other applications. However, to date there has no work using evolutionary to explore the latent space of a GAN to generate images according to feature measures. Evolutionary search has frequently been used to generate artistic images [4,15,16]. In previous work [2,4], have either reduced the dimensionality of the search space through programmatic encodings or have constrained the images with priors [3,21]. In this work, we employed GANs to generate novel images by evolving the latent vector to maximise and minimise single and two-dimensional image feature values for two datasets, faces and butterflies. In our system, images are created by optimising the latent space of the GAN in order to create images that score high or low on feature measures. We show that the generation of images in this space requires the use of carefully constructed constraints on image realism. We also show that GANs trained on different image sets appear to impose different bounds on the values of features measures that can be evolved. We show that GANs can be successful in generating of photorealistic images and used nature-inspired methods in order to create

© Springer Nature Switzerland AG 2018
L. Cheng et al. (Eds.): ICONIP 2018, LNCS 11306, pp. 452–465, 2018.
https://doi.org/10.1007/978-3-030-04224-0_39

diverse images. By training GANs we can apply our approach for the discovery of new images that can be used in field of designing video games.

The paper is structured as follows. Section 2 outlines related work. Section 3 presents the methodology used for evolving images. Section 4 presents our results on single dimensional feature experiments with constraint. The main approach is described in Sect. 5. Finally, Sect. 6 discusses results and future work.

2   Related Work

Aesthetic feature measures have been often applied to the creation of new artistic images using evolutionary search [4,15,16,22]. There has also been work in the evolution of existing images [20,23]. This work differs from previous work in the use of a GAN as a mapper from latent search vector to the image feature space and also in the use of the discriminator network and feature metrics to constrain these images.

In terms of deep learning, Gatys [9–11] used a convolutional network to transfer artistic style into existing image. These new approaches in network architectures and training methods enabled the generation of realistic images [7,26]. Recently Dosovitskiy and Brox [6] trained networks of generating images from feature vector and combining an auto-encoder-style approach with Deep Convolutional Generative Adversarial Networks training. Furthermore, Nguyen [24] used priors from Deep Generative Network to generate image variants that look close to natural within a preferred inputs for neurons.

The nature-inspired algorithm for diversity optimisation was introduced in [8], based on optimisation for Traveling Salesman Problem (TSP) which is a NP-hard combinatorial optimisation problem with real world applications. For our investigation we consider recent work [3] on feature based diversity optimisation for images.

3   Methodology

In this section, the methods used to evolve images are discussed. Follows, the descriptions of technical system, features, and features optimization applied in this paper.

3.1   Our System

Now, we describe our system that is based on Generative Adversarial Networks (GANs) [13]. In the nutshell, GANs are based on a two-player game in which the generator network produces sample, competes against the discriminator network, that has to distinguish between the training data and generator samples [12]. Figure 1 shows the structure of our system. We train two

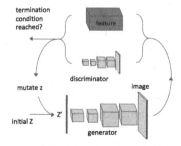

Fig. 1. The final setup of the system. The latent vector Z is randomly seeded and sent through the system, mutating until it reaches an optimal solution or the termination condition is reached.

networks: (1) a generator, to generate images from a latent vector Z of 100 real numbers, and (2) a discriminator which scores the images in the generator for realness as both networks are trained. We make two copies of the GAN formed by these networks and train one with two different image datasets: the Celebfaces attributes (*CelebA*) dataset containing more than 150k celebrity faces [17] and the other with imagenet [5] class of butterflies containing over 45k butterfly images. In this paper, we present our experiments as a proof-of-concept.

Our GAN implementation is build on a modified version of the PyTorch GAN code [1]. The generator component of the network consists of 5 deconvolutional layers. The activation functions for the first 4 layers are a ReLU activation [19]. The last deconvolutional layer uses Tanh. The discriminator uses LeakyReLU activation in all layers. The generator takes 100 elements of vector Z as input and generates a 128 × 128 pixels image. The discriminator takes an image and generates a normally distributed *realness* score with the most real images scoring zero. As a final step, Fig. 1 (at the top) shows, we applied the feature function, as described in [3].

The GAN and the necessary feature functions are linked together to drive evolution. The combined system works as follows. A randomly initialised latent feature vector is sent into the generator, which outputs an image. This image is run through both the chosen feature function and the discriminator, and both contribute to a score. The evolutionary process, guided by the score, mutates Z with the goal of optimising both the realness and the desired feature of the output image.

3.2   Features

This section describes in more detail the features used in our experiments. We denote a function f for an image I, representing feature. This function maps an image I to a scalar value $f(I)$. The features are neutral measures of the properties of an image. We intend to use features that taken from the literature [25], and have been empirically derived from surveys studies or from databases of popular images. For our experiments we use the following features: hue, mean-saturation, smoothness, Reflectional Symmetry [15] and Global Contrast Factor [18]. The features are defined as follows.

Hue is the value of the hue of every pixels in the image I. The range of *Hue* is $[0, 1]$. Note that both 0 and 1 represent the colour red and hue has a periodic quantity.

Mean-saturation is the mean value of saturation for all pixels of image I. The range is $[0, 1]$ with 0 representing low saturation and 1 representing high saturation. *Smoothness* of an image is computed, for a given image I with N pixels as:

$$1 - \sum_{i=1}^{N} \sum_{c=1}^{3} gradient(I_{ic})/3N,$$

where *gradient* is the gradient magnitude image produced by MATLAB *intermediate* image gradient method, which calculated gradients between adjoining pixel values on each colour channel. We perceive from this thus $smoothness(I)$ is the disparity of colour between adjacent pixels and also lies within the range $[0, 1]$. *Reflectional Symmetry* is a measure based on den Heijer's work [15] to measure the degree which an image reflects itself. Symmetry divides an image into four quadrants and measures horizontal, vertical, and diagonal symmetry. Note Symmetry is defined for image I as:

$$Symm(I) = S_h(I) + S_v(I) + S_d(I)/3.$$

Global Contrast Factor (GCF) is a measure of mean contrast between neighbouring pixels at different image resolutions. GCF is determined by calculating the local contrast at each pixel at resolution r:

$$lc_r(I_{ij}) = \sum\nolimits_{I_{kl} \in N(I_{ij})} |lum(I_{kl}) - lum(I_{ij})|$$

where $lum(P)$ is the perceptual luminosity of pixel P and $N(I_{ij})$ are the four neighbouring pixels of I_{ij} at resolution r. The mean local contrast at the current resolution is defined as:

$$C_r = (\sum\nolimits_{i=1}^{m} \sum\nolimits_{j=1}^{n} lc_r(I_{ij}))/(mn).$$

From these local contrasts, GCF is calculated as:

$$GCF = \sum\nolimits_{r=1}^{9} w_r \cdot C_r.$$

The pixel resolutions correspond to different *superpixel* sizes of 1,2,4,8,16,25,50, 100, and 200. Each superpixel is set to the average luminosity of the pixel's it contains. The w_r are empirically derived weights of resolutions from [18] giving highest weight to moderate resolutions.

3.3   Feature Optimisation

In this work we investigated the use of single and two-dimensional features optimization. We explore this optimisation space with respect to feature values. For a single feature our system optimise particular feature. We define the minimization process $feature$ and maximization process $(1.0 - feature)^1$. For two-dimensional features (f, g) we have four optimisation targets representing the combinations of minimizing and maximizing f and g.

---

[1] Note that for feature GCF maximisation is achieved through 1/GCF and scaling in the range $[0, 1]$.

To maintain image realness we penalise the combined score with a measure for realness from the GAN *discriminator*. Thus our fitness functions for single features are shown in Eqs. (1–2) and for two-dimensional features in (3–6).

$$feature \times discriminator \tag{1}$$
$$(1.0 - feature) \times discriminator \tag{2}$$
$$feature\ 1 \times feature\ 2 \times discriminator \tag{3}$$
$$feature\ 1 \times (1.0 - feature\ 2) \times discriminator \tag{4}$$
$$(1.0 - feature\ 1) \times feature\ 2 \times discriminator \tag{5}$$
$$(1.0 - feature\ 1) \times (1.0 - feature\ 2) \times discriminator \tag{6}$$

Fig. 2. The first two images obtained by using evolutionary algorithms: *(1+1) EA.* Two following images obtained without realness constraints by minimizing saturation and hue.

In two-dimensional feature experiments we use 6 feature combinations: hue-saturation, hue-symmetry, saturation-symmetry, smoothness-saturation, GCF-smoothness and GCF-saturation. These combinations were chosen to produce potentially interesting outputs. GCF-smoothness and GCF-saturation were selected due to related work indicating GCF-smoothness would constrain each other [3], resulting in lower image diversity.

4   Preliminary Experiments

We refined the methodology through an experimental process. Initial experimentation used *(1+1) EA* and Covariance Matrix Adaptation Evolutionary Strategy (CMA-ES) [14] frameworks to determine the performance of both algorithms. CMA-ES is well-know for evolving vectors of real numbers. Because of this useful property for optimizing non-linear, non convect problem in the continuous space we applied CMA-ES for 2000 mutations (the equivalent of 80000 iterations), and *(1+1) EA* for 80000 iterations. As illustrated in Fig. 2 CMA-ES was able to achieve more extreme feature value. This superiority of optimisation applied to all feature metrics. We ran all of our experiments on single nodes Intel® Core^{TM} i7-6700 series with 4 core processors. CMA-ES ran for 80 minutes and *(1+1) EA* ran 240 minutes for an optimisation run.

4.1   Feature Experiments Without Realness Constraint

It was initially assumed that the GAN would be able to create face-like images with some input vector. The tests were performed which did not incorporate a constraint on realness as part of the optimisation process. As Fig. 2 demonstrates optimizing features without the discriminator produces abstract images. To constrain images to be more realistic three constraining methods were tested: (1) reducing the degrees of freedom in the covariance matrix, (2) discarding images that failed, according to the discriminator, a certain given realness threshold, and (3) incorporating the discriminator's return value into the optimisation function. It was found that the third option of integrating realness as a variable *discriminator* gave the most visually interesting results. The option of discarding images resulted in CMA-ES failing to progress.

4.2   Single Dimensional Feature Experiments with Constraint

Single feature experiments require the fewest variables to optimise and as such, could be expected to evolve the image with the least difficulty. The results of each feature values are shown in Table 1 (col. 1-2). As can be seen the ranges of features above are very small, with the exception of symmetry. In these runs the *discriminator* term has strong effect on constraining the feature values. For symmetry, the larger range might be explained by presence of both symmetric and asymmetric faces in the training dataset. In line with the small feature ranges, the constrained images only showed small variations as seen in Fig. 3(a)–(b) corresponding to the hue and symmetry measures in Table 1.

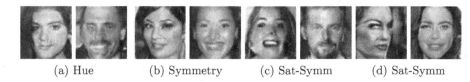

(a) Hue   (b) Symmetry   (c) Sat-Symm   (d) Sat-Symm

Fig. 3. Images (a) and (b) obtained by single features. The first image corresponds to minimizing features hue (0.0841) and symmetry (0.7985). The second image corresponds to maximizing features hue (0.1275) and symmetry (0.9198). Images (c) and (d) obtained by two-dimensional features. The first image corresponds to minimizing and the second image corresponds to maximizing features saturation (Sat) and symmetry (Symm).

Table 1. Single features value with cut-off of 0.2 for faces and butterfly datasets. Single dimensional feature values obtained from experiments with constraint for butterfly datasets.

Feature	MIN-F const	MAX-F const	Min-Face	Max-Face	Min-Butter	Max-Butter
Hue	0.0841	0.1275	**0.0337**	**0.4886**	**0.1083**	**0.5282**
Saturation	0.3306	0.3543	0.2176	0.4954	0.1205	0.5918
Smoothness	0.9737	0.9843	0.9582	0.9843	0.9462	0.9887
Symmetry	**0.7985**	**0.9198**	0.5904	0.9198	0.5568	0.9428
GCF	0.0276	0.0286	0.0106	0.0348	0.0090	0.0417

4.3 Two-Dimensional Feature Experiments with Constraint

Running the experiments on two-dimensional features gave similarly constrained images to the single features. Figure 3(c) and (d) shows images evolved to minimise and maximise saturation and symmetry. As can be seen, there is some success in evolving different amounts of symmetry but not a particularly strong difference in saturation. It appears, for all feature combinations, the realness constraint is preventing strong exploration of the feature space.

4.4 Impact of Cut-Off Function

In order to maintain balance between image realness and exploration we adjusted the discriminator term by passing the raw result of *discriminator* $= x$ for an image. The cut-off function f are defined as follows:

$$f(x) = \begin{cases} x & \text{if } x \geq c \\ s & \text{if } x < c \end{cases} \tag{7}$$

with a cutoff c and stable value s. In the experiments that follow we set s to the value close to zero, returning maximum realness, and adapted c to test its effect. With the cut-off function, the search is unaffected until the image reaches a certain threshold of realness. When the threshold is reached the system has no variation in respect to the realness value, thus giving priority to aesthetic features over realness values. A subjective analysis of possible cut-off values needed to be performed in order to determine the optimal value for future experiments. Figure 4 demonstrates the effect of different cut-off values on both image realness and aesthetic feature value.

A relationship can be observed from the above images. As the cut-off decreases, so does the degree we are able to evolve the feature value. However, it can be noted that even the 0.02 cut-off was able to create a far lower hue compared to the constrained results – while still being realistic enough to be called a face. The 0.02 cut-off was used in the remaining single feature experiments.

5   Results

This section presents the results of evolution in single and two-dimensional image features. The previous experiments were all carried on faces. In the following, GANs generated from both the faces and the butterfly datasets were used.

5.1   Single Dimensional Feature Experiments with Cut-Off Function

We conducted single feature dimension experiments using faces and butterfly datasets and optimised the following features: hue, saturation, smoothness, symmetry and GCF. For these experiments, we use a cut-off of 0.02 on the discriminator output for both GANs. The results were obtained for the minimum and maximum feature values from each experiment. Figure 5 shows the results of the experiments for the single dimensional feature with corresponding to an image minimising (left) and maximising (right) the feature for faces and butterflies respectively.

Fig. 4. The results of minimizing hue with cut-off at 0.2 (left), 0.05 (middle) and 0.02 (right).

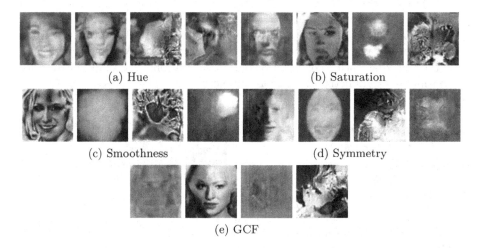

(a) Hue   (b) Saturation

(c) Smoothness   (d) Symmetry

(e) GCF

Fig. 5. All single feature optimisations with 0.02 cut-off.

Table 1 (col. 3 − 6) shows the minimum and maximum value for each feature for the faces (Min-Face, Max-Face) and butterfly datasets (Min-Butter, Max-Butter), respectively. We observe that hue has the highest range with respect to feature value. The use of the butterfly dataset provides good way to see how evolution with aesthetic measure responds to the priors embedded in GAN. For the single dimensional experiments we observe in Fig. 5 that images generated with the faces dataset appear more real than the images generated with butterfly dataset. This is likely to be due to the more diverse nature of the butterfly dataset. The images shown in Fig. 5(a) have the most variance in the hue dimension. Images with the lowest value for hue appear most realistic. In contrast the image with higher value for hue appears less realistic. We observe that image generated from the butterfly dataset achieves a higher feature range. Figure 5(b) shows that in spite of the saturation feature for faces extending over a narrow range the resulting faces are not very realistic. The butterfly dataset is able to produce higher values of saturation, resulting in a realistic and colorful image. In Fig. 5(c) we observe that minimization produces realistic images with superimposed darker shadows. In contrast maximizing smoothness produces less realistic images. The images shown in Fig. 5(d) produce high values for symmetry for both datasets. These images appear symmetrical and less real. Images with lower symmetry value are more real. Finally, the images shown in Fig. 5(e) exhibit less realness for faces and butterfly dataset in minimization and more realism in maximization.

5.2   Two-Dimensional Feature Experiments with Cut-Off Function

In our next experiment, we evolve images using the GAN to minimise and maximise in two feature dimensions. These experiments aim to give us insight into how features interact with each other and also the impact of the image priors as embedded in the GAN on the extent to which features can be optimised. After training our GAN models on the faces and butterfly dataset, We run experiments with the following feature combinations: GCF-saturation; GCF-smoothness; hue-saturation; hue-symmetry; saturation-symmetry; and smoothness-saturation.

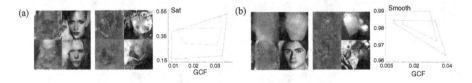

Fig. 6. Images obtained by two-dimensional features with a 0.008 cut-off constraint. Images correspond to 2 feature pairs, GCF, saturation and GCF, smoothness for faces and butterfly dataset, from top left, respectively. Note the images follow their positions on the graph.

The feature pair values resulting from these experiments are shown in Tables 2 (for faces) and 3 (for butterflies). The images corresponding to these values are shown in Figs. 6 and 8. The first column of Figs. 6 and 8, show images, in clockwise order from top-left, for min-max, max-max, max-min, min-min combinations of features for faces. The corresponding pictures butterflies dataset is shown in the second column. The last column, plots the positions in feature-space, of the four face images from the first column (in red) and the four butterfly images from the second column (in blue). The shape of the quadrilateral in these plots provides an indication of how feature values are constrained with respect to each other and by the GAN used to generate them. Figure 7 shows the trace of GCF and Saturation for faces as the evolutionary run proceeds.

Based on our findings from the previous experiments with single dimension diversity, we reduced the cut-off to 0.008 for the two-dimensional feature experiments to try to maintain the realism of the images. The impact of this smaller cut-off can be observed in Figs. 6 and 8 in terms of the relatively small areas of feature space contained by the plots. Looking at two-dimensional features in turn. The images shown in Fig. 6(a) have the highest GCF values and for max-min and max-max optimisation appear most realistic.

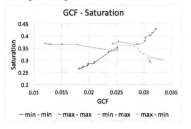

Fig. 7. The trace of GCF and Saturation for the four face images shown in Fig. 6(a).

We observe in Fig. 6(a) that image generated on the butterflies dataset achieve a higher score for GCF and permit a higher range of saturation. The feature plot in Fig. 6 shows that GCF and Saturation can vary independently. In contrast, Fig. 6(b)'s plot indicates some difficulty in minimising both smoothness and GCF. From the plot it also appears to be relatively difficult to simultaneously maximise GCF and smoothness. This result is in concordance with the observations in [3] which found that GCF and smoothness, being spatial features, appeared to be in conflict with each other. More specifically the high contrast required for high GCF scores is in direct conflict with the low contrast required for high smoothness scores. Also notable from Fig. 6(b) is a relative lack of realism in the faces as compared to those in Fig. 6(a). The images shown in Fig. 8(a) illustrate the relationship between hue and saturation. The butterfly pictures show the most variance in the saturation dimension and the face pictures show marginally more variance in hue. Images high in saturation seem to appear sharper – with the face image that maximises both features having quite harsh colour, more contrast, and a mask-like appearance. For Fig. 8(b) both sets of images have similar ranges of symmetry but faces have a much narrower range of hue. Highly symmetric images seem to be less realistic, tending to ovoid shapes, with detail seemly sacrificed in order to maximise symmetry. In contrast asymmetric images appear to have more realistic textures and more intense colours.

Figure 8(c) combines saturation and symmetry. As before, highly symmetric images appear less realistic. The evolutionary process seems to have difficulty maximising both saturation and symmetry for both GANs. Clearly it is possible to create artificial images that score highly on both feature dimensions so this difficulty may be reflective of the rarity of this feature combination in the training sets for these GANs. As a final observation, the butterfly picture maximising both features resembles an insect's face, perhaps an interesting consequence of having diverse images in the training set.

Finally, the images in Fig. 8(d) show difficulty in minimising both smoothness and saturation. There is a smaller corresponding problem in maximising both features. In both data sets the most realistic images are produced by the minimisation of smoothness and the maximisation of saturation, perhaps indicating that the priors in the dataset are biased toward rougher and more colourful images.

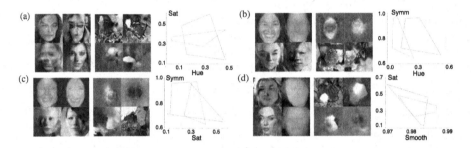

Fig. 8. Images obtained by two-dimensional features with a 0.008 cut-off constraint. The first four images correspond to features hue and saturation, hue and symmetry, saturation and symmetry, smoothness and saturation, from top, respectively. The images follow their positions on the graph.

Table 2. Two-dimensional features with cut-off for faces dataset.

Feature pairs	Min.f1–Min.f2	Min.f1–Max.f2	Max.f1–Min.f2	Max.f1–Max.f2
Hue-Saturation	0.0426–0.3318	0.0457–0.3671	0.2900–0.2727	0.4651–0.4257
Hue-Symmetry	0.0480–0.7233	0.0549–0.9577	0.2523–0.6448	0.0935–0.9722
Saturation-Symmetry	0.2573–0.7020	0.3190–0.9654	0.4256–0.6336	0.3582–0.9679
Smoothness-Saturation	0.9814–0.2378	0.9751–0.4876	0.9901–0.2912	0.9903–0.3670
GCF-Saturation	0.0164–0.2930	0.0126–0.4048	0.0310–0.2796	0.0332–0.4160
GCF-Smoothness	0.0166–0.9890	0.0166–0.9890	0.0326–0.9762	0.0290–0.9893

Table 3. Two-dimensional features with cut-off for butterfly dataset.

Feature pairs	Min.f1–Min.f2	Min.f1–Max.f2	Max.f1–Min.f2	Max.f1–Max.f2
Hue-Saturation	0.1622–0.1956	0.1218–0.4772	0.4659–0.1402	0.3458–0.5912
Hue-Symmetry	0.1286–0.7232	0.1744–0.9558	0.5067–0.6701	0.3075–0.9614
Saturation-Symmetry	0.1447–0.7133	0.1594–0.9554	0.5760–0.6512	0.3014–0.9647
Smoothness-Saturation	0.9840–0.1469	0.9706–0.6177	0.9852–0.1291	0.9868–0.4661
GCF-Saturation	0.0100–0.1743	0.0102–0.3820	0.0334–0.1983	0.0357–0.5205
GCF-Smootness	0.0094–0.9826	0.0093–0.9840	0.0378–0.9638	0.0319–0.9838

6   Discussion and Future Work

In this work, we have shown how to apply GAN in order to generate images scoring high or low for given feature values. We used evolutionary search to maximise and minimise single features and pairs of features for two datasets, faces and butterflies. We have shown that GANs known for their successful generation of photorealistic images with combination of evolutionary search can be a powerful technique for creating novel images. We have presented a novel latent variable approach based on nature-inspired methods and have shown how to explore the latent space of a GAN to create semi-realistic images that sample different regions of feature spaces. Additionally, we studied the effects of different values of the cut-off function on the appearance of the images. Finally, our experimental results on 2 datasets demonstrated that proposed approach is promising image transformation in regards to maximizing diversity in single- and two-dimensional spaces applications.

For future research, it would be interesting to explore intermediate points in the feature space to gain more insight into the relationships between features and to explore additional constraints and their effect on the process of generating novel images. The potential are of the future work is to use Multi-Objective Optimization Algorithms to evolve the latent vector in terms of choosing multiple-criteria. The work can be extend by exploring different aesthetic features and can be use in fields of industrial design, entertainment (video games) or architecture that can beneficial to the exploration of variant solutions.

References

1. PyTorch tutorial: Dcgan. https://github.com/yunjey/pytorch-tutorial
2. McCormack, J., d'Inverno, M. (eds.): Computers and Creativity. Springer, Heidelberg (2012). https://doi.org/10.1007/978-3-642-31727-9
3. Alexander, B., Kortman, J., Neumann, A.: Evolution of artistic image variants through feature based diversity optimisation. In: GECCO, pp. 171–178 (2017)
4. Correia, J., Machado, P., Romero, J., Carballal, A.: Evolving figurative images using expression-based evolutionary art. In: ICCC, p. 24 (2013)

5. Deng, J., Dong, W., Socher, R., Li, L.J., Li, K., Fei-Fei, L.: Imagenet: a large-scale hierarchical image database. In: CVPR, pp. 248–255. IEEE Computer Society (2009)
6. Dosovitskiy, A., Brox, T.: Generating images with perceptual similarity metrics based on deep networks. In: NIPS, pp. 658–666 (2016)
7. Dosovitskiy, A., Springenberg, J.T., Brox, T.: Learning to generate chairs with convolutional neural networks. In: CVPR, pp. 1538–1546. IEEE Computer Society (2015)
8. Gao, W., Nallaperuma, S., Neumann, F.: Feature-based diversity optimization for problem instance classification. In: Handl, J., Hart, E., Lewis, P.R., López-Ibáñez, M., Ochoa, G., Paechter, B. (eds.) PPSN 2016. LNCS, vol. 9921, pp. 869–879. Springer, Cham (2016). https://doi.org/10.1007/978-3-319-45823-6_81
9. Gatys, L.A., Ecker, A.S., Bethge, M.: Texture synthesis using convolutional neural networks. In: NIPS, pp. 262–270 (2015)
10. Gatys, L.A., Ecker, A.S., Bethge, M.: Image style transfer using convolutional neural networks. In: CVPR, pp. 2414–2423. IEEE Computer Society (2016)
11. Gatys, L.A., Ecker, A.S., Bethge, M., Hertzmann, A., Shechtman, E.: Controlling perceptual factors in neural style transfer. In: CVPR, pp. 3730–3738. IEEE Computer Society (2017)
12. Goodfellow, I., Bengio, Y., Courville, A.: Deep Learning. MIT Press, Cambridge (2016)
13. Goodfellow, I., et al.: Generative adversarial nets. In: Advances in Neural Information Processing Systems, pp. 2672–2680 (2014)
14. Hansen, N., Müller, S.D., Koumoutsakos, P.: Reducing the time complexity of the derandomized evolution strategy with covariance matrix adaptation (CMA-ES). Evol. Comput. $11(1)$, 1–18 (2003)
15. den Heijer, E., Eiben, A.E.: Investigating aesthetic measures for unsupervised evolutionary art. Swarm Evol. Comput. 16, 52–68 (2014)
16. Kowaliw, T., Dorin, A., McCormack, J.: Promoting creative design in interactive evolutionary computation. IEEE Trans. Evol. Comput. $16(4)$, 523–536 (2012)
17. Liu, Z., Luo, P., Wang, X., Tang, X.: Deep learning face attributes in the wild. In: ICCV (2015)
18. Matkovic, K., Neumann, L., Neumann, A., Psik, T., Purgathofer, W.: Global contrast factor-a new approach to image contrast. In: Computational Aesthetics, pp. 159–168 (2005)
19. Nair, V., Hinton, G.E.: Rectified linear units improve restricted Boltzmann machines. In: ICML, pp. 807–814 (2010)
20. Neumann, A., Alexander, B., Neumann, F.: The evolutionary process of image transition in conjunction with box and strip mutation. In: Hirose, A., Ozawa, S., Doya, K., Ikeda, K., Lee, M., Liu, D. (eds.) ICONIP 2016. LNCS, vol. 9949, pp. 261–268. Springer, Cham (2016). https://doi.org/10.1007/978-3-319-46675-0_29
21. Neumann, A., Alexander, B., Neumann, F.: Evolutionary image transition using random walks. In: Correia, J., Ciesielski, V., Liapis, A. (eds.) EvoMUSART 2017. LNCS, vol. 10198, pp. 230–245. Springer, Cham (2017). https://doi.org/10.1007/978-3-319-55750-2_16
22. Neumann, A., Neumann, F.: Evolutionary computation for digital art. In: GECCO, pp. 937–955. ACM (2018)

23. Neumann, A., Szpak, Z.L., Chojnacki, W., Neumann, F.: Evolutionary image composition using feature covariance matrices. In: GECCO, pp. 817–824 (2017)
24. Nguyen, A., Dosovitskiy, A., Yosinski, J., Brox, T., Clune, J.: Synthesizing the preferred inputs for neurons in neural networks via deep generator networks. In: NIPS, pp. 3387–3395 (2016)
25. Nixon, M., Aguado, A.S.: Feature Extraction & Image Processing, 2 edn. Academic Press, Boston (2008)
26. Radford, A., Metz, L., Chintala, S.: Unsupervised representation learning with deep convolutional generative adversarial networks. arXiv preprint arXiv:1511.06434 (2015)

Efficient Single Image Super Resolution Using Enhanced Learned Group Convolutions

Vandit Jain[⊠], Prakhar Bansal, Abhinav Kumar Singh, and Rajeev Srivastava

Indian Institute of Technology (Banaras Hindu University), Varanasi, India
{vandit.jain.cse15,prakhar.bansal.cse15,abhinavkr.singh.cse15,
rajeev.cse}@iitbhu.ac.in

Abstract. Convolutional Neural Networks (CNNs) have demonstrated great results for the single-image super-resolution (SISR) problem. Currently, most CNN algorithms promote deep and computationally expensive models to solve SISR. However, we propose a novel SISR method that uses relatively less number of computations. On training, we get group convolutions that have unused connections removed. We have refined this system specifically for the task at hand by removing unnecessary modules from original CondenseNet. Further, a reconstruction network consisting of deconvolutional layers has been used in order to upscale to high resolution. All these steps significantly reduce the number of computations required at testing time. Along with this, bicubic upsampled input is added to the network output for easier learning. Our model is named SRCondenseNet. We evaluate the method using various benchmark datasets and show that it performs favourably against the state-of-the-art methods in terms of both accuracy and number of computations required.

Keywords: Convolutional Neural Networks · Deep learning
Image super resolution · Learned group convolutions

1 Introduction

Super Resolution (SR) problem is defined as recovering a high resolution image from a low resolution image. This is a highly ill-posed problem with multiple solutions possible for a single input image. This problem finds many applications such as medical imaging, security and surveillance among others.

In recent years, deep learning methods have performed better as compared to interpolation-based [2], reconstruction-based [6,7] or other example-based methods [3–5,8] that have been used in the past. This is proved by the fact that the first effort in the direction of deep learning for solving the problem of single image super resolution [9] performed better than several previous models not using deep learning algorithms.

V. Jain and P. Bansal—Authors contributed equally.

© Springer Nature Switzerland AG 2018
L. Cheng et al. (Eds.): ICONIP 2018, LNCS 11306, pp. 466–475, 2018.
https://doi.org/10.1007/978-3-030-04224-0_40

This lead to development of several other methods that used deep learning [11–19]. However, all these methods in order to get a slight performance improvement (in terms of PSNR) promoted use of deep, computationally heavy CNN models. It would be objectively correct to say that such heavy resources are not available at all situations for such lengthy periods of time. In order to solve this problem, it is required to build a model that uses less number of multiplication-addition operations (FLOPs) to come up with a high resolution image.

In this work, we present a novel super resolution model termed SRCondeseNet that uses the concept of removing unused connections in the network to form group convolutions. Normal group convolutions also help in reducing the number of connections but the latter method comes with a huge loss in accuracy. Once the features are extracted using this reduced model, reconstruction is done using deconvolutional layers with 1×1 kernels to produce a high resolution image. Also applied is the concept of residual learning i.e. the bicubic upsampled input is added to network output so that the model only has to learn the difference [11]. Our contributions through this work are:

1. Our model incorporates the use of group convolutions and pruning in the field of super resolution thereby producing a lightweight CNN model for this problem.
2. Setting state-of-the-art in terms of performance metrics such as PSNR and SSIM along with using less number of FLOPs as compared to current light weight SISR methods.

2 Related Work

Here we focus on various deep learning methods that have been used to solve the SISR problem. Also we go through various methods that have been proposed to come up with efficient, lightweight CNNs.

2.1 Single Image Super Resolution

Various deep learning methods have been applied in the past, to solve the SISR problem, many of which have been summarized in [23]. First, Dong et al. proposed in [9] the replacement of all steps to produce a high resolution image - feature extraction then mapping then reconstruction - by a single neural network. The deep learning model performed better than other example-based methods. However, it was proposed in [9] that deeper networks may not be effective for SISR. This was proved wrong by Kim et al. in [11]. They used a very deep CNN model that performed better than [9]. Kim et al. in [11] used residual learning proposed by He et al. in [30] to combat the problem of vanishing gradients that arises in deep models. Since then, the concept of residual learning has been used by many CNN models [13,15,17–20]. Hence, we also include the feature of global residual learning in order to avoid the vanishing gradient problem that is bound to happen in a deep model like ours. Moreover, some models [12,17,19,20] advocate the use of recursive layers in the CNN. This helps in reducing the number of effective parameters required. However, these models require heavy computation power, significantly higher than our model. To achieve real time performance, [21] proposed use of sub-pixel convolutions instead of bicubic upsampled image taken as the input. Similarly, [10,15,18] start with low-resolution image as input to the network. This way the model works on low resolution image thereby helping in reducing the number of computations. We also use this idea for the aforementioned purpose. Some methods [13,14] have advocated the use of GAN (Generative Adverserial Networks) to produce visually-pleasing images along with promising results on quantitative metrics like PSNR and SSIM.

2.2 Efficient Convolutional Neural Networks

Many attempts have been made to build CNNs that use less computation power without compromising accuracy. One such method is weight pruning. Weight pruning is removing unwanted connections in a neural network. CondenseNets [1] which are explained below use weight pruning.

CondenseNet. Our model employs blocks that are modified version of CondenseNet blocks [1]. Original CondenseNet blocks use learned group convolutions. In this method, the model goes through two kinds of stages: condensing stages and optimization stage. In the former, using sparsity inducing regularization, unimportant filters are removed. The convolutional layers used here have 1×1 kernels. Thus, number of connections depend on number of input channels and number of output channels only. The condensation is done by calculating L_1-norm over every incoming feature and every filter group. Then we remove those columns that have L_1-norm value lesser than other columns. The number of feature map connections that are left after every condensing stage depends on the condensing factor C. Once we get the lighter model, it goes through optimization stage where it is trained. Every block contains several *denselayers* and structure of each original *denselayer* is described in Fig. 1 (left).

3 Proposed Method

In this section, we describe our proposed method, SRCondenseNet, in detail. First we take as input the low resolution(LR) image and pass it into an input convolutional layer. The output of this layer is fed into modified CondenseNet that contains *denselayers* that are stacked into four blocks. The output of the last block is sent to what is called as the reconstruction network. It comprises of a bottleneck layer, a set of deconvolution layer, whose number depends on the scaling factor. Next comes the reconstruction layer with one output channel to get the final image.

3.1 Modified CondenseNet Blocks

In Sect. 2.2, we explained original CondenseNet blocks. However, CondenseNet has been designed for classification task. Hence, several modifications have been

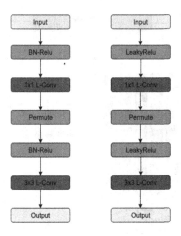

Fig. 1. Comparison between structure of *denselayer* of CondenseNet [1] (left) and SRCondenseNet (right). BN layer is removed and Relu is replaced by LeakyRelu.

done to suit it for the SISR problem. SRCondenseNet contains *denselayer* structure, which has been depicted in Fig. 1 (right).

Every block contains many *denselayer* named structures. We have removed Batch-Normalization layers as suggested by Nah et al. [22] and Lim et al. [16]. This removes unnecessary computations. Also Relu activation has been replaced by LeakyRelu to combat the "dying ReLU" problem. We stack up four blocks each containing 7 *denselayers* (blue) as shown in Fig. 2. Only one out of four blocks (black dashed line) is shown in the figure to avoid clutter. Number of input channels in every *denselayer* depends on growth rate and increase in number linearly according to it. Every block has its own growth rate. After testing several values for trade-off between model size and accuracy, we set growth rate of all the blocks to 20. Thus, every subsequent *denselayer* has 20 more input channels than the previous one.

Moreover, original CondenseNet contains transition layers between blocks comprising of average pooling layers. For SISR problem, there was no need of pooling layers. We have skipped these transition layers in our model. Thus, width and height of input into first block is equal to width and height of output from last block.

3.2   Reconstruction Network

The reconstruction network comes after the modified CondenseNet blocks as shown in Fig. 2. It starts with the bottleneck layer (green) which is a 1×1 layer to reduce the output feature maps to a very less number thereby reducing the number of computations in further layers. 1×1 kernel also helps in the purpose. Number of output feature maps are set to 128. Next, this is followed by a set of deconvolutional layers (pink). Their number depends on the scaling factor(r).

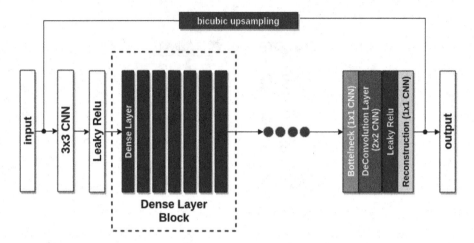

Fig. 2. Our model (SRCondenseNet) structure. We have four such blocks (black dashed line) each containing seven *denselayer* (blue) structures. Only one block is shown for clarity. This is followed by the reconstruction network containing bottleneck, deconvolution and reconstruction CNN at the end. (Color figure online)

With r equal to 2, we have a single deconvolutional layer with stride 2. Deconvolutional layers help in reducing the number of parameters and computational complexity by a factor of r^2 throughout the model. This is because, by using bicubic interpolated image as an input, instead of upscaling it at the last using deconvolutional layers, increases the size of input to all feature extraction layers by a factor of r^2. This method of upscaling also improves performance in reconstruction. Again, number of feature maps are set to 128 for all deconvolutional layers. Finally, we end up the model with a convolutional layer (yellow) with one channel as output to get the final YCbCr image.

3.3   Global Residual Learning

Deep CNN models with high number of layers tend to suffer from vanishing gradient problem. Hence, as proposed by He et al. in [30], this problem is solved by adding a global residual connection. In our model, we add a residual connection in which we add a bicubic interpolated image to the output received from the model. This makes the learning easier and more and more layers can be stacked.

4   Experiments

4.1   Datasets

We have used 91 images from Yang et al. [24] and 200 images from the Berkeley Segmentation Dataset (BSD)[25] for training. We cut out several patches of the original images with a stride of 64, size of which depends on the scaling factor.

For every case, during training, input size of the image to the network is 32 × 32. Hence, for scaling factor of 2, we cut out patches of 64 × 64. Further, we have performed data augmentation on these patches. Eventually, we get five patches for a single original patch. These are converted to YCbCr image and only Y-channel is processed.

We test our model on standard datasets: SET5 [26], SET14 [27] and Urban100 [28].

4.2   Implementation Details

We set the initial learning rate to 0.0001 and keep the cosine learning rate method as used by Huang et al. in [1]. We run the network for 180 epochs with both condensing factor and number of groups set to 4 to have every condensing stage with 30 epochs. LeakyReLUs have negative slope set to 0.1. We train our network on a Tesla P40 GPU. All networks were optimized using Adam [29]. We have used a robust Charbonnier loss function instead of $L1$ or $L2$ function that is generally used [9,11,12,16] to aid high-quality reconstruction performance [18].

4.3   Comparison with State-of-the-Art Methods

Comparison on the Basis of Accuracy. Peak signal-to-noise Ratio (PSNR) and structures similarity (SSIM) are the two standard metrics for comparison. Table 1 shows comparisons for SISR results for various models (scale = x2). Clearly, our method performs handsomely when compared to current state-of-the-art models using similar computation power. Various standard testing datasets have been used. Figure 3 shows a qualitative comparison of images from various testing datasets.

Table 1. Average PSNR/SSIM values for x2 scale factor for various models on different models. Red indicates best value and blue indicates second best value.

Dataset	SRCNN [9]	VDSR [11]	LapSRN [18]	DRRN [19]	SRCondenseNet(ours)
Set5	36.66/0.9542	37.53/0.9587	37.52/0.9590	37.74/0.9591	37.79/0.9594
Set14	32.45/0.9067	33.03/0.9124	33.08/0.9130	33.23/0.9136	33.23/0.9137
Urban100	29.50/0.8946	30.76/0.9140	30.41/0.9100	31.23/0.9188	31.24/0.9190

Comparison on the Basis of FLOPs. SRCondenseNet uses the concept of learned group convolutions. Thus, it requires relatively less computation power to produce better results. Here, for comparison, we have used the definition of FLOPs(number of multiplications and additions) to compare computational complexity. Similar method was used in the original CondenseNet [1] paper. We have used the same method to calculate FLOPs for all models. Scale is taken as 2 here as well. SRCNN [9], VDSR [11] & DRRN [19] take bicubic interpolated

Fig. 3. Qualitative comparison. The first row shows low resolution input, bicubic interpolation of LR input, output from our model, original HR image (left to right) of image from set5. Similarly second row are images for img_013 from set14.

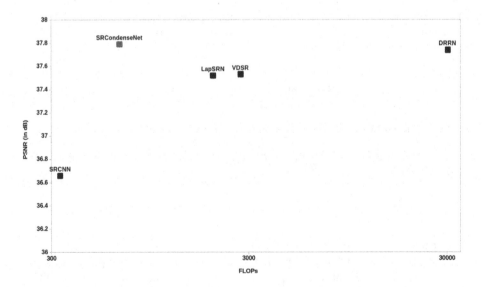

Fig. 4. Graph showing average PSNR vs FLOPs(SET5 [26] scale x2) trade-off.

input, hence we take input image size as 64 × 64 for these models. Whereas, we take 32 × 32 input image size for LapSRN [18] & our model as these models use original low resolution image. Table 2 shows that our model is lighter than most models. There is a trade-off between computational complexity and PSNR as can be seen in Fig. 4. SRCNN [9] contains only three parameterised convolutional layers and thus is unable to learn good enough mapping between a low resolution image and its coressponding high resolution image. Number of layers in SRCNN [9] is very less as compared to all other models mentioned in Tables 1 and 2 which makes it computationally less expensive (without explicitly applying any technique to reduce number of parameters) than other models (including ours). However, it should also be noted that it produces significantly poorer results than all other models. On the other hand, rest of all the models are computationally heavier than our model.

Table 2. FLOPs count (x1e6) for various models with suitable input to produce a size 64 × 64 output image and scale factor of 2. Red indicates best value and blue indicates second best value.

Model	SRCNN [9]	VDSR [11]	LapSRN [18]	DRRN [19]	SRCondenseNet(ours)
FLOPs(x1e6)	332.32	2727.61	1988.38	30235.17	668.88

5   Conclusion and Future Work

In this paper, we propose a single image super resolution method that uses pruned CNNs to solve the problem using less number of computations. The proposed method outperforms state-of-the-art by a considerable margin in terms of PSNR and SSIM while maintaining less number of FLOPs than comparable methods. Learned group convolutions after our modifications are found to be performing well for the SISR task. This work promotes the use of efficient CNNs that have been used widely in high-level computer vision tasks into low-level vision tasks such as SISR.

In this work, Charbonnier loss has been used throughout the process. We intend on integrating perceptual loss in the proposed method in order to produce visually pleasing images as claimed by Ledig et al. [13] and Sajjadi et al. [14] in future.

Acknowledgements. The authors are grateful to HP Inc. for their support to the Innovations Incubator Program. They are thankful to other stakeholders of this program including Leadership, and Faculty Mentors at IIT-BHU, Drstikona and Nalanda Foundation. Authors are also grateful to Dr. Prasenjit Banerjee, Nalanda Foundation for his mentoring and support

References

1. Huang, G., Liu, S., Maaten, L.V.D., Weinberger, K.Q.: CondenseNet: an efficient DenseNet using learned group convolutions. In: IEEE Conference on Computer Vision and Pattern Recognition (2018)
2. Duchon, C.E.: Lanczos filtering in one and two dimensions. J. Appl. Meteorol. **18**(8), 1016–1022 (1979)
3. Schulter, S., Leistner, C., Bischof, H.: Fast and accurate image upscaling with super-resolution forests. In: Conference on Computer Vision (2015)
4. Timofte, R., De Smet, V., Van Gool, L.: A+: adjusted anchored neighborhood regression for fast super-resolution. In: Cremers, D., Reid, I., Saito, H., Yang, M.-H. (eds.) ACCV 2014. LNCS, vol. 9006, pp. 111–126. Springer, Cham (2015). https://doi.org/10.1007/978-3-319-16817-3_8
5. Perez-Pellitero, E., Salvador, J., Ruiz-Hidalgo, J., Rosenhahn, B.: PSyCo: manifold span reduction for super resolution. In: IEEE Conference on Computer Vision and Pattern Recognition (2016)
6. Glasner, D., Bagon, S., Irani, M.: Super-resolution from a single image. In: International Conference on Computer Vision (2009)
7. Yang, J., Wright, J., Huang, T., Ma, Y.: Image superresolution via sparse representation. IEEE Trans. Image Process. **19**(11), 2861–2873 (2010)
8. Salvador, J., Perez-Pellitero, E.: Naive bayes superresolution forest. In: International Conference on Computer Vision (2015)
9. Dong, C., Loy, C.C., He, K., Tang, X.: Learning a deep convolutional network for image super-resolution. In: Fleet, D., Pajdla, T., Schiele, B., Tuytelaars, T. (eds.) ECCV 2014. LNCS, vol. 8692, pp. 184–199. Springer, Cham (2014). https://doi.org/10.1007/978-3-319-10593-2_13
10. Dong, C., Loy, C.C., Tang, X.: Accelerating the super-resolution convolutional neural network. In: Leibe, B., Matas, J., Sebe, N., Welling, M. (eds.) ECCV 2016. LNCS, vol. 9906, pp. 391–407. Springer, Cham (2016). https://doi.org/10.1007/978-3-319-46475-6_25
11. Kim, J., Lee, J.K., Lee, K.M.: Accurate image super-resolution using very deep convolutional networks. In: Computer Vision and Pattern Recognition, pp. 1646–1654 (2016)
12. Kim, J., Lee, J.K., Lee, K.M.: Deeply-recursive convolutional network for image SuperResolution. In: Computer Vision and Pattern Recognition, pp. 1637–1645 (2016)
13. Ledig, C., et al.: Photo-realistic single image super-resolution using a generative adversarial network. In: IEEE Conference on Computer Vision and Pattern Recognition (CVPR), pp. 105–114 (2017)
14. Sajjadi, M.S.M., Schlkopf, B., Hirsch, M.: EnhanceNet: single image super-resolution through automated texture synthesis. In: IEEE International Conference on Computer Vision (ICCV), Venice, pp. 4501–4510 (2017)
15. Tong, T., Li, G., Liu, X., Gao, Q.: Image super-resolution using dense skip connections. In: The IEEE International Conference on Computer Vision (ICCV) (2017)
16. Lim, B., Son, S., Kim, H., Nah, S., Lee, K. M.: Enhanced deep residual networks for single image super-resolution. In: IEEE Conference on Computer Vision and Pattern Recognition Workshops (CVPRW), pp. 1132–1140 (2017)
17. Cheng, X., Li, X., Tai, Y., Yang, J.: SESR: single image super resolution with recursive squeeze and excitation networks. In: International Conference on Pattern Recognition (ICPR) (2018)

18. Lai, W.S., Huang, J.B., Ahuja, N., Yang, M.H.: Deep Laplacian pyramid networks for fast and accurate super-resolution. In: IEEE Conference on Computer Vision and Pattern Recognition (2017)
19. Tai, Y., Yang, J., Liu, X.: Image super-resolution via deep recursive residual network. In: IEEE Conference on Computer Vision and Pattern Recognition (CVPR), pp. 2790–2798 (2017)
20. Tai, Y., Yang, J., Liu, X., Xu, C.: MemNet: a persistent memory network for image restoration. In: IEEE International Conference on Computer Vision (ICCV), Venice, pp. 4549–4557 (2017)
21. Shi, W., et al.: Real-time single image and video super-resolution using an efficient sub-pixel convolutional neural network. In: IEEE Conference on Computer Vision and Pattern Recognition (CVPR), pp. 1874–1883 (2016)
22. Nah, S., Kim, T.H., Lee., K.M.: Deep multi-scale convolutional neural network for dynamic scene deblurring. arXiv preprint arXiv:1612.02177 (2016)
23. Hayat, K.: Super-Resolution via Deep Learning (2017). arXiv:1706.09077
24. Yang, J., Wright, J., Huang, T.S., Ma, Y.: Image super resolution via sparse representation. IEEE Trans. Image Process. **19**(11), 2861–2873 (2010)
25. Arbelaez, P., Maire, M., Fowlkes, C., Malik, J.: Contour detection and hierarchical image segmentation. IEEE Trans. Pattern Anal. Mach. Intell. **33**(5), 898–916 (2011)
26. Bevilacqua, M., Roumy, A., Guillemot, C., Alberi-Morel, M.L.: Low-complexity SingleImage super-resolution based on nonnegative neighbor embedding. In: British Machine Vision Conference (2012)
27. Zeyde, R., Elad, M., Protter, M.: Low-complexity SingleImage super-resolution based on nonnegative neighbor embedding. In: International Conference on Curves and Surfaces, pp. 711–730 (2012)
28. Huang, J.B., Singh, A., Ahuja, N.: Single image superresolution from transformed self-exemplars. In: IEEE Conference on Computer Vision and Pattern Recognition (2015)
29. Kingma, D., Adam, J.B.: A method for stochastic optimization. In: International Conference on Learning Representations, ICLR (2015)
30. He, K., Zhang, X., Ren, S., Sun, J.: Deep residual learning for image recognition. In: Computer Vision and Pattern Recognition, pp. 770–778 (2016)

Linear Periodic Discriminant Analysis of Multidimensional Signals

Dounia Mulders[1,2](✉), Cyril de Bodt[1], Nicolas Lejeune[2], André Mouraux[2], and Michel Verleysen[1]

[1] ICTEAM Institute, Université catholique de Louvain,
Place du Levant 3, 1348 Louvain-la-Neuve, Belgium
dounia.mulders@uclouvain.be
[2] IONS Institute, Université catholique de Louvain,
Avenue Mounier 53, 1200 Woluwe-Saint-Lambert, Belgium

Abstract. Extracting relevant information from noisy multidimensional signals has tremendous impacts in numerous applications, ranging from audio separation to electrophysiological recording analysis. Linear filters are often considered to reconstruct and interpret the latent sources generating the data. Known properties of the sources can be used to guide their separation. In neuroscience, the cortical processes underlying perception in different modalities (visual, auditory, ...) is often studied using electroencephalography (EEG) during periodic stimulation, eliciting periodic activity in neural sources, some of which being specific to the considered modality. Whereas current approaches extract sources either periodic or discriminative, none of them accounts for both aspects at once. This paper proposes several methods extracting periodic sources specific between two classes, hence termed as Linear Periodic Discriminant Analysis methods. They are validated on synthetic data and EEG recordings of subjects to whom periodic stimulation from two modalities is applied. The methods highlight modality-specific periodic responses.

Keywords: Linear filtering · Specific periodic components
Generalized Rayleigh Quotient · Source separation · Steady-states
EEG

1 Introduction

Multidimensional signals are encountered in a broad range of situations including music editing, speech recognition or physiological monitoring. In many cases, such observed signals result from the noisy mixing of several components or sources of interest [6]. So-called source separation algorithms are therefore considered to recover the latent sources. The mixing is often linearly approximated, allowing an easy interpretation of the sources in the original sensor space. Independent Component Analysis (ICA) methods are probably the most well-known approaches to this kind of problems, only relying on the minimal assumption of

© Springer Nature Switzerland AG 2018
L. Cheng et al. (Eds.): ICONIP 2018, LNCS 11306, pp. 476–487, 2018.
https://doi.org/10.1007/978-3-030-04224-0_41

independence between the source signals [8]. ICA has proven to be very useful in many contexts, including the extraction of single speech signal from mixed recordings [7] or artifacts removal from electrophysiological signals [12]. In several applications however, specific structures of the latent sources are expected, which can be used to guide the definition of source separation algorithms. In audio signal processing for instance, diverse forms of guidance have already been considered, using either priors on the mixing processes or the temporal dynamics of the sources [20]. Such temporal structures are also exploited in the study of biological signals, for instance to enhance periodic components of a given multidimensional signal [14]. Other works have extended this kind of methods to extract quasi-periodic signals such as the electrocardiogram [17].

In this line, structured, and in particular periodic, signals also often arise in neuroscience. For instance, periodic stimulation is increasingly proposed to probe sensory perception in humans [3]. This kind of stimulus elicit a periodic activity at the stimulation frequency in some neural populations, known as a steady-state response (SSR), which can be partly recorded with electroencephalography (EEG). Each source contributing to the SSR associated to stimuli from one modality (e.g. visual, auditory, tactile,...) can be common to several studied modalities, or specific to a particular one. As an example, if cool and warm periodic stimuli are applied to a subject, part of the SSR can be expected to be common to the two classes, warm and cool, since both stimuli share some characteristics (they both activate the spinothalamic system). To study the specific cortical processing of each class of stimulus however, one is more interested in finding the most discriminant components of the SSR. Although linear filtering methods aiming to extract either periodic or discriminant components exist and are well-documented in the literature, the optimization of filters defining periodic components which are specific to one class of signals among two has surprisingly not yet been studied. Such a method would better capture the mechanisms governing specific sensory perception in humans, thereby helping to understand and subsequently diagnose sensory disorders. In this context, this paper presents methods to extract such sources. The approach is inspired, on the one hand, by the source separation methods aiming to maximize periodicity of the extracted components [14] and, on the other hand, by the Common Spatial Pattern (CSP) algorithm largely studied and adapted in Brain-Computer Interfacing (BCI) researches [2,11]. The specific periodic components are obtained by optimizing interpretable linear filters, hence the name of the presented approach: Linear Periodic Discriminant Analysis (LPDA). Several objective functions are proposed, compared and assessed on synthetic and real data. In particular, results on both simulated and real data are convincing, the presented methods enabling the discovery of modality-specific periodic structures. Our methodology, combining periodicity and discrimination maximization, outperforms state-of-the-art approaches accounting for one of both aspects at once.

The paper is organized as follows. Section 2 introduces and motivates the definition of several LPDA objective functions. Experiments are conducted in Sect. 3, both on synthetic and EEG data. Section 4 finally concludes the work.

2 Linear Periodic Discriminant Analysis Methods

This section defines and justifies the proposed approaches to extract periodic components specific to one among two multidimensional signals, belonging to two different classes. The zero-mean (over time) signals from the first and second classes, respectively denoted by $\mathbf{x}_1(t)$ and $\mathbf{x}_2(t) \in \mathbb{R}^C$, are assumed to result from the linear mixing of latent periodic sources. For instance, they can be C-channel EEG signals recorded on one subject in two different conditions, e.g. exposed to periodic warm vs. cool stimulation. The sources, of period $T_\star$ and fundamental frequency $f_\star = 1/T_\star$ for both classes 1 and 2, can either be common to both signals or specific to $\mathbf{x}_1$ or $\mathbf{x}_2$. This leads to the noiseless latent model

$$\mathbf{x}_i(t) = A_u \mathbf{z}_u(t) + A_{s,i} \mathbf{z}_{s,i}(t), \tag{1}$$

where $i \in \{1,2\}$ and s (resp. u) indexes the *specific* (resp. *unspecific*) mixing matrices $A_{s,i} \in \mathbb{R}^{C \times N_{s,i}}$ (resp. $A_u \in \mathbb{R}^{C \times N_u}$) and periodic sources $\mathbf{z}_{s,i}(t) \in \mathbb{R}^{N_{s,i}}$ (resp. $\mathbf{z}_u(t) \in \mathbb{R}^{N_u}$), $N_{s,i}$ (resp. N_u) being the number of specific (resp. unspecific) sources. In practice, different sources of noise will also affect the observed signals. Also, each specific source has a distinct level of specificity expressed by its variance. For convenience, the sources in a vector are ordered in decreasing order of specificity; for instance, $\mathbf{z}_{s,1}^1(t)$ is the most specific source for class 1, where the exponent indicates the component of the vector $\mathbf{z}_{s,1}(t)$. Each column of a mixing matrix A is a spatial pattern of the associated source [16]. Under this framework, both the specific spatial patterns $A_{s,i}$ and the specific source time courses $\mathbf{z}_{s,i}(t)$ should be recovered using $\mathbf{x}_1$ and $\mathbf{x}_2$. To this end, spatial filters $\mathbf{w} \in \mathbb{R}^C$ are optimized to obtain filtered signals $s_i(t) := \mathbf{w}^T \mathbf{x}_i(t)$ such that $s_1(t) \approx \mathbf{z}_{s,1}^1(t)$. Similarly, the filters can be optimized for class 2 to obtain $s_2(t) \approx \mathbf{z}_{s,2}^1(t)$ instead. Once a first optimal filter $\mathbf{w}_1$ for class i is found by optimizing a cost function F, a second optimal filter can be obtained, leading to a component in the orthogonal subspace of the first one. A matrix $W \in \mathbb{R}^{C \times C}$ is hence recursively built, whose columns $\mathbf{w}_k$ are the filters ranked according to F. The first $N_{s,i}$ columns of the pseudo-inverse of W^T then estimate the spatial patterns of the extracted components for class i: $\left[(W^T)^{-1}\right](:,1:N_{s,i}) = \widehat{A_{s,i}}$.

In order to obtain optimized filters, one must first define relevant cost functions characterizing the class-specific periodicity of the filtered signals. In the following, we propose several such objective functions, based on generalized Rayleigh quotients capturing periodicity differences, to obtain the filters for the specific activity of class 1, without loss of generality; both classes can be swapped to obtain the specific activity of class 2.

• **Method 1 (M1).** Periodic Component Analysis [14,19] aims to extract periodic sources from a single data set based on the periodicity measure $(\sum_t (s(t+T_\star) - s(t))^2 / (\sum_t (s(t))^2)$ of the signal $s(t)$. In the studied two-classes setting, a ratio of periodicity inspired by this measure can be optimized to find components $s_i(t) = \mathbf{w}^T \mathbf{x}_i(t)$, being periodic in class 1 and not necessarily in class 2:

$$\mathbf{w}_1 = \arg\min_{\mathbf{w}} \left\{ F_1(\mathbf{w}) = \frac{\sum_t (s_1(t+T_\star) - s_1(t))^2}{\sum_t (s_2(t+T_\star) - s_2(t))^2} = \frac{\mathbf{w}^T R_1(T_\star) \mathbf{w}}{\mathbf{w}^T R_2(T_\star) \mathbf{w}} \right\}, \quad (2)$$

where $R_i(T_\star) = \mathbb{E}_t\{(\mathbf{x}_i(t+T_\star) - \mathbf{x}_i(t))(\mathbf{x}_i(t+T_\star) - \mathbf{x}_i(t))^T\}$. It can be noted that $F_1(\mathbf{w}) \geq 0$ and that $F_1(\mathbf{w}) = 0$ for a perfectly periodic signal $s_1(t)$. However, $F_1(\mathbf{w}) = 0$ as well when $s_1(t) = 0$ even though a flat signal may probably not be related to the sources of interest. This objective should hence be used with caution when some channels are correlated, in which case $s_1(t)$ could be 0.

- **M2.** Alternatively, the periodicity of the filtered signals can be emphasized by a simple averaging over the fundamental periods: $s_{\mu i}(t) := \frac{1}{K} \cdot \sum_{k=0}^{K-1} s_i(t+kT_\star) := \mathbf{w}^T \mathbf{x}_{\mu i}(t)$, with K the number of periods in the available signals. As in the CSP algorithm [2], the variance ratio of these averaged signals can be optimized as

$$\mathbf{w}_1 = \arg\max_{\mathbf{w}} \left\{ F_2(\mathbf{w}) = \frac{\sum_t (s_{\mu 1}(t))^2}{\sum_t (s_{\mu 2}(t))^2} = \frac{\mathbf{w}^T C_{\mu 1} \mathbf{w}}{\mathbf{w}^T C_{\mu 2} \mathbf{w}} \right\}, \quad (3)$$

where $C_{\mu i} = \mathbb{E}_t\{\mathbf{x}_{\mu i}(t)\mathbf{x}_{\mu i}(t)^T\}$.

- **M3.** Another approach is based on Canonical Correlation Analysis (CCA), which was already proposed to extract periodic components [15]. In the present work however, the method maximizes the correlation between the differences of the filtered signals in both classes and an arbitrary periodic signal $\mathbf{w}_y^T \mathbf{y}(t)$ of the same length as s_i. This periodic signal is defined from the Fourier series of a periodic signal of fundamental frequency $f_\star$: $\mathbf{y}(t) = (\sin(2\pi f_\star t) \ \cos(2\pi f_\star t) \ \sin(2\pi 2 f_\star t) \ \ldots \sin(2\pi N_h f_\star t) \ \cos(2\pi N_h f_\star t))^T$, where N_h indicates the number of accounted harmonics. Since the correlation $\mathbb{E}_t\{\mathbf{w}^T(\mathbf{x}_1(t) - \mathbf{x}_2(t))\mathbf{y}(t)^T \mathbf{w}_y\}$ must be maximized with scale invariance, the optimization problem is formulated as

$$(\mathbf{w}_1, \mathbf{w}_{1,y}) = \arg\max_{\mathbf{w}, \mathbf{w}_y} \left\{ F_3(\mathbf{w}, \mathbf{w}_y) = \frac{\mathbf{w}^T (C_{1;y} - C_{2;y}) \mathbf{w}_y}{\sqrt{\mathbf{w}^T C_{12} \mathbf{w} \cdot \mathbf{w}_y^T C_\mathbf{y} \mathbf{w}_y}} \right\}, \quad (4)$$

where $C_{i;y} = \mathbb{E}_t\{\mathbf{x}_i(t)\mathbf{y}(t)^T\}$, $C_{12} = \mathbb{E}_t\{(\mathbf{x}_1(t) - \mathbf{x}_2(t))(\mathbf{x}_1(t) - \mathbf{x}_2(t))^T\}$ and $C_\mathbf{y} = \mathbb{E}_t\{\mathbf{y}(t)\mathbf{y}(t)^T\}$. This method is only well suited when the common activity is assumed to be highly similar (in terms of amplitude and phase) in both classes, leading to its cancellation through the point-by-point difference $\mathbf{x}_1(t) - \mathbf{x}_2(t)$.

- **M4.** Inspired by the Spectral Contrast Maximization (SCM) approach [18], the ratio of spectral content at the fundamental frequency $f_\star$ and its harmonics in both classes can be optimized. With $S_i(f) := \mathcal{F}_f\{s_i(t)\} = \mathbf{w}^T \mathcal{F}_f\{\mathbf{x}_i(t)\} = \mathbf{w}^T X_i(f)$ the Fourier transform of the filtered signals at frequency f, a fourth objective can be formulated in the following way:

$$\mathbf{w}_1 = \arg\max_{\mathbf{w}} \left\{ F_4(\mathbf{w}) = \frac{\mathbb{E}_{f \in \nu}\{|S_1(f)|^2\}}{\mathbb{E}_{f \in \nu}\{|S_2(f)|^2\}} = \frac{\mathbf{w}^T S_{\mathbf{x}_1} \mathbf{w}}{\mathbf{w}^T S_{\mathbf{x}_2} \mathbf{w}} \right\}, \quad (5)$$

with $\nu := \{\pm f_\star, \pm 2f_\star, \ldots, \pm N_h f_\star\}$ the set of frequencies of interest and $S_{\mathbf{x}_i} := \mathbb{E}_{f \in \nu}\{X_i(f)X_i(f)^*\}$. This objective is expected to be maximized by concentrating the frequency content of $s_1(t)$ to these frequencies. It should however be stressed that it can also be maximized if its denominator tends toward 0 while keeping the numerator bounded.

• **M5.** The quantity $\sum_t (s_i(t+T_\star) - s_i(t))^2$ is used in (2) to describe the periodicity of the signal $s_i(t)$. An alternative periodicity measure which can be considered is $\sum_t s_i(t+T_\star) \cdot s_i(t)$ [17], leading to the optimization problem

$$\mathbf{w}_1 = \arg\max_{\mathbf{w}} F_5(\mathbf{w}), \text{ with} \tag{6}$$

$$F_5(\mathbf{w}) = \frac{\sum_t s_1(t+T_\star) \cdot s_1(t) - s_2(t+T_\star) \cdot s_2(t)}{\sum_t (s_1(t))^2 + (s_2(t))^2} = \frac{\mathbf{w}^T(C_1(T_\star) - C_2(T_\star))\mathbf{w}}{\mathbf{w}^T(C_1(0) + C_2(0))\mathbf{w}},$$

where $C_i(\tau) = \mathbb{E}_t\{\mathbf{x}_i(t+\tau)\mathbf{x}_i(t)^T\}$. Note that a difference of periodicity between the two classes is considered instead of a ratio similar to (2), because the matrices $C_i(T_\star)$ are not necessarily positive definite contrarily to $R_i(T_\star)$.

• **M6.** Another variant of (2) can be defined using a difference instead of a ratio of periodicity. A distinct normalization ensures the scale-invariance of the filters:

$$\mathbf{w}_1 = \arg\min_{\mathbf{w}}\left\{F_6(\mathbf{w}) = \frac{\mathbf{w}^T(R_1(T_\star) - R_2(T_\star))\mathbf{w}}{\mathbf{w}^T(C_1(0) + C_2(0))\mathbf{w}}\right\}. \tag{7}$$

• **M7.** A similar change can be applied to (5), leading to a last formulation:

$$\mathbf{w}_1 = \arg\max_{\mathbf{w}}\left\{F_7(\mathbf{w}) = \frac{\mathbf{w}^T(S_{\mathbf{x}_1} - S_{\mathbf{x}_2})\mathbf{w}}{\mathbf{w}^T(C_1(0) + C_2(0))\mathbf{w}}\right\}. \tag{8}$$

All the objective functions above, except the third one, are expressed as generalized Rayleigh quotients of the form $(\mathbf{w}^T A \mathbf{w})/(\mathbf{w}^T B \mathbf{w})$ and can be solved by the generalized eigenvalue decomposition (GEVD) of the matrix pair (A, B), yielding a set of filters $\{\mathbf{w}_k\}_{k=1}^C$ corresponding to the generalized eigenvectors ranked in decreasing or increasing order of cost function values, depending on whether the objective function is maximized or minimized. When A is symmetric and B positive-definite, the decomposition is real and hence interpretable [5]. The formulation (4) is also solved analytically, as for classical CCA [9].

It is here assumed that the filters are ranked in decreasing order of specificity for class 1, hence with $\mathbf{w}_1$ being the most specific filter for class 1. The successive extracted sources are denoted by $s_i^k(t) := \mathbf{w}_k^T \mathbf{x}_i(t)$.

3   Experimental Results

This section assesses the methods introduced in Sect. 2. Their behavior and performances are first studied and compared in Sect. 3.1 on synthetic data, for which the latent sources are known. Then, preliminary results on a real EEG data set are presented in Sect. 3.2, for which the specific patterns are obtained. As to the quality assessment, different criteria are proposed in Sect. 3.1.

3.1   Synthetic Data

As stated in Sect. 1, the studied methods aim to extract class-specific periodic components from two multidimensional signals. Therefore, synthetic data are generated by linearly mixing specific and common components for the two classes, with additional noise. Denoting by $X_i(t) \in \mathbb{R}^C$ the simulated data from class i at time sample t, $Z_{u,i}(t) \in \mathbb{R}^{N_u}$ and $Z_{s,i}(t) \in \mathbb{R}^{N_s}$ the common (i.e. unspecific to the class) and specific periodic factors respectively, for $i \in \{1, 2\}$, data are simulated according to the model

$$X_i(t) = \underbrace{A_u \cdot Z_{u,i}(t)}_{\text{periodic interference}} + \underbrace{A_s \cdot Z_{s,i}(t)}_{\text{signal}} + \underbrace{N_i(t)}_{\text{noise}}, \tag{9}$$

with the common and specific mixing matrices $A_u \in \mathbb{R}^{C \times N_u}$ and $A_s \in \mathbb{R}^{C \times N_s}$, $N_i(t) \in \mathbb{R}^C$ being additive random noise. The common activity can be seen as interferences in this problem. Although the mixing matrices are common to both classes, the factor (or source) time courses are not. In particular:

• the specific factors $Z_{s,1}$ and $Z_{s,2}$ are mutually independent, with distinct specificity levels expressed by their variances which are proportional to $(N_s - 1, \ N_s - 2, \ \ldots, \ 1, \ 0)$ and $(0, \ 1, \ \ldots, \ N_s - 2, \ N_s - 1)$ [21]. These sources are ordered in decreasing order of specificity for class 1, the first (resp. last) $\lfloor \frac{N_s}{2} \rfloor$ being specific to class 1 (resp. 2). If $N_s = 1$, there is no specific factor for class 2.

• The common factors $Z_{u,i}$ are equal in both classes, up to an arbitrary time lag, as the common activity can have distinct latencies, e.g. due to differences in fiber conduction velocities in the case of EEG signals.
For each simulation, the entries of the mixing matrices are uniformly distributed in $[0, 1]$ and the factors are random periodic signals of period $T_\star$. However, since the performances of the optimized filters highly depend on the relative position in $\mathbb{R}^C$ of the columns of A_u with respect to the ones of A_s, the columns of A_s can be altered to reach some pre-defined angles with respect to the ones of A_u.
  Regarding the assessment, different types of metrics can be considered.

• The reconstruction quality of the mixing matrix is evaluated using the canonical angle (CA) between the subspace spanned by the columns of A_s and its estimate $\widehat{A_s}$ computed from the optimized filters. The CA is the maximum angle found between orthogonal bases of these subspaces [5], and is therefore invariant to permutation and scaling of the columns of each matrix. Focusing on the most discriminant periodic source, the angle between the first true and estimated spatial patterns ($A_s^1 := A_s(:, 1)$ and $\widehat{A_s^1} := \widehat{A_s}(:, 1)$) is also evaluated.

• The time-correlation $\mathrm{corr}(s_1^1(t), \mathbf{z}_{s,1}^1(t))$ between the first true and estimated sources, $\mathbf{z}_{s,1}^1(t)$ being the first component of $\mathbf{z}_{s,1}(t)$, is insensitive to the pseudo-inversion required to obtain the spatial patterns from the optimized filters.

• The periodicity difference between $s_1^1(t)$ and $s_2^1(t)$ assesses the relevance of the first filter $\mathbf{w}_1$ in terms of periodicity and discrimination. The following measure is employed to quantify the periodicity of a signal $y(t) \in \mathbb{R}$ at a frequency $f_\star$:

$$M_\pi(y) = 100 \cdot \frac{\sum_{k=1}^{\lfloor f_s/(2 \cdot f_\star) \rfloor} Y_{NS}(k \cdot f_\star)}{\sum_f |Y(f)|}, \tag{10}$$

where f_s is the sampling frequency, $Y(f) = \mathcal{F}_f\{y(t)\}$ and Y_{NS} is the *noise-subtracted spectrum*. At frequency f, $Y_{NS}(f) \in \mathbb{R}$ is obtained by subtracting the average amplitude at 10 neighboring frequencies (5 higher and 5 lower) from the frequency amplitude $|Y(f)|$. Additional details are provided in [13,14]. The components extracted using the methods from Sect. 2 should maximize $M_\pi(s_1^1) - M_\pi(s_2^1)$, while providing topographical patterns of the extracted components which can be interpreted thanks to the linear filtering. Only this criterion can be used on real data as it does not rely on the 'true' sources.

In the following, each signal is generated with $C = 20$ channels, a sampling rate of 500 Hz, $f_\star = 0.5$ Hz and a duration of 20 seconds. The signal variances are defined over the channels and time samples [18]. Denoting $\sigma_u^2 := \mathrm{var}(A_u \cdot \mathbf{z}_{u,i}(t))$, $\sigma_s^2 := \mathrm{var}(A_s \cdot \mathbf{z}_{s,i}(t))$ and $\sigma_N^2 := \mathrm{var}(\mathbf{n}_i(t))$, the two first quantities control the signal-to-interference ratio (SIR), while the signal-to-noise ratio (in dB) is defined as $\mathrm{SNR} = 10 \cdot \log_{10}(\sigma_s^2/\sigma_N^2)$. In order to simulate realistic EEG data, which are studied in Sect. 3.2, pink additive noise is chosen for $N_i(t)$, i.e. with a frequency power spectrum proportional to $1/f$ [1]. Unless stated otherwise, there are $N_u = 10$ unspecific factors and the SIR is fixed with $\sigma_u = 1$ and $\sigma_s = 1$. The canonical angles between A_u and A_s are $\geq 70°$ and there is a time lag of 0.3 seconds between $Z_{u,1}(t)$ and $Z_{u,2}(t)$. The performances of the methods are commented with either 1 or 4 specific pattern(s) and then with varying unspecific activity according to N_u and σ_u.

• The performances of the methods with $N_s = 1$ specific factor are depicted in Figs. 1a to d. The angular position of the first specific pattern A_s^1 is also compared to the C-length vector of the Fourier frequency coefficients of the mixed signals at $f_\star$ in Fig. 1a (curve labeled 'FFT'). Although this pattern is usually employed in neuroscience studies to describe the periodic cortical activity elicited by periodic stimuli, it can be highly different from the spatial pattern of the source of interest. All the quality measures considered indicate the superiority of M7 to recover the specific pattern, followed by M2 and M5. Comparing Figs. 1b and c indicates that M3 probably amplifies common instead of specific activity, leading to both high $M_\pi(s_1^1)$ and $M_\pi(s_2^1)$. As this method uses a point-by-point difference between the two classes of signals, it is sensitive to a non-zero time lag between the common activity in both data sets.

• When $N_s = 4$, the results in Figs. 1e to h confirm the overall previous conclusions. The canonical angles between all the true and estimated specific patterns shown in Fig. 1f are larger than the angles computed only using the most specific pattern (Fig. 1e), indicating that the methods indeed recover more easily the first specific pattern. All methods are affected by the increased number of specific sources, and this effect is more pronounced for M1.

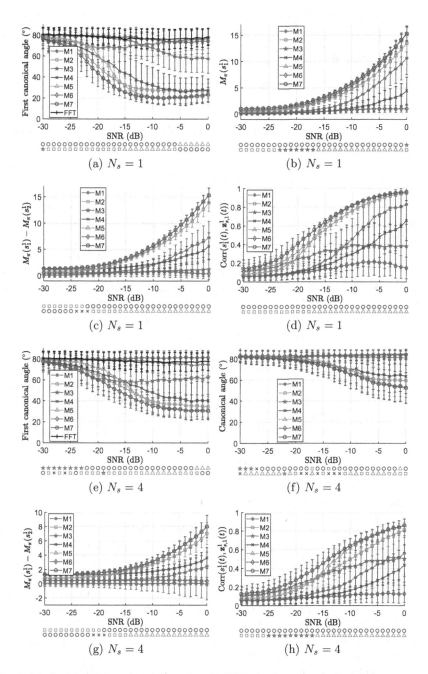

Fig. 1. Methods performances (mean ± std of 500 runs) on synthetic data as a function of the SNR. The two rows of markers below each figure indicate (1) the best method on average for each SNR and (2) the best method which is significantly outperformed by the best one (t-tests, Holm-Bonferroni corrected).

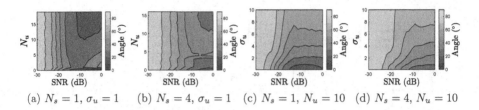

(a) $N_s = 1, \sigma_u = 1$ (b) $N_s = 4, \sigma_u = 1$ (c) $N_s = 1, N_u = 10$ (d) $N_s = 4, N_u = 10$

Fig. 2. Angle between the first specific true and estimated patterns, for M7 on synthetic data as a function of the SNR and either N_u (a, b) or σ_u (c, d). Results are averaged over 500 signal generations and $\sigma_s = 1$.

- The effects of σ_u and N_u on the M7 performances (the seemingly most efficient method) are shown in Fig. 2. For a fixed SIR, the number of common factors N_u only affects the performances when the SNR gets closer to 0, leading to improved results as N_u increases. On the other hand, increasing σ_u deteriorates the recovery of the most specific pattern, especially when $N_s = 4$. Let us note that, unlike the angle, the correlation between the first true and estimated sources is almost independent of σ_u (not shown due to a lack of space), suggesting that increasing σ_u does not affect the first optimal filter but rather the following ones, which in turn influences the first pattern after inverting the filter matrix.

The results presented in this section are obtained on data with 20 channels. Similar conclusions can be drawn with different dimensions, although on average all the angles tend to increase with the dimension, as a consequence of the curse of dimensionality [10]. Results are also similar with different canonical angles between A_u and A_s, with overall decreasing performances when they tend to be small as it becomes impossible for a filter to be orthogonal to the interference patterns A_u. M2 is especially affected by the reduction of these angles, while M5 and M7 are more robust (results not shown due to a lack of space).

3.2   Electroencephalogram (EEG) Data

This section shows brief results of the LPDA methods and of state-of-the-art filters on a real EEG data set, with 64 channels and sampled at 1000 Hz, which was recorded on 15 healthy subjects to whom periodic stimulation from four classes was applied: painful warm or non-painful cool, each applied on a variable or fixed skin surface (denoted resp. by w1, w2, c1 and c2). Each subject received each kind of stimulus with a frequency $f_\star = 0.2$ Hz and a duration $T = 75$ s (15 cycles). From a neuroscientific point of view, interest lies in finding the specific components of the condition w1 compared to c1 and w2 compared to c2. It would allow to disentangle the neural processing of warm and cool stimuli, thereby providing cues on the specific brain responses to painful stimuli.

As to the compared methods, to the best of our knowledge, there is no existing approach optimizing both the periodicity and the specificity of an extracted component at once. The results of our methods are hence compared to one method maximizing the periodicity (SCM [14]), another one optimizing the discrimination (CSP [11]) and Principal Component Analysis (PCA), a widely used non-guided filtering method.

LPDA methods are applied on the EEG data of each subject to obtain, for each ordered pair of conditions among w1–c1, w2–c2, c1–w1, c2–w2, the most specific source for the first condition of the pair and its associated spatial pattern and filter. Each optimal filter $\mathbf{w}_1$ can be applied to both signals of the pair to obtain $s_1^1(t)$ and $s_2^1(t)$. Table 1 confirms the superiority of method M7, followed by M5, in terms of difference of periodicity in any two compared classes, as it was observed on synthetic data, with the periodicity measure (10). M2, M3 and M6 results are of the same order of magnitude as the ones of M4 and are not shown due to a lack of space. Except for M7, the standard deviations are roughly of the same order as the mean results, indicating a high variability across subjects. It should be stressed that the performances of all methods vary coherently with respect to the subjects since M7 always statistically significantly outperforms the other schemes. These results are encouraging as they show that periodic components specific to one class can be extracted, M7 overcoming the three baselines. Further validations should be undertaken to disentangle physiologically interpretable patterns from potentially slightly overfitted results.

Table 1. For each pair of conditions (in row), mean(std) for the 15 subjects of the difference of periodicity M_π (10) between the filtered signals in both conditions using the most specific filter for the first condition of the pair. The best performances per row are in bold. Non-italic characters indicate that the corresponding signal is significantly less periodic than the best one of the same row (paired t-tests with Holm-Bonferroni correction). For M4 and M7, $N_h = 10$.

	LPDA methods				Baseline methods		
	M1	M4	M5	M7	SCM	CSP	SVD
w1–c1	0.55(0.42)	0.35(0.43)	0.55(0.94)	**3.25(1.04)**	2.84(1.08)	0.25(0.56)	0.22(0.32)
w2–c2	0.31(0.46)	0.09(0.15)	0.68(1.23)	**3.25(1.05)**	2.98(0.98)	0.15(0.29)	0.17(0.37)
c1–w1	0.10(0.53)	0.30(0.27)	0.31(0.39)	**2.43(0.67)**	1.59(0.86)	−0.12(0.28)	−0.19(0.28)
c2–w2	0.11(0.46)	0.19(0.14)	0.42(0.47)	**2.44(0.55)**	1.78(0.69)	−0.01(0.36)	−0.12(0.43)

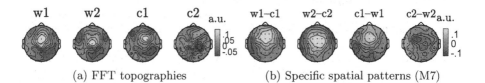

(a) FFT topographies   (b) Specific spatial patterns (M7)

Fig. 3. Group-level average spatial patterns of (a) the frequency amplitudes at $f_\star = 0.2$ Hz and (b) the most specific components extracted by M7 ($N_h = 10$) for the first class in each pair.

Finally, Fig. 3 shows the spatial pattern of the frequency spectrum at $f_\star$, usually analyzed in neuroscience, and the most specific spatial pattern extracted for each class of signals. The specific patterns of the two warm conditions are relatively similar but different form those of the cool conditions. The specific warm patterns have a larger central positive cluster than the corresponding FFT topographies, while the specific cool patterns are more diffuse, reflecting a larger inter-subjects variability. Cold receptors being known to habituate rapidly to repeated stimuli [4], this could affect the stability of the SSR.

4   Conclusions and Perspectives

This paper proposes several linear filtering methods for multidimensional signals aiming to extract periodic components which are specific to one class among two. Seven objective functions are defined, combining discrimination and periodicity maximization in simple formulations. The methods are discussed and their results compared on synthetic data sets, allowing to study their behavior and reliability in a controlled context. They are also applied on real EEG recordings to highlight modality-specific spatial patterns of brain activity elicited by periodic cool and warm stimulation. The results are encouraging, as the extracted patterns are consistent with the nature of the considered stimuli and indeed exhibit a markedly increased periodicity for the considered class.

Further investigations will be carried regarding the proper validation of the extracted specific patterns to enable their fair interpretation in the neuroscience context. This validation is however not trivial due to the very limited number of trials available for each subject and condition. While the current results are assessed by quantifying the difference of periodicity between each compared pair of modalities, the discriminative power of the filters could also be analyzed.

Acknowledgments. DM and CdB are Research Fellows of the FNRS. The authors gratefully thank Prof. Christian Jutten for insightful discussions.

References

1. Bedard, C., Kroeger, H., Destexhe, A.: Does the 1/f frequency scaling of brain signals reflect self-organized critical states? Phys. rev. let. **97**(11), 118102 (2006)
2. Blankertz, B., Tomioka, R., Lemm, S., Kawanabe, M., Muller, K.R.: Optimizing spatial filters for robust EEG single-trial analysis. IEEE Signal Process. Mag. **25**(1), 41–56 (2008)
3. Colon, E., Legrain, V., Mouraux, A.: Steady-state evoked potentials to study the processing of tactile and nociceptive somatosensory input in the human brain. Clin. Neurophysiol. **42**(5), 315–323 (2012)
4. De Keyser, R., van den Broeke, E.N., Courtin, A., Dufour, A., Mouraux, A.: Event-related brain potentials elicited by high-speed cooling of the skin: a robust and non-painful method to assess the spinothalamic system in humans. Clin. Neurophysiol. **129**(5), 1011–1019 (2018)
5. Golub, G.H., Van Loan, C.F.: Matrix Computations, vol. 3. JHU Press, Baltimore (2012)

6. Hyvärinen, A.: Independent component analysis: recent advances. Phil. Trans. R. Soc. A **371**(1984), 20110534 (2013)
7. Hyvärinen, A., Oja, E.: Independent component analysis: algorithms and applications. Neural Netw. **13**(4–5), 411–430 (2000)
8. Jutten, C., Herault, J.: Blind separation of sources, part I: an adaptive algorithm based on neuromimetic architecture. Signal Process. **24**(1), 1–10 (1991)
9. Krzanowski, W.: Principles of Multivariate Analysis, vol. 23. OUP Oxford, Oxford (2000)
10. Leskovec, J., Rajaraman, A., Ullman, J.D.: Mining of Massive Datasets. Cambridge University Press, New York (2014)
11. Lotte, F., Guan, C.: Regularizing common spatial patterns to improve BCI designs: unified theory and new algorithms. IEEE TBME **58**(2), 355–362 (2011)
12. Makeig, S., Bell, A.J., Jung, T.P., Sejnowski, T.J.: Independent component analysis of electroencephalographic data. In: Advances in Neural Information Processing Systems, pp. 145–151 (1996)
13. Mouraux, A., Iannetti, G.D., Colon, E., Nozaradan, S., Legrain, V., Plaghki, L.: Nociceptive steady-state evoked potentials elicited by rapid periodic thermal stimulation of cutaneous nociceptors. J. Neurosci. **31**(16), 6079–6087 (2011)
14. Mulders, D., de Bodt, C., Lejeune, N., Mouraux, A., Verleysen, M.: Spatial filtering of EEG signals to identify periodic brain activity patterns. In: Deville, Y., Gannot, S., Mason, R., Plumbley, M.D., Ward, D. (eds.) LVA/ICA 2018. LNCS, vol. 10891, pp. 524–533. Springer, Cham (2018). https://doi.org/10.1007/978-3-319-93764-9_48
15. Nakanishi, M., Wang, Y., Wang, Y.T., Mitsukura, Y., Jung, T.P.: A high-speed brain speller using steady-state visual evoked potentials. Int. J. Neural Syst. **24**(06), 1450019 (2014)
16. Samadi, S., Amini, L., Cosandier-Rimélé, D., Soltanian-Zadeh, H., Jutten, C.: Reference-based source separation method for identification of brain regions involved in a reference state from intracerebral EEG. IEEE Trans. Biomed. Eng. **60**(7), 1983–1992 (2013)
17. Sameni, R., Jutten, C., Shamsollahi, M.B.: Multichannel electrocardiogram decomposition using periodic component analysis. IEEE Trans. Biomed. Eng. **55**(8), 1935–1940 (2008)
18. Sameni, R., Jutten, C., Shamsollahi, M.B.: A deflation procedure for subspace decomposition. IEEE Trans. Signal Process. **58**(4), 2363–2374 (2010)
19. Saul, L.K., Allen, J.B.: Periodic component analysis: an eigenvalue method for representing periodic structure in speech. In: Advances in Neural Information Processing Systems, pp. 807–813 (2001)
20. Vincent, E., Bertin, N., Gribonval, R., Bimbot, F.: From blind to guided audio source separation: how models and side information can improve the separation of sound. IEEE Signal Process. Mag. **31**(3), 107–115 (2014)
21. Wu, W., Chen, Z., Gao, S., Brown, E.N.: A probabilistic framework for learning robust common spatial patterns. In: Annual International Conference of the IEEE Engineering in Medicine and Biology Society, EMBC 2009, pp. 4658–4661 (2009)

Correlation Filter Tracking
with Complementary Features

Wei Wang[1,2] ⓘ, Weiguang Li[1,2], and Mingquan Shi[1(✉)]

[1] Chongqing Institute of Green and Intelligent Technology,
Chinese Academy of Sciences, Chongqing, China
shimq@cigit.ac.cn
[2] University of Chinese Academy of Sciences, Beijing, China

Abstract. Although Correlation Filters (CF) tracking algorithms have inherent capability to tackle various challenging scenarios individually, none of them are robust enough to handle all the challenges simultaneously. For any online tracking based on Correlation Filters, feature is one of the most important factors due to its representation power of target appearance. In this paper, we proposed a new tracking framework by integrating the advantage of complementary features to achieve robust tracking performance. The key issue of this work lies in the fact that different features respond to different tracking challenges, which also applies to deep learning features and hand-craft features. Moreover, for the tracking speed balance, we train a light-weight deep CNN features by using end-to-end learning method, which has the same Parameter magnitude as the hand-crafted features. Experimental results on OTB-2013, OTB-2015 large benchmarks datasets show that the proposed tracker performs favorably against several state-of-the-art methods.

Keywords: Visual tracking · End-to-end learning · Correlation filters

1 Introduction

In recent years, tracking based on correlation filter (CF) [1–6] has greatly improved the tracking speed and performance. The popularity of correlation filters has the following important properties: First, CF-based tracking algorithms can significantly improve tracking speed due to computing the spatial correlation in the Fourier domain as element-wise product. Second, correlation filters provide a more power discriminative model by taking the surrounding context information into account. In addition, it is also well known that feature is one of the more important and effective factors in visual tracking frame-work [7]. With the rapid development of deep Neural Network [8–10], CF-based tracking using deep learning features [11–13] has achieved remarkable successes because those features have good invariance in image translation, rotation and deformation.

Despite above advantages, due to the huge number of parameters, tracking using deep learning features are limited in the low speed, also have some failure cases. For example, in some extreme light conditions, deep learning features tracker is often vulnerable to tracking failures. As shown in Fig. 1(d) in the singer2 of OTB benchmark

L. Cheng et al. (Eds.): ICONIP 2018, LNCS 11306, pp. 488–500, 2018.
https://doi.org/10.1007/978-3-030-04224-0_42

tracking using deep learning features has already fallen to tracking failure from frame 50, but tracking using hand-crafted features has been successful tracking in Fig. 1(d). These scenarios can also be found in more benchmark sequences. On the other hand, due to the lack of priori knowledge of the tracked target, such CNN trained for image recognition task often lead to over-fitting on tracking.

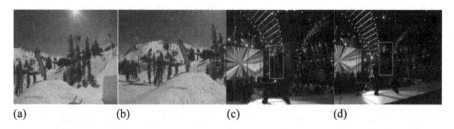

(a)   (b)   (c)   (d)

Fig. 1. On video sequence with deformation and rotation (a, b), tracking with HOG feature fall to tracking failure, while on video with extreme illumination condition (c, d), CNN feature fall to the tracking failure. (red: tracking result with CNN feature; green: tracking result with HOG feature) (Color figure online)

There are two methods to tackle this shortcoming: The first is using an end-to-end training of deep architectures in which CF is treated as a special correlation filter layer added in a Siamese network. This network is a lightweight CNN architecture [14, 15] with only thousands of parameters, which make CF-based tracker using CNN features achieve similar speed to that using hand-crafted features. Second, multiple features has shown its power in tracking task and performs favorably against the state-of-the-art [16–18]. Motivated by above discussion, we propose a novel tracker using lightweight CNN features and HOG features to get the CF response maps which are chosen by comparing their Peak to Side-lobe Ratio (PSR) [19] to estimate target location. Extensive evaluations demonstrate that the proposed tracker achieves the state-of-the-art results.

In brief, the main contributions of this paper are as follows

(1) We propose a new tracking framework which jointly considers the advantage of two different kinds of features. This framework not only improves the whole tracking performance but also achieves real-time speed.
(2) A lightweight CNN feature is trained by end-to-end learning method for tracking features fusing. This makes these two CF trackers using complementary features work synchronously.

2   Related Works

2.1   Tracking Based on Correlation Filter

Tracking based on correlation filters have attracted wide attention due to its high computing speed by using fast Fourier transforms [6, 19, 20]. Correlation filters tracking establish a model using cyclic sampling which is a dense sampling method and obviously superior to sparse sampling for tracking algorithm. Early correlation filtering tracking algorithms, like Minimum Output Sum of Squared Error filter tracker (MOSSE) and cycle kernel structure tracker (CSK), only use GRAY features for tracking, achieve extremely high tracking speed. By adopting the HOG features, kernelized correlation filter (KCF) using circularly generated samples, further improve the tracking performance and remained a high tracking speed. Based on KCF, several improvements have been proposed through a variety of different methods, like Discriminative Scale Space Tracking (DSST) [21], Context-Aware Correlation Filter Tracking (CACF) [22], Learning Spatially Regularized Correlation Filters (SRDCF) [23] and Long-term Correlation (LCT) [24]. These algorithms using single hand-crafted features are mainly improved on ridge regression model. Recently, some trackers use multiple features to improve tracking performance. Staple [25] combine HOG and global colour histogram (CN) features for target tracking, reach tracking speed of 80 FPS. [26] proposed a co-trained KCF tracker which fuse different convolutional features and handcrafted features. Moreover, [28] propose a novel incremental oblique random forest tracker which uses a more powerful proximal SVM to obtain oblique hyper planes to capture the geometric structure of the data better. Although using multiple features, the different between our work and [26] are as follows: (1) we use a lightweight convolutional neural network as the feature extractor; (2) The features we use to train the CF are completely unrelated.

2.2   CNN Feature Based Tracker

In recent years, deep convolutional neural networks (CNN) have been widely used in target tracking. Early CNN based trackers mainly use a pre-trained CNN as a feature extractor and build upon discriminative or regression models. Based on particle filter framework, FCNT [12] fuses the high-level and low-level CNN feature to track target. MDNET [11] proposed a multi domain network tracking algorithm which add convolutional neural network to target classification layer. HCF [27] proposed a coarse-to-fine tracking framework which use a hierarchical convolutional feature, further improves the tracking precision. The promotion [18] of HCF interprets correlation filters as the counterparts of convolution filters in deep neural networks. ECO [29] use feature dimensionality reduction to achieve a peak performance with high tracking speed. More recently, several tracking methods use end-to-end training networks to run at high frame rates while achieving state-of-the-art performance. SiamFC [30] developed a three-layers CNN architecture to learn the mapping between two consecutive frames as a spatial correlation response. As an extension of SiamFC, CFnet [14] treats the CF as a special layer of the whole end-to-end learning architecture. [15] presents an end-to-end lightweight network architecture to learn the convolutional features and perform the correlation tracking process simultaneously. By using this lightweight CNN features, CF-based tracker achieves almost the same tracking speed as those trackers using hand-crafted features, additionally with relatively high tracking performance.

3 The Proposed Tracker

In this section, we introduce the framework of our tracker which consists of Correlation Filter Tracking, End-to-end representation learning and complementary features ensemble tracking.

3.1 Correlation Filter Tracking

A typical correlation filtering tracker is to learn a discriminative classifier and locate the target by searching for the maximum response value of the correlation map. Denote x with size M × N as the input feature of the correlation filtering tracker, where M, N, indicates the width and height. At first, the ideal response $y \in R^{\wedge}(M \times N)$ need to be given by the following equation:

$$y(m, n) = e^{-\frac{(m-M/2)^2 + (n-N/2)^2}{2\sigma^2}} \tag{1}$$

where σ is the kernel width. The correlation filter w can be learned by minimizing the ridge regression loss:

$$w^* = \arg\min \sum_{m,n} \|w \cdot x(m, n) - y(m, n)\|^2 + \lambda \|w\|^2 \tag{2}$$

where λ is a regularization parameter used to reduce over-fitting, • is an inner product symbol. As a ridge regression formula, (2) has a close-form. So the w can be given by solving (2) using least- square method:

$$W = \frac{X^T X + \lambda I}{X^T Y} \tag{3}$$

Since X is circulant, it can be diagonalized so that Eq. (3) can be rewritten in the Fourier domain [10]:

$$\hat{w} = \frac{\hat{x}^* \odot \hat{y}}{\hat{x}^* \odot \hat{x} + \lambda} \tag{4}$$

where $\hat{x}$ means $\hat{x} = \text{FFT}(x)$ as well as other symbols, $\odot$ indicates the element-wise product, $*$ means the complex conjugate of $\hat{x}$. Let z be the feature vector of the next frame image which has a circulant matrix Z, its correlation response can be calculated by:

$$\hat{f} = \hat{z} \odot \hat{w} \tag{5}$$

The target can be located by searching for the position of maximum value of the correlation response map $\hat{f}$.

3.2   End-to-End Representation Learning

In many existing deep learning DCF tracker, correlation filter only connect to a pre-train CNN feature extractor without any deep integration. The feature extractor has been learned off-line by using a huge image detection dataset, which are not necessarily fit for tracking tasks. In this section, we introduce an end-to-end training of deep architectures with a few thousand parameters which is generally preferable to training individual components separately. As shown in Fig. 2, in the end-to-end training architecture, CF is treated as a special correlation filter layer added in a Siamese network so that back-propagate gradients through an entire SGD learning procedure.

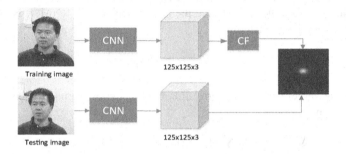

Fig. 2. The end-to-end learning architecture

This end-to-end learning architecture is a Siamese convolutional framework which has a training image pairs (x', z') representing the object of interest and the search area. For example, in a CF training explanation, the image pairs can be considered as the video frame T and T + 1 according to the Training image and Testing image of in Fig. 2. Each input image is processed by a CNN to extract features. The response maps of these two features are cross-correlated by equal:

$$L(\theta) = \| f - \tilde{f} \|^2 + \gamma \|\theta\|^2$$
$$\text{s.t.} \quad f = F^{-1}(\hat{z} \odot \hat{w}) \tag{6}$$

where F^{-1} is inverse discrete Fourier transform, $\tilde{f}$ is the real response map of z. Here, $z = \varphi(x')$ is the feature extracted from the CNN. To avoid non-convergence condition, we also use the weight decay method in the CNN parameter optimization.

In the proposed end-to-end framework, the derivative of the Correlation Filter has a closed-form expression so that the gradient can be back-propagated. Since the computation in CF tracking is converted in Fourier domain, the gradient of discrete Fourier transform and inverse discrete Fourier transform can be formulated as:

$$\hat{f} = F(f), \frac{\partial L}{\partial \hat{f}^*} = F\left(\frac{\partial L}{\partial f}\right), \frac{\partial L}{\partial f} = F^{-1}\left(\frac{\partial L}{\partial \hat{f}^*}\right) \tag{7}$$

The whole propagated backwards is given by:

$$\frac{\partial L}{\partial z} = \frac{\partial L}{\partial \varphi(x)} = F^{-1}\left(\frac{\partial L}{\partial(\hat{\varphi}(x))^*} + \left(\frac{\partial L}{\partial \hat{\varphi}(x)}\right)^*\right) \tag{8}$$

The back-propagation in correlation filter layer still can be computed in Fourier frequency domain so that the CNN connected a CF layers apply the offline training on large-scale datasets with. Since this ultra-lightweight CNN feature has approximately the same number of parameters as hand-crafted features, we use the two kinds of feature to improve the whole tracking performance.

3.3 Complementary Features Ensemble Tracking

Tracking with single feature may be not powerful enough to handle extreme illumination and complex background. The obvious way to tackle this is using multiple features. Many existed works have used multiple features such as combining pre-trained CNN or hand-crafted features to improve the tracking performance. However, due to the huge number of parameter, CNN based tracker suffers from high computational time which makes them unsuitable for real-time applications.

In this work, we train CF by using lightweight CNN and HOG feature respectively, comparing the two CF response maps with PSR to locate the target. The lightweight CNN has been trained by end-to-end representation learning method mentioned in Sect. 3.2. As shown in Fig. 3, the architecture of the proposed tracker has two parallel correlation filters, which are trained in each frame by lightweight CNN features and HOG features respectively. The filters of the two CFs can be solved by formula:

$$\begin{cases} W_{CNN} = \dfrac{X_{CNN}^T X_{CNN} + \lambda I}{X_{CNN}^T Y} \\[2mm] W_{HOG} = \dfrac{X_{HOG}^T X_{HOG} + \lambda I}{X_{HOG}^T Y} \end{cases} \tag{9}$$

Both formulas in (10) can be calculated in Fourier domain to get the response maps by (6).

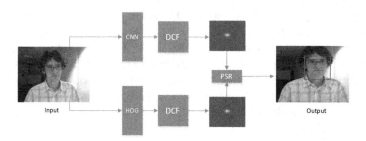

Fig. 3. The proposed training architecture

In general, CF-based trackers evaluate the tracking result by estimating on the highest peak of the response map. However, this tracking result may be disturbed by similar objects or background clutters leading to incorrect detection. As a result, incorrect detection would further affect the CF model duo to inaccurate training samples. Consequently, we use PSR as the criterion to measure the response map generated by the two kinds of features. PSR is defined by formula:

$$PSR = \frac{f_{\max} - mean(f)}{\delta} \qquad (10)$$

Where f is the response map, δ is the standard deviation of f. The measure formula is defined as:

$$P = \max(PSR_{CNN}, PSR_{HOG}) \qquad (11)$$

Thus, we use the response map with higher PSR to estimate the target location and update the training samples for the next frame. Note that our method does not only use CF maximum response value, like LCT [24], to determine the tracking result. That is due to that CF maximum response value may not really reflect the tracking result when the CF model is contaminated in case of hard tracking challenge, e.g., heavy occlusion, deformation and rotation.

4 Experiments and Analysis

In this section, we first describe the implement details of the proposed tracker. Then, we introduce favorable benchmark datasets and their evaluation metrics used in the experiment. Last, we evaluate the tracker from overall and sub attribute performance compared to other state-of-the-art trackers.

4.1 Implement Details

The overall procedure of the proposed tracker is presented in Algorithm 1. The lightweight CNN have two convolutional layers partly come from VGG [31] without pooling layers and the output forced to 32 channels. The training sets come from the videos of ImageNet [32] that overlap with the test set. Each pair of frames is chosen in the nearest 10 frames and cropped 1.5× padding of target patches. We set stochastic gradient descent (SGD) with momentum of 0.9, learning rate of R = 0.0001 to train the network from the scratch.

The proposed tracker framework is based on KCF with three target scales. According to the discussion in KCF, We use linear kernel instead of Gauss kernel, to map the feature space. For combining features of the CF tracker, we propose a parallel tracking structure in which CNN feature extraction is applied in GPU and HOG feature is processed in CPU at the same time.

Algorithm 1:

Input: the first frame of the image and the lightweight CNN.

Output: target position.

1   Extract the lightweight CNN and HOG features from the crop patch of the target in the first frame.

2 Train the CF model W_{CNN}, W_{HOG} by Eq. (4)

3 From 2 to N

4   Extract the lightweight CNN and HOG features from the crop patch of the target in the current frame.

5 get the response map f_{CNN}, f_{HOG} by Eq. (5) with the current features

6 Get the optimal response map by comparing the PSR with Eq. (11)

IF PSR (f_{CNN}) >= PSR (f_{HOG})

$f = f_{CNN}$

ELSE

$f = f_{HOG}$

7 Estimate the target position with response map f.

8 Update the CF model.

9 End

10 End

4.2   Data Sets and Evaluation Metrics

OTB [33] is a popular tracking benchmark that contains a large number of annotated videos with substantial variations. Its video sequences are categorized with 11 attributes based on different challenging factors [33]. To better analyze the performance of our approach, we evaluated our tracker on all above the 11 challenges.

OTB provides three evaluation metrics: one pass evaluation (OPE), time robustness evaluation (TRE) and spatial robustness evaluation (SRE). Each metrics use Precision plot and Success plot to evaluate the tracker performance. The area under curve (AUC) of success plot is the average of the success rates according to the sampled overlap thresholds. Given a tracked bounding box K_t and the ground-truth bounding extent K_0 of a target object, the overlap score is defined as:

$$\varphi_K = \left| \frac{K_0 \cap K_t}{K_0 \cup K_t} \right| \qquad (12)$$

where $\cap$ and $\cup$ are regional intersection and union operation, while $|*|$ is definite for the number of pixels in the frame. In the success plot, the x-axis depicts a set of thresholds for the overlap to indicate the tracking success, while in the precision plot the x-axis depicts a set of thresholds for the error to indicate the tracking precision.

4.3 Comparisons with More State-of-the-Arts

Evaluation on OTB2013

Our tracker is compared with recent proposed trackers including CF2 [27], MEEM [34], KCF [20], DSST [21] and Staple [25] on the 50 sequences of OTB2013. Among the compared trackers, CF2 use deep CNN features and the others use hand-crafted features. It needs to be emphasized that Staple is a complementary features tracker which combines the response map of HOG and CN features with their weighted sum to track target. The distance precision rate and overlap success rate of OPE and SRE are shown in Fig. 4.

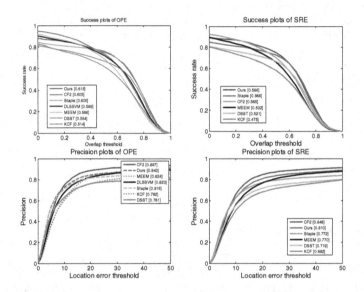

Fig. 4. Precision and Success plots on OTB 2013 of the compared tracker

The larger the area of the curve, the better the performance of the algorithm is. Overall, our algorithm performs favorably against the compared state-of-the-art tracker in two metrics: OPE and SRE. Our complementary feature tracker leads to 3.5% and 6.5% increases in success plots of OPE compared with CF2 using hierarchical deep features and DSST using HOG features respectively.

Evaluation on OTB2015

We also evaluate our tracker on a larger benchmark OTB2015 comparing with 5 state-of-the-art methods. As shown in Fig. 5, our tracker also has superior annotated with overall performance than the compared methods.

In order to demonstrate the feasibility of the proposed complementary feature Methods, we also provide a baseline tracker which only uses the lightweight CNN or HOG features as the feature extractor. The accuracy of OPE of the ablation comparison on OTB2015 are shown on Fig. 6. It is easy to see that the performance of our proposed tracker is superior to the baseline due to the use of complementary features.

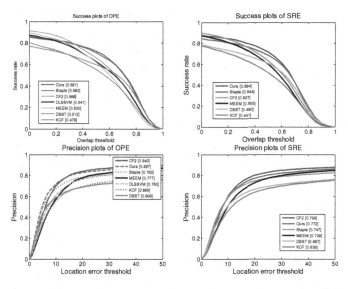

Fig. 5. Precision and Success plots on OTB 2015 of the compared tracker

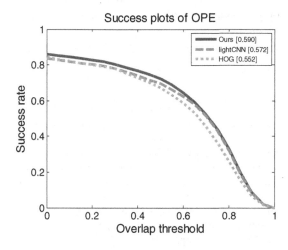

Fig. 6. Success plots of OPE of the compared method on OTB 2015

Tracking Speed Analysis

Speed is another important factor in tracking task. It is obvious that tracking with deep CNN features often have a low speed due to the expensive computing of feature extracting. For the speed analysis, we compare the feature extracting speed of three kinds of feature: deep feature (using VGG4-3), end-to-end light-weight CNN feature and HOG feature in Table 1.

Notice that the two CNN features and HOG feature extraction are computed in GPU and CPU respectively. The compared tracker CF2 using VGG-net as the feature

Table 1. The feature extracting speed using deep feature (VGG4-3), lightweight CNN feature and HOG feature (the images is from *Basketball* in OTB)

	Deep CNN feature	Lightweight CNN feature	HOG feature
Feature dimension	$20 \times 50 \times 512$	$125 \times 125 \times 32$	$125 \times 125 \times 31$
Speed	98 ms	19 ms	18 ms

extractor runs at a speed of 10 FPS. The average tracking speed of ours is 48 FPS and nearly 5 times faster than CF2. The main reason is that the lightweight CNN features only have several thousand parameters, far less than traditional deep CNN feature.

5   Conclusions

In this paper, we proposed a new simple tracker based on complementary features, which use end-to-end light-weight CNN feature and HOG feature to train the correlation filter respectively. The proposed tracker also uses PSR instead peak value to select the appropriate response map to locate the target. Evaluation on large-scale benchmark dataset demonstrates that our tracker achieves favorable performance against the compared state-of-the-art methods. It should be pointed out that our proposed tracker not only has superior performance, but also runs at a fast speed which is enough for real-time applications.

References

1. Junseok, K., Leek, M.L.: Visual tracking decomposition. In: IEEE Conference on Computer Vision and Pattern Recognition, pp. 1269–1276. IEEE Press, San Francisco (2010)
2. Hare, S., Saffari, A., Torr, P.H.S.: Struck: structured output tracking with kernels. In: International Conference on Computer Vision, pp. 263–270. IEEE Computer Society (2011)
3. Avidan, S.: Ensemble tracking. IEEE Trans. Pattern Anal. Mach. Intell. **29**(2), 261–271 (2007)
4. Bolme, D.S., Beveridge, J.R.: Visual object tracking using adaptive correlation filters. Comput. Vis. Pattern Recognit. **119**(5), 2544–2550 (2010)
5. Galoogahi, H. K., Sim, T., Lucey, S.: Multi-channel correlation filters. In: IEEE International Conference on Computer Vision, pp. 3072–3079, IEEE (2014)
6. Henriques, J.F., Caseiro, R., Martins, P., Batista, J.: Exploiting the circulant structure of tracking-by-detection with kernels. In: Fitzgibbon, A., Lazebnik, S., Perona, P., Sato, Y., Schmid, C. (eds.) ECCV 2012. LNCS, vol. 7575, pp. 702–715. Springer, Heidelberg (2012). https://doi.org/10.1007/978-3-642-33765-9_50
7. Wang, N., Yeung, D.Y.: Learning a deep compact image representation for visual tracking. In: Proceedings of the 26th International Conference on Neural Information Processing Systems, pp. 809–817. Curran Associates Inc., Lake Tahoe (2013)
8. Zhang, K., Liu, Q., Wu, Y., Yang, M.H.: Robust visual tracking via convolutional networks without training. IEEE Trans. Image Process. **25**(4), 1779–1792 (2016)
9. Krizhevsky, A., Sutskever, I., Hinton, G.E.: Imagenet classification with deep convolutional neural network. Adv. Neural. Inf. Process. Syst. **60**(1), 1097–1105 (2012)

10. Girshick, R., Donahue, J., Darrell, T., Malik, J.: Rich feature hierarchies for accurate object detection and semantic segmentation. In: IEEE Conference on Computer Vision and Pattern Recognition, pp. 580–587. IEEE Computer Society (2014)

11. Nam, H., Han, B.: Learning multi-domain convolutional neural networks for visual tracking. In: IEEE Conference on Computer Vision and Pattern Recognition, pp. 4293–4302. IEEE (2016)

12. Wang, L., Ouyang, W., Wang, X., Lu, H.: Visual tracking with fully convolutional networks. In: IEEE International Conference on Computer Vision, pp. 3119–3127. IEEE (2015)

13. Wang, L., Ouyang, W., Wang, X., Lu, H.: STCT: sequentially training convolutional networks for visual tracking. In: IEEE Conference on Computer Vision and Pattern Recognition, pp. 1373–1381. IEEE (2016)

14. Valmadre, J., Bertinetto, L., Henriques, J., Vedaldi, A., Torr, P.H.S.: End-to-end representation learning for correlation filter based tracking. In: IEEE Conference on Computer Vision and Pattern Recognition, pp. 5000–5008. IEEE (2017)

15. Wang, Q., Gao, J., Xing, J., Zhang, M., Hu, W.: Dcfnet: discriminant correlation filters network for visual tracking. arXiv:2017.1704.04057, http://cn.arxiv.org/abs/1704.04057

16. Bertinetto, L., Valmadre, J., Golodetz, S., Miksik, O., Torr, P.: Staple: complementary learners for real-time tracking. Comput. Vis. Pattern Recognit. 38(2), 1401–1409 (2016)

17. Zhang, L., Suganthan, P.N.: Robust visual tracking via co-trained kernelized correlation filters. Pattern Recognit. 69(1), 82–93 (2017)

18. Ma, C., Xu, Y., Ni, B., Yang, X.: When correlation filters meet convolutional neural networks for visual tracking. IEEE Signal Process. Lett. 23(10), 1454–1458 (2016)

19. Bolme, D.S., Beveridge, J.R., Draper, B.A., Lui, Y.M.: Visual object tracking using adaptive correlation filters. In: IEEE Conference on Computer Vision and Pattern Recognition, vol. 119, pp. 2544–2550. IEEE (2010)

20. Henriques, J.F., Caseiro, R., Martins, P., Batista, J.: High-speed tracking with kernelized correlation filters. IEEE Trans. Pattern Anal. Mach. Intell. 37(3), 583–596 (2014)

21. Danelljan, M., Häger, G., Khan, F.S., Felsberg, M.: Accurate scale estimation for robust visual tracking. In: British Machine Vision Conference, vol. 65, pp. 1–11 (2014)

22. Mueller, M., Smith, N., Ghanem, B.: Context-aware correlation filter tracking. In: IEEE Conference on Computer Vision and Pattern Recognition, pp. 1387–1395. IEEE (2016)

23. Danelljan, M., Häger, G., Khan, F. S., Felsberg, M.: Adaptive decontamination of the training set: a unified formulation for discriminative visual tracking. In: IEEE Conference on Computer Vision and Pattern Recognition, pp. 1430–1438. IEEE (2016)

24. Ma, C., Yang, X., Zhang, C., Yang, M.H.: Long-term correlation tracking. In: IEEE Conference on Computer Vision and Pattern Recognition, pp. 5388–5396. IEEE (2015)

25. Bertinetto, L., Valmadre, J., Golodetz, S., Miksik, O., Torr, P.: Staple: complementary learners for real-time tracking. In: IEEE Conference on Computer Vision and Pattern Recognition, vol. 38, no. 2, pp. 1401–1409. IEEE (2015)

26. Zhang, L., Suganthan, P.N.: Robust visual tracking via co-trained kernelized correlation filters. Pattern Recognit. 69, 82–93 (2017)

27. Ma, C., Huang, J.B., Yang, X., Yang, M.H.: Hierarchical convolutional features for visual tracking. In: IEEE International Conference on Computer Vision, pp. 3074–3082. IEEE Computer Society (2015)

28. Zhang, L., Varadarajan, J., Suganthan, P.N., Ahuja, N., Moulin, P.: Robust visual tracking using oblique random forests. In: IEEE Conference on Computer Vision and Pattern Recognition, pp. 5825–5834. IEEE (2017)

29. Danelljan, M., Bhat, G., Khan, F.S., Felsberg, M.: Eco: efficient convolution operators for tracking. In: IEEE Conference on Computer Vision and Pattern Recognition, pp. 6931–6939. IEEE (2017)
30. Bertinetto, L., Valmadre, J., Henriques, João F., Vedaldi, A., Torr, Philip H.S.: Fully-convolutional siamese networks for object tracking. In: Hua, G., Jégou, H. (eds.) ECCV 2016. LNCS, vol. 9914, pp. 850–865. Springer, Cham (2016). https://doi.org/10.1007/978-3-319-48881-3_56
31. Simonyan, K., Zisserman, A.: Very deep convolutional networks for large-scale image recognition. arXiv 2014.1409.1556. http://cn.arxiv.org/abs/1409.1556
32. Russakovsky, O., Deng, J., Su, H.: imagenet large scale visual recognition challenge. Int. J. Comput. Vis. **115**(3), 211–252 (2015)
33. Wu, Y., Lim, J., Yang, M.H.: Object tracking benchmark. IEEE Trans. Pattern Anal. Mach. Intell. **37**(9), 1834–1848 (2015)
34. Zhang, J., Ma, S., Sclaroff, S.: MEEM: robust tracking via multiple experts using entropy minimization. In: Fleet, D., Pajdla, T., Schiele, B., Tuytelaars, T. (eds.) ECCV 2014. LNCS, vol. 8694, pp. 188–203. Springer, Cham (2014). https://doi.org/10.1007/978-3-319-10599-4_13

Fast Image Recognition with Gabor Filter and Pseudoinverse Learning AutoEncoders

Xiaodan Deng, Sibo Feng, Ping Guo$^{(\boxtimes)}$, and Qian Yin$^{(\boxtimes)}$

Image Processing and Pattern Recognition Laboratory,
Beijing Normal University, Beijing, China
dengxiaodancyc@163.com, sibofeng@mail.bnu.edu.cn,
pguo@ieee.org, yinqian@bnu.edu.cn

Abstract. Deep neural network has been successfully used in various fields, and it has received significant results in some typical tasks, especially in computer vision. However, deep neural network are usually trained by using gradient descent based algorithm, which results in gradient vanishing and gradient explosion problems. And it requires expert level professional knowledge to design the structure of the deep neural network and find the optimal hyper parameters for a given task. Consequently, training a deep neural network becomes a very time consuming problem. To overcome the shortcomings mentioned above, we present a model which combining Gabor filter and pseudoinverse learning autoencoders. The method referred in model optimization is a non-gradient descent algorithm. Besides, we presented the empirical formula to set the number of hidden neurons and the number of hidden layers in the entire training process. The experimental results show that our model is better than existing benchmark methods in speed, at same time it has the comparative recognition accuracy also.

Keywords: Pseudoinverse learning autoencoder · Gabor filter
Image recognition · Handcraft feature

1 Introduction

Neural network has attracted many researchers to study, and it has been used in many fields successfully. Currently the most used model for image recognition is convolutional neural networks (CNN). In 1998, Yann LeCun and Yoshua Bengio published a paper on the application of neural networks in handwriting recognition and optimization with back propagation algorithm, and presented model LeNet5 in [1] is considered as the beginning of CNN. Its network structure includes the convolutional layer, the pooling layer and the full connection layer, which are the basic components of the modern CNN network. In 2012, Alex used AlexNet [2] in the contest of ImageNet to refresh the record of image classification and set the position of deep learning in computer vision. AlexNet uses five convolution layers and three fully connected layers for classification. Subsequently, there are many other successful CNN models, which became deeper and more complex. In 2015, He *et al.* proposed the ResNet [3] model which reached 152 layers.

© Springer Nature Switzerland AG 2018
L. Cheng et al. (Eds.): ICONIP 2018, LNCS 11306, pp. 501–511, 2018.
https://doi.org/10.1007/978-3-030-04224-0_43

Recently, successful CNN models usually have complex structure and need to set many hyper-parameters. Those parameters are related to the performance of the CNN models and they are difficult to tune. Many research groups have presented their research results, however it is difficult to repeat them. On the other hand, because there are too many hyper-parameters, the training of the CNN model is a time-consuming process. Moreover, most deep neural networks are trained by the gradient descent (GD) based algorithms and variations [1, 4]. Also, it is found that the gradient descent based algorithm in deep neural networks has inherent instability. This instability blocks the learning process of the previous or later layers. In addition, gradient descent method is easy to be stuck in vanishing gradient problem. Though CNN has good performing result, it need much professional knowledge to use and it takes a lot of time to train.

In order to reduce the training time and improve the generalization ability of neural network, we present a model, which combines the Gabor filter [5] and pseudoinverse learning autoencoders (PILAE) [6], to deal with image recognition problem. Gabor transformation belongs to the window Fourier transformation, and the Gabor function can extract relevant characteristics in different scales and directions in the frequency domain. In addition, Gabor function is similar to the biological function of human eyes, so it is often used for texture recognition and has achieved good results. The main advantage of using CNN based deep nets is the features are leant from images, while the advantage of using traditional handcraft - features is that the feature extraction speed. Therefore, to extract features using Gabor filter (GF) is much easier than that of using CNN. In Feng et al.'s work [7], histogram of oriented gradient (HOG) is used to extract features. However, the generation processing of the HOG descriptor is tedious, resulting in slow speed and poor real-time performance. Besides, due to the nature of the gradient, the descriptor is quite sensitive to noise. Hence, we choose Gabor filter to extract features first. Then, PILAE based feed forward neural net is adopted to extract independent feature vectors and perform image recognition.

Our proposed GF + PILAE model optimization does not need gradient descent based algorithms. The learning procedure of our model is forward propagating and the whole structure of network is determined with a given strategy in the process of propagation, including the depth of the network and the number of neurons in the hidden layer. It is a completely quasi-automatic learning procedure, so even users without professional knowledge they can easily use it. It is our efforts to prompt democratized artificial intelligence development.

2   Related Work

2.1   Gabor Filter

The class of Gabor functions was presented by Gabor [8]. The basic idea of Gabor function is to add a small window to the signal. The Fourier transform of the signal is mainly concentrated in the small window, so it can reflect the local characteristics of the signal. Daugman [9] extended Gabor function to two-dimensional cases. Gabor wavelet function was regarded as the best model for simulating visual sensory cells in the cerebral cortex [10]. Each visual cell can be viewed as a Gabor filter with a certain direction and scale. When an external stimulus such as image signal inputs visual cells, the output response of visual cells is the convolution of image with Gabor filter, and the output signal is further processed by the brain to form the final impression of cognition. This model can better explain human vision's tolerance to scale and direction change. The two-dimensional Gabor kernel function is defined as follows [11]:

$$G_{\lambda,\theta,\varphi,\sigma,\gamma}(x,y) = \exp\left(-\frac{x'^2 + \gamma^2 y'^2}{2\sigma^2}\right)\cos\left(2\pi\frac{x'}{\lambda} + \varphi\right),$$
$$x' = (x - x_0)cos\theta + (y - y_0)sin\theta,$$
$$y' = -(x - x_0)sin\theta + (y - y_0)cos\theta.$$

(1)

Equation (1) is obtained by the multiplication of a Gaussian function and a cosine function. The arguments x and y specify the position of a light impulse, where (x_0, y_0) is the center of the receptive field in the spatial domain. Θ is the orientation of parallel bands in the kernel of Gabor filter, and the valid values are real numbers from 0 to 360. φ is the phase parameter of cosine function in Gabor kernel function, and the valid values is from -180 to $180°$. γ is the space aspect ratio, which represents the ellipticity of the Gabor filter. λ is the wavelength parameter of the cosine function in the Gabor kernel function. σ is the standard deviation of Gaussian function in the Gabor kernel function. This parameter determines the size of acceptable area in the Gabor filter. Its value is related to the Bandwidth b and the value of λ. The Bandwidth b indicates the difference in high and low frequency. Equation (2) presents the relationship of b, σ and λ:

$$b = log_2 \frac{\frac{\sigma}{\lambda}\pi + \sqrt{\frac{ln2}{2}}}{\frac{\sigma}{\lambda}\pi - \sqrt{\frac{ln2}{2}}}\frac{\sigma}{\lambda} = \frac{1}{\pi}\sqrt{\frac{ln2}{2}}\cdot\frac{2^b + 1}{2^b - 1}.$$

(2)

Usually we use the Gabor filter in 8 directions, 5 scales, and these parameters can be adjusted. Figure 1 is a sample of Gabor filter bank with forty different Gabor filters. Feature extraction is performed using Gabor filter, as shown in Eq. (3).

$$\mathbf{I}_G = \mathbf{I} \oplus \mathbf{G}.$$

(3)

where I is the grayscale distribution of the image, I_G is the feature extracted from I, "$\oplus$" stands for 2D convolution operator, G is the defined Gabor filter. Equation (3) can

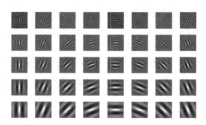

Fig. 1. Gabor filter bank. These filters are in different scales and orientations [12]

be efficiently computed by fast Fourier transform, $\mathbf{I}_G = F^{-1}(F(\mathbf{I})F(\mathbf{G}))$, where F^{-1} is the inverse Fourier transform.

Gabor filters are sensitive to the edge information of images and able to adapt to the obvious environments with different light. Studies have found that Gabor wavelet transformation is very suitable for texture expression and separation. Gabor filter needs less data and can meet the real-time requirements compared with other methods. On the other hand, it can tolerate a certain degree of image rotation and deformation.

2.2   Autoencoders

An autoencoder [13] was first proposed by Rumelhart *et al.* in 1986. The autoencoding neural network is an unsupervised learning scheme, which uses the back propagation algorithm and tries to encode input vectors into hidden vectors, and decode hidden vectors into input vectors. Autoencoders are usually used to reduce dimension and feature learning task. The autoencoder consists of two parts: encoder and decoder. The autoencoder can compress the input into potential space representation and the decoder can reconstruct the input from potential space representation. The loss function of autoencoder can be defined as reconstruction error function in Eq. (4),

$$E = \sum\nolimits_{i=1}^{N} \left\| \left(W_d \left(f \left(W_e x^i + b_e \right) \right) + b_d \right) - x^i \right\|^2, \qquad (4)$$

where W_e and W_d is respectively the weights of encoder and decoder, b_e and b_d is the bias of the encoder and decoder respectively.

A stacked autoencoder is a feed forward neural network in which the outputs of each encoder layer are the inputs of the successive layer. A way to obtain good parameters for a stacked autoencoder is to use greedy layer-wise training [14]. This method trains the weight parameters of each layer individually while freezing parameters for the remains of the model. To produce the better results, after this phase of training is complete, fine-tuning with backpropagation algorithm can be used to improve the results by tuning the parameters of all layers at the same epoch.

2.3   Pseudoinverse Learning Autoencoder

Pseudoinverse learning algorithm (PIL) [15–17] was originally proposed by Guo *et al.*, which is a fast algorithm for training feedforward neural networks. The whole training process of the PIL just needs simple matrix inner product operation and pseudoinverse operation. It improves the learning accuracy by adding layers, without iteration optimization like other gradient descent based algorithms. Moreover, it is convenient to use and does not require users to set various hyper-parameters. The depth parameter in the training process are automatically adjusted.

For a classification task, suppose that the data set $D = \{X^i, O^i\}_{i=1}^{N}$ denotes N samples, where $X^i = (x_1, x_2, \ldots, x_d) \in R^d$ and $O^i = (o_1, o_2, \ldots, o_m) \in R^m$ denotes the i-th input sample and the corresponding expected output respectively. For a single hidden layer forward network, the most used sum-of-square objective function is as follows,

$$E = \frac{1}{2N}\sum_{i=1}^{N}\sum_{j=1}^{m}\left\|g_j\left(x^i, \hat{\theta}\right) - o_j^i\right\|^2, \tag{5}$$

where $g_j\left(x^i, \hat{\theta}\right)$ is the j-th output neuron, which shows the map from input value into predicted value, and it is defined as follows,

$$g_j\left(x, \hat{\theta}\right) = \sum_{i=1}^{p} w_{i,j}^1 f\left(\sum_{k=1}^{d} w_{k,i}^0 x_k + b\right). \tag{6}$$

To simplification, we can represent the map in matrix form. The hidden layer is defined as follows,

$$H = f(XW_0 + b), \quad X \in R^{N \times d}, \quad W_0 \in R^{d \times p}, \tag{7}$$

where H is a matrix representing the output of hidden layer, X is the input matrix which has N vectors with d dimension, W_0 is the weights matrix between input and hidden layer which has d rows and p columns, b is a bias parameter in the input layer, $f(\cdot)$ is an activation function. Details of the PIL can be found in Refs. [15–17].

3   Proposed Methodology

3.1   Proposed Classification Model

We proposed a classification model combining Gabor filter and pseudoinverse learning autoencoders (PILAE), which is a forward network without needing iterative optimization by gradient descent algorithm. The structure of the model is shown in Fig. 2.

The input image is first filtered by Gabor filter bank, then we can obtain feature maps. Gabor feature map is a kind of Handcraft feature, and it is easy to obtain. The

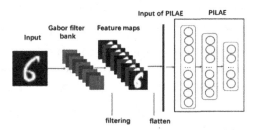

Fig. 2. The proposed method combining Gabor filter and PILAE. Gabor filters extract features from input image to form feature maps, then the PILAE extract features further and make classification.

feature maps are fused into a vector as the input of PILAE. PILAE is used to further extract features and make classification.

3.2 Feature Extraction

In our model, feature extraction is set to two parts. First, image features is extracted by Gabor filter. With different scales and different orientations, Gabor filters can extract different features from original image input. Moreover, the biological model function of human eyes is similar to Gabor function, so Gabor filter performs well in extract features. Second, PILAE can extract features from input vector. PILAE is consist of several layers, in which there are different number of neurons. If the number of neurons in hidden layer is less than the dimension of the input, it is equivalent to extracting features from the input vectors.

3.3 Training Model

The Gabor filter part can be considered as data preprocessing, which transforms input data to a first feature data space, while the training focuses on the PILAE section. For a single autoencoder that uses pseudoinverse algorithm to train, the encoder part can be represented as follows,

$$H = f(W_e X), \tag{8}$$

where X is the input, which can be considered as the succeed feature data input. H is the hidden output, which can be considered as the succeed feature data output. $f(\cdot)$ is the activation function, W_e is the weight between input and hidden layer. And the decoder part can be represented as follows,

$$G = g(W_d H), \tag{9}$$

where G is a vector mapped from vector H, $g(\cdot)$ is the activation function, W_d is the weight between hidden layer and output layer. Our objective function is as follows,

$$J = \frac{1}{2} \|W_d H - X\|^2 + \frac{\hat{\lambda}}{2} \|W_d\|^2, \tag{10}$$

where $\hat{\lambda}$ is a regularization parameter which can be selected with a formula in Ref. [19]. This is an error function with weight decay. The goal of training the auto-encoder is to find the weight parameter to minimize the error function. By using pseudoinverse learning algorithm, we can get the weight of decoder is as follows,

$$W_d = XH^T \left(HH^T + \hat{\lambda}\right)^{-1}. \tag{11}$$

The weights W_e and W_d can be set as $W_d = W_e^T$, which is termed as tied weights. The basic low-rank approximation is adopted to avoid identity mapping [18]. To get the optimal result, we can add layer to the model, which is called as stacked autoencoders in the literatures. The number of first hidden layer neuron is set to equal to the rank of input matrix. For succeeding hidden layer the number of neuron is set to be $p = \beta Dim(x)$, $\beta \in (0, 1]$, where $Dim(x)$ is the dimension of input vectors. Because the autoencoder is trained with PIL algorithm, we name it as PILAE [18]. When the autoencoder training is finished, we discard the decoder parts and cascade the encoders to form a stacked autoencoder network. For the task of classification or regression, in the final network output layer, a classifier at the end of the network is used to get the final result of classification. The classifiers can be a PIL as well as its variants [20], a support vector machine (SVM), a multilayer neural network (MLP), or a radial basis function network, and so on. In this work, a Softmax classifier is used to get final output results.

4   Performance Evaluation

To evaluate the performance of the proposed methods, we conduct experiments to compare our method with other methods based on benchmark data set. In the experiments, the parameters of Gabor filter are set as follow: the scale of Gabor filter is 2, 4 and 6, respectively, the wave length λ is $\pi/2$, the orientation θ is set as 8 different orientations from 0 to $7\pi/8$, and the difference between the two adjacent orientations is $\pi/8$, standard deviation of the Gaussian function σ is 1.0, spatial aspect ratio γ is 0.5, phase parameter φ is 0. We conduct experiments to generate different numbers of Gabor feature maps, which are fused into a feature vector input to the following network layer. The MNIST dataset and the CIFAR-10 dataset are used in the experiments. All the experiments are conducted on the same hardware computer with Core i7 3.20 GHz processors.

4.1   MNIST Dataset

In deep learning and pattern recognition, MNIST is the most widely used database. MNIST is a handwritten digits images recognition data set including 70,000 handwritten digital images of 0–9, 60,000 images out of which are used as training samples and the rest 10,000 images are the test samples. Each image in the dataset is $28 \times 28 = 784$ pixels.

Table 1 shows the comparison results of our method and other benchmark methods using MNIST dataset. In our method, we use four Gabor feature maps and three layers of PILAE to get the result. The structure of PILAE is 705-635-571 in GF + PILAE. 5 and 10 convolution kernels are used in different layers in LeNet5. MLP uses one hidden layer and the number of hidden neuron is 300. PILAE has one encoder layer in this experiment. We use 4 HOG (Histogram of oriented gradient) feature maps in the method of HOG + PILAE. From Table 1, we observe that our method can obtain comparable accuracy to other baseline method, while the training speed of our method is fast than others.

Figure 3(a) shows that the highest accuracy was obtained as the number of Gabor features was four, and the accuracy decreases as the number of Gabor features continued to increase. The reason may be the fact that the more feature maps are used the

Table 1. Performance comparison of MNIST dataset.

Model	Training accuracy (%)	Testing accuracy (%)	Training time (s)
GF + PILAE	98.86	98.42	103.25
LeNet5	98.51	98.49	1270.8
PILAE	93.88	93.78	33.54
MLP	97.87	97.80	411.68
SVM	98.72	96.46	2593.28
HOG + PILAE	98.36	98.02	112.58

more noise will be learned. From Fig. 3(b) we observe that the training time increases as the number of Gabor features increases. With the number of Gabor features increasing, the dimension of the network's input increases, so the training time will become longer. In the future, we will continue to study how to fuse the features to get better performance.

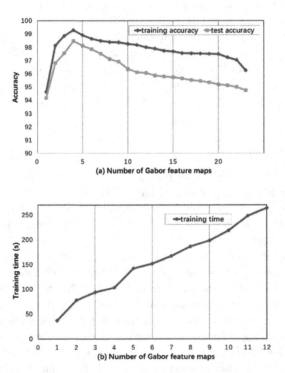

Fig. 3. (a) The accuracy with different numbers of Gabor feature maps for MNIST. (b) The training time with different numbers of Gabor feature maps for MNIST.

4.2   CIFAR-10 Dataset

CIFAR-10 image database includes 60,000 32 × 32 color images. The images are divided into 10 categories, and there are 6000 images in each category. The whole database is divided into five training packages and one test package. Each package contains 10,000 images, so there are 50,000 training images and 10,000 test images in total. We use different numbers of Gabor filters in this experiment, and the structure of PILAE is 3012-2955-2524.

Figure 4 shows that the best test accuracy is obtained at the 6 feature maps. As the number of feature maps increasing from 1 to 6, the training accuracy and the test accuracy is increasing. When the number of feature maps is greater than 6, the accuracy decreases, but the change is small. We made experiments using other methods, such as MLP with one hidden layer and 2000 neurons, PILAE with 3 layers, HOG + PILAE with 4 HOG features and LeNet5 with 20 and 50 kernels in different layers. The results are shown in Table 2. All the results show that when getting the same accuracy, our method is faster than other methods. However, our method is not perfect on the accuracy on CIFAR-10. The reason maybe the color information is lost when filtered by Gabor. In the future, we will pay more attention on processing color images.

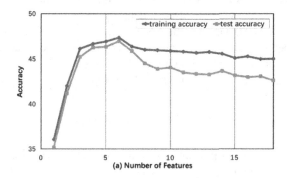

Fig. 4. The accuracy with different numbers of Gabor feature maps for CIFAR-10.

4.3   Discussion

From experimental results presented in Tables 1 and 2, we can know that training speed of our model is fast compared with other models. In the proposed model, Gabor filter is used to extract features, and extracted feature is a kind of handcraft feature. Compared with features learned with CNN, the Gabor feature is easier to obtain and less time consumption. With Gabor features, it can meet the real-time processing requirements compared with other learning feature methods. On the other hand, the training processing of the PILAE is a completely forward propagation without iteration. The connecting weights of PILAE are computed by using pseudoinverse learning algorithm directly. In additions, the depth of the network is dynamically increasing, and the number of hidden layers in PILAE is data dependent. That is, relative simple data set will generate relative simple network structure. While complex data set will require

Table 2. Performance comparison of CIFAR10 dataset.

Model	Training accuracy (%)	Testing accuracy (%)	Training time (s)
GF + PILAE	48.34	47.02	388.23
LeNet5	64.31	63.02	6743.82
PILAE	45.16	44.08	151.47
MLP	38.98	38.32	765.68
HOG + PILAE	46.57	46.05	436.58

the network architecture has more hidden layer to learn better data representation, and consequently reach good performance for given task.

In the training process, compared with gradient descent based learning algorithm, we need not set those learning optimization related hyper parameters, such as learning rate, momentum, and iterative epoch number. Those hyper parameters are difficult to tune if without rich experiences and professional knowledge. The transfer learning based on convolution neuron network (CNN) may be fast in inference stage, but it cannot be faster in training stage. As we known, CNN based network such as Alexnet is trained by gradient descent based algorithm, which is iterative and needs to adjust the learning hyper parameters also. So it is time consuming in training stage.

Our method performs well on MNIST data set, however, it does not obtain good test accuracy on CIFAR10 data set. The reason is that we only use one color channel to conduct experiments, this will lose the color information about the RGB image in CIFAR10. In the future, we will design the more complicated network architecture to improve the classification accuracy to the color image.

5   Conclusions

In this paper, we proposed a fast image recognition model that combines Gabor filter and pseudoinverse learning autoencoder. This model integrates the advantages of both Gabor filter and PILAE. Gabor filters extract features from input in different scales and orientations, and then the feature maps are sent to PILAE. The training of our model is fast, because it does not need back propagation or iterative optimization. Moreover, the number of layers is automatically determined, and we give the method to set the number of hidden neurons. We estimate the performance of our network using some benchmark datasets such as MNIST, CIFAR-10. The results show that on classification tasks, our model has a better performance than other models especially in learning speed. Because our model has no empirical parameters, it is easy to use even for person without professional knowledge. This is our effort to prompt the development of automatic machine learning and expect to democratize artificial intelligence.

Acknowledgements. The research work described in this paper was fully supported by the grants from the National Natural Science Foundation of China (Project No. 61472043), the Joint Research Fund in Astronomy (U1531242) under cooperative agreement between the NSFC and CAS, and Natural Science Foundation of Shandong (ZR2015FL006). Prof. Ping Guo and Qian Yin are the authors to whom all correspondence should be addressed.

References

1. LeCun, Y., Bottou, L., Bengio, Y., Haffner, P.: Gradient-based learning applied to document recognition. Proc. IEEE **86**(11), 2278–2324 (1998)
2. Krizhevsky, A., Sutskever, I., Hinton, G.E.: ImageNet classification with deep convolutional neural networks. In: Advances in Neural Information Processing Systems, pp. 1097–1105 (2012)
3. He, K., Zhang, X., Ren, S., Sun, J.: Deep residual learning for image recognition. In: Proceedings of the IEEE Conference on Computer Vision and Pattern Recognition, pp. 770–778 (2016)
4. Karen, S., Andrew, Z.: Very deep convolutional networks for large-scale image recognition. ArXiv:1409.1556[cs.CV] (2014)
5. Tai, S.L.: Image representation using 2D Gabor wavelets. IEEE Trans. Pattern Anal. Mach. Intell. **18**(10), 959–971 (1996)
6. Wang, K., Guo, P., Yin, Q., et al.: A pseudoinverse incremental algorithm for fast training deep neural networks with application to spectra pattern recognition. In: 2016 International Joint Conference on Neural Networks (IJCNN), pp. 3453–3460. IEEE (2016)
7. Feng, S., Li, S., Guo, P., Yin, Q.: Image recognition with histogram of oriented gradient feature and pseudoinverse learning autoencoders. In: 24th International Conference on Neural Information Processing (ICONIP 2017), pp. 740–749. Springer, Cham (2017)
8. Gabor, D.: Theory of communication. J. Inst. Electr. Eng. I Gen. **93**(26), 429–441 (1946)
9. Daugman, J.D.: Two dimensional spectral analysis of cortical receptive field profiles. Vision Res. **20**(10), 847–856 (1980)
10. Jones, J., Palmer, L.: An evaluation of the two-dimensional Gabor filter model of simple receptive fields in cat striate cortex. J. Neurophysiol. **58**(6), 1233–1258 (1987)
11. Kruizinga, P., Petkov, N.: Nonlinear operator for oriented texture. IEEE Trans. Image Process. **8**(10), 1395–1407 (1999)
12. Fazli, S., Afrouzian, R., Seyedarabi, H.: High-performance facial expression recognition using gabor filter and probabilistic neural network. In: 2009 IEEE International Conference on Intelligent Computing and Intelligent Systems, pp. 93–96 (2009)
13. Rumelhart, D.E., Hinton, G.E., Williams, R.J.: Learning representations by back-propagating errors. Nature **323**(6088), 533–536 (1986)
14. Hinton, G.E., Osindero, S., Teh, Y.W.: A fast learning algorithm for deep belief nets. Neural Comput. **18**(7), 1527–1554 (2006)
15. Guo, P., Chen, P.C.L., Sun, Y.: An exact supervised learning for a three-layer supervised neural network. In: Second International Conference on Neural Information Processing (ICONIP 1995), pp. 1041–1044 (1995)
16. Guo, P., Lyu, M.R., Mastorakis, N.E.: Pseudoinverse learning algorithm for feedforward neural networks. In: Advances in Neural Networks and Applications, pp. 321–326 (2001)
17. Guo, P., Lyu, M.R.: A pseudoinverse learning algorithm for feedforward neural networks with stacked generalization applications to software reliability growth data. Neurocomputing **56**, 101–121 (2004)
18. Wang, K., Guo, P., Xin, X., Ye, Z.: Autoencoder, low rank approximation and pseudoinverse learning algorithm. In: 2017 IEEE International Conference on Systems, Man, and Cybernetics (SMC), pp. 948–953. IEEE (2017)
19. Guo, P., Lyu, M., Chen, P.: Regularization parameter estimation for feedforward neural networks. IEEE Trans. Syst. Man Cybern. B **33**(1), 35–44 (2003)
20. Guo, P.: A VEST of the pseudoinverse learning algorithm. Preprint arXiv:1805.07828 (2018)

TiedGAN: Multi-domain Image Transformation Networks

Mohammad Ahangar Kiasari, Dennis Singh Moirangthem, Jonghong Kim, and Minho Lee[(⊠)]

School of Electronics Engineering, Kyungpook National University,
Daegu, South Korea
ahangar100@gmail.com, mdennissingh@gmail.com,
jonghong89@gmail.com, mholee@gmail.com

Abstract. Recently, domain transformation has become a popular challenge in deep generative networks. One of the recent well-known domain transformation model named CycleGAN, has shown good performance in transformation task from one domain to another domain. However, CycleGAN lacks the capability to address multi-domain transformation problems because of its high complexity. In this paper, we propose TiedGAN in order to achieve multi-domain image transformation with reduced complexity. The results of our experiment indicate that the proposed model has comparable performance to CycleGAN as well as successfully alleviates the complexity issue in the multi-domain transformation task.

Keywords: Generative models · Generative adversarial networks
Image domain transformation

1 Introduction

Recently, deep generative models have been proposed to learn any complex distribution of a large dataset. [1] proposed generative adversarial networks (GAN), which consists of a generator and a discriminator. The discriminator learns to distinguish between the generated data and real data while the generator tries to fool the discriminator by generating data similar to the real data. By training the paired discriminator and generator alternatively, the network is able to imitate the real data distribution. GANs have found various applications in recent times [5–7]. In the conventional GAN, the input of the generator is usually random samples z drawn from a simple known distribution such as a uniform distribution. However, in recent past, the definition of generative models have changed in order to utilize it for domain transformation tasks. In these models, multiple generators are able to produce its outputs in multiple domains in the dataset with different modalities.

[2] introduced a multi domain transformation based on the variational autoencoder (VAE). The distribution of the latent variables in the bottle-neck of

© Springer Nature Switzerland AG 2018
L. Cheng et al. (Eds.): ICONIP 2018, LNCS 11306, pp. 512–518, 2018.
https://doi.org/10.1007/978-3-030-04224-0_44

the variational autoencoder is confined to a normal Gaussian distribution with the help of Kullback–Leibler (KL) divergence loss. However, the results show that the generated images lack high frequency information. In other words, the generated images are suffering from blurriness. To our knowledge, it is because of the higher constraints set in the bottle-neck of the autoencoders.

Researchers in [3,8] concurrently proposed a novel domain transformation method based on GAN named CycleGAN. The model has a superior performance on a wide range of image categories. However, CycleGAN is originally designed to transfer images between two domains. But, in real-world applications, a multi-domain transformation algorithm is much more desirable. The aim of this paper is to find a way to improve CycleGAN to make it applicable for multi-domain transformation tasks. If CycleGAN in the current form is used for a multi-domain transformation task, the number of generators required in the network will increase exponentially due to the current structure of CycleGAN. To this end, we propose TiedGAN where we introduce one auxiliary loss function to the networks which can efficiently reduce the number of generators required in the presence of multiple domains. Our experiments and results show that the proposed model is able to simultaneously and efficiently translate images between multiple domains.

2   Related Works

2.1   Generative Adversarial Networks (GANs)

GAN is a deep model for generating a complex dataset from a prior distribution in order to mimic a target data distribution. GAN consists of a generator and a discriminator. The discriminator is a logistic regression network that gradually learn to distinguish the fake data (generated by the generator) and the real data while the generative model learns to fool the discriminator by generating fake data more similar to the target distribution.

Let x represent the training dataset and z be randomly sampled from a prior distribution. In the framework, the generator provides fake data $G(z)$ given the input z and the discriminator tries to distinguish the fake data $G(z)$ from the training data $x, D(G(\mathbf{z}))$.

The parameters of the $G(\cdot)$ and $D(\cdot)$ are updated simultaneously to minimize $E_{\mathbf{z} \sim p_z}[\log(1 - D(G(\mathbf{z})))]$ and maximize $E_{\mathbf{x} \sim p_x}[\log D(\mathbf{x})]$.

Generally speaking, the optimization process is based on a minimax optimization as shown in Eq. (1) [1].

$$\min_G \max_D V(G, D) = E_{\mathbf{x} \sim p_x}[\log D(\mathbf{x})] + E_{\mathbf{z} \sim p_z}[\log(1 - D(G(\mathbf{z})))] \qquad (1)$$

where p_x and p_z indicate the distribution of original and the random data, respectively.

2.2   CycleGAN

Recently, the concept of generative models has been adapted to domain transformation tasks in which the generator gets one of the domains in the dataset as an input and then translate it to another domains. One of the well-known domain transformation models is CycleGAN proposed by [8]. In the case of two domains, CycleGAN consists of two generators G_{12}, G_{21} and two discriminators D_1, D_2. G_{12} tries to fool D_1 in order to transfer images from the first domain to the second one. The generator G_{21} and the discriminator D_2 are defined similarly. In addition, cycleGAN applied "*cycle consistency*" to find the existing inherent similarities between the domains.

3   Proposed TiedGAN

The proposed model has been inspired by the CycleGAN. Figure 1 illustrates the structure of the proposed model named TiedGAN. In Fig. 1, x_i, where $i = 1, 2, 3$, presents the data in i^{th} domain in the dataset. The model consists of multiple generators G_{ij} that present transfer functions from i^{th} domain to j^{th} domain. Corresponding to Fig. 1, in the case of three domains, G_{12} and G_{13} transfer images from first domain to the second and third domains, respectively. G_{21} and G_{31} are defined similarly. All the generators G_{ij} have one encoder E_{ij} and one decoder Dec_{ij} each.

In the model, one of the domains is set as a mediator between all the other domains. In the sense that the model is trained to learn the proper transfer functions from each domain to the mediator, and from the mediator to any other domains. To this purpose, we apply adversarial, *consistency* and *tie* losses together in the proposed model. The adversarial loss is to match the distribution of the generated samples with the target distribution. The consistency loss is to reconstruct the same image while we are transforming that particular image from one domain to another one and then transforming it back to the original domain [8]. In addition, in order to reduce the number of generators in the model, we applied *tie* loss between the second and third domains. As shown in Fig. 1, *tie* loss equation is elucidated in Eq. (2).

$$\tilde{d}_2 = Dec_{12}(E_{31}(Dec_{13}(E_{21}(x))))$$
$$L_{tie} = \mathbb{E}_{\mathbf{x} \sim p_2}[||\tilde{d}_2 - x||_2^2] \tag{2}$$

where p_2 represents the second domain distribution. In the case of transformation between three different domains, the original CycleGAN needs six autoencoders to transfer images between the domains while our model needs only four autoencoders.

The total loss of the model including adversarial, consistency, and tie losses are as follows:

$$L_{cycle1} = \mathbf{E}_{\mathbf{x}\sim p_1}[||x - G_{21}(G_{12}(x))||_2^2]$$
$$L_{cycle2} = \mathbf{E}_{\mathbf{x}\sim p_2}[||x - G_{31}(G_{13}(x))||_2^2]$$
$$L_{G_{12}} = \mathbf{E}_{\mathbf{x}_2\sim p_2}[log(D_1(x_2))] + \mathbf{E}_{\mathbf{x}_1\sim p_1,\mathbf{x}_2\sim p_2}[1 - log(D_1(G_{12}(x_1)))] \quad (3)$$
$$L_{G_{13}} = \mathbf{E}_{\mathbf{x}_3\sim p_3}[log(D_2(x_3))] + \mathbf{E}_{\mathbf{x}_1\sim p_1,\mathbf{x}_3\sim p_3}[1 - log(D_3(G_{13}(x_1)))]$$
$$L_{total} = L_{G_{12}} + L_{G_{13}} + L_{cycle1} + L_{cycle2} + L_{tie}$$

where x_1, x_2 and x_3 represent the input data of the first, second, and third domains, respectively. L_{cycle1}, and L_{cycle2} are the consistency losses. $L_{G_{12}}$ and $L_{G_{21}}$ are the adversarial losses for G_{12} and G_{13}, respectively.

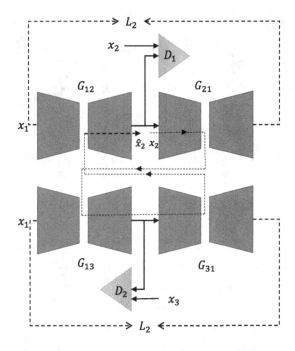

Fig. 1. Schematic of the model for three domains. L_1 and L_2 represent the consistency losses. The x_2 and $\hat{x}_2$ are the input and the target for computing the *tie* loss.

4   Experiments and Results

To evaluate the effectiveness of the proposed model, we conduct our experiments on the Large-scale CelebFaces Attributes (CelebA) [4] dataset. This dataset comprises of 202,599 images, each with various characteristics. The images have 40

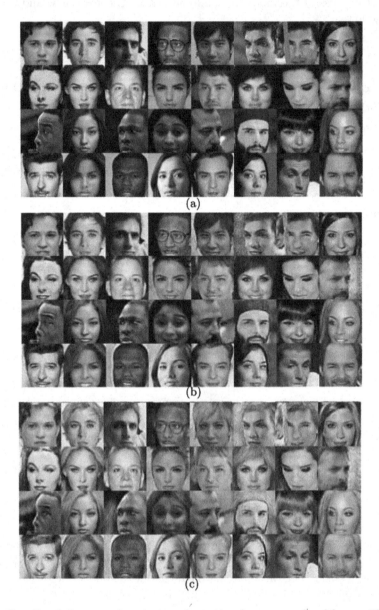

Fig. 2. Results of the cross-domain transformation from a couple of faces with black hair to the second and third domains corresponding to smiling faces and blond-haired, respectively. The figure has three panels of celebrity faces. (a) The first panel shows the original black-haired faces. (b) The second panel shows the corresponding generated smiling faces. (c) The last panel illustrates the corresponding generated blond-haired faces.

different attributes based on their characteristics and modalities. We can categorize the dataset into different domains as per the attributes of the images, but the domains have no paired images between each other. In our work, the images were separated into various styles depending on the attributes, for example, faces with blond or black hair, non-smiling and smiling faces, etc. The size of the original images in the dataset was 178×218. However, the original images have been cropped to 64×64 by selecting and aligning only the facial areas of the images in order to focus on the relevant parts as well as decrease the computation load.

In the experiment, we used the images with back hair, blond hair, and smiling face domains. Here, we would like to emphasize that there are no paired images between domains in the dataset, and hence supervised learning is not possible. The results show the promising performance in the multi-domain transformation. The results of the proposed model for the CelebA dataset is illustrated in Fig. 2. Figure 2(a), shows the original black-haired faces. The translated images to the blond-haired and smiling faces are shown in Figs. 2(b) and (c), respectively. The results show that the model not only has less complexity compared to the CycleGAN, but also can generate plausible images in different domains.

5   Conclusion

In this paper we introduced TiedGAN, an enhanced version of CycleGAN, to efficiently address multi domain image transformation problem. We performed experiments on three domains of the popular CelebA dataset including black haired, blond-haired and smiling faces domains. The results of our experiments showed that the proposed model illustrates a good performance in the multi domain transformation task with much lesser complexity.

Acknowledgements. This work was partly supported by Institute for Information & Communications Technology Promotion (IITP) grant funded by the Korea government (MSIT) (2016-0-00564, Development of Intelligent Interaction Technology Based on Context Awareness and Human Intention Understanding) (50%) and by the National Research Foundation of Korea (NRF) grant funded by the Korea government (MSIP) (No. NRF-2016R1A2A2A05921679) (50%).

References

1. Goodfellow, I., et al.: Generative adversarial nets. In: Advances in Neural Information Processing Systems, pp. 2672–2680 (2014)
2. Liu, M.Y., Breuel, T., Kautz, J.: Unsupervised image-to-image translation networks. In: Advances in Neural Information Processing Systems, pp. 700–708 (2017)
3. Liu, M.Y., Tuzel, O.: Coupled generative adversarial networks. In: Advances in Neural Information Processing Systems, pp. 469–477 (2016)
4. Liu, Z., Luo, P., Wang, X., Tang, X.: Deep learning face attributes in the wild. In: Proceedings of International Conference on Computer Vision (ICCV) (2015)
5. Makhzani, A., Shlens, J., Jaitly, N., Goodfellow, I.: Adversarial autoencoders. In: International Conference on Learning Representations (2016)

6. Radford, A., Metz, L., Chintala, S.: Unsupervised representation learning with deep convolutional generative adversarial networks. In: International Conference on Learning Representations (2016)
7. Tolstikhin, I.O., Gelly, S., Bousquet, O., Simon-Gabriel, C.J., Schölkopf, B.: Adagan: boosting generative models. In: Advances in Neural Information Processing Systems, pp. 5424–5433 (2017)
8. Zhu, J.Y., Park, T., Isola, P., Efros, A.A.: Unpaired image-to-image translation using cycle-consistent adversarial networks. In: IEEE International Conference on Computer Vision (2017)

Graph Matching Based on Fast Normalized Cut

Jing Yang[1,2], Xu Yang[2(✉)], Zhang-Bing Zhou[1,3], and Zhi-Yong Liu[2,4,5]

[1] School of Information Engineering, China University of Geosciences (Beijing),
Beijing 100083, China
`zhangbing.zhou@gmail.com`
[2] State Key Laboratory of Management and Control for Complex Systems,
Institute of Automation, Chinese Academy of Sciences, Beijing 100190, China
`xu.yang@ia.ac.cn`
[3] Computer Science Department, TELECOM SudParis, Evry 91011, France
[4] Center for Excellence in Brain Science and Intelligence Technology,
Chinese Academy of Sciences, Shanghai 200031, China
[5] University of Chinese Academy of Sciences, Beijing 100049, China

Abstract. Graph matching is important in pattern recognition and computer vision which can solve the point correspondence problems. Graph matching is an NP-hard problem and approximate relaxation methods are used to solve this problem. But most of the existing relaxation methods solve graph matching problem in the continues domain without considering the discrete constraints. In this paper, we propose a fast normalized cut based graph matching method which takes the discrete constraints into consideration. Specifically, a regularization term which is related to the discrete form of the permutation matrix is added to the objective function. Then, the objective function is transformed to a form which is similar to the fast normalized cut framework. The fast normalized cut algorithm is generalized to get the permutation matrix iteratively. The comparisons with the state-of-the-art methods validate the effectiveness of the proposed method by the experiments on synthetic data and image sequences.

Keywords: Graph matching · Fast normalized cut
Discrete constraints

1 Introduction

Graph matching is a fundamental problem in computer vision, pattern recognition and image understanding, which can be used to find the correspondence between two point sets extracted from two images. Graph matching further lays the foundation for many computer vision tasks, such as feature tracking [12], object recognition [17] and shape matching [15]. In this paper, we mainly focus on solving the point correspondence problems by graph matching.

© Springer Nature Switzerland AG 2018
L. Cheng et al. (Eds.): ICONIP 2018, LNCS 11306, pp. 519–528, 2018.
https://doi.org/10.1007/978-3-030-04224-0_45

The problem of finding the correspondence between two point sets can be solved by RANSAC [7], Iterative Closet Point (ICP) [20] and other methods which based on the appearance descriptors. But these methods only consider the relationships between feature points and may fail in the presence of local appearance ambiguities. Many recent graph matching methods incorporate both the relationships between points and edges which explore pairwise consistency in frames. Since these methods encode the geometric information when construct a graph and match two graphs, they can develop a more satisfactory correspondence than these methods only based on feature points. But graph matching with pairwise constraints is an NP-hard problem which is not easy to find an exactly result and an approximate relaxation method is needed to find the approximate result. An important kind of approximate methods is based on spectral decomposition. Umeyama's [18] method is regarded as the first spectral method in this domain. Leoedeanu and Hebert [13] proposed the Spectral Matching (SM) method and introduced the notation of affinity matrix which can encode the similarity of feature points and edges extracted from two images. This method solved the graph matching problem by calculating the optimal rank-one approximation of the affinity matrix. Cour et al. [6] incorporated the mapping constraints within the relaxation scheme when formulated the permutation matrix and this method was called the Spectral Matching with Affine Constraint (SMAC). The rank-one approximation of the affinity matrix was also used to get the final result. Cho et al. [4] introduced a method called Reweighted Random Walks for graph Matching (RRWM). Graph matching problem is solved by simulating random walks with reweighted jumps which enforced the matching constraints on the affinity graph. Another kind of methods tries to gain the discrete solution of graph matching problem. The Integer Projected Fixed Point (IPFP) method [14] was proposed by Leodeanu et al., which optimized the objective function in discrete domain and can get the discrete permutation matrix directly. In [11], Jiang et al. also proposed a new graph matching algorithm which can get the discrete solution. They added the $\mathcal{L}_2$-norm as a binary constraint of the objective function and the final discrete matrix was calculated by the increase of the $\mathcal{L}_2$-norm.

Normalized Cut is a widely used graph-cut technique in many applications, such as image segmentation [16] and clustering [9]. Since finding the optimized cut is an NP-hard problem, several spectral methods are used to obtain the minimize normalized cut approximately by eigen-computation. Xu et al. [1] proposed a Fast Normalized Cut with Linear Constraints method based on the original graph cut method [16], which incorporated the prior knowledge as a linear constraint into normalized cut problems. Their also designed an algorithm which can be used to optimized a general problem. The approximate relaxation methods which are presented above usually optimized graph matching problem in a continuous domain. These methods relaxed the constraint to continuous domain and defined a new problem firstly. Then they solved the continuous problem and got a global optimum solution. The continues optimal solution would be mapped back to discrete domain in the end. But when these methods redefined the graph matching problem, they only considered the continuous constraints which may result in a local optimum solution.

In this paper, we propose a graph matching method based on the fast normalized cut method and discrete constraint. First, we take the discrete constraint into consideration when redefine the problem in continuous domain and a graph matching model is set up. Second, the fast normalized cut method which has the similar model with this method is utilized to obtain the solution. Last, the Hungarian Algorithm is used to get the discrete solution.

2 Problem Formulation

Given two graphs $\mathbf{G}(\mathbf{V}_g, \mathbf{E}_g)$ and $\mathbf{H}(\mathbf{V}_h, \mathbf{E}_h)$, where $\mathbf{V}_g(\mathbf{V}_h)$ and $\mathbf{E}_g(\mathbf{E}_h)$ represent the node set and edge set respectively. Each node $\mathbf{v}_i \in \mathbf{V}_g$ and $\mathbf{v}_j \in \mathbf{V}_g$ are connected by an edge $\mathbf{e}_{ij} \in \mathbf{E}_g$. Each edge is assigned an attribute which could be a real number or an attribute vector. Graph $\mathbf{H}$ has the same definition. Graph matching problem aims at finding a correspondence between node sets $\mathbf{V}_g$ and $\mathbf{V}_h$. We construct an affinity matrix $\mathbf{W}$ which can measure the compatible between node $\mathbf{v}_i$ and node $\mathbf{v}_j$ by two kinds of attributes. The diagonal element of $\mathbf{W}$ represents the similarity between node $\mathbf{v}_i \in \mathbf{V}_g$ and $\mathbf{v}'_i \in \mathbf{V}_h$. The non-diagonal element of this matrix represents the comparability of edge $\mathbf{e}_{ij} \in \mathbf{E}_g$ and edge $\mathbf{e}_{i'j'} \in \mathbf{E}_h$. A permutation matrix $\mathbf{X}$ is used to denote the one-to-one correspondence solution. If the element in $\mathbf{X}_{ij}$ equals to 1, it means that node $\mathbf{v}_i$ in graph $\mathbf{G}$ corresponds to node $\mathbf{v}_j$ in graph $\mathbf{H}$. Else, the $\mathbf{X}_{ij} = 0$ presents the opposite situation. In this paper, we only consider the equal-sized graph matching problem.

In most recent research [4,6,8,14], the graph matching problem is calculated by the following model:

$$\max \ \mathbf{x}^{\mathbf{T}}\mathbf{W}\mathbf{x}$$
$$\text{s.t.} \quad \mathbf{A}\mathbf{x} = 1, \ \mathbf{x}_i \in \{0, 1\}, \tag{1}$$

where $\mathbf{x}$ is the row-wise vector of permutation matrix $\mathbf{X}$ and $\mathbf{A}$ is a matrix of size $2\mathbf{N} \times \mathbf{N}^2$. The alphabet $\mathbf{N}$ is the number of nodes. And the matrix $\mathbf{A}$ is used to ensure the doubly stochastic constraint of $\mathbf{X}$. In most occasions, affinity matrix $\mathbf{W}$ is not a positive definite matrix and the objective function is usually non-convex. The above function defines the permutation matrix in discrete domain which had been proved to be an NP-hard problem. Some research proposed the relaxation methods to solve this problem [5,6,20,21]. They relaxed the constraint from discrete domain to continuous domain and constructed the following function:

$$\max \ \mathbf{x}^{\mathbf{T}}\mathbf{W}\mathbf{x}$$
$$\text{s.t.} \quad \mathbf{A}\mathbf{x} = 1, \ \mathbf{x}_i \geq 0. \tag{2}$$

The solution of this problem can be obtained by approximate methods in continuous domain firstly. Then this solution will be mapped back to the discrete domain by Hungarian Algorithm or other methods. But these existing graph matching methods solved the problem without consideration of discrete constraints which may get a weak local optimization solution in continuous domain.

3 Graph Matching Based on Fast Normalized Cut

In this section, we propose a graph matching based on fast normalized cut method [1]. The discrete constraints are taken into consideration when construct the objective function. The new objective function has a positive definite matrix and it is similar to the graph cut model.

3.1 Graph Matching Model

The original graph matching problem is a discrete problem and the permutation matrix $\mathbf{X}$ only can contain number 1 or 0. This matrix also has the double stochastic constraint. So when the right discrete solution is obtained, the permutation matrix has the following algebraic properties:

$$\|\mathbf{x}\|_2^2 = \mathbf{N}. \tag{3}$$

The operation $\|\cdot\|_2$ represents the $\mathcal{L}_2$-norm. From the mathematical theory of $\|\cdot\|_2$ we know that as the number of 1 increases, the value of $\mathcal{L}_2$-norm increases when it is less than $\mathbf{N}$. The maximum value of $\|\mathbf{x}\|_2^2$ is obtained when $\mathbf{x}$ only contains 1 and 0 and the value is equals to $\mathbf{N}$. The number of 1 is fixed to $\mathbf{N}$ when the optimal discrete solution is obtained. The vector $\mathbf{x}$ is a column vector and it has the property that:

$$\mathbf{x}^{\mathbf{T}}\mathbf{x} = \sum_i \mathbf{x}_i^2, \tag{4}$$

which is equal to the square number of $\|\mathbf{x}\|_2$. So we used $\mathbf{x}^{\mathbf{T}}\mathbf{x}$ represents $\|\mathbf{x}\|_2^2$ to encode the discrete attribute of the solution. The objective function can be rewrote as follows:

$$\max \mathbf{x}^{\mathbf{T}}\mathbf{W}\mathbf{x}$$
$$\text{s.t.}\ \ \mathbf{A}\mathbf{x} = 1, \mathbf{x}^{\mathbf{T}}\mathbf{x} = N,\ \mathbf{x}_i \in [0,1]. \tag{5}$$

The above function incorporates the discrete constraint into graph matching. But the discrete constraint is added as a part of the function's constraints and this function may not easy to be solved directly. The Lagrange Multiplier Method is used to optimize this problem. And the objective function is presented as follows:

$$\max \mathbf{x}^{\mathbf{T}}\mathbf{W}\mathbf{x} + \alpha\mathbf{x}^{\mathbf{T}}\mathbf{x}$$
$$\text{s.t.}\ \ \mathbf{A}\mathbf{x} = 1,\ \mathbf{x}_i \in [0,1], \tag{6}$$

where $\alpha \mathbf{x}^T \mathbf{x}$ is a regularization part and α is a control parameter which can affect the solution of this function. If permutation vector $\mathbf{x}$ is more discrete, the value of $\mathbf{x}^T \mathbf{x}$ is lager and the objective function is closer to the discrete case. Equation (6) can be transformed to the following form:

$$\max \mathbf{x}^T \widetilde{\mathbf{W}} \mathbf{x}$$
$$\text{s.t. } \mathbf{A}\mathbf{x} = 1, \ \mathbf{x}_i \in [0,1], \tag{7}$$

where $\widetilde{\mathbf{W}} = \mathbf{W} + \alpha \mathbf{I}_1$. The notation $\mathbf{I}_1$ denotes a identity matrix with size $\mathbf{N}^2 \times \mathbf{N}^2$. We find that as the α is equal or greater than a number β, matrix $\widetilde{\mathbf{W}}$ becomes a positive semidefinite matrix. The number is:

$$\beta = |\min\{\lambda_{\min}, 0\}|, \tag{8}$$

where $\lambda_{\min}$ is the smallest eigenvalue of affinity matrix $\mathbf{W}$.

3.2 Fast Normalized Cut

Normalized cut is an important method in graph cut problem. This method usually defines a weighted graph and the weight of the graph edges denotes the similarity between two nodes. But finding the real cut of the nodes is a discrete problem which is also an NP-hard problem. The spectral relaxation methods had been applied to get the approximate solution of this problem based on eigen-computations [16]. In [1], the authors proposed a new algorithm called Fast Normalized Cut with Linear Constraints. They incorporated the prior knowledge of the graph cut problem into the original framework by adding the linear constraints. Then they designed a iterative algorithm to find the feasible solution. Our graph matching mode has the resemble part with this framework. In this paper, we are inspired by the graph cut method and the solution of the graph matching problem is calculated iteratively. The algorithm based on Eq. (7) is showed in Algorithm 1.

The main idea of the Algorithm is similar to the Fast Normalized Cut algorithm and SMAC. According the constraints of the objective function, the vector $\mathbf{v}$ can be decomposed into two part $\mathbf{n}_0$ and $\mathbf{u}$. The matrix $\mathbf{P}$ is a projection matrix which can encode the constraints [16]. Notation $\mathbf{d}_v$ is a parameter matrix which measures the difference of each iterative result. In each iteration step, $\mathbf{v}_k$ is closer to the real result. This vector has the similar significance with the transfer matrix in the PageRank Algorithm [2]. It can accelerate the convergence procedure. From research [1], we know that $\mathbf{x}^T \mathbf{W} \mathbf{x}$ is an increasing function when $\mathbf{W}$ is a semidefinite positive matrix. So this function can obtain a maximum value at a fixed point. In our method, the affinity matrix is a semidefinite positive matrix. So the proposed algorithm will converge and return a result. The final result of Algorithm 1 is continues, we choose the Hungarian algorithm [10] which is a linear assignment method to map the result back to discrete domain. Then the satisfactory permutation matrix is obtained.

Algorithm 1 Graph Matching based on Fast Normalized Cut

Input: Affinity matrix $\widetilde{\mathbf{W}}$, graph size N, initialized vector $\mathbf{n_0}$, constraint matrix $\mathbf{A}$

$\mathbf{P} = \mathbf{I} - \mathbf{A}^{\mathbf{T}}(\mathbf{A}\mathbf{A}^{\mathbf{T}})^{-1}\mathbf{A}$, k=0

$\mathbf{n_0} = \mathbf{A}^{\mathbf{T}}(\mathbf{A}\mathbf{A}^{\mathbf{T}})^{-1}\mathbf{1}$

$\gamma = \sqrt{1 - \|\mathbf{n_0}\|^2}$

$\mathbf{v_0} = \gamma\dfrac{\mathbf{P}\widetilde{\mathbf{W}}\mathbf{n_0}}{\|\mathbf{P}\widetilde{\mathbf{W}}\mathbf{n_0}\|}$

repeat

   $\mathbf{u}_{k+1} = \gamma\dfrac{\mathbf{P}\widetilde{\mathbf{W}}\mathbf{v}_k}{\|\mathbf{P}\widetilde{\mathbf{W}}\mathbf{v}_k\|}$

   $\mathbf{v}_{k+1} = \mathbf{u}_{k+1} + \mathbf{n_0} \times \mathbf{d}_v$

   $\mathbf{d}_v = |\mathbf{v}_{k+1} - \mathbf{v}_k|$

   $k = k + 1$

until $\mathbf{v}$ converges

Output: A permutation vector $\mathbf{v}$

---

4   Experiments

In this section, the proposed method is applied to both synthetic point matching and the CMU house dataset matching tasks. This methods is compared with the state-of-the-art methods, including the Spectral Matching(SM) [13], the Spectral Matching with Affine Constraint (SMAC) [6], the Reweighted Random Walks for graph Matching(RRWM) [4], the Probabilistic graph and hypergraph matching(PGM) [19] and the Integer Projected Fixed Point (IPFP) [14].

4.1   Synthetic Graph Matching

The setting of this experiments is following [6]. We need to construct two random graphs $\mathbf{G}$ and $\mathbf{H}$ firstly. The first graph $\mathbf{G}$ which contains $\mathbf{N}$ nodes and $\mathbf{M}$ edges is generated in the 2D surface randomly and uniformly. The sparsity of this graph is set to be 0.5. A weighted matrix $\mathbf{W}^{\mathbf{G}}$ is generated uniformly in $[0,1]$. Each edge of the graph will be assigned a weight $\mathbf{W}^{\mathbf{G}}_{ij}$. The second graph $\mathbf{H}$ is generated from the first graph. The graph $\mathbf{G}$ is permutated by a permutation matrix $\mathbf{P}$ firstly. Then the Gaussian noise $\mathcal{N}(0,\sigma)$ is added to disturbed the graph $\mathbf{H}$. The affinity matrix $\mathbf{W}$ is computed as follows:

$$\mathbf{W}_{ii',jj'} = \exp(-\left|\mathbf{W}^{\mathbf{G}}_{ij} - \mathbf{W}^{\mathbf{H}}_{i'j'}\right|^2/\varepsilon). \tag{9}$$

The notation ε is the scaling factor which is set to be 0.15.

Since we focus on the equal-sized graph matching in this paper, two comparisons are performed with respect to the number of nodes $\mathbf{N}$ and the Gaussian noise level σ respectively. The nodes number $\mathbf{N}$ is increased from 30 to 50 by step size 1 when σ is set 0.1. The noise level is increased from 0 to 0.2 by step size 0.025, with the node number $\mathbf{N}$ is set to be 40.

The comparison results are showed in Fig. 1. The left sub-figure depicts the comparison results with respect to the nodes number, from which we can observe that as the number of nodes increases the matching accuracy decreases and the proposed method performs better than other methods. The right sub-figure is respect to the noise level σ, and we can obtain the similar observation as the left sub-figure generally.

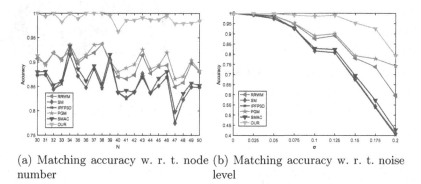

(a) Matching accuracy w. r. t. node number

(b) Matching accuracy w. r. t. noise level

Fig. 1. Synthetic graph matching result. The matching accuracy is compared with respect to node number and noise level.

4.2   CMU House Matching

In this experiment, the feature point matching tasks on the CMU House sequence is performed [3,4]. This dataset contains 101 images of a toy house in different viewpoints. Each image in the dateset is marked 30 landmark points. We match all possible image pairs spaced by 0, 10, 20, 30, 40, 50, 60, 70, 80, 90,100 and 110 frames. Average matching accuracy of per sequence gap is calculated. Matrixes $\mathbf{W^G}_{ij}$ and $\mathbf{W^H}_{i'j'}$ denote the Euclidean distance between two points respectively in this experiment. The affinity matrix is then computed by the same format as the Synthetic graph matching.

The edge density is also set 0.5 and $\varepsilon = 0.15$. Figure 2 shows the results of this dataset. The left sub-figure illustrates the results related to the sequence gap. We can observe that as the sequence gap increases, the accuracy decreases. Since a constraint which can bring an discrete solution is added in our method, our method obtain a better correspondence than other methods. The result of this experiment is consistent with the synthetic graph matching. A sample of house matching is exhibited on the right (We choose the sequence gap which equals to 70). In the sample graph, the green lines represent right matching results.

5   Complexity Analysis and Future Work

The complexity of the proposed work is $\mathcal{O}(\mathbf{N}^4)$ when the graph is connected by the complete method. In the outside loop of the algorithm, the complexity

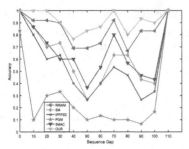

(c) Matching accuracy w. r. t. sequence gap   (d) *CMU house* matching sample

Fig. 2. The CMU House sequence matching result and sample

is $\mathcal{O}(\mathbf{N}^2)$ in order to compute $\mathbf{P}$ and $\mathbf{n}_0$. It costs $\mathcal{O}(\mathbf{N}^4)$ in the inside loop. Since $\mathcal{O}(\mathbf{N}^4)$ is bigger than $\mathcal{O}(\mathbf{N}^2)$, $\mathcal{O}(\mathbf{N}^2)$ is neglected in this algorithm. But in some real experiments, graph is usually constructed by the sparse connection method. In that way, the complexity of the proposed method is related to the sparsification technique which will reduce the complexity.

Graphs which are compared in this paper are supposed to be equal-sized. In real tasks, there exists graphs with different number of nodes. In the future work, we will extend the proposed method to the unequal-sized graph matching. Since the inside loop costs the most complexity, some modification will be done to save the computations in the future.

6   Conclusion

In this paper, we propose a graph matching algorithm based on fast normalized cut algorithm. It incorporates the discrete constrain into the objective function. The discrete constraint is added as a part of the function and can be seen as a parameter which can control the discreteness of the function. The proposed framework of objective function is resemble to the fast normalized cut and an effective algorithm is designed to calculate the result iteratively. The experiments both in synthetic graph and CMU house dataset demonstrate the effectiveness of the proposed method.

Acknowledgments. This work is supported partly by the National Natural Science Foundation (NSFC) of China (grants 61772479, 61662021, 61773047, 61503383, 61633009, U1613213, 61627808, 61502494, and U1713201), partly by the National Key Research and Development Plan of China (grant 2016YFC0300801 and 2017YFB1300202), and partly by the Development of Science and Technology of Guangdong Province Special Fund project (grant 2016B090910001).

References

1. Achuurmans, D., Li, W., Xu, L.: Fast normalized cut with linear constraints. In: 22th IEEE Conference on Computer Vision and Pattern Recognition, pp. 2866–2873. Miami (2009)
2. Bryan, K., Leise, T.: The 25,000,000,000 eigenvetor: the linear algebra behind google. Siam Rev. **48**(3), 569–581 (2006)
3. Cetano, T.S., MeAuley, J.J., Cheng, L., Le, Q.V., Smola, A.J.: Learning graph matching. IEEE Trans. Pattern Anal. Mach. Intell. **31**(6), 1048–1058 (2009)
4. Cho, M., Lee, J., Lee, K.M.: Reweighted random walks for graph matching. In: 20th Conference and Workshop on Neural Information Processing Systems, pp. 313–320. Vancouver (2006)
5. Conte, D., Foggia, P., Sansone, C., Vento, M.: Thirty years of graph matching in pattern recognition. Int. J. Pattern Recognit. Artifical Intell. **18**(3), 265–298 (2004)
6. Cour, M., Srinivasan, P., Shi, J.: Balanced graph matching. In: 20th Conference and Workshop on Neural Information Processing Systems, pp. 313–320. Vancouver (2006)
7. Fischler, M.A., Bolles, R.C.: Random sample consensus: a paradigm for model fitting with applications to image analysis and automated cartography. Commun. Assoc. Comput. Mach. **24**(6), 381–395 (1981)
8. Gold, S., Rangarajan, A.: A graduated assignment algorithm for graph matching. IEEE Trans. Pattern Anal. Mach. Intell. **18**(4), 377–388 (1996)
9. Hagen, L., Kahng, A.: New spectral methods for ratio cut partioning and clustering. IEEE Trans. Comput. Aided Des. Intergrated Circuits Syst. **11**(9), 1074–1085 (1992)
10. James, M.: Algorithms for the assignment and transportation problems. SIAM J. Appl. Math. **5**(1), 32–38 (1957)
11. Jiang, B., Tang, J., Ding, C., Luo, B.: Binary constraint preserving graph matching. In: 30th IEEE Conference on Computer Vision and Pattern Recognition, pp. 550–557. Hawaii (2017)
12. Jiang, H., Yu, X.S., Martin, D.R.: Linear scale and rotation invariant matching. IEEE Trans. Pattern Anal. Mach. Intell. **33**(7), 1339–1355 (2011)
13. Leoedeanu, M., Hebert, M.: A spectral technique for correspondence problems using pairwise constraints. In: 10th IEEE Conference on International Conference on Computer Vision, pp. 1482–1489. Beijing (2005)
14. Leoedeanu, M., Hebert, M., Sukthankar, R.: An integer projected fixed point method for graph matching and map inference. In: 23th Conference and Workshop on Neural Information Processing Systems, pp. 1114–1122. Vancouver (2009)
15. Michel, D., Oikonomidis, I., Argyros, A.A.: Scale invariant and deformation tolerant partial shape matching. Image Vis. Comput. **29**(7), 459–469 (2011)
16. Shi, J.B., Jitendra, M.: Normalized cuts and image segmentation. IEEE Trans. Pattern Anal. Mach. Intell. **22**(8), 888–905 (2000)
17. Sullivan, J., Carlsson, S.: Recognizing and tracking human action. In: Heyden, A., Sparr, G., Nielsen, M., Johansen, P. (eds.) ECCV 2002. LNCS, vol. 2350, pp. 629–644. Springer, Heidelberg (2002). https://doi.org/10.1007/3-540-47969-4_42
18. Umeyama, S.: An eigendecomposition approach to weighted graph matching problems. IEEE Trans. Pattern Anal. Mach. Intell. **10**(5), 695–703 (1988)
19. Zass, R., Shashua, A.: Probabilistic graph and hypergraph matching, In: 21th IEEE Conference on Computer Vision and Pattern Recognition, pp. 1-8. Anchorage(2008)

20. Zhang, Z.Y.: Iterative point matching for registration of free-form curves and surfaces. Int. J. Comput. Vis. **13**(2), 119–152 (1994)
21. Zhou, F., De la Torre, F.: Factorized graph matching. In: 25th IEEE Conference on Computer Vision and Pattern Recognition, pp. 127–134. Providence (2012)

Two-Phase Transmission Map Estimation for Robust Image Dehazing

Qiaoling Shu[1], Chuansheng Wu[1], Ryan Wen Liu[2,3(✉)], Kwok Tai Chui[4], and Shengwu Xiong[3]

[1] Department of Mathematics, Wuhan University of Technology, Wuhan, China
[2] School of Navigation, Wuhan University of Technology, Wuhan, China
wenliu@whut.edu.cn
[3] Hubei Key Laboratory of Transportation Internet of Things, School of Computer Science and Technology, Wuhan University of Technology, Wuhan, China
[4] Department of Electronic Engineering, City University of Hong Kong, Kowloon Tong, Hong Kong

Abstract. A robust two-phase transmission map estimation framework is proposed in this paper for single image dehazing. The proposed framework first estimates the coarse transmission map through the statistical assumption of dark channel prior (DCP). To refine the coarse transmission map, a novel image-gradient-guided high-order variational method is then proposed in the second phase. The resulting L1-regularized high-order nonsmooth optimization problem will be effectively solved using the primal-dual algorithm. Once the fine transmission map is accurately obtained, the final haze-free image could be restored based on the haze imaging model of Koschmieder. To further enhance dehazing performance, an improved tolerance mechanism is incorporated into the proposed method to suppress the undesirable artifacts usually produced by DCP in large sky regions. Numerous experiments on both synthetic and realistic images were performed to compare our proposed method with several state-of-the-art dehazing methods. Dehazing results have illustrated the superior performance of the proposed method.

Keywords: Image dehazing · Image restoration · Dark channel prior Total generalized variation · Primal-dual algorithm

1 Introduction

Images captured under hazy imaging condition often suffer from noticeable visibility degradation and apparent contrast reduction. The low vision quality could easily produce negative effects in practical applications, such as intelligent transporation, remote sensing and video surveillance, etc [1,2]. To enhance image quality, current dehazing methods can be mainly classified into four types [1–4]: multi-scale fusion methods, contrast enhancement methods, Retinex-based

© Springer Nature Switzerland AG 2018
L. Cheng et al. (Eds.): ICONIP 2018, LNCS 11306, pp. 529–541, 2018.
https://doi.org/10.1007/978-3-030-04224-0_46

methods and physics-based methods. This paper mainly focuses on the physics-based dehazing methods since only these methods take the essential principles of haze-degraded images into consideration.

The dehazing performance of physics-based methods is strongly dependent on the image features and priors adopted to estimate clean images. It has been observed that small image patches in haze-free images exhibit a uniformly colored surface and the same depth [5]. As a consequence, the contrast color-lines prior [5,6] has been proposed to enhance imaging quality in different applications. Recently, the non-local haze-lines prior [7,8] generated the reliable dehazing quality based on the assumption that colors of a haze-free image are well approximated by a few hundred distinct colors leading to tight clusters in RGB space. Current research [1,9] has shown that the accurate estimation of transmission map, which is inversely proportional to scene depth, can generate high-quality dehazing results. He *et al.* [10] discovered the simple but effective dark channel prior (DCP) to reliably estimate the transmission map. To make dehazing more desirable, many improvements were considered to refine the DCP-based coarse transmission map, e.g., standard median filtering [11], anisotropic diffusion [12], guided image filtering [13], and weighted guided image filtering [14,15], etc. The popular total variation (TV) regularizer [16] has also been introduced to optimize the transmission map. However, the resulting map often suffers from unwanted staircase-like artifacts since TV encourages piecewise constant solutions. Under DCP assumption, variational methods [17,18] were proposed to perform simultaneous depth map estimation and haze-free image restoration. It is obvious that DCP-based methods perform well in dehazing due to the assumption that most local non-sky patches in clean images have pixel values close to zero [19]. The restored images with large sky regions easily suffer from block effects or serious color distortion because of the underestimated transmission map [14]. Correction mechanism [20] or sky detection (or segmentation) [19,21] should be incorporated into traditional dehazing methods to enhance dehazing performance.

The latest generation of deep neural networks (DNN) [22,23] has recently achieved impressive results in the field of image dehazing. For example, convolutional neural networks (CNN) [24] and its modified version with coarse-to-fine strategy [25] have been adopted to estimate the medium transmission map which is subsequently used to restore the latent haze-free image through atmospheric scattering model. If the transmission map is not estimated accurately, it would bring negative effects on haze-free image restoration. To address this problem, CNN-based learning method [26,27] has been proposed to directly estimate the clean image from its degraded version. However, the dehazing performance is usually dependent on the volume and diversity of training datasets. If the training datasets donot include the similar geometrical features existed in images to be restored, it is difficult to generate satisfactory image quality. Generative adversarial network (GAN) [28,29] could be used to generate realistic-looking synthetic images and improve dehazing performance. To make dehazing easier and more flexible, we still tend to propose a robust dehazing method based on traditional but effective *two-step* framework. In particular, it first estimates the transmission map, and then restores the latent haze-free image.

In this work, we propose to develop a two-phase transmission map estimation method for single image dehazing. The proposed method follows the general *two-step* dehazing framework (i.e., transmission map estimation and haze-free image restoration) and mainly generates the contribution in the first step. In this step, the coarse transmission map is pre-estimated under DCP assumption [10]. To obtain the refined transmission map, an image-gradient-guided high-order variational method is proposed to optimize the transmission map. Compared with previous refinement methods, our proposed method is able to yield more natural-looking transmission map, which generates homogeneous appearance within surfaces/objects and significant edges at depth discontinuities. The optimized transmission map could guarantee satisfactory dehazing performance. The quality of restored haze-free image in the second step will be further enhanced using an improved tolerance mechanism.

The remainder of this paper is organized as follows: Sect. 2 briefly overviews the DCP-based coarse transmission map estimation. The image-gradient-guided high-order variational method for refining transmission map is proposed in Sect. 3. The latent haze-free image is restored in Sect. 4. Experiments on different types of images are performed in Sect. 5. Finally, we conclude this paper with our main contributions in Sect. 6.

2   Coarse Transmission Map Estimation

The coarse transmission map estimation in our proposed two-phase framework is pre-estimated through the widely-used local prior, i.e., DCP [10]. Here we will briefly overview the image degradation model and the DCP-based estimation method. The Koschmieder's physical model [30,31] commonly adopted to describe the formation of hazy image is given by

$$I(\mathbf{x}) = J(\mathbf{x})t(\mathbf{x}) + A\left(1 - t(\mathbf{x})\right), \tag{1}$$

where $\mathbf{x} \in \Omega$ is the pixel coordinate with Ω being image domain, I is the observed hazy image, J is the latent sharp image (i.e., *haze-free image*) to restore, A is the global atmospheric light, and t denotes the scene transmission map. The motivation of DCP [10] is that at least one color channel in local non-sky patches has very low intensity values, even close to zero. Under the homogeneous atmosphere assumption, the transmission map t is given by $t(\mathbf{x}) = \exp\left(-\mu d(\mathbf{x})\right)$ with μ being the scattering coefficient of atmosphere and $d(\mathbf{x})$ being the depth of scene. Let J be an outdoor haze-free image. The dark channel $J^{\mathrm{dark}}(\mathbf{x})$ can be formulated as follows

$$J^{\mathrm{dark}}(\mathbf{x}) = \min_{\mathbf{y} \in \Omega^l(\mathbf{x})} \left(\min_c \left(J_c\left(\mathbf{y}\right)\right)\right) \to 0, \tag{2}$$

where $c \in \{r, g, b\}$ denotes the color channel, $\Omega^l(\mathbf{x})$ is the local patch centered at $\mathbf{x} \in \Omega$ with Ω being image domain. He *et al.* [10] proposed to roughly estimate the scene transmission map based on DCP assumption. The atmospheric scattering model (1) can be transformed into the following version

$$\min_{y \in \Omega^l(x)} \left(\min_c \left(\frac{I_c(y)}{A_c} \right) \right) = t_0(x) \min_{y \in \Omega^l(x)} \left(\min_c \left(\frac{J_c(y)}{A_c} \right) \right) + (1 - t_0(x)).$$

It is easy to estimate the coarse transmission map t_0, i.e.,

$$t_0(x) = 1 - v \min_{y \in \Omega^l(x)} \left(\min_c \left(\frac{I_c(y)}{A_c} \right) \right), \tag{3}$$

where v is a constant parameter used to keep some amount of haze for distant object. Through a large number of experiments, we empirically set the parameter to be $v = 0.90$ in this work. The global atmospheric light A is obtained by selecting the top 0.1% brightest pixels in the dark channel (see [10] for details).

3 Refined Transmission Map Estimation

3.1 TGV-Regularized Variational Model

The coarse transmission map gained by Eq. (3) is assumed to be constant in local image regions. It is easy to cause the unwanted blocking effects in final restored images [10]. The transmission map should have homogeneous appearance within surfaces/objects and significant edges at depth discontinuities. To achieve a good balance between surface/object-smoothing and edge-preserving, a second-order total generalized variation (TGV)-regularized variational model [32,33] with a guided image will be proposed to optimize the coarse transmission map. Compared with commonly-used TV regularizer [16,17,34] which favors piecewise constant solutions, TGV encourages piecewise smooth solutions from both theoretical [32] and practical aspects [18]. It means that TGV essentially has the capacity of generating more natural-looking transmission map leading to satisfactory dehazing results. Therefore, the nonsmooth TGV-regularized variational model is proposed for refining the coarse transmission map t_0, i.e.,

$$E(t) = \frac{1}{2} \int_\Omega |t - t_0|^2 + \frac{\lambda}{2} \int_\Omega |\nabla t - \nabla I|^2 + \text{TGV}_\alpha^2(t), \tag{4}$$

where t is the refined transmission map to estimate, λ is a positive parameter, and I represents the grayscale of the input color hazy image I. The first two terms in Eq. (4) are both convex data-fidelity terms. In particular, the first part denotes a measure of the distance between the refined transmission map t and its coarse version t_0. Thus, the optimized transmission map could preserve the main structural details of the input image. Inspired by guided image filtering [13], we tend to utilize the gradient of hazy image ∇I in the second term as a guided image constraint to sharpen the edge of refined transmission map. This is similar as He et al.'s guided image filtering [13]. This data-fidelity term ensures that t has an edge only if I has an edge. The second-order regularizer $\text{TGV}_\alpha^2(t)$

in Eq. (4) is able to suppress undesirable staircase-like artifacts in estimated transmission map. Consequently, the proposed variational model can smooth the objects in smooth regions and preserve the sharp edges in discontinuous regions. Mathematically, the discrete formulation of $\mathrm{TGV}_\alpha^2(t)$ can be defined as follows

$$\mathrm{TGV}_\alpha^2(t) = \min_{\omega \in \mathcal{C}_c^2(\Omega, R^2)} \alpha_1 \int_\Omega |\nabla t - \omega| + \alpha_0 \int_\Omega |\mathcal{E}(\omega)|$$

where α_1 and α_0 are positive parameters, $\mathcal{C}_c^2(\Omega, R^2)$ denotes the space of the vector field, and $\mathcal{E}$ represents the symmetrized derivative operator. To further enhance the accuracy of transmission map estimation, the local geometrical features should be considered to assist in improving the estimation performance. A 2-D binary mask M is incorporated into the TGV-regularized variational model (4) in this work. Thus, the original second-order variational model (4) for generating refined transimission map could be rewritten as follows

$$E(t) = \frac{1}{2} \int_\Omega |t - t_0|^2 + \frac{\lambda}{2} \int_\Omega |\nabla t - \nabla I|^2 + \alpha_1 \int_\Omega |(\nabla t - \omega) \circ \mathrm{M}| + \alpha_0 \int_\Omega |\mathcal{E}(\omega)|$$

$$\propto \frac{1}{2} \int_\Omega \left| t - \frac{t_0 + \lambda \nabla^T \nabla I}{1 + \lambda \nabla^T \nabla} \right|^2 + \alpha_1 \int_\Omega |(\nabla t - \omega) \circ \mathrm{M}| + \alpha_0 \int_\Omega |\mathcal{E}(\omega)|, \qquad (5)$$

where $\circ$ denotes the element-wise multiplication operator. It is necessary to select the proper mask M using the local geometrical features of transmission map. In this work, an experience-guided threshold K_t is simply determined based on image gradients ∇t, i.e., $K_t = \mathrm{mean}(|\nabla t|)$. In particular, for any element $m_i \in \mathrm{M}$, $m_i = 0$ if the corresponding value $\nabla t_i < K_t$, and $m_i = 1$ otherwise. To compensate some missing elements, we tend to further optimize mask M by directly using the morphology dilation operation. Although the mask M is selected based on the manually-selected threshold K_t, it is still worthy of consideration because the dehazing results have illustrated the satisfactory imaging performance.

3.2 Numerical Optimization

The unconstrained optimization problem (5) related to image dehazing in this work is convex but nonsmooth. To guarantee solution stability and efficiency, the primal-dual algorithm of Chambolle-Pock [35–37] is introduced to handle the TGV-regularized variational model (5). The related dual formulation of the primal problem (5) can be defined as follows

$$\min_{t,\omega} \max_{u \in U, v \in V} \frac{1}{2} \int_\Omega |t - \hat{t}|^2 + \langle (\nabla t - \omega) \circ \mathrm{M}, u \rangle + \langle \mathcal{E}(\omega), v \rangle, \qquad (6)$$

with $\hat{t} = (t_0 + \lambda \nabla^T \nabla I)/(1 + \lambda \nabla^T \nabla)$. In Eq. (6), u and v are dual variables. The convex variable sets U and V are respectively given by

$$U = \{u = (u_1, u_2) \mid \|u\|_\infty \le \alpha_1\}, \qquad (7)$$

Algorithm 1. Primal-dual algorithm for model (5)

1: **Input:** M, I, α_1, α_0, δ, τ and λ.
2: **while** $j \leq J_{\max}$ **do**
3:      $u^{j+1} = \text{proj}_U \left(u^j + \delta \left(\left(\nabla \bar{t}^j - \bar{w}^j \right) \circ M \right) \right),$
4:      $v^{j+1} = \text{proj}_V \left(v^j + \delta \left(\mathcal{E} \left(\bar{w}^j \right) \right) \right),$
5:      $t^{j+1} = \frac{\tau \left(\left(t_0 + \lambda \nabla^T \nabla I \right) / \left(1 + \lambda \nabla^T \nabla \right) \right) + t^j + \tau \text{div} \left(u^{j+1} \right)}{1 + \tau},$
6:      $w^{j+1} = w^j + \tau \left(u^{j+1} + \text{div}^h \left(v^{j+1} \right) \right),$
7:      $\bar{t}^{j+1} = 2t^{j+1} - t^j$, $\bar{w}^{j+1} = 2w^{j+1} - w^j$.
8: **end while**
9: **Output:** $t \leftarrow t^{J_{\max}}$.

$$V = \left\{ v = \begin{pmatrix} v_{11} & v_{12} \\ v_{21} & v_{22} \end{pmatrix} \mid \|v\|_\infty \leq \alpha_0 \right\}. \tag{8}$$

The primal-dual algorithm for solving the TGV-regularized variational model (5) is detailedly summarized in Algorithm 1. The Euclidean projectors $\text{proj}_U (u)$ and $\text{proj}_V (v)$ are defined as follows

$$\text{proj}_U (u) = \frac{u}{\max \left(1, |u| / \alpha_1 \right)}, \tag{9}$$

$$\text{proj}_V (v) = \frac{v}{\max \left(1, |v| / \alpha_0 \right)}. \tag{10}$$

The divergence operators $\text{div} (u)$ and $\text{div}^h (v)$ in Algorithm 1 are defined as $\text{div} (u) = \partial_x^{-1} u_1 + \partial_y^{-1} u_2$ and $\text{div}^h (v) = \left(\partial_x^{-1} v_{11} + \partial_y^{-1} v_{12}, \partial_x^{-1} v_{21} + \partial_y^{-1} v_{22} \right)^T$. In our experiments, the parameters $\delta = \tau = 1/\sqrt{12}$ are pre-selected to guarantee the convergence of the primal-dual algorithm.

4   Latent Haze-Free Image Restoration

After generating the refined transmission map t, we can restore the latent haze-free image according to the atmospheric scattering model (1), i.e.,

$$J(\mathrm{x}) = \frac{I(\mathrm{x}) - A}{\max \left(t(\mathrm{x}), t_{lb} \right)} + A, \tag{11}$$

where t_{lb} denotes the lower bound of transmission map t. As suggested in [10], selecting t_{lb} as 0.1 is available for most practical applications. It is generally known that DCP-based dehazing models often fail to restore hazy images with large sky regions. To overcome this limitation, a tolerance-based correction mechanism is incorporated into the image restoration model (11) as follows

$$J(\mathrm{x}) = \frac{I(\mathrm{x}) - A}{\min \left(\max \left(\frac{K}{|I(\mathrm{x}) - A|}, 1 \right) \cdot \max \left(t(\mathrm{x}), t_{lb} \right), 1 \right)} + A, \tag{12}$$

where K is a predefined tolerance. It means that for any pixel in hazy image I, the areas with $|I(\mathrm{x}) - A| < K$ will be considered as sky areas. The corresponding transmission map should be corrected. The constant parameter K used in Eq. (12) limits the improvement of image dehazing. To further enhance imaging performance, an adaptive method for tolerance selection is proposed as follows

$$K = \min \left(\text{median} \left(|I(\mathrm{y}) - A| \right), K_{ub} \right), \quad \mathrm{y} \in \Omega^l(\mathrm{x}), \tag{13}$$

with K_{ub} being the upper bound of K. According to our experimental experience, the manual selection $K_{ub} = (0.15\ 0.15\ 0.15)$ is satisfactory for most applications without further modification. The whole procedure of our two-phase transmission map estimation method for image dehazing is visually summarized in Fig. 1.

Fig. 1. Framework of our proposed dehazing method: (a) input hazy image, (b) coarse transmission map t_0 using DCP, (c) dehazing result with (b), (d) refined transmission map t using our two-phase estimation method, (e) initial dehazing result with (d), (f) final dehazing result with a tolerance mechanism implemented for (e).

5   Experiment Results and Discussion

To verify the effectiveness of the proposed method, we first present several dehazing results to illustrate the influence of transmission map estimation on dehazing performance. Numerous experiments are then conducted on realistic images to compare our proposed method with several state-of-the-art dehazing methods, e.g., Tarel-09 [31], He-13 [13], Meng-13 [38] and Berman-16 [7]. The experience-dependent parameters $\alpha_1 = 1.0$, $\alpha_0 = 2.0$ and $\lambda = 5.0$ were exploited throughout all experiments. The size of local patch $\Omega^l(\mathrm{x})$ in Eq. (3) is set to be 15×15 for all DCP-based dehazing methods (including He-13 [13], Meng-13 [38] and our proposed method).

(a) Hazy   (b) Coarse Map t_0 (c) Results with (b)   (d) Refined Map t (e) Results with (d)

Fig. 2. Dehazing results on different realistic hazy images: (a) input hazy images, (b) estimated coarse transmission maps t_0 using DCP, (c) recovered haze-free images with (b), (d) refined transmission maps t using our two-phase estimation method, (e) recovered haze-free images with (d).

5.1   Influence of Transmission Map on Image Dehazing

Experiments are performed on six different realistic images to investigate the influence of transmission map on dehazing performance, shown in Fig. 2. It can be found that the DCP-based coarse transmission maps t_0 in Fig. 2(b) contain obvious textures of different scales. To guarantee high-quality dehazing, the estimated transmission maps should contain few or (ideally) no visible textures since there is essentially no relationship between the textures and the amount of haze. As a consequence, the dehazing results in Fig. 2(c) suffer from the unwanted halo effects around image edges leading to visual quality degradation. In contrast, our two-phase estimation method could effectively oversmooth the textures while generating refined transmission maps with piecewise smooth property in Fig. 2(d). The final dehazing results in Fig. 2(e) could eliminate the undesirable halo effects and benefit from more natural-looking appearance. The visual image quality is improved accordingly.

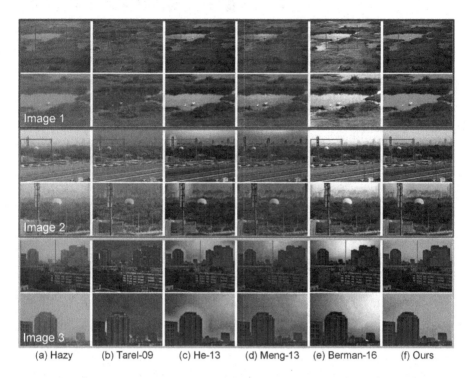

(a) Hazy   (b) Tarel-09   (c) He-13   (d) Meng-13   (e) Berman-16   (f) Ours

Fig. 3. Comparison of dehazing results on three different natural images. From left to right: (a) hazy image, restored images generated by (b) Tarel-09 [31], (c) He-13 [13], (d) Meng-13 [38], (e) Berman-16 [7] and (f) Ours.

5.2   Experimental Results on Natural Images

To further assess the dehazing performance, we implement the comparative experiments on several natural hazy images. The comparison results are visually illustrated in Fig. 3. We find that Tarel-09 [31] generates the worst dehazing results. He-13 [13] is able to yield satisfactory imaging performance for "Image 1" but fails to work for "Image 2" and "Image 3" due to the presence of sky regions. The undesirable halo effects have seriously degraded the image quality. Meng-13 [38] and Berman-16 [7] could effectively reduce the haze effects. However, the resulting artifacts or color distortion bring negative impacts on final imaging performance. In contrast, the proposed two-phase estimation method performs well in haze removal as well as artifacts suppression. High-quality images could be guaranteed with more natural appearances.

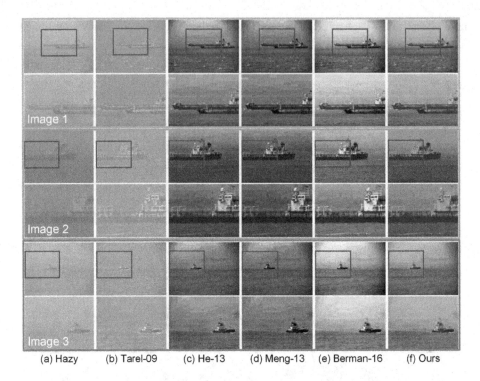

Fig. 4. Comparison of dehazing results on three different maritime images. From left to right: (a) hazy image, restored images generated by (b) Tarel-09 [31], (c) He-13 [13], (d) Meng-13 [38], (e) Berman-16 [7] and (f) Ours.

5.3   Experimental Results on Maritime Images

We extend our proposed method to handling maritime hazy images in this subsection. Due to the specific imaging conditions, maritime images commonly contain different geometrical structures compared with other natural images. In particular, they sometimes include large sky regions leading to the ineffectiveness of widely-used DCP in practical applications. The dehazing results are visually illustrated in Fig. 4. It could be observed that Tarel-09 [31] fails to effectively remove the haze. In contrast, the other competing methods could significantly reduce the haze and enhance the image quality. However, He-13 [13], Meng-13 [38] and Berman-16 [7] still suffer from color distortion or severely block effect in sky regions. By contrast, the proposed method can generate a good balance between haze reduction and artifacts suppression. This is due to the superior properties of TGV regularizer and correction mechanism. The visual quality has been significantly enhanced.

6   Conclusion

A novel two-phase transmission estimation framework is proposed for robust image dehazing. The DCP-based coarse transmission map was estimated in the first phase. The coarse version was then refined using the TGV-regularized variational method in the second phase. The proposed method benefits much from the image-gradient-guided data-fidelity term, which effectively sharpens the edge of refined transmission map. In addition, because of the detail-preserving property of TGV regularizer, the final restored images contain more fine details without unwanted artifacts. Numerous experiments have demonstrated that our proposed method is comparable or superior to current state-of-the-art single image dehazing methods. It is well known that TGV-regularized variational methods have the capacity of removing undesirable noise while preserving important image structures. Therefore, there is a huge potential to extend the proposed method to image dehazing in the presence of noise in our future work.

Acknowledgments. This work was supported by the National Natural Science Foundation of China (No.: 51609195), and the Fund of Hubei Key Laboratory of Transportation Internet of Things (No.: WHUTIOT-2017B003).

References

1. Li, Y., You, S., Brown, M.S., Tan, R.T.: Haze visibility enhancement: a survey and quantitative benchmarking. Comput. Vis. Image Understand. **165**, 1–16 (2017)
2. Wang, W., Yuan, X.: Recent advances in image dehazing. IEEE/CAA J. Autom. Sinica **4**(3), 410–436 (2017)
3. Ancuti, C.O., Ancuti, C.: Single image dehazing by multi-scale fusion. IEEE Trans. Image Process. **22**(8), 3271–3282 (2013)
4. Kim, J.H., Jang, W.D., Sim, J.Y., Kim, C.S.: Optimized contrast enhancement for real-time image and video dehazing. J. Vis. Commun. Image Represent. **24**(3), 410–425 (2013)
5. Fattal, R.: Dehazing using color-lines. ACM Trans. Graph. **34**(1), 13 (2014)
6. Ono, S., Yamada, I.: Color-line regularization for color artifact removal. IEEE Trans. Comput. Imaging **2**(3), 204–217 (2016)
7. Berman, D., Treibitz, T., Avidan, S.: Non-local image dehazing. In: IEEE CVPR, pp. 1674–1682 (2016)
8. Berman, D., Treibitz, T., Avidan, S.: Air-light estimation using haze-lines. In: IEEE ICCP, pp. 1–9 (2017)
9. Fattal, R.: Single image dehazing. ACM Trans. Graph. **27**(3), 72 (2008)
10. He, K., Sun, J., Tang, X.: Single image haze removal using dark channel prior. IEEE Trans. Patt. Anal. Mach. Int. **33**(12), 2341–2353 (2011)
11. Gibson, K.B., Vo, D.T., Nguyen, T.Q.: An investigation of dehazing effects on image and video coding. IEEE Trans. Image Process. **21**(2), 662–673 (2012)
12. Tripathi, A.K., Mukhopadhyay, S.: Single image fog removal using anisotropic diffusion. IET Image Process. **6**(7), 966–975 (2012)
13. He, K., Sun, J., Tang, X.: Guided image filtering. IEEE Trans. Patt. Anal. Mach. Int. **35**(6), 1397–1409 (2013)

14. Li, Z., Zheng, J., Zhu, Z., Yao, W., Wu, S.: Weighted guided image filtering. IEEE Trans. Image Process. **24**(1), 120–129 (2015)
15. Li, Z., Zheng, J.: Edge-preserving decomposition-based single image haze removal. IEEE Trans. Image Process. **24**(12), 5432–5441 (2015)
16. Li, L., Feng, W., Zhang, J.: Contrast enhancement based single image dehazing via TV-L1 minimization. In: IEEE ICME, pp. 1–6 (2014)
17. Fang, F., Li, F., Zeng, T.: Single image dehazing and denoising: a fast variational approach. SIAM J. Imaging Sci. **7**(2), 969–996 (2014)
18. Liu, R.W., Xiong, S., Wu, H.: A second-order variational framework for joint depth map estimation and image dehazing. In: IEEE ICASSP, pp. 1433–1437 (2018)
19. Wang, W., Yuan, X., Wu, X., Liu, Y.: Dehazing for images with large sky region. Neurocomputing **238**, 365–376 (2017)
20. Gao, Y., Hu, H.M., Wang, S., Li, B.: A fast image dehazing algorithm based on negative correction. Signal Process. **103**, 380–398 (2014)
21. Liu, Y., Li, H., Wang, M.: Single image dehazing via large sky region segmentation and multiscale opening dark channel model. IEEE Access **5**, 8890–8903 (2017)
22. Krizhevsky, A., Sutskever, I., Hinton, G.E.: ImageNet classification with deep convolutional neural networks. In: NIPS, pp. 1097–1105 (2012)
23. Ancuti, C., Ancuti, C.O., Timofte, R., et al.: NTIRE 2018 challenge on image dehazing: methods and results. In: CVPR, pp. 1004–1014 (2018)
24. Cai, B., Xu, X., Jia, K., Qing, C., Tao, D.: DehazeNet: an end-to-end system for single image haze removal. IEEE Trans. Image Process. **25**(11), 5187–5198 (2016)
25. Ren, W., Liu, S., Zhang, H., Pan, J., Cao, X., Yang, M.-H.: Single image dehazing via multi-scale convolutional neural networks. In: Leibe, B., Matas, J., Sebe, N., Welling, M. (eds.) ECCV 2016. LNCS, vol. 9906, pp. 154–169. Springer, Cham (2016). https://doi.org/10.1007/978-3-319-46475-6_10
26. Li, B., Peng, X., Wang, Z., Xu, J., Feng, D.: AOD-Net: All-in-one dehazing network. In: ICCV, pp. 4770–4778 (2017)
27. Zhang, H., Sindagi, V., Patel, V.M.: Multi-scale single image dehazing using perceptual pyramid deep network. In: CVPR, pp. 902–911 (2018)
28. Goodfellow, I., Pouget-Abadie, J., Mirza, M., et al.: Generative adversarial nets. In: NIPS, pp. 2672–268 (2014)
29. Li, R., Pan, J., Li, Z., Tang, J.: Single image dehazing via conditional generative adversarial network. In: CVPR, pp. 8202–8211 (2018)
30. Tan, R.: Visibility in bad weather from a single image. In: CVPR, pp. 1–8 (2008)
31. Tarel, J.P., Hautiere, N.: Fast visibility restoration from a single color or gray level image. In: ICCV, pp. 2201–2208 (2009)
32. Bredies, K., Kunisch, K., Pock, T.: Total generalized variation. SIAM J. Imaging Sci. **3**(3), 492–526 (2010)
33. Liu, R.W., Shi, L., Yu, C.H., Wang, D.: Box-constrained second-order total generalized variation minimization with a combined L1,2 data-fidelity term for image reconstruction. J. Electron. Imaging **24**(3), 033026 (2015)
34. Liu, R.W., Shi, L., Huang, W., Xu, J., Yu, C.H., Wang, D.: Generalized total variation-based MRI Rician denoising model with spatially adaptive regularization parameters. Magn. Reson. Imaging **32**(6), 702–720 (2014)
35. Chambolle, A., Pock, T.: A first-order primal-dual algorithm for convex problems with applications to imaging. J. Math. Imaging Vision **40**(1), 120–145 (2011)
36. Condat, L.: A primal-dual splitting method for convex optimization involving Lipschitzian, proximal and linear composite terms. J. Optimiz. Theor. Appl. **158**(2), 460–479 (2013)

37. Lu, W., Duan, J., Qiu, Z., Pan, Z., Liu, R.W., Bai, L.: Implementation of high-order variational models made easy for image processing. Math. Methods Appl. Sci. **39**(14), 4208–4233 (2016)
38. Meng, G., Wang, Y., Duan, J., Xiang, S., Pan, C.: Efficient image dehazing with boundary constraint and contextual regularization. In: ICCV, pp. 617–624 (2013)

An Effective Single-Image Super-Resolution Model Using Squeeze-and-Excitation Networks

Kangfu Mei[1], Aiwen Jiang[1(✉)], Juncheng Li[2], Jihua Ye[1], and Mingwen Wang[1]

[1] School of Computer Information Engineering, Jiangxi Normal University,
Nanchang, China
{meikangfu,jiangaiwen,jhye,mwwang}@jxnu.edu.cn
[2] Department of Computer Science and Technology, East China Normal University,
Shanghai, China
51164500049@stu.ecnu.edu.cn

Abstract. Recent works on single-image super-resolution are concentrated on improving performance through enhancing spatial encoding between convolutional layers. In this paper, we focus on modeling the correlations between channels of convolutional features. We present an effective deep residual network based on squeeze-and-excitation blocks (SEBlock) to reconstruct high-resolution (HR) image from low-resolution (LR) image. SEBlock is used to adaptively recalibrate channel-wise feature mappings. Further, short connections between each SEBlock are used to remedy information loss. Extensive experiments show that our model can achieve the state-of-the-art performance and get finer texture details.

Keywords: Single image super resolution
Squeeze-and-excitation block · Channel-wise recalibrate
Deep residual learning · Image restoration

1 Introduction

Single-image super-resolution (SISR) is a popular computer vision problem, which aims to reconstruct a high-resolution (HR) image from a low-resolution (LR) image. However, SISR is still considered as an ill-posed inverse problem due to high-level information loss during image downsampling. To solve this problem, many algorithms have been proposed.

Early methods [15,17,19–21], besides bicubic and bilinear interpolation, learned the mapping from LR to HR pairs directly by sacrificing certain accuracy or speed for improvements. Super-Resolution Convolutional Neural Network (SRCNN) proposed by Dong et al. [3] was the first successful model that adopted CNN structure to solve SISR problem and obtained great performance improvement. In SRCNN, convolutional neural network was used to learn non-linear

© Springer Nature Switzerland AG 2018
L. Cheng et al. (Eds.): ICONIP 2018, LNCS 11306, pp. 542–553, 2018.
https://doi.org/10.1007/978-3-030-04224-0_47

mapping from each LR vector to a set of HR vector. Due to the outstanding performance of SRCNN, several deeper and more complicated models has been proposed to follow it, such as VDSR proposed by Kim et al. [8]. Though VDSR achieved excellence performance, its speed remained slow speed as it use a very deep residual convolutional network and an upscale image preprocess.

To avoid the complexities of feature extraction network and upscale prepro-cess, Shi et al. [16] replaced upscale preprocess with sub-pixel convolution layers. The sub-pixel layers could produce HR image from feature maps directly with a set of up-scaling filters. This architecture greatly improved the speed of networks. Therefore, following the strategy of up-sampling layer, Ledig et al. [12] further proposed a SRResNet with a very deep ResNet [5] architecture. Lai et al. [11] proposed the LapSRN, which use learned kernel as up-sampling unit to direct produced SR images.

In spite of great success achieved in the above architectures, the main issue that how to model mapping from LR to HR images better in a fast and flexible way remained unsolved. In this paper, we have proposed a Super-Resolution Squeeze-and-Excitation Network (SrSENet) for SISR. The concept of SEBlock [6] is employed to better modeling interdependencies between chan-nels. Short connections from input to each SEBlock are used to remedy informa-tion lost. And different deconvolution layers are used for different scales under the same feature extraction architecture. The proposed method is evaluated on some popular publicly available benchmarks. Extensive experiments show that our proposed model can achieves competitive accuracy in a more accurate and flexible way. It can greatly reduce models complexity by using less layers and allow designing more flexible applications.

The contributions of this paper are two folds:

- We have introduced an effective super-resolution network with SEBlock. It performs dynamic channel-wise feature recalibration to provide a new power-ful architecture to improve the representational ability of information extrac-tion part from low-resolution images.
- We have set up a new state-of-the-art super-resolution method with fast running speed and accurate result in the measurement of PSNR and SSIM without increasing the complexity of the network, especially in case of large upscale rate.

2 Related Work

2.1 Single-Image Super-Resolution

In this section, we are mainly concentrated on reviewing mainstream deep learn-ing based single-image super-resolution methods. Typically, a SISR network could be approximately divided into two parts. The first part could be seen as a feature extraction block, which is composed of many stacked convolutional layers. The second part records up-scaling information from LR images to HR images. Recent works are concentrated on improving the first part by changing the way of skip connections between inputs of each layer. In other words, they focus on changing the proportion of information captured by initial layers.

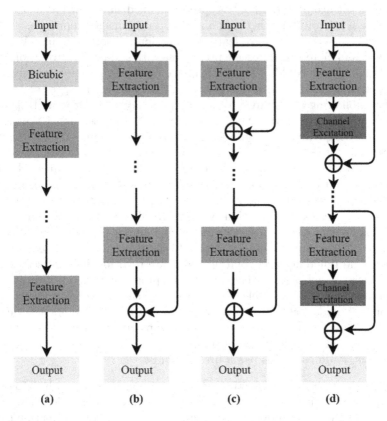

Fig. 1. Comparisons on network architectures of four typical deep learning based SISR categories.

We group mainstream deep learning based SISR models into four categories, as shown in Fig. 1. The (a) category contains feature extraction, such as network in [3]. The (b) category like [8] introduces short connection as residual-learning. The (c) category like accepts input in each feature extraction layer. Our proposed model could be categorized into the last category (d). The difference from the other three categories is that each extraction layer block receives input before channel-wise modeling. In this way, network could better learn mapping between LR-HR images.

2.2   Squeeze-and-Excitation Channel

Different from works on enhancing spatial encoding, SENet [6] was proposed to fully capture channel-wise dependencies through adaptive recalibration. The

SENet was separated into two steps, squeeze and excitation, to explicitly model channel interdependencies.

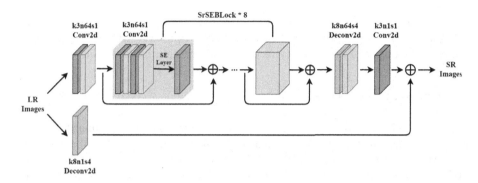

Fig. 2. Our proposed Network architectures of SrSENet in upscale of 4x. Blue blocks represent a Convolutional layer. Yellow blocks represent a LeakRelu layer. Green Blocks represent a Transposed Convolutional layer. (Color figure online)

After initial images were input into the first convolution layer, the output feature $U \in \mathbb{R}^{W \times H \times C}$ was passed to a SEBlock to do squeeze and excitation operator. The squeeze operator was used to embed information from global receptive field into a channel descriptor in each layer. Then a sigmoid activation function and FC layer were later used to gain nonlinear interaction between each layers. The squeeze operator produced a sequence S in $1 \times 1 \times C$ which represented the correlations of each layer. The excitation operator later was employed to perform feature recalibration through reweighting the original feature mappings

$$\tilde{U} = F_{scale}(U, S) = u_c \times s_c,$$

where u_c refers to the parameters of the c-th filter and s_c denotes the element of c-th channel descriptor. This architecture can help feature extraction parts better capture the information from input to output. In our work, we combined SEBlock with ResNet for feature extraction.

2.3   Transposed Convolutional Layer

In order to obtain super-resolution images, a simple idea is to upscale original image first, then final HR image is directly generated from the resulted scaled image. It is not difficult to find that this kind of strategy wastes much time on preprocessing without any obvious advantage.

Shi et al. [16] first proposed to use sub-pixel convolution layer to produce HR images directly. It upscale a LR image by periodic shuffling the elements of a $W \times H \times C \cdot r^2$ tensor to a tensor of shape $rH \times rW \times rC$. However, it didn't make full use of the correspondence information from LR to HR. LapSRN [11] was proposed by to use a multiple transposed convolutional layer to deal with different upscale rate in a progressive way. Without any preprocessing step like upscale, LapSRN achieved more accurate information between LR and HR in a fast way.

Following previous works, we use transposed convolutional layer with different parameters for different upscale rate, which can keep network simple and improve the power of networks to record reconstruction information.

3   Proposed Method

The proposed method aims to extract information from the LR image I_L and learn mapping function F from I_L feature maps to HR images I_H. We describe I_L with C channels in size of $W \times H$. With upscale rate r, I_H is in size of $rW \times rH$. Our ultimate goal is to minimize the loss between the reconstructed images and the corresponding ground truth HR images. In the following, we will describe the details of the proposed method.

3.1   Network Architecture

Our proposed method is inspired from SRResNet [12] and LapSRN [11]. Following LapSRN, our model contains two parts: residual learning stage and image reconstruction stage, as shown in Fig. 2.

Unlike SRResNet and LapSRN, in the residual learning stage, we introduced SrSEBlock to extract features from LR images. The SrSEBlock structure integrates ResNet and SENet, which can better capture information from inputs and better modeling interdependencies between channels.

As VDSR [8] suggested, in the SR ill-posed problem, surrounding pixels were useful to correctly infer center pixel. With larger receptive field a SR model has, it could use more contextual information from LR to better learn correspondences from LR to HR. In our proposed network, the filters of SrSEBlock is in size of $3 \times 3 \times 64$. Therefore, in case of depth D layer, its receptive field could be seen as $(2 \times D + 1) \times (2 \times D + 1)$ in the original image space. The bigger receptive field means our network can use more context to reconstruct images.

As we know, with the increase of network depth, gradient disappearance or explosion will occur during training and the high-frequency information will also disappear. So, we introduce a short connection between SrSEBlocks which can receive input information before channel-wise modeling.

In the proposed network, we employ 8 SrSEBlocks to generate a feature mapping, and then we employ a transposed layer to transform the resulted mapping directly into a residual image by applying a deconvolutional layer. On different upscale rates, we don't increase the number of deconvolution layers, just directly change parameters such as the kernel size, stride and padding steps, to obtain corresponding residual image. In image reconstruction stage, the up-sampled LR image feature mappings and the learned residual feature mappings are added together to reconstruct HR image. By using residual image learning, network converges efficiently. The final feature mapping is output directly as the SR image.

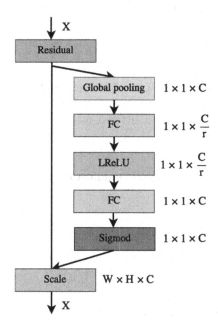

Fig. 3. The architecture of SELayer.

3.2 Channels Excitation in SrSEBlock

Different from recent work that focus on enhancing spatial encoding, we use SrSEBlock to model correlations between channels. In this section, we will describe how the SrSEBlock work in our network.

In details, feature maps are input into a SELayer as Fig. 3 shows. The corresponding excitations to each channel are output to scale original feature map. Taking a feature maps U in size of $W \times H \times C$ as input, we first do a global average pooling to generate channel-wise statistics z in size of $1 \times 1 \times C$., as show in below

$$z_c = \frac{1}{W \times H} \sum_{i=1}^{W} \sum_{j=1}^{H} u_c(i,j).$$

In order to learn nonlinear interaction between each channels, we use two FC layers with non-linear activations to form a bottleneck, as done in He et al. [5]. This architecture could limit model complexity and benefit for generalization. The reduction ratio r at 16 is accepted to do dimensionality reduction. The final output s of SELayer is use to scale corresponding channels of residual feature mappings.

In this way, noise information in previous feature mappings could be reduced. And channels that contain useful information will be highly activated, helping to boost feature's discriminative abilities. In the later ablation experiment, we will show its effectiveness.

4   Experiments

In our experiment settings, given a set of HR images $\{Y_i\}$ and the corresponding down-sampled LR images $\{X_i\}$ through bicubic, our goal is to minimize the Charbonnier Penalty Function [2] defined as below, which is a differentiable variant of L_1 norm

$$\rho(z) = \sqrt{z^2 + \varepsilon^2}.$$

The loss is minimized using stochastic gradient descent with the standard back-propagation. We solve:

$$G^* = arg \min_G \frac{1}{n} \sum_{i=1}^{n} \rho(Y_i - G(X_i)),$$

where G represents our SR image networks.

4.1   Datasets for Training and Testing

Different from previous work, we use DIV2K [1] to train our model for more realistic modeling. DIV2K is a newly distributed high quality image dataset for image super resolution. Its training data has 800 high definition, high resolution images. In our experiments, we find different image processing framework will produce different bicubic downscale results. So for fair comparison, we all use the bicubic downsampling algorithm in Matlab image processing tool to generate LR-HR image pairs for our network training. For each pair, we crop HR sub image in 96×96 size and downscale it to LR images by different downscale factors. We export the pairs as MAT variable in HDF5 type.

4.2   Experiment Setup

We compare our proposed SrSENet with several state-of-the-art methods such as SRCNN [3], FSRCNN [4], SelfExSR [7], VDSR [8], DRCN [8] and LapSRN [11] on five common used benchmark datasets Set5, Set14 [22], BSDS100 [13], Urban100 [7] and Manga109 [14]. The restoration quality of the resulted SR is evaluated by using PSNR and SSIM [18].

Three scaling cases $\{2\times, 4\times, 8\times\}$ are considered. On each case, the architecture of feature extraction part of our network is kept the same, and the transposed convolutional layer size is changed according to different up-scale rate. The source code of our method is available on GitHub[1].

4.3 Training Details

We use 8 SrSEBlocks to do feature extraction. For each upscale deconvolution layer, we use respective convolutional kernels [4, 2, 1], [8, 4, 2], [16, 8, 4] for 2x, 4x, 8x rate up-scaled super-resolution image respectively. Here in the format $[*, *, *]$, the first represents kernel size, the second represents stride steps, and the last is padding size in transposed layer. If dealing with odd multiples of magnification, we can also easily achieve an odd magnification by modifying the kernel size of the convolutional network to an odd number (e.g., [3, *, *]). During the training, we set the initial learning rate at 0.0001. We use Adma optimizer [10] with $\beta_1 = 0.9$ to let network convergence and the training batches is 64. It roughly takes half day on a machine using four TitanX GPUs for a single upscale training. For illustration, the respective PSNR testing curves of our SrSENet on Set14 are shown in Fig. 4.

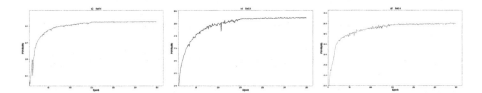

Fig. 4. Respective PNSR testing curves of SrSENet on dataset Set14 for three scaling cases. Left: scale 2×, Middle: scale 4×, Right: scale 8×.

The quantitative performance comparisons are shown in Table 1. From the experiment results, we can easily find that our proposed method obtains competitive performance in all datasets in different upscale rates. Especially in larger scale case, the advantages of our method are more obvious. Our method can achieve top performance with less network depth. In Fig. 5, we further show some realistic results for visual comparison. We can find that the fine texture of images in our method are recovered better.

In order to verify the effectiveness of the introduced SEblock, we additionally have set up an ablation experiment. We construct a reduced version network by removing SElayers out of the proposed SrSENet, while keeping other parts remained. We have compared the reduced version with our SrSENet. The performance comparisons on 4× scale are shown in Table 2. From the results, we can

---

[1] Source code: https://github.com/MKFMIKU/SrSENet.

Table 1. Quantitative comparisons of state-of-the-art methods. Red text indicates the best performance and blue italics text indicates the second best performance. We use results from LapSRN to do comparation, and attention that Layers in the table include convolution and deconvolution.

Algorithm	Scale	Set5 PSNR/SSIM	Set14 PSNR/SSIM	BSDS100 PSNR/SSIM	Urban100 PSNR/SSIM	Manga109 PSNR/SSIM
Bicubic	2x	33.65/0.930	30.34/0.870	29.56/0.844	26.88/0.841	30.84/0.935
SelfExSR [7]	2x	36.49/0.954	32.44/0.906	31.18/0.886	29.54/0.897	35.78/0.968
SRCNN [3]	2x	36.65/0.954	32.29/0.903	31.36/0.888	29.52/0.895	35.72/0.968
FSRCNN [4]	2x	36.99/0.955	32.73/0.909	31.51/0.891	29.87/0.901	36.62/0.971
VDSR [8]	2x	37.53/0.958	32.97/0.913	31.90/0.896	30.77/0.914	37.16/0.974
DRCN [9]	2x	37.63/0.959	32.98/0.913	31.85/0.894	30.76/0.913	37.57/0.973
LapSRN [11]	2x	37.52/0.959	33.08/0.913	31.80/0.895	30.41/0.910	37.27/0.974
SrSENet	2x	37.56/0.958	33.14/0.911	31.84/0.896	30.73/0.917	37.43/0.974
Bicubic	4x	28.42/0.810	26.10/0.704	25.96/0.669	23.15/0.659	24.92/0.789
SelfExSR [7]	4x	30.33/0.861	27.54/0.756	26.84/0.712	24.82/0.740	27.82/0.865
SRCNN [3]	4x	30.49/0.862	27.61/0.754	26.91/0.712	24.53/0.724	27.66/0.858
FSRCNN [4]	4x	30.71/0.865	27.70/0.756	26.97/0.714	24.61/0.727	27.89/0.859
VDSR [8]	4x	31.35/0.882	28.03/0.770	27.29/0.726	25.18/0.753	28.82/0.886
DRCN [8]	4x	31.53/0.884	28.04/0.770	27.24/0.724	25.14/0.752	28.97/0.886
LapSRN [11]	4x	31.54/0.885	28.19/0.772	27.32/0.728	25.21/0.756	29.09/0.890
SrSENet	4x	31.40/0.881	28.10/0.766	27.29/0.720	25.21/0.762	29.08/0.888
Bicubic	8x	24.40/0.657	23.19/0.568	23.67/0.547	20.74/0.515	21.47/0.649
SelfExSR [7]	8x	25.52/0.704	24.02/0.603	24.18/0.568	21.81/0.576	22.99/0.718
SRCNN [3]	8x	25.33/0.689	23.85/0.593	24.13/0.565	21.29/0.543	22.37/0.682
FSRCNN [4]	8x	25.41/0.682	23.93/0.592	24.21/0.567	21.32/0.537	22.39/0.672
VDSR [8]	8x	25.72/0.711	24.21/0.609	24.37/0.576	21.54/0.560	22.83/0.707
LapSRN [11]	8x	26.14/0.738	24.44/0.623	24.54/0.586	21.81/0.581	23.39/0.735
SrSENet	8x	26.10/0.703	24.38/0.586	24.59/0.539	21.88/0.571	23.54/0.722

easily find that the introduced SEblocks indeed plays great importance on final excellence performance. On the other scales, we could achieve similar conclusions as well. We owe the its effectiveness to it introducing channel-wise attention mechanism, which makes channel information of each pixel on SR image adaptively learnable.

Table 2. Ablation experiment: quantitative comparisons on 4× scale.

Algorithm	Set5 PSNR/SSIM	Set14 PSNR/SSIM	BSDS100 PSNR/SSIM	Urban100 PSNR/SSIM	Manga109 PSNR/SSIM
Reduced version	31.30/0.880	28.10/0.766	27.16/0.720	25.08/0.760	28.84/0.886
SrSENet	31.40/0.881	28.10/0.766	27.29/0.720	25.21/0.762	29.08/0.888

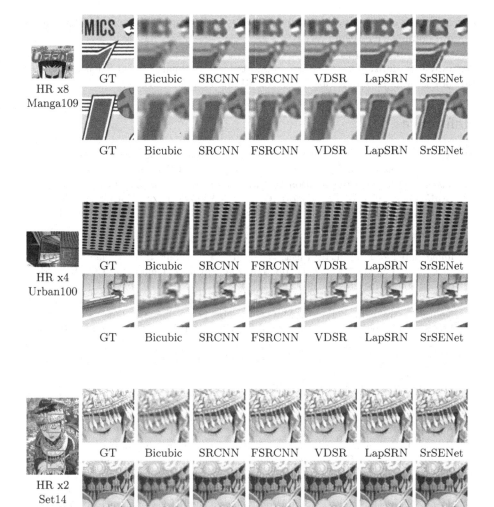

Fig. 5. Visual comparisons on Bicubic, SRCNN, FSRCNN, VDSR, LapSRN and SrSENet on upscale rate of 8×, 4×, 2×.

5   Conclusions

In this paper, we have proposed a new effective super-resolution model by using a deep residual network with SrSEBlock. Our method focuses on modeling channels correlations between feature mappings from the LR image. By modeling channel wise, we have confirmed that our method could produce more realistic texture on realistic images. We set a new state-of-the-art super-resolution method without increasing the complexities of the network. We believe that our approach can be applied to other real-world computer vision problems and achieve competitive results.

Acknowledgment. This work was supported by National Natural Science Foundation of China under Grant Nos. 61365002, 61462042 and 61462045.

References

1. Agustsson, E., Timofte, R.: NTIRE 2017 challenge on single image super-resolution: dataset and study. In: Proceedings of the IEEE Conference on Computer Vision and Pattern Recognition Workshops, pp. 1110–1121. IEEE, Hawaii (2017)
2. Bruhn, A., Weickert, J., Schnörr, C.: Lucas/Kanade meets Horn/Schunck: combining local and global optic flow methods. Int. J. Comput. Vis. **61**(3), 211–231 (2005)
3. Dong, C., Loy, C.C., He, K., Tang, X.: Learning a deep convolutional network for image super-resolution. In: Fleet, D., Pajdla, T., Schiele, B., Tuytelaars, T. (eds.) ECCV 2014. LNCS, vol. 8692, pp. 184–199. Springer, Cham (2014). https://doi.org/10.1007/978-3-319-10593-2_13
4. Dong, C., Loy, C.C., Tang, X.: Accelerating the Super-Resolution Convolutional Neural Network. In: Leibe, B., Matas, J., Sebe, N., Welling, M. (eds.) ECCV 2016. LNCS, vol. 9906, pp. 391–407. Springer, Cham (2016). https://doi.org/10.1007/978-3-319-46475-6_25
5. He, K., Zhang, X., Ren, S., Sun, J.: Deep residual learning for image recognition. In: Proceedings of the IEEE Conference on Computer Vision and Pattern Recognition, pp. 770–778. IEEE, Las Vegas (2016)
6. Hu, J., Shen, L., Sun, G.: Squeeze-and-excitation networks. In: Proceedings of the IEEE Conference on Computer Vision and Pattern Recognition. IEEE, Salt Lake (2018)
7. Huang, J.B., Singh, A., Ahuja, N.: Single image super-resolution from transformed self-exemplars. In: Proceedings of the IEEE Conference on Computer Vision and Pattern Recognition, pp. 5197–5206. IEEE, Boston (2015)
8. Kim, J., Kwon Lee, J., Mu Lee, K.: Accurate image super-resolution using very deep convolutional networks. In: Proceedings of the IEEE Conference on Computer Vision and Pattern Recognition, pp. 1646–1654. IEEE, Las Vegas (2016)
9. Kim, J., Kwon Lee, J., Mu Lee, K.: Deeply-recursive convolutional network for image super-resolution. In: Proceedings of the IEEE Conference on Computer Vision and Pattern Recognition, pp. 1637–1645. IEEE, Las Vegas (2016)
10. Kingma, D., Ba, J.: Adam: a method for stochastic optimization. In: Proceedings of the International Conference on Learning Representations, San Diego (2015)
11. Lai, W.S., Huang, J.B., Ahuja, N., Yang, M.H.: Deep laplacian pyramid networks for fast and accurate super-resolution. In: Proceedings of the IEEE Conference on Computer Vision and Pattern Recognition, pp. 5835–5843. IEEE, Hawaii (2017)
12. Ledig, C., et al.: Photo-realistic single image super-resolution using a generative adversarial network. In: Proceedings of the IEEE Conference on Computer Vision and Pattern Recognition, pp. 105–114. IEEE, Hawaii (2017)
13. Martin, D., Fowlkes, C., Tal, D., Malik, J.: A database of human segmented natural images and its application to evaluating segmentation algorithms and measuring ecological statistics. In: Proceedings of the IEEE International Conference on Computer Vision, pp. 416–423. IEEE, Vancouver (2001)

14. Matsui, Y., Ito, K., Aramaki, Y., Fujimoto, A., Ogawa, T., Yamasaki, T., Aizawa, K.: Sketch-based manga retrieval using manga109 dataset. Multimed. Tools Appl. **76**(20), 21811–21838 (2017)
15. Schulter, S., Leistner, C., Bischof, H.: Fast and accurate image upscaling with super-resolution forests. In: Proceedings of the IEEE Conference on Computer Vision and Pattern Recognition, pp. 3791–3799. IEEE, Boston (2015)
16. Shi, W., et al.: Real-time single image and video super-resolution using an efficient sub-pixel convolutional neural network. In: Proceedings of the IEEE Conference on Computer Vision and Pattern Recognition, pp. 1874–1883. IEEE, Las Vegas (2016)
17. Timofte, R., De Smet, V., Van Gool, L.: A+: adjusted anchored neighborhood regression for fast super-resolution. In: Cremers, D., Reid, I., Saito, H., Yang, M.-H. (eds.) ACCV 2014. LNCS, vol. 9006, pp. 111–126. Springer, Cham (2015). https://doi.org/10.1007/978-3-319-16817-3_8
18. Wang, Z., Bovik, A.C., Sheikh, H.R., Simoncelli, E.P.: Image quality assessment: from error visibility to structural similarity. IEEE Trans. Image Process. **13**(4), 600–612 (2004)
19. Yang, C.Y., Yang, M.H.: Fast direct super-resolution by simple functions. In: Proceedings of the IEEE International Conference on Computer Vision, pp. 561–568. IEEE, Sydney (2013)
20. Yang, J., Wright, J., Huang, T., Ma, Y.: Image super-resolution as sparse representation of raw image patches. In: Proceedings of the IEEE Conference on Computer Vision and Pattern Recognition, pp. 1–8. IEEE, Anchorage (2008)
21. Yang, J., Wright, J., Huang, T.S., Ma, Y.: Image super-resolution via sparse representation. IEEE Trans. Image Process. **19**(11), 2861–2873 (2010)
22. Zeyde, R., Elad, M., Protter, M.: On single image scale-up using sparse-representations. In: Boissonnat, J.-D., et al. (eds.) Curves and Surfaces 2010. LNCS, vol. 6920, pp. 711–730. Springer, Heidelberg (2012). https://doi.org/10.1007/978-3-642-27413-8_47

Image Denoising Using a Deep Encoder-Decoder Network with Skip Connections

Raphaël Couturier, Gilles Perrot, and Michel Salomon$^{(\boxtimes)}$

FEMTO-ST Institute, CNRS - Univ. Bourgogne Franche-Comté (UBFC),
Belfort, France
{raphael.couturier,gilles.perrot,michel.salomon}@univ-fcomte.fr

Abstract. In many areas images can be corrupted by various types of noise and therefore image denoising is a prerequisite. For example, medical images like the 4D-CT or ultrasound ones, are prone to noise and artifacts that can affect diagnostic confidence. Remote sensing is another field for which image preprocessing is mandatory to improve the quality of source images. Synthetic Aperture Radar (SAR) images are typically corrupted by multiplicative speckle noise. In this paper, a deep neural network able to deal with both additive white Gaussian and multiplicative speckle noises is developed, showing also some blind denoising capacity. The experiments on noisy images show that the proposal, which consists in a encoder-decoder, is efficient and competitive in comparison with state-of-the-art methods.

Keywords: Image denoising · Additive and multiplicative noises
Deep learning · Encoder-decoder

1 Introduction

In today's digital world, an increasing amount of digital images is produced every day. Nevertheless, the visual quality of an image is not guaranteed, since different sources of noise can influence the pixel values. A main source is the acquisition process and particularly the presence of defaults in the capturing device: noise can be produced by the sensor, misaligned lenses, and so on, but noise can also be added during its edition, storage or transmission. As a result, different types of noise can appear in a digital image, such as Gaussian noise, Salt-and-pepper noise, etc., and at different levels. For an observer, the impact of noise can range from isolated speckles up to images that seem to show nothing but noise.

To recover as precisely as possible a clean image y from a noisy version x that is the outcome of an arbitrary stochastic corruption process n: $x = n(y)$, an efficient image denoising method is needed. Formally, the goal of image denoising is thus to find a function f that approximates as well as possible the inverse function of n:

$$f = \underset{f}{\operatorname{argmin}} \operatorname{E}_y \parallel f(x) - y \parallel_2^2 . \tag{1}$$

© Springer Nature Switzerland AG 2018
L. Cheng et al. (Eds.): ICONIP 2018, LNCS 11306, pp. 554–565, 2018.
https://doi.org/10.1007/978-3-030-04224-0_48

It should be noticed that additive white Gaussian noise is often targeted, in which case the corruption process can be rewritten as $x = y + \mathcal{N}(0, \sigma)$ where σ is the standard deviation.

To solve this problem, there are two main categories of methods: model-based optimization methods and discriminative learning methods. The objective of the former methods is to directly solve the optimization problem, but, as this problem is usually complex, they are time consuming. On the other hand, discriminative learning methods try to learn a set Θ of parameters defining a nonlinear function $\hat{f}$ that approximates n^{-1} by minimizing a loss function according to a data set that consists of clean-noisy images pairs. In that case, the previous problem can be expressed as follows:

$$\Theta = \underset{\theta}{\mathrm{argmin}} \frac{1}{N} \sum_{i=1}^{N} \| \hat{f}(x_i) - y_i \|_2^2 \tag{2}$$

where x_i is the noisy version of y_i and N is the size of the data set. Compared to model-based methods, discriminative ones are less flexible since they are usually trained to deal for a specific underlying model of corruption.

Typical examples of model-based methods are BM3D [4] and WNNM [7], while neural networks are representatives of the discriminative family. Obviously, even if the MLP has been investigated [2], with the current rise of deep learning, deep neural networks are now the most actively studied discriminative methods. One of the first deep network proposal was made by Xie *et al.* [18] in 2012, it consisted in a stacking of auto-encoders where each auto-encoder was trained one after the other. In 2014, Long *et al.* [10] introduced the Fully Convolutional Networks (FCN) for semantic segmentation, an architecture that allows to produce segmentation maps whatever the image size and faster than with patch classification approaches. A work that has led to the widespread use of deep networks in which the fully connected part is dropped for dense predictions.

Image denoising is such a dense prediction task, whose objective is to recover for each pixel its original gray level value. Consequently, among the various deep networks that have recently been investigated to tackle the image denoising problem, almost all of them have adopted the FCN paradigm. However, even if these networks belong to the same family, differences among them can be observed. First, a FCN can be trained to recover directly the clean image or to predict a residual image that is subtracted from the noisy input one [19]. Second, a central problem when using a CNN for image denoising (or segmentation) is due to pooling layers. Indeed, a pooling layer usually performs a spatial downsampling and as the input and output images must have the same size, it means that an upsampling process is needed. On the one hand the pooling permits to enlarge the field of view, but on the other hand the aggregation throws away useful spatial information. To address this issue, several different architectures have emerged: architectures without pooling layers and the encoder-decoder architecture.

The proposal is presented thereafter throughout the following sections. Section 2 starts with a discussion on related existing discriminative denoising methods and more specifically deep learning ones. An overview of the proposed deep neural network design with its main characteristics is given in the following section. Section 4 is dedicated to the experiments, showing the relevance of the proposed approach. Finally, some concluding remarks are given in Sect. 5.

2   Related Works

A first example of architecture for image denoising that only has convolutional layers is the deep network called DnCNN (Deep network CNN) proposed by Zhang *et al.* [19] considering a residual learning formulation. The CNN is composed of layers with three different convolutional blocks using a unique convolution kernel size of 3 × 3: Convolution+ReLU for the first layer, Convolution+BatchNorm+ReLU in the intermediate layers, and only Convolution in the final layer. It has a receptive field whose size depends on the network depth and which is correlated with the effective patch size of other denoising methods. In fact, most of the denoising methods such as BM3D, WNNM, MLP, and so on operate on patches. The authors have thus chosen to increase the receptive field through a large depth. The networks of 17 and 20 layers that they trained for additive Gaussian denoising, respectively for a specific noise level and for blind denoising, outperformed slightly both BM3D and WNNM on the BSD68 data set of grayscale images. A recently proposed alternative to an increased depth or to increasing filter sizes is the use of dilated kernels, also known as atrous convolution. Indeed, convolutional layers that only use 3 × 3 kernels but with multiple atrous rates perform an analysis of the image at multiple scales without needing a large depth. This approach has also been studied by Zhang *et al.* in another work [21], leading to similar denoising performances.

Zhang *et al.* have finally introduced another architecture [20] to handle a wide range of noise levels and spatially variant noise. This architecture, called FFDNet for fast and flexible denoising convolutional neural network, consists of a CNN similar to the one of DnCNN, but that does not predict the noise. The CNN receives as input four sub-images obtained from the initial input image using a reversible downsampling operator (factor is set to 2) and a tunable noise level map. As output it produces four denoised sub-images which are then upsampled to recover the final output image. For Additive White Gaussian Noise (AWGN) removal, the experiments show that DnCNN is better for low noise levels ($\sigma \leq 25$), whereas for larger values FFDNet becomes gradually slightly better with the increase of noise level. This result is all the more interesting as it is the version of DnCNN trained for a specific noise level that is considered, whereas FFDNet is trained in a blind context with noise level $\sigma \in [0; 75]$.

An encoder-decoder is quite different. The encoder consists of convolutional layers that successively downsample the input image into small abstraction maps from which the noise is removed as the process goes deeper. The decoder is then fed by the final abstraction map in order to reconstruct a clean image thanks to deconvolutional layers. The reconstruction by the decoder is clearly the most difficult part since image details might be lost during the features extraction performed by the encoder. To mitigate this problem, a common approach is to adopt the skip connection method. In this work we have considered an encoder-decoder with such connections. A similar architecture, but considering a residual learning pattern, has been investigated by Gu *et al.* in [6] for SAR image despeckling. Compared to SAR-BM3D and DnCNN, this Residual Encoder-Decoder NETwork (RED-NET) has given improved denoising performances.

3 The Proposed Deep Network for Image Denoising

3.1 Network Architecture

Our network is similar to a generator architecture introduced by Isola *et al.* [8] in their investigation of conditional adversarial networks to solve image-to-image translation problems, an architecture which is itself an adaptation of one issued from [14]. The generator we consider is the U-Net [15] version corresponding to an encoder-decoder having skip connections between mirrored layers in both encoder and decoder stacks. The encoder extracts salient features preserving the detailed underlying structure of the image, while simultaneously removing the noise, whereas the decoder produces a clean version of the input image by recovering successive image details as it progresses through its layers in a bottom-up way from the bottleneck layer of the encoder. Each skip connection allows to directly shuttle the information from an encoder layer to its corresponding decoder one, and this is appealing since the input noisy image and output clean version share large parts of the low-level information like the location of prominent edges. In fact, skip connections allow to remember different levels of details that are useful to reconstruct the final output image.

Symmetric skip connections are very interesting because they facilitate the training and improve image recovery. On the one hand, skip connections allow to solve the vanishing gradient problem by backpropagating the signal directly and, on the other hand as both input and output images have the same content, the recovery of the clean version can benefit from the details appearing in the corrupted one. Thus, better results are usually obtained with skip connections.

From an architecture point of view, following the specification given in [8], the encoder is almost exclusively composed of Convolution-BatchNorm-ReLU layers, a typical choice in CNN, except the first layer that does not undergo the batch normalization. Let us notice that, as illustrated in the previous section, this kind of convolutional block is also the one used in many FCN that do not have an encoder-decoder architecture. For its part, the decoder consists of a mixing of this kind of layer and a variant of it integrating a dropout rate of 50% before the ReLU activation. Neither pooling nor unpooling operations are used, because aggregation induces some losses of details which, in the context of image denoising, can be awkward. Since convolutions and deconvolutions use 4×4 kernels with stride 2, each encoder and decoder layer will produce feature maps which are downsampled and upsampled, respectively, by a factor of 2. The last top decoder layer is mapped back to the output clean image with a deconvolution followed by a tanh activation. Our code is available on GitHub[1].

---

[1] https://github.com/rcouturier/ImageDenoisingwithDeepEncoderDecoder.

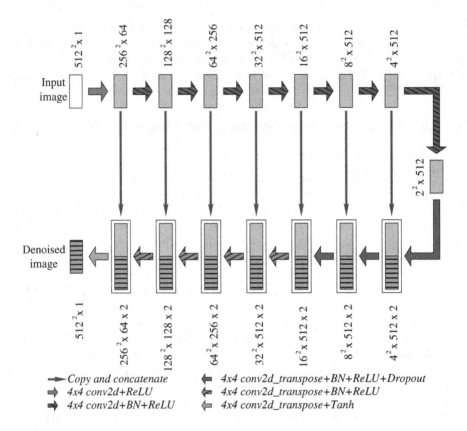

Fig. 1. Schematic diagram of the proposed deep encoder-decoder network architecture.

Figure 1 shows the detailed structure of the proposed encoder-decoder. The symmetric u-shape of the network can be in particular noticed, with the contracting path which consists of the encoding layers (left-right arrows) on the top part, while the expansive path on the bottom part is made of the same number of decoding layers (right-left arrows). Each light blue box represents a set of feature maps issued from an encoding layer and each greenish-blue box with horizontal line pattern one from a decoding layer. Obviously, the size of the input image defines an upper bound on the number of layers in the encoder and decoder. The x-y size of the feature maps, as well as their number, is provided on the top (encoding part) or bottom (decoding part) of each box. For example, in the case of the decoder, $4^2 \times 512 \times 2$ means that there are 1024 maps of size 4^2, where 512 maps are the result of the decoding of the bottleneck layer and the other 512 ones the higher resolution features map copied from encoder. The different arrows denote the different operations. On the one hand convolution operations are represented by left-right arrows and on the other hand deconvolution ones by right-left arrows. The respective TensorFlow module implementing each operation, namely *conv2d* and *conv2d_transpose*, are used in the labels.

3.2   Loss Function

A factor that has a major impact on the obtained neural network is obviously the loss function used to drive the training process. Despite its importance, the choice of this function is hardly ever discussed in most research works. Usually the choice simply consists in deciding whether to use the $L1$–norm or the $L2$–norm, the latter being the most popular option. However, even if properties of the $L2$–norm explain why it is the default choice, in the case of image restoration tasks and particularly for image denoising it is disputable. First, the key objective of image denoising is to improve the visual quality from a human observer's point of view and the $L2$–norm is clearly not correlated with this desirable objective. Second, it is known that the Euclidean metric is optimal when white Gaussian noise is encountered, but for other noise schemes alternative metrics should be considered [5]. Therefore, a loss function that is based on a metric reflecting the visual quality should be investigated.

Such an investigation can be found in [22], a paper in which the authors compared several losses considering two state-of-the-art metrics for image quality: the Structural SIMilarity (SSIM) index [16] and the MultiScale Structural SIMilarity (MS-SSIM) index [17]. They compared both norms, SSIM, MS-SSIM, and their own loss function that is a combination of MS-SSIM and $L1$–norm on different image restoration tasks, among which joint denoising and demosaicking of color image patches (31×31 pixels) using a FCN of three layers with PReLU activation in the first two ones. We independently came up with the same idea to investigate a loss function that combines the losses $\mathcal{L}^{L1}$ and $\mathcal{L}^{SSIM}$ denoted by $\mathcal{L}^{L1+SSIM}$ in the following. It should be noted that the work presented in [22] focuses on the analysis of loss functions and not on the design of a FCN for image denoising. Formally, $\mathcal{L}^{L1}$ and $\mathcal{L}^{SSIM}$ losses are defined by:

$$\mathcal{L}^{L1}(x,y) = \frac{1}{|x|} \sum_{p \in x} |x - y|, \mathcal{L}^{SSIM}(x,y) = 1 - \frac{1}{|x|} \sum_{p \in x} \frac{2\mu_x \mu_y + C_1}{\mu_x^2 + \mu_y^2 + C_1} \cdot \frac{2\sigma_{xy} + C_2}{\sigma_x^2 + \sigma_y^2 + C_2}$$

$$(3)$$

where x is the noisy version of the clean image y and μ, σ, are means and standard deviations that depend on pixel p. Both are computed using a Gaussian filter with standard deviation σ_G. $C_1 = (K_1 L)^2$ and $C_2 = (K_2 L)^2$ are two constants, where L is the dynamic range of the pixel values (1 in our case due to normalization in $[0;1]$), $K_1 \ll 1$ and $K_2 \ll 1$. In fact, the SSIM index is a similarity measure that combines three comparison functions measuring different kinds of changes between images: luminance, contrast, and structure, but thanks to a simplification it can be expressed as a product of two terms.

4   Experimental Results

4.1   Data Set and Network Training

For image denoising, images from the Berkeley Segmentation Data Set (BSD or BSDS)[2] [13] are widely used for training and testing. For example, in [12] they used the 300 images from BSDS300 to generate patches for training and 200 images for testing (the 200 fresh images from BSDS500). In [19], Zhang *et al.* followed [3] and hence trained their image denoising model DnCNN using 400 BSD images considering three different noise levels. More precisely, for each noise level they cropped $128 \times 1,600$ patches of size 40×40. In [20], they used a similar approach to train FFDNet for AWGN denoising in grayscale images.

However, it is admitted that the training of deep networks can benefit from a large data set and therefore the question of extending the routinely used small BSD training set arises. Hence, in their most recent works [20,21], Zhang *et al.* not only considered 400 images from BSD, but also selected 400 images from the validation set of ImageNet database and $4,744$ images of Waterloo Exploration Database [11]. According to their experimental study in [21], the training with an enlarged data set does not improve the denoising performance. In a first evaluation we do not consider the BSD data set, neither for training, nor for testing. We used as data set a subset of the $10,000$ gray images of 512×512 pixels provided by the BOSS database [1]: the first $3,000$ images of the database are used, with $2,800$ images for training and the 200 remaining ones for testing.

A network is trained during 50 epochs for a specific type of noise, using the Adam optimizer [9]. The traditional SGD is replaced due to the observation of Mao *et al.* [12] that Adam provides a faster training convergence of their encoder-decoder networks. The computations have been completed on a NVIDIA Tesla Titan X GPU, with a training time for a given noise level of about 5 h.

4.2   Denoising Performance

A measure used to assess the denoising performance of an approach is the Peak Signal-to-Noise Ratio (PSNR), even if it is known to be a poor quality metric when the purpose is to compare the images as perceived by the human visual system. Indeed, a high PSNR value and good visual quality do not necessarily go together [16]. It is rather the simultaneous taking into account of PSNR and mean SSIM index which is a good indicator of the visual quality: when both metrics have high values, the quality is regarded as high.

Quantitative Results. Table 1 shows the quantitative results gained for AWGN, including noise levels $\sigma \in \{10, 30, 50, 70, 80\}$, and speckle reduction on the test set of 200 images. The speckle noise is modeled as a multiplicative noise that follows a Gamma distribution $\Gamma(L, 1)$ of unit mean and variance $\frac{1}{L}$,

---

[2] https://www2.eecs.berkeley.edu/Research/Projects/CS/vision/grouping/ segbench/.

where $L = 1$. In each case the average PSNR and SSIM values of the noisy input images are given, as well as the corresponding outcomes produced by BM3D, and those issued by the proposed encoder-decoder and DnCNN. The results of DnCNN have been obtained by using directly network models provided by the authors in the GitHub[3] of their Matlab implementation. For the speckle case, the BM3D values have not been computed since the corresponding SAR version should have been used, while DnCNN is dropped due its focus on Gaussian denoising. As can be seen, the encoder-decoder can achieve satisfactory denoising results and outperforms almost systematically BM3D and DnCNN. Indeed, except for AWGN with $\sigma = 10$, in which case the encoder-decoder gives high values but slightly lower than those of BM3D and DnCNN, better PSNR and SSIM results are obtained. Moreover, the noisier the images, the more advantageous it is to use the encoder-decoder to recover clean images. For the speckle case, a comparison with the RED-NET results [6] shows that the proposal achieves a nearly similar performance for $L = 1$. Finally, we can notice that the encoder-decoder is able to deal with AWGN and speckle noise, once trained on the targeted noise, an important feature which is looked for in the perspective of blind denoising. Overall, the residual learning strategy adopted by DnCNN seems interesting for low noise levels, but as the noise increases the reconstruction of a clean image as performed by the proposed network is clearly more appropriate.

Table 1. Average PSNR (dB)/SSIM obtained for AWGN and speckle.

AWGN	Noisy input images	$\mathcal{L}^{L1+SSIM}$		
σ		BM3D	Encoder decoder	DnCNN
10	28.37/0.5798	36.97/0.9282	36.07/0.9273	37.23/0.9304
30	19.17/0.2002	31.02/0.8284	32.06/0.8626	31.11/0.8334
50	15.10/0.1052	27.56/0.7591	29.97/0.8181	27.45/0.7613
70	12.64/0.0664	24.97/0.7091	28.48/0.7865	24.55/0.7041
80	11.69/0.0545	23.76/0.6882	27.99/0.7743	
Speckle $L = 1$	10.24/0.1441		27.86/0.7852	

To further highlight the suitability of the encoder-decoder, it might be interesting to have an idea of the denoising performance given by other methods. Therefore, in Table 2 are shown the behaviors observed by Zhang *et al.* on BSD68 set [20] for AWGN removal with BM3D [4], WNNM [7], MLP [2], DnCNN, FFD-Net. It can be seen that DnCNN and FFDNet outperform other methods. Even if these results are obtained on a different data set, considering the performances of BM3D and DnCNN in both tables as reference, the proposed encoder-decoder appears as a valuable competitor for state-of-the-art approaches.

We have also completed a preliminary evaluation of the encoder-decoder blind denoising ability for AWGN. To train the network, the first thousand image from

---

[3] https://github.com/cszn/DnCNN.

Table 2. Average PSNR (dB) obtained by Zhang *et al.* [20] for AWGN on BSD68.

AWGN σ	BM3D	WNNM	MLP	DnCNN	FFDNet
15	31.07	31.37	-	31.72	31.63
25	28.57	28.83	28.96	29.23	29.19
35	27.08	27.30	27.50	27.69	27.73
50	25.62	25.87	26.03	26.23	26.29
75	24.21	24.40	24.59	24.64	24.79

data set are used, where for each image its corresponding noisy versions with $\sigma = 10, 30, 50$, and 70 are computed. The training set size is thus increased by 43% (4,000 images), as is the computation time (7.2 h). Once trained, this blind denoiser yields the following results for $\sigma = 80$: a PSNR of 27.50 dB and 0.7564 for SSIM value. Obviously, these values are inferior to the ones obtained with the network trained specifically for $\sigma = 80$ shown in Table 1. But they are slightly better than those given by the network trained only for $\sigma = 70$: a PSNR of 27.33 dB and 0.7542 for SSIM value. These results are encouraging but a deeper investigation is needed to confirm that the proposed network can be suitably trained to deal simultaneously with different unknown noise levels.

Visual Results. Images (a) to (g) of Fig. 2 illustrate the visual results of BM3D and the proposed deep network, considering a same image, for AWGN with $\sigma = 30$ and 70. It can be seen that the encoder-decoder preserves sharp edges and finer details as the noise level increases. This point is clearly highlighted through the comparison of images (f) and (g), since the clock on the left shaded part of Big Ben appears far blurrier with BM3D. Furthermore, even if for $\sigma = 30$ the PSNR result is better for BM3D, the neural network yields an image with a better visual quality: a look on the cloudy upper left part in the images (c) and (d) is convincing in our opinion. This observation is further supported by the SSIM value which is equal to 0.9021 for the clean image recovered by the encoder-decoder, whereas the one for BM3D is equal to 0.8929.

Figure 2 also shows the denoising results on two different images with noise level 70 given by DnCNN. In both cases the proposed network recovers a clean image with far more details and a better visual quality. This is again confirmed by the higher values of PSNR and SSIM: for Big Ben the values obtained from the image recovered by the encoder-decoder are, respectively, 25.01 dB and 0.8204 *versus* 24.11 dB and 0.7875 for DnCNN, while for the image with the bird they are 21.86 dB and 0.7162 *versus* 21.24 dB and 0.6547. In the case of the speckle noise presented in Fig. 3, despite the huge corruption interesting details are brought out, especially in the shaded part of the building.

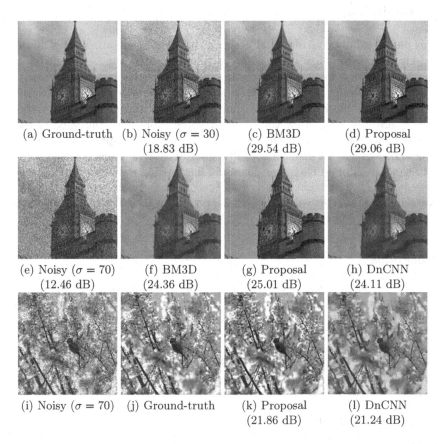

(a) Ground-truth (b) Noisy ($\sigma = 30$)   (c) BM3D   (d) Proposal
                     (18.83 dB)   (29.54 dB)   (29.06 dB)

(e) Noisy ($\sigma = 70$)   (f) BM3D   (g) Proposal   (h) DnCNN
      (12.46 dB)   (24.36 dB)   (25.01 dB)   (24.11 dB)

(i) Noisy ($\sigma = 70$) (j) Ground-truth   (k) Proposal   (l) DnCNN
                                              (21.86 dB)   (21.24 dB)

Fig. 2. AWGN denoising results (PSNR) of an image with noise level $\sigma = 30$: (a)–(d) and two images with noise level $\sigma = 70$: (e)–(h) for Big Ben and (i)–(l) for the bird.

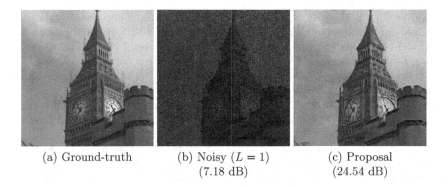

(a) Ground-truth   (b) Noisy ($L = 1$)   (c) Proposal
                          (7.18 dB)   (24.54 dB)

Fig. 3. Speckle denoising results (PSNR) of one image with $L = 1$.

5   Conclusion

In this paper, a fully convolutional network that consists in an encoder-decoder with skip connections has been proposed for image denoising. The great lines of the network have been presented and the choice of the loss function used to carry out the training discussed. The results obtained on grayscale images show that the network can remove AWGN and multiplicative speckle noise, provided that it is suitably trained for the targeted noise. Moreover, compared to some competing approaches for image denoising, the network appears to be able to produce state-of-the-art denoising results. Finally, a preliminary evaluation of its ability to address blind Gaussian denoising has yielded favorable performance.

Acknowledgment. This work has been supported by the EIPHI Graduate School (contract "ANR-17-EURE-0002").

References

1. Bas, P., Filler, T., Pevný, T.: "Break Our Steganographic System": the ins and outs of organizing BOSS. In: Filler, T., Pevný, T., Craver, S., Ker, A. (eds.) IH 2011. LNCS, vol. 6958, pp. 59–70. Springer, Heidelberg (2011). https://doi.org/10.1007/978-3-642-24178-9_5
2. Burger, H.C., Schuler, C.J., Harmeling, S.: Image denoising: can plain neural networks compete with BM3D? In: 2012 IEEE Conference on Computer Vision and Pattern Recognition (CVPR), pp. 2392–2399. IEEE (2012)
3. Chen, Y., Pock, T.: Trainable nonlinear reaction diffusion: a flexible framework for fast and effective image restoration. IEEE Trans. Pattern Anal. Mach. Intell. **39**(6), 1256–1272 (2017)
4. Dabov, K., Foi, A., Katkovnik, V., Egiazarian, K.: Image denoising by sparse 3-D transform-domain collaborative filtering. IEEE Trans. Image Process. **16**(8), 2080–2095 (2007)
5. François, D., Wertz, V., Verleysen, M., et al.: Non-euclidean metrics for similarity search in noisy datasets. In: 13th European Symposium on Artificial Neural Networks (ESANN), pp. 339–344 (2005)
6. Gu, F., Zhang, H., Wang, C., Zhang, B.: Residual encoder-decoder network introduced for multisource SAR image despeckling. In: 2017 SAR in Big Data Era: Models, Methods and Applications (BIGSARDATA), pp. 1–5. IEEE (2017)
7. Gu, S., Zhang, L., Zuo, W., Feng, X.: Weighted nuclear norm minimization with application to image denoising. In: 2014 IEEE Conference on Computer Vision and Pattern Recognition (CVPR), pp. 2862–2869 (2014)
8. Isola, P., Zhu, J., Zhou, T., Efros, A.A.: Image-to-image translation with conditional adversarial networks. In: 2017 IEEE Conference on Computer Vision and Pattern Recognition (CVPR), pp. 5967–5976. IEEE (2017)
9. Kingma, D.P., Ba, J.: Adam: a method for stochastic optimization. In: 3rd International Conference for Learning Representations (ICLR) (2015)
10. Long, J., Shelhamer, E., Darrell, T.: Fully convolutional networks for semantic segmentation. In: 2015 IEEE Conference on Computer Vision and Pattern Recognition (CVPR), pp. 3431–3440. IEEE (2015)

11. Ma, K., Duanmu, Z., Wu, Q., Wang, Z., Yong, H., Li, H., Zhang, L.: Waterloo exploration database: new challenges for image quality assessment models. IEEE Trans. Image Process. **26**(2), 1004–1016 (2017)
12. Mao, X., Shen, C., Yang, Y.: Image restoration using very deep convolutional encoder-decoder networks with symmetric skip connections. In: Lee, D.D., Sugiyama, M., Luxburg, U.V., Guyon, I., Garnett, R. (eds.) Advances in Neural Information Processing Systems 29 (NIPS 2016), pp. 2802–2810. Curran Associates, Inc. (2016)
13. Martin, D., Fowlkes, C., Tal, D., Malik, J.: A database of human segmented natural images and its application to evaluating segmentation algorithms and measuring ecological statistics. In: 8th IEEE International Conference on Computer Vision (ICCV), vol. 2, pp. 416–423. IEEE (2001)
14. Radford, A., Metz, L., Chintala, S.: Unsupervised representation learning with deep convolutional generative adversarial networks. In: 4th International Conference for Learning Representations (ICLR) (2016)
15. Ronneberger, O., Fischer, P., Brox, T.: U-Net: convolutional networks for biomedical image segmentation. In: Navab, N., Hornegger, J., Wells, W.M., Frangi, A.F. (eds.) MICCAI 2015. LNCS, vol. 9351, pp. 234–241. Springer, Cham (2015). https://doi.org/10.1007/978-3-319-24574-4_28
16. Wang, Z., Bovik, A.C., Sheikh, H.R., Simoncelli, E.P.: Image quality assessment: from error visibility to structural similarity. IEEE Trans. Image Process. **13**(4), 600–612 (2004)
17. Wang, Z., Simoncelli, E.P., Bovik, A.C.: Multiscale structural similarity for image quality assessment. In: 37th Asilomar Conference on Signals, Systems Computers, vol. 2, pp. 1398–1402. IEEE (2003)
18. Xie, J., Xu, L., Chen, E.: Image denoising and inpainting with deep neural networks. In: Pereira, F., Burges, C.J.C., Bottou, L., Weinberger, K.Q. (eds.) Advances in Neural Information Processing Systems 25 (NIPS 2012), pp. 341–349. Curran Associates, Inc. (2012)
19. Zhang, K., Zuo, W., Chen, Y., Meng, D., Zhang, L.: Beyond a gaussian denoiser: residual learning of deep CNN for image denoising. IEEE Trans. Image Process. **26**(7), 3142–3155 (2017)
20. Zhang, K., Zuo, W., Zhang, L.: FFDNet: toward a fast and flexible solution for CNN based image denoising. IEEE Trans. Image Process. **27**(9), 4608–4622 (2018)
21. Zhang, K., Zuo, W., Gu, S., Zhang, L.: Learning deep CNN denoiser prior for image restoration. In: 2017 IEEE Conference on Computer Vision and Pattern Recognition (CVPR), pp. 2808–2817. IEEE (2017)
22. Zhao, H., Gallo, O., Frosio, I., Kautz, J.: Loss functions for image restoration with neural networks. IEEE Trans. Image Process. **3**(1), 47–57 (2017)

MSCE: An Edge-Preserving Robust Loss Function for Improving Super-Resolution Algorithms

Ram Krishna Pandey[1]([✉]), Nabagata Saha[2], Samarjit Karmakar[2],
and A. G. Ramakrishnan[1]

[1] Department of Electrical Engineering, Indian Institute of Science, Bangalore, India
{ramp,agr}@iisc.ac.in
[2] Department of Computer Science and Engineering, NIT Warangal, Warangal, India
{snabagata,ksamarjit}@student.nitw.ac.in

Abstract. With the recent advancement in the deep learning technologies such as CNNs and GANs, there is significant improvement in the quality of the images reconstructed by deep learning based super-resolution (SR) techniques. In this work, we propose a robust loss function based on the preservation of edges obtained by the Canny operator. This loss function, when combined with the existing loss function such as mean square error (MSE), gives better SR reconstruction measured in terms of PSNR and SSIM. Our proposed loss function guarantees improved performance on any existing algorithm using MSE loss function, without any increase in the computational complexity during testing.

Keywords: Loss function · CNN · GAN · Super-resolution · Mean square error · Mean square Canny error · Edge preservation · PSNR SSIM

1 Introduction

Super-resolution is the process of obtaining a high resolution (HR) image from one or more low resolution (LR) images. Classical reconstruction based image super-resolution requires multiple low-resolution images with sub-pixel misalignment at the same scale, whereas single image super-resolution requires a database of LR and HR matched pairs to learn a mapping function between the patch pairs at different scales [1]. Given a low resolution image during testing, this learned function or representation can be used to reconstruct the corresponding HR image.

Since the advent of deep learning technologies in the past decade, super-resolution algorithms have shown remarkable improvement in the quality of the reconstructed image. Most of the work reported in the literature have used mean square error (MSE) loss function to minimize the error between the reconstructed

© Springer Nature Switzerland AG 2018
L. Cheng et al. (Eds.): ICONIP 2018, LNCS 11306, pp. 566–575, 2018.
https://doi.org/10.1007/978-3-030-04224-0_49

model output and the ground truth image [2–7]. Minimizing this loss function may reduce the high frequency content in the reconstructed image and thus may blur the edges in it. Also, the reconstructed image may not lie precisely in the manifold of the HR image. Researchers have endeavored to find ways to solve this problem to a good extent as can be seen in SRGAN [8], where the authors claim that the reconstructed output lies precisely in the manifold of HR images, even if the reconstructed images have less peak signal to noise ratio (PSNR) and structural similarity (SSIM). Ledig et al. [8] have used a weighted combination of MSE loss, content loss [9] and adversarial loss to reconstruct the HR image. This approach requires a deep architecture, such as the VGG net [10], to obtain the local covariance structure in the image. Most of the image transformation tasks use mean square error as loss function, which provides smooth transformed images.

Our main contributions in this paper are as follows:

- We have performed a large number of experiments to obtain a robust loss function that improves the performance of the existing algorithms that employ MSE loss function.
- While training, we apply Canny edge detector [11] the reconstructed output (in batches) and also separately on the corresponding ground truth image to compute the proposed mean square Canny error (MSCE) and assign weights (convex combination) based on our experiments i.e. the loss function can be given as: $\mu \times MSE + (1 - \mu) \times MSCE$.
- Our approach guarantees performance improvement in terms of PSNR and SSIM over the existing approaches, if the model is trained on one dataset and tested on different datasets as mentioned in Tables 1 and 2.
- Our model does not incur additional overhead in terms of computation during testing to obtain the performance gain reported in Tables 1 and 2 due to our proposed MSCE loss function.

2   Related Work

Super-resolution and image denoising can be assumed as image transformation tasks. In super-resolution, a LR image is fed to a transformation network such as a multilayer neural network to generate a HR image. Most of the image processing tasks such as image denoising and super-resolution minimize a per-pixel loss function to obtain reconstruction. In this work, our focus is on improving the quality of existing super-resolution algorithms such as SRCNN [2] and ESPCN [3] that use per-pixel loss function. Recently proposed perceptual loss function has shown significant improvement in the perceptual quality of the images. Simonyan et al. [12] use perceptual loss for feature visualization. Gatys et al. [13] and [14] use perceptual loss for texture synthesis and style transfer, respectively. These approaches solve an optimization problem and hence, are slower.

Johnson et al. [9] and Pandey et al. [15] use the benefits of per-pixel as well as perceptual loss functions and propose a computationally efficient, optimization-free approach that provides results for image transformation tasks that are qualitatively similar to those of the above optimization-based approaches. The super-resolution algorithm SRGAN [8] uses a weighted combination of three different loss functions, namely mean square error, perceptual and adversarial loss to obtain a sharper reconstruction. The images reconstructed by these methods perceptually look sharper, even if they have low values of PSNR and SSIM.

In this work, our focus is on improving the perceptual quality, PSNR and SSIM without incurring any additional computational overhead during testing by the addition of a new, robust loss function that aims to preserve the edge information.

3   The Proposed Edge-Preserving MSCE Loss Function

We employ Canny edge detector [11] to detect the edges in the reconstructed and ground truth images. We have chosen this algorithm, since Canny operator provides the most reliable edges amongst all the edge detection algorithms in the literature, and also satisfies all the general edge detection criteria.

Most of the recent papers on image super-resolution and denoising use mean square error as the loss function. This loss function may smooth the edge components in an image. We thought of preserving the edges by defining the loss function as a convex combination of mean square error loss and our edge preserving loss as follows:

Suppose the training set consists of image pairs $\{L_i, H_i\}$; $i = 1...N$, where N is the total number of training examples. The model Θ, parameterized by λ, predicts the output O_j for a given input L_j. Let C denote the Canny operator. Let $C(\Theta_\lambda(L_j)))$ be the resultant image obtained by applying Canny operator on the predicted output image, $O_j = \Theta_\lambda(L_j)$. The proposed edge preserving loss function, called the mean square Canny error - (MSCE) is given by:

$$Loss = \mu \times \underbrace{\frac{1}{N} \sum_{j=1}^{N} \parallel \Theta_\lambda(L_j) - H_j \parallel_F}_{MSE\,Loss\,(l_{mse})} + (1 - \mu) \times \underbrace{\frac{1}{N} \sum_{j=1}^{N} \parallel C(\Theta_\lambda(L_j)) - C(H_j) \parallel_F}_{Edge\,preserving\,loss\,(l_{edge})} \qquad (1)$$

The first term in the equation above is the mean square loss function used to minimize the error between the reconstructed output and the ground truth image. The second term in the loss function is the edge preserving loss function. After a large number of experiments, the weighing factor μ has been fixed to lie in the range $0.8 \leq \mu \leq 0.99$. To minimize this loss function, Adam optimizer [16] is used with learning rate $(lr) = 0.001$, $\beta_1 = 0.999$ and $\beta_2 = 0.99$.

3.1   Choosing the Value of μ

- **Exhaustive Experimentation:** We performed experiments varying μ in the range $0.8 \leq \mu \leq 0.99$ by incrementing its value by 0.01 each time. We found that the models were consistently giving better results for the particular values of $\mu = 0.84$, 0.85 and 0.86. For the results reported in the Figs. 1, 2, 3 and 4 and Tables 1 and 2, the value of μ used is 0.85.

– **Dynamic Choice of** μ: While performing the experiments, we found that sometimes, values of μ (still in the range $0.8 \leq \mu \leq 0.99$) other than the three specific ones mentioned above, gave better results. We made a list of those different values of μ and tried each of them parallely in each epoch. For the subsequent epoch, we select the model corresponding to the least value of the loss function. Let l_{mse} and l_{edge} denote the mean square error loss and our edge-preserving loss, respectively, as mentioned in Eq. 1. Let $\{\hat{\lambda}, \hat{\mu}\}$ be the optimal model parameters and μ be the weighing parameter currently chosen during training. In each epoch, we selected the value of μ that minimized the loss function in the right hand side of Eq. 2:

$$\hat{\mu} = \underset{\mu}{\operatorname{argmin}}\{\mu \times l_{mse} + (1 - \mu) \times l_{edge}\} \qquad (2)$$

We used the earlier approach for calculating the loss in our experiments, results for which have been reported in the Tables. A dynamic choice of the value of μ gives similar results in less number of epochs. It can be experimented further to possibly achieve still better results.

4 Datasets Used for Training and Testing

The models are trained on DIV2K [17] training dataset with the original architecture (without changing the architectural details of the existing model) proposed in the respective papers. We have performed testing on different datasets such as Set5 [18], Set14 [19], BSD [20] and URBAN100 [21] for the different scale factors of 2, 3, 4 and 8. We have found that there is consistent performance gain over the original models, in terms of PSNR and SSIM, on all the datasets on which our MSCE loss function has been tested so far. These results are seen quantitatively in Tables 1 and 2.

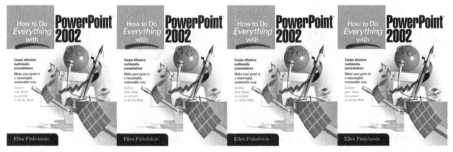

(a) SRCNN original   (b) SRCNN MSCE   (c) ESPCN original   (d) ESPCN MSCE

Fig. 1. Qualitative comparison of the results for an upscale factor of 2, when the ppt image from Set14 is directly fed to the original model and the model modified with MSCE loss trained by us. (a) The output image reconstructed by the original SRCNN model. (b) The output image reconstructed by SRCNN model modified with MSCE loss function. (c) Image reconstructed by the original ESPCN model. (d) Output image reconstructed by ESPCN model modified with MSCE loss function.

5   Experiments and Discussion

We have performed extensive experiments on different super-resolution algorithms proposed recently, by augmenting the original loss function with our proposed mean square Canny error loss function.

We have validated the effectiveness of our proposed MSCE loss function on the recent techniques of SRCNN [2] and ESPCN [3]. We found that the addition of our MSCE loss leads to better results and the improvement is consistent on both methods across different upscaling factors of 2, 3, 4 and 8.

(a) SRCNN original (b) SRCNN MSCE   (c) ESPCN original   (d) ESPCN MSCE

Fig. 2. Comparison of the results for an upscale factor of 3, when the comic image from Set14 is directly fed to the original model and the model modified with MSCE loss trained by us. (a) Output image reconstructed by the original SRCNN model. (b) Output image reconstructed by SRCNN model modified by MSCE loss function. (c) Output image reconstructed by the original ESPCN model. (d) Output image reconstructed by ESPCN model modified by MSCE loss function.

(a) SRCNN original (b) SRCNN MSCE   (c) ESPCN original   (d) ESPCN MSCE

Fig. 3. Comparison of the results for an upscale factor of 4, when the baby input image from Set5 is directly fed to the original model trained by us and the model modified with MSCE loss trained by us. (a) Output image reconstructed by the original SRCNN model. (b) Output image reconstructed by SRCNN model modified by MSCE loss function. (c) Output image reconstructed by the original ESPCN model. (d) Output image reconstructed by ESPCN model modified by MSCE loss function.

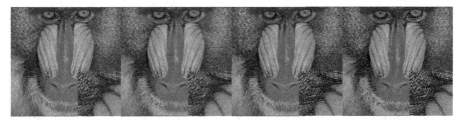

(a) SRCNN original (b) SRCNN MSCE (c) ESPCN original (d) ESPCN MSCE

Fig. 4. Comparison of the results for an upscale factor of 8, when the baboon input image from Set14 is directly fed to the original model trained by us and the model modified with MSCE loss trained by us. (a) Output image reconstructed by the original SRCNN model. (b) Output image reconstructed by SRCNN model modified by MSCE loss function. (c) Output image reconstructed by the original ESPCN model. (d) Output image reconstructed by ESPCN model modified by MSCE loss function.

Table 1. P_* and S_* are the PSNR and SSIM values obtained by the method $*$ for the upscaling factors of 2, 3, 4 and 8, whereas P_*^c and S_*^c are the corresponding PSNR and SSIM values obtained after augmenting the loss function by the MSCE loss function designed by us. All the models other than bicubic (non-learnable) have been trained on DIV2K training dataset. For testing, we have used 4 datasets, namely Set5, Set14, Urban and BSD.

Dataset		$P_{bicubic}$	$S_{bicubic}$	P_{srcnn}	S_{srcnn}	$\mathbf{P^c_{srcnn}}$	$\mathbf{S^c_{srcnn}}$	P_{espcn}	S_{espcn}	$\mathbf{P^c_{espcn}}$	$\mathbf{S^c_{espcn}}$
Set5	2x	27.02	0.92	28.44	0.93	28.57	0.93	26.48	0.92	26.59	0.92
	3x	25.41	0.89	26.59	0.90	26.75	0.91	25.882	0.91	25.888	0.91
	4x	21.96	0.79	23.22	0.82	23.37	0.83	22.35	0.81	22.49	0.82
	8x	18.10	0.61	18.740	0.63	18.743	0.63	18.33	0.62	18.43	0.62
Set14	2x	24.10	0.86	25.22	0.88	25.32	0.88	23.50	0.87	23.56	0.87
	3x	22.65	0.81	23.62	0.84	23.68	0.84	23.06	0.84	23.06	0.84
	4x	20.01	0.70	20.96	0.73	21.04	0.73	20.12	0.71	20.32	0.72
	8x	17.13	0.53	17.57	0.56	17.58	0.56	17.20	0.56	17.27	0.56
Urban	2x	20.66	0.84	22.26	0.87	22.44	0.87	21.38	0.87	21.42	0.87
	3x	20.22	0.79	21.47	0.83	21.53	0.83	21.18	0.83	21.18	0.83
	4x	16.92	0.65	17.81	0.69	17.84	0.69	17.54	0.70	17.59	0.70
	8x	14.63	0.48	15.04	0.50	15.04	0.50	14.94	0.507	14.99	0.509
BSD	2x	25.88	0.89	25.96	0.90	26.18	0.90	23.36	0.87	23.41	0.88
	3x	21.86	0.77	22.49	0.81	22.54	0.81	22.34	0.81	22.35	0.81
	4x	21.43	0.73	22.08	0.77	22.13	0.77	21.28	0.76	21.41	0.77
	8x	18.43	0.57	18.78	0.59	18.81	0.59	18.47	0.587	18.58	0.589

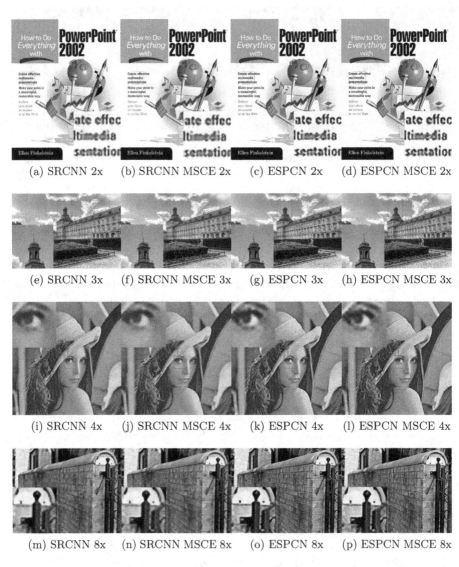

(a) SRCNN 2x   (b) SRCNN MSCE 2x   (c) ESPCN 2x   (d) ESPCN MSCE 2x

(e) SRCNN 3x   (f) SRCNN MSCE 3x   (g) ESPCN 3x   (h) ESPCN MSCE 3x

(i) SRCNN 4x   (j) SRCNN MSCE 4x   (k) ESPCN 4x   (l) ESPCN MSCE 4x

(m) SRCNN 8x   (n) SRCNN MSCE 8x   (o) ESPCN 8x   (p) ESPCN MSCE 8x

Fig. 5. Comparison of the results obtained on down-sampled (by bicubic interpolation without blurring) images on different upscaling factors. (a), (b), (c) and (d) have been down-sampled by a factor of 2 and reconstructed. (e), (f), (g) and (h) have been down-sampled by a factor of 3 and reconstructed. (i), (j), (k) and (l) have been down-sampled by a factor of 4 and reconstructed. (m), (n), (o) and (p) have been down-sampled by a factor of 8 and reconstructed.

6   Results

Figures 1, 2, 3 and 4 show both the results qualitatively: one obtained by passing the input image directly to the original models SRCNN [2] and ESPCN [3] with the loss functions used in the original papers, and the other obtained by augmenting the loss function with our MSCE loss function. Comparison of the results obtained on down-sampled (by bicubic interpolation without blurring) images are shown in Fig. 5, for different upscaling factors.

Tables 1 and 2 list the quantitative results obtained by the two superresolution methods on the datasets Set5, Set14, URBAN and BSD for different upscaling factors and the corresponding values obtained after they are modified by our MSCE loss function.

Note 1: Table 1 lists the results obtained from the LR images created by downsampling using normal bicubic interpolation. Whereas, the results reported in Table 2 are obtained by blurring the image by a Gaussian filter with radius 2 and then downsampling by bicubic interpolation to obtain the LR images at different scales.

Table 2. P_* and S_* are the PSNR and SSIM values obtained by the method $*$ at the different upscaling factors of 2, 3, 4 and 8, whereas P_*^c and S_*^c are the corresponding PSNR and SSIM values obtained by augmenting the loss function by the MSCE loss function designed by us. All the models other than bicubic (non-learnable) are trained on DIV2K (blurred by Gaussian blurring, then downsampled by bicubic) training dataset. For testing, we use 4 datasets, namely Set5, Set14, Urban and BSD.

Dataset		$P_{bicubic}$	$S_{bicubic}$	P_{srcnn}	S_{srcnn}	P_{srcnn}^c	S_{srcnn}^c	P_{espcn}	S_{espcn}	P_{espcn}^c	S_{espcn}^c
Set5	2x	21.10	0.77	23.96	0.83	24.06	0.84	21.21	0.75	21.87	0.79
	3x	21.63	0.79	22.38	0.85	24.75	0.86	22.50	0.80	22.88	0.83
	4x	20.12	0.72	21.92	0.78	21.94	0.78	21.53	0.77	21.92	0.78
	8x	17.72	0.59	18.17	0.61	18.34	0.61	18.48	0.613	18.56	0.614
Set14	2x	19.35	0.67	21.75	0.76	21.78	0.76	19.50	0.67	20.08	0.70
	3x	19.84	0.69	20.64	0.77	22.25	0.78	20.63	0.73	20.99	0.74
	4x	18.67	0.62	20.02	0.69	20.07	0.69	19.75	0.68	20.04	0.69
	8x	16.84	0.52	17.16	0.54	17.29	0.54	17.35	0.55	17.43	0.54
Urban	2x	16.57	0.63	18.87	0.74	18.86	0.74	16.93	0.63	17.33	0.66
	3x	17.63	0.66	18.96	0.76	20.03	0.76	18.68	0.71	18.87	0.72
	4x	15.85	0.58	17.05	0.65	17.07	0.65	16.96	0.64	17.05	0.64
	8x	14.42	0.47	14.74	0.49	14.81	0.49	14.96	0.49	14.97	0.48
BSD	2x	21.00	0.71	23.27	0.80	23.28	0.80	20.92	0.71	21.67	0.75
	3x	19.82	0.66	20.27	0.75	21.59	0.75	20.30	0.70	20.60	0.72
	4x	20.13	0.67	21.32	0.73	21.35	0.73	21.06	0.73	21.38	0.73
	8x	18.18	0.56	18.38	0.58	18.51	0.58	18.70	0.58	18.70	0.58

7 Conclusion

A large number of research papers have been published in the recent past by designing different models or algorithms that work reasonably well. The unique contribution of our work is that it improves the performance of any existing method, rather than proposing another technique. In this paper, we have proposed a robust edge-preserving loss function that adds performance gain in terms of PSNR and SSIM to any existing model, without increasing the computational cost involved in testing. We train the existing model by adding weighted Canny edge based loss. Minimizing this loss function helps to preserve the edges by giving more weightage to the edges. As shown by the Tables of results, the PSNR and SSIM values obtained after including our MSCE loss function are consistently better.

References

1. Glasner, D., Bagon, S., Irani, M.: Super-resolution from a single image. In: 12th IEEE International Conference on Computer Vision, pp. 349–356 (2009)
2. Dong, C., Loy, C.C., He, K., Tang, X.: Learning a deep convolutional network for image super-resolution. In: Fleet, D., Pajdla, T., Schiele, B., Tuytelaars, T. (eds.) ECCV 2014. LNCS, vol. 8692, pp. 184–199. Springer, Cham (2014). https://doi.org/10.1007/978-3-319-10593-2_13
3. Shi, W., Caballero, J., Husz, F., Totz, J., Aitken, A. P., Bishop, R., Rueckert, D., Wang, Z.: Real-time single image and video super-resolution using an efficient sub-pixel convolutional neural network. In: IEEE Conference on Computer Vision and Pattern Recognition, pp. 1874–1883 (2016)
4. Pandey, R.K., Ramakrishnan, A.G.: A Hybrid Approach of Interpolations and CNN to Obtain Super-Resolution. arXiv preprint arXiv:1805.09400 (2018)
5. Pandey, R.K., Ramakrishnan, A.G.: Language independent single document image super-resolution using CNN for improved recognition. arXiv preprint arXiv:1701.08835 (2017)
6. Pandey, R.K., Ramakrishnan, A.G.: Sadhana **43**, 15 (2018). https://doi.org/10.1007/s12046-018-0794-1
7. Pandey, R.K., Maiya, S.R., Ramakrishnan, A.G.: A new approach for upscaling document images for improving their quality. In: 14th International IEEE India Conference INDICON 2017, IIT Roorkee, 15–17 December (2017)
8. Ledig, C., et al.: Photo-realistic single image super-resolution using a generative adversarial network. In: Proceedings of the IEEE Conference on Computer Vision and Pattern Recognition (2017)
9. Johnson, J., Alahi, A., Fei-Fei, L.: Perceptual losses for real-time style transfer and super-resolution. In: Leibe, B., Matas, J., Sebe, N., Welling, M. (eds.) ECCV 2016. LNCS, vol. 9906, pp. 694–711. Springer, Cham (2016). https://doi.org/10.1007/978-3-319-46475-6_43
10. Simonyan, K., Zisserman, A.: Very deep convolutional networks for large-scale image recognition. In: International Conference on Learning Representations (ICLR) (2015)
11. Canny, J.: A computational approach to edge detection. IEEE Trans. Pattern Anal. Mach. Intell. **6**, 679–698 (1986)

12. Simonyan, K., Vedaldi, A. Zisserman, A.: Deep inside convolutional networks: visualising image classification models and saliency maps. arXiv preprint arXiv:1312.6034 (2013)
13. Gatys, L.A., Ecker, A.S., Bethge, M.: Texture synthesis using convolutional neural networks. In: Advances in Neural Information Processing Systems 28 (2015)
14. Gatys, L.A., Ecker, A.S., Bethge, M.: a neural algorithm of artistic style. arXiv preprint arXiv:1508.06576 (2015)
15. Pandey, R.K., Karmakar, S., Ramakrishnan, A.G.: Computationally efficient approaches for image style transfer. arXiv preprint arXiv:1807.05927 (2018)
16. Kinga, D., Ba, J.: Adam: a method for stochastic optimization. In: Proceedings of the International Conference on Learning Representations (ICLR) (2015)
17. Agustsson, E., Timofte, R.: NTIRE 2017 challenge on single image super-resolution: dataset and study. In: Proceedings of the IEEE Conference on Computer Vision and Pattern Recognition (CVPR) Workshops (2017)
18. Bevilacqua, M., Roumy, A., Guillemot, C., Alberi-Morel, M.L.: Low-complexity single-image super-resolution based on nonnegative neighbor embedding. In: Proceedings of the BMVC (2012)
19. Zeyde, R., Elad, M., Protter, M.: On single image scale-up using sparse-representations. In: Proceedings of the International Conference Curves and Surfaces, Avignon-France (2010)
20. Martin, D., Fowlkes, C., Tal, D., Malik, J.: A database of human segmented natural images and its application to evaluating segmentation algorithms and measuring ecological statistics. In: Proceedings of the 8th International Conference on Computer Vision, vol. 2, pp. 416–423 (2001)
21. Huang, J.B., Singh, A., Ahuja, N.: Single image super-resolution from transformed self-exemplars. In: Proceedings of the IEEE Conference on Computer Vision and Pattern Recognition, pp. 5197–5206 (2015)

Author Index

Printed in the United States
By Bookmasters